ROYAL HISTORICAL SOCIETY

ANNUAL
BIBLIOGRAPHY
OF
BRITISH AND IRISH
HISTORY

ROYAL HISTORICAL SOCIETY

ANNUAL BIBLIOGRAPHY OF BRITISH AND IRISH HISTORY

PUBLICATIONS OF 1992

GENERAL EDITORS:

BARBARA ENGLISH & J.J.N. PALMER

J.S.A. Adamson   Andrew Ayton   Virginia Davis
C.C. Eldridge   Barbara English   A.C. Howe
James Kirk   V.J. Morris   Mary O'Dowd
D.J. Orton   J.J.N. Palmer   Michael S. Partridge
Huw Pryce   Stephen Taylor   John Turner   A. Williams

For the Royal Historical Society and
in association with the Institute of
Historical Research

OXFORD UNIVERSITY PRESS

1993

Oxford University Press, Walton Street, Oxford OX2 6DP
Oxford New York Toronto
Delhi Bombay Calcutta Madras Karachi
Kuala Lumpur Singapore Hong Kong Tokyo
Nairobi Dar es Salaam Cape Town
Melbourne Auckland Madrid
and associated companies in
Berlin Ibadan

Oxford is a trade mark of Oxford University Press

Published in the United States
by Oxford University Press Inc., New York

British Library Cataloguing in Publication Data
Data available

Library of Congress Cataloging in Publication Data
Data available
ISBN 0–19–820461–2

1 3 5 7 9 10 8 6 4 2

Typeset by Pentacor PLC, High Wycombe, Bucks
Printed and bound in
Great Britain by Biddles Ltd.
Guildford and King's Lynn

CONTENTS

Preface   ix

Abbreviations   x

A Auxiliary   V.J. MORRIS AND D.J. ORTON

   a. Archives and Bibliography   1
   b. Works of Reference   9
   c. Historiography   12

B General   BARBARA ENGLISH

   a. Long Periods: National   17
   b. Long Periods: Local   26
   c. Collective Volumes   36
   d. Genealogy and Heraldry   48

C Roman Britain   J.J.N. PALMER

   a. Archaeology   51
   b. History   53

D England 450–1066   A. WILLIAMS

   a. General   56
   b. Politics and Institutions   57
   c. Religion   58
   d. Economic Affairs and Numismatics   60
   e. Intellectual and Cultural   62
   f. Society and Archaeology   64

E England 1066–1500   ANDREW AYTON

   a. General   67
   b. Politics   69
   c. Constitution, Administration and Law   70
   d. External Affairs   71
   e. Religion   72
   f. Economic Affairs   76
   g. Social Structure and Population   79
   h. Naval and Military   82
   i. Intellectual and Cultural   83
   j. Visual Arts   87
   k. Topography   90

F England and Wales 1500–1714   J.S.A. ADAMSON

   a. General   93

Contents

b.	Politics	94
c.	Constitution, Administration and Law	98
d.	External Affairs	99
e.	Religion	101
f.	Economic Affairs	105
g.	Social History (General)	107
h.	Social Structure and Population	109
i.	Naval and Military	110
j.	Political Thought and History of Ideas	112
k.	Arts and Cultural History	114
l.	Science and Technology	118

G   Britain 1714–1815   STEPHEN TAYLOR

a.	General	120
b.	Politics	121
c.	Constitution, Administration and Law	123
d.	External Affairs	124
e.	Religion	126
f.	Economic Affairs	128
g.	Social Structure and Population	131
h.	Naval and Military	133
i.	Intellectual and Cultural	135
j.	Science	139

H   Britain 1815–1914   A.C. HOWE AND MICHAEL S. PARTRIDGE

a.	General	140
b.	Politics	141
c.	Constitution, Administration and Law	145
d.	External Affairs	146
e.	Religion	147
f.	Economic Affairs	151
g.	Social Structure and Population	156
h.	Social Policy	160
i.	Education	162
j.	Naval and Military	164
k.	Science and Medicine	165
l.	Intellectual and Cultural	168

I   Britain since 1914   JOHN TURNER

a.	General	174
b.	Politics	176
c.	Constitution, Administration and Law	180
d.	External Affairs	181
e.	Religion	186
f.	Economic Affairs	187
g.	Social Structure and Population	192
h.	Social Policy	194

Contents

i.	Naval and Military	196
j.	Intellectual and Cultural	199

J Medieval Wales HUW PRICE

a.	General	203
b.	Politics	203
c.	Constitution, Administration and Law	203
d.	External Affairs	204
e.	Religion	204
f.	Economic Affairs	204
g.	Social Structure and Population	205
h.	Naval and Military	205
i.	Intellectual and Cultural	206

K Scotland Before the Union JAMES KIRK

a.	General	206
b.	Politics	207
c.	Constitution, Administration and Law	207
d.	External Affairs	208
e.	Religion	208
f.	Economic Affairs	209
g.	Social Structure and Population	210
h.	Naval and Military	210
i.	Intellectual and Cultural	211

L Ireland to c.1640 VIRGINIA DAVIS

a.	General	211
b.	Politics	212
c.	Constitution, Administration and Law	212
d.	External Affairs	212
e.	Religion	212
f.	Economic Affairs	213
g.	Social Structure and Population	214
h.	Naval and Military	215
i.	Intellectual and Cultural	215

M Ireland since c.1640 MARY O'DOWD

a.	General	216
b.	Politics	216
c.	Constitution, Administration and Law	218
d.	External Affairs	218
e.	Religion	218
f.	Economic Affairs	219
g.	Social Structure and Population	220
h.	Naval and Military	221
i.	Intellectual and Cultural	221

Contents

j. Local History   222
k. Science and Medicine   222

N Empire and Commonwealth   C.C. ELDRIDGE

a. General   223
b. Politics   224
c. Constitution, Administration and Law   226
d. Religion   229
e. Economic Affairs   230
f. Social Structure and Population   231
g. Naval and Military   234
h. Intellectual and Cultural   235

Author Index   238
Personal Name Index   280
Place Index   296
Subject Index   307

PREFACE

The Annual Bibliography of the Royal Historical Society is designed in the first place to serve the urgent needs of scholars, which has meant subordinating total coverage and refinements of arrangement to speed of production. Nevertheless, it aims to be comprehensive as possible: this year's volume includes 1340 books and 3675 articles by 4307 authors. There is some variety in the subdivisions of sections, which are those approved by section editors. Searchers are advised to use these subdivisions in conjunction with the four indices. The indices form an important part of the work and cover, as far as is practicable, all major subjects of the book and articles listed, whether or not they figure in the titles. The counties used in identifying places are, in general, the pre-1974 counties. A list of the county abbreviations employed is given at the head of the Index of Places.

Within each subsection (except Bc) the items are arranged in alphabetical order of authors. All collective works, including collections of reprints, are placed in section Bc. Reprinted articles and brief reports (in, for example, archaeological reports or exhibition catalogues) are not normally itemised separately, and where this is the case the fact is noted in the Bc section. All other articles contained in these collective works are individually listed in the appropiate subsection and there referred to by their Bc numbers in lieu of the usual publication details. Apart from the short list of standard abbreviations which follows this preface, these Bc numbers are the only abbreviated form of publication details.

Items covering more than two sections are listed in Ba or Bb; any items that extend over two sections may appear in the first section and be cross-referenced at the head of the second. Items in the section 'Empire and Commonwealth post 1783' are included only if they contribute substantially to British or Irish history: for example, 'British teaplanters in Ceylon' would be included whereas 'The social structure of Ceylon' would not.

The editors are especially indebted to the director and the library staff of the Institute of Historical Research for undertaking the searches without which the compilation of this Bibliography would not be practicable. The editors also wish to express their gratitude to George Slater of the Computer Centre and to Matthew Palmer for programming assistance, and to Lynn Foster for her invaluable work in entering data.

University of Hull
April 1993

Arch.	Archaeological
B.	Bulletin
BAR	British Archaeological Reports
c.	circa
(comp.)	compilers
CBA	Council for British Archaeology
(ed.)	editor(s)
edn.	edition
fl.	floruit
HMSO	Her Majesty's Stationery Office
Inst.	Institute
J.	Journal
ns	new series
no.	number
P.	Proceedings
Q.	Quarterly
R.	Review
rev.	revised
ser.	series
Soc.	Society
T.	Transactions
UP	University Press

A. AUXILIARY

See also Fk1–4,16,21,64,71,75,77,84; Ge32; Hi20,l28,89; Kb3; La10; Ma4

(a) Archives and Bibliography

1. Adams, J.N.; Davies, M.J. (comp.). *A bibliography of nineteenth-century legal literature, vol. 1: A-G.* London; Avero and Chadwyck-Healey; 1992. Pp 1046 + microfiche.
2. Adams, Simon. 'The papers of Robert Dudley, earl of Leicester, 1: the Browne-Evelyn collection', *Archives* 20 (1992), 63–85.
3. Anon. *A handlist of the Inner Temple archives*, parts 1–2. London; Inner Temple; 1992. 2 vols. Unpaginated.
4. Anon. *Accessions to repositories and reports added to the National Register of Archives, 1990.* London; HMSO (Royal Commission on Historical Manuscripts); 1991. Pp 72.
5. Anon. *Annual review, 1990–1991.* London; HMSO (Royal Commission on Historical Manuscripts); 1991. Pp 40.
6. Anon. *Foreign Office correspondence: political, 1955 (FO 371/ 113453–118673).* Kew; List and Index Soc. vol. 245; 1991. Pp 326.
7. Anon. *Memoranda rolls 16–17 Henry III preserved in the Public Record Office: Michaelmas 1231-Michaelmas 1233.* London; HMSO; 1991. Pp 311.
8. Anon. *Record repositories in Great Britain: a geographical directory.* London; HMSO (Royal Commission on Historical Manuscripts); 9th edn. 1991. Pp 46.
9. Anon. *Select classified guide to the holdings of the Churchill Archives Centre, January 1992.* Cambridge; Churchill Archives Centre; 1992. Pp 48.
10. Anon. *The British Library catalogue of additions to the manuscripts: the Cecil of Chelwood papers (Additional MSS. 51071–51204).* London; British Library; 1991. Pp ix, 171.
11. Anon. *Treasury Board papers, 1765–1770 (T1/437–480).* Kew; List and Index Soc. vol. 244; 1991. Pp 551.
12. Anon. *ZLIB: British Transport Historical Records Office Library, 1694–1982.* Kew; List and Index Soc. vol. 246; 1991. Pp 493.
13. Anon. 'A bibliography of the works of Michael Howard', Bc150, 303–04.
14. Anon. 'A list of essays on Scottish history published during the year 1990', *Scottish Historical R.* 70 (1991), 181–87.
15. Anon. 'Bibliography' [church in Devon and Cornwall], Bc148, 223–29.
16. Anon. 'Bibliography of Christopher Brooke', Bc70, 333–39.
17. Anon. 'Bibliography of Victorian historicism and medievalism', Bc225, 229–86.

18. Anon. 'Bibliography; religion; sexuality', *J. of Women's History* 3/3 (1992), 141–92.
19. Anon. 'Bibliography' [the early church in Wales], Bc96, 150.
20. Anon. 'John G. Hurst: a ceramic bibliography', Bc191, 7–15.
21. Anon. 'Oliver MacDonagh: writings, 1947–1989', Bc143, 241–45.
22. Anon. 'Periodical literature in Wales, 1990', *Archaeologia Cambrensis* 140 (1991), 176–80.
23. Anon. 'Publications of D.T. Whiteside', Bc87, xiii-xvi.
24. Anon. 'Reports on historical manuscripts', *Royal Commission on Historical Manuscripts Annual Report* (1991–2), 25–40.
25. Anon. 'Select bibliography' [portraiture], Bc93, 422–30.
26. Anon. 'Select bibliography' [Royal Mint], Bc226, 759–74.
27. Axton, Marie (ed.). *Calendar and catalogue of Sark Seigneurie archive, 1526–1927.* Kew; List and Index Soc. (Special ser. 26); 1991. Pp 266.
28. Baker, P.A. 'The Moran papers', *Medical History* 36 (1992), 455–59.
29. Baxter, Colin F. *The Normandy campaign, 1944: a selected bibliography.* London; Greenwood; 1992. Pp 184.
30. Beadle, Richard. 'Dated and datable manuscripts in Cambridge libraries', Bc220, 238–45.
31. Bendall, A. Sarah. *Maps, land and society: a history, with a carto-bibliography of Cambridgeshire estate maps, c.1600–1836.* Cambridge; Cambridge UP; 1992. Pp xxiv, 404.
32. Bennett, John; Chatterton, Matthew; Smallbone, Linda. 'British Labour history publications 1991', *Labour History R.* 57/2 (1992), 21–42.
33. Bennett, Shelley M. 'A checklist of pre-Raphaelite works of art in the Huntington library and art collections', *Huntington Library Q.* 55 (1992), 225–51.
34. Berridge, Virginia. 'The archives of addiction: the Society for the Study of Addiction', *Archives* 20 (1992), 286–95.
35. Bird, Stephen. 'Classified archive deposits 1990', *Labour History R.* 57/2 (1992), 16–20.
36. Black, Jeremy. 'The papers of British diplomats, 1689–1793', *Archives* 20 (1992), 225–53.
37. Blackwell, Kenneth; Spadoni, Carl. *Detailed catalogue of the second archive of Bertrand Russell.* London; Thoemmes; 1992. Pp 460.
38. Bowler, Catherine; Brimblecombe, Peter. 'Archives and air pollution history', *J. of the Soc. of Archivists* 13 (1992), 136–42.
39. Braches, Ernst. 'The first years of the Fagel collection in Trinity college, Dublin', Bc133, 189–96.
40. Britnell, R.H.; *et al.* (comp.). 'Review of periodical literature', *Economic History R.* 2nd ser. 45 (1992), 154–87.
41. Bromley, Allan. *The Babbage papers in the Science Museum library: a cross-referenced list.* London; Science Museum; 1991.
42. Brown, C.E. 'Conservation of waterlogged wood: a review', Bc41, 121–123.
43. Camp, Anthony J. 'Society of Genealogists', *Local Historian* 22 (1992), 68–73.
44. Cassidy, Cheryl M. (comp.). 'Victorian periodicals 1990: an annotated bibliography', *Victorian Periodicals R.* 24 (1991), 202–23.

45. Catterall, Peter (ed.). *Bibliography of British history, 1945–1987.* Oxford; Blackwell; 1991. Pp 900.

46. Chamberlaine-Brothers, R.J. 'List of publications on the history of Warwickshire published in 1990, with addenda for 1989', *Warwickshire History* 8 (1991), 61–68.

47. Clapinson, Mary; Rogers, T.D. (ed.). *Summary catalogue of post-medieval western manuscripts in the Bodleian library, Oxford: acquisitions, 1916–1975 (SC 37300–55936).* Oxford; Oxford UP; 1991. 3 vols. Pp 705; 692; 749.

48. Codignola, Luca. *Guide to documents relating to French and British north America in the archives of the Sacred Congregation 'De propaganda fide' in Rome, 1622–1799.* Ottawa; National Archives of Canada; 1991. Pp xiii, 250.

49. Cohen, Edward H. (ed.). 'Victorian bibliography for 1990', *Victorian Studies* 34 (1990–1), 520–659.

50. Cohen, I. Bernard. 'The review of the first edition of Newton's *Principia* in the *Acta eruditorum*, with notes on the other reviews', Bc87, 323–53.

51. Colson, Jean; Middleton, Roger; Wardley, Peter. 'Annual review of information technology developments for economic and social historians, 1991', *Economic History R.* 2nd ser. 45 (1992), 378–412.

52. Coppens, Chris; de Schepper, Marcus. 'Printer to town and university: Henrick van Haestens at Louvain, with a check-list (1621–1628)', Bc133, 107–26.

53. Cornish, Rory T. *George Grenville: a bibliography.* London; Greenwood; 1992. Pp xxxiv, 227.

54. Cowie, Leonard W. *William Wilberforce, 1759–1833: a bibliography.* London; Greenwood; 1992. Pp 160.

55. Cox, R.W. *Sport in Britain: a bibliography of publications, 1800–1988.* Manchester; Manchester UP; 1991. Pp 272.

56. Craig, B.L. 'The role of records and of record-keeping in the development of the modern hospital in London, England and Ontario, Canada, c.1890–1940', *B. of the History of Medicine* 65 (1991), 376–97.

57. Creaton, Heather J.; Trowles, T. (comp.). 'Periodical articles on London history, 1990', *London J.* 16 (1991), 174–91.

58. Donnachie, Ian; Whatley, Christopher A. 'Bibliography' [Scottish history], Bc6, 171–80.

59. Dyson, Brian. 'Archives at the university of Hull: a research resource', *History of Education Soc. B.* 47 (1991), 56–63.

60. Edwards, Francis. 'Les archives de la province anglaise de la Compagnie de Jesus à Mount Street (Londres)', *L'histoire des croyants: memoire vivante des mommes melanges Charles Molette,* 2 vols. Pp 950. (Abbeville; Paillart; 1989), v. 2: 443–57.

61. Ellis, Joyce; Luckin, Bill; Hogg, Malcolm 'Review of periodical articles', *Urban History Yearbook* 18 (1991), 127–52.

62. Elrington, Christopher. 'The Victoria County History', *Local Historian* 22 (1992), 128–37.

63. English, Barbara; Palmer, J.J.N. (ed.). *Royal Historical Society annual*

bibliography of British and Irish history: publications of 1991. Oxford; Oxford UP; 1992. Pp x, 314.

64. Erskine, Audrey. 'The documentation of Exeter cathedral: the archives and their application', Bc104, 1–9.

65. Evans, A.K.B. 'The custody of Leicester's archives from 1273 to 1947', *Leicestershire Arch. & Historical Soc. T.* 66 (1992), 105–20.

66. Fellows-Jensen, Gillian. 'Scandinavian place-names of the Irish sea province', Bc108, 31–42.

67. Figala, Karin; Harrison, John; Petzold, Ulrich. *'De scriptoribus chemicis*: sources for the establishment of Isaac Newton's (al)chemical library', Bc87, 135–79.

68. Forster, G.C.F. 'Review of periodical literature and occasional publications', *Northern History* 28 (1992), 248–62.

69. Forster, Geoffrey. 'Libraries for the few: the members of the Association of Independent Libraries and their archives', *Library History* 9 (1991), 15–26.

70. Foster, Stewart. 'The archives of the English Province of the Servite Fathers', *Catholic Archives* 12 (1992), 17–20.

71. Fowler, Simon. *Army records for family historians.* London; Public Record Office; 1992. Pp 97.

72. Gelling, Margaret. 'A chronology for Suffolk place-names', Bc113, 53–64.

73. Gibbs, Graham; McGurk, Patrick; Stroud, Patricia. 'Bibliography of the writings of Michael Wilks', Bc12, 507–13.

74. Gibson, Jeremy; Medlycott, Mervyn (comp.). *Local census listings, 1522–1930: holdings in the British Isles.* Birmingham; Federation of Family History Societies; 1992. Pp 60.

75. Gibson, William T. 'The papers of Bishop Connop Thirlwall of St Davids', *J. of Welsh Ecclesiastical History* 9 (1992), 64–67.

76. Goodall, Francis. 'Bibliography' [business history], *Business Archives* 62 (1991), 55–63.

77. Goose, Nigel. 'Current bibliography of urban history', *Urban History* 19 (1992), 316–402.

78. Graham, T.W. 'A list of articles on Scottish history published during the year 1990', *Scottish Historical R.* 70 (1991), 172–80.

79. Griffiths, Rhidian; O'Leary, Paul (comp.). 'Articles relating to the history of Wales published mainly in 1990', *Welsh History R.* 16 (1992–3), 277–88.

80. Hale, Matthew; Hawkins, Richard; Partridge, Michael (comp.). 'List of publications on the economic and social history of Great Britain and Ireland published in 1991', *Economic History R.* 2nd ser. 45 (1992), 755–93.

81. Harris, Kate. *Glastonbury abbey records at Longleat house: a summary list.* Taunton; Somerset Record Soc. vol. 81; 1991. Pp xiii, 105.

82. Hart, Elizabeth. 'Bibliography of the published writings of Frank Baker', *P. of the Wesley Historical Soc.* 47 (1989–90), 232–39.

83. Harvey, Charles; Press, Jon. 'Researching a business biography: the career of William Morris, 1861–1896', *Business Archives* 62 (1991), 17–27.

84. Harvey, P.D.A. 'The documents of landscape history: snares and delusions', *Landscape History* 13 (1991), 47–52.
85. Hawkings, David T. *Criminal ancestors: a guide to historical criminal records in England and Wales.* Stroud; Sutton; 1992. Pp xiv, 458.
86. Haworth, Richard; Davies, K.M. 'A bibliography of the writings of J.H. Andrews', Bc118, 425–40.
87. Herren, Michael W. 'Hiberno-Latin lexical sources of Harley 3376, a Latin-Old English glossary', Bc222, 371–80.
88. Higgs, Edward. 'The 1891 census: continuity and change', *Local Historian* 22 (1992), 184–90.
89. Hodson, Sara S. 'A checklist of pre-Raphaelite manuscripts in the Huntington library', *Huntington Library Q.* 55 (1992), 147–223.
90. Hollis, Stephanie; Wright, Michael (comp.). *Old English prose of secular learning.* Woodbridge; Boydell (Annotated bibliographies of Old and Middle English literature, vol. 4); 1992. Pp 400.
91. Hough, Brenda. 'The work of the Church of England Record Centre', *Archives* 19 (1991), 351–57.
92. Howard-Hill, T.H. (ed.). *British literary bibliography, 1970–1979: a bibliography.* Oxford; Oxford UP; 1992. Pp xix, 912.
93. Hughes, Peter. 'Sorting religious archives', *Catholic Archives* 12 (1992), 3–16.
94. Huws, Gwilym; Roberts, D. Hywel E. *Wales.* Oxford; Clio (World bibliographical series vol. 122); 1991. Pp 247.
95. James, David. *Guide for family historians: list of parish, nonconformist and other related records held by the West Yorkshire Archive Service.* Wakefield; West Yorkshire Archive Service; 1992. Pp 140.
96. Jarvis, A.E. 'An attempt at a bibliography of Samuel Smiles', *Industrial Archaeology R.* 13 (1991), 162–71.
97. Jewitt, A. Crispin. *Maps for empire: the first 2,000 War Office maps.* London; British Library; 1992. Pp xxii, 511.
98. Jones, Peter (ed.). *Sir Isaac Newton: a catalogue of the manuscripts and papers collected and published on microfilm by Chadwyck-Healey.* Cambridge; Chadwyck-Healey; 1991. Pp 148.
99. Kain, Roger J.P.; Wilmot, Sarah. 'Tithe surveys in national and local archives', *Archives* 20 (1992), 106–17.
100. Kaiser-Lahme, Angela. 'Westalliierte Archivpolitik während und nach dem Zweiten Weltkrieg: die Beschlagnahmung, Sicherung und Auswertung deutscher Archive und Dokumente durch die Amerikaner und Briten, 1943–1946' [The archive policy of the Western allies during and after the second world war: the seizure, protection and analysis of German archives and documents by the Americans and British, 1943–1946], *Der Archivar* 45 (1992), 397–410.
101. Karsor, Eugene L. (ed.). *The battle of Jutland: a bibliography.* London; Greenwood; 1991. Pp 192.
102. Kavanagh, Gaynor. 'Mangles, muck and myths: rural history museums in Britain', *Rural History* 2 (1991), 187–204.
103. Keen, Rosemary. 'The Church Missionary Society Archives: or thirty years work in the basement', *Catholic Archives* 12 (1992), 21–30.

104. Kenney, Alison. 'Catalogue of the archives of the Manchester Society of Architects', *B. of the John Rylands University Library of Manchester* 74/2 (1992), 37–63.

105. Ker, N.R.; Piper, Alan J. *Medieval manuscripts in British libraries, vol. 4: Paisley-York.* Oxford; Oxford UP; 1992. Pp xl, 826.

106. Kushner, Tony. 'A history of Jewish archives in the United Kingdom', *Archives* 20 (1992), 3–16.

107. Land, Andrew; Lowe, Rodney; Whiteside, Noel (ed.). *The development of the welfare state, 1939–1951: a guide to the documents in the Public Record Office.* London; HMSO (Public Record Office handbooks 25).; 1992. Pp xii, 241.

108. Llewellyn, Howard (comp.). *A bibliography of Cardiff directories, 1795–1978.* Cardiff; Survey of Cardiff (occasional paper no. 4); 1990. Pp 47.

109. Lloyd, Andrew. 'Early atlas and printed books from the Manchester Geographical Society collection: a catalogue', *B. of the John Rylands University Library of Manchester* 73/2 (1991), 37–158.

110. Lynn, Martin. 'British business archives and the history of business in nineteenth century West Africa', *Business Archives* 62 (1991), 28–39.

111. Malcolmson, A.P.W. (ed.). *Eighteenth century Irish official papers in Great Britain: private collections, 2.* Belfast; Public Record Office of Northern Ireland; 1990. Pp 494.

112. Maxted, Ian. 'Joyce Youings: a bibliography of her work', Bc218, 227–50.

113. McCloskey, Donald N.; Hersh, George K. (ed.). *A bibliography of historical economics to 1980.* Cambridge; Cambridge UP; 1991. Pp 516.

114. McKinlay, Alan. 'British business history: a review of the periodical literature for 1990', *Business History* 34/2 (1992), 1–11.

115. McLelland, D. 'The Jordanhill bibliography of Scottish education', *Scottish Educational R.* 23 (1991), 104–11.

116. McNiven, Peter. '*The Guardian* archives in the John Rylands university library of Manchester', *B. of the John Rylands University Library of Manchester* 74/2 (1992), 65–84.

117. Meyer, W.R. 'The archives of the Leeds Education Authority, (1903–1960)', *Local Historian* 21 (1991), 162–67.

118. Mitchell, L.M.; Sampson, K.J.; Woolgar, C.M. (ed.). *A summary catalogue of the papers of Earl Mountbatten of Burma.* Southampton; University of Southampton; 1991. Pp vi, 309.

119. Morgan, Kevin. 'Theses and dissertations on British Labour history 1991', *Labour History R.* 57/2 (1992), 5–8.

120. Morgan, Nicholas. 'Born 1900, still going strong: setting up the United Distillers' Archive', *Business Archives* 63 (1992), 1–11.

121. Morgan, Paul (comp.). *Select index of manuscript collections in Oxford libraries outside the Bodleian.* Oxford; Bodleian Library; 1991. Pp 50.

122. Morgan, Raine (comp.). 'Annual list and brief review of articles on agrarian history, 1990', *Agricultural History R.* 40 (1992), 71–80.

123. Morris, V.J.; Orton, D.J. (comp.). 'List of books and pamphlets on agrarian history, 1990', *Agricultural History R.* 39 (1991), 171–75.

124. Morris, V.J.; Orton, D.J. (comp.). 'List of books and pamphlets on agrarian history 1991', *Agricultural History R.* 40 (1992), 168–72.

125. Mullett, Michael. *Sources for the history of English nonconformity, 1660–1830.* London; British Records Assocation (Archives and the user 8); 1991. Pp 116.

126. Myat-Price, E.M. 'List 3 of theses in history and archaeology relating to Surrey', *Surrey Arch. Collections* 81 (1991–2), 133–37.

127. Nenadic, Stana. 'Research in urban history: a review of recent theses', *Urban History Yearbook* 18 (1991), 153–72.

128. Newsome, Eileen Betty (comp.). *An index to the archbishop of York's marriage bonds and allegations, 1780–1789.* York; Borthwick Inst.; 1991. Pp iv, 326.

129. Nicol, Alexandra (ed.). *Curia regis rolls, 1242–1243, preserved in the Public Record Office, vol. 17: 26–27 Henry III.* London; HMSO; 1991. Pp xxi, 650.

130. Nixon, Malcolm I. 'Business records deposited in 1989', *Business Archives* 62 (1991), 64–77.

131. Ó Catháin, Séamus. 'The Irish folklore archive', *History Workshop J.* 31 (1991), 145–48.

132. O'Day, Alan; Stevenson, John. *Irish historical documents since 1800.* Dublin; Gill & Macmillan; 1992. Pp xi, 252.

133. O'Donnell, E.E. 'My Browne Heaven: the Father Browne collection', *Catholic Archives* 12 (1992), 32–38.

134. Paisey, David. 'Printed books in English and Dutch in early printed catalogues of German university libraries', Bc133, 127–48.

135. Perks, Rob. 'Audio and oral history collections', *Labour History R.* 57/1 (1992), 12–16.

136. Perks, Rob. (ed.). *Oral history: an annotated bibliography.* London; British Library National Sound Archive; 1990. Pp 183.

137. Pickering, Oliver. 'BCMSV: a database of manuscript English verse in Leeds university library', Bc220, 258–67.

138. Ponko, Vincent. *Britain in the middle east, 1921–1956: an annotated bibliography.* New York; Garland; 1990. Pp 542.

139. Pugh, Patricia (comp.). *A catalogue of the papers of Frederick Dealtry Lugard, baron Lugard of Abinger, 1858–1945 in Rhodes House library, Oxford.* Oxford; Bodleian Library; 1989. Pp 236.

140. Purcell, Mary. 'Dublin diocesan archives: Hamilton papers (2)', *Archivium Hibernicum* 45 (1990), 3–37.

141. Quinn, Dermot A. 'A bibliography of the published works of Geoffrey Barraclough', Bc81, 199–22.

142. Rasor, Eugene L. (comp.). *The Falklands/Malvinas campaign: a bibliography.* London; Greenwood; 1991. Pp 208.

143. Raymond, Stuart. *British genealogical periodicals: a bibliography of their contents, vol. 1: Collectanea topographica et genealogica.* Birmingham; Federation of Family History Societies; 1991. Pp 83.

144. Raymond Stuart. *British genealogical periodicals: a bibliography of*

their contents, vol. 2: The Genealogist, part 1. Birmingham; Federation of Family History Societies; 1991. Pp 44.

145. Raymond, Stuart. *Gloucestershire and Bristol: a genealogical bibliography.* Birmingham; Federation of Family History Societies (British genealogical bibliographies); 1992. Pp 88.

146. Raymond, Stuart (ed.). *Dorset: a genealogical bibliography.* Birmingham; Federation of Family History Societies; 1991. Pp vii, 113.

147. Raymond, Stuart (ed.). *Somerset: a genealogical bibliography.* Birmingham; Federation of Family History Societies; 1991. Pp 107.

148. Rendel, Rosemary. 'The Records of the Converts' Aid Society', *Catholic Archives* 12 (1992), 39–41.

149. Rhodes, Dennis E. 'Bibliography of the published works of Anna Simoni to the end of 1990', Bc133, 211–15.

150. Rhodes, Dennis E. 'The published writings of John Claud Trewinard Oates (1912–1990)', Bc88, 397–409.

151. Roberts, Phyllis B. *Thomas Becket in the medieval Latin preaching tradition: an inventory of sermons about St Thomas Becket, c.1170-c.1400.* The Hague; Nijhoff; 1992. Pp 270.

152. Rogal, Samuel J. (ed.). *Medicine in Great Britain from the Restoration to the nineteenth century, 1660–1800: an annotated bibliography.* London; Greenwood; 1992. Pp 272.

153. Rogers, David. *The Bodleian Library and its treasures, 1320–1700.* Henley-on-Thames; Ellis; 1991. Pp 176.

154. Rose, Louise; Rowe, Margery. 'The new north Devon Record Office', *Archives* 19 (1991), 289–96.

155. Rouse, Richard H.; Rouse, Mary A. (ed.). *Registrum Anglie de libris doctorum et auctorum veterum.* London; British Library with British Academy (Corpus of British medieval library catalogues); 1991. Pp clxix, 34.

156. Rudoe, Judy. *Decorative arts, 1850–1950: a catalogue of the British Museum collection.* London; British Museum; 1991. Pp 368.

157. Ruston, Alan. 'Richard Nelson Lee and Nelson Lee Junior, authors of Victorian pantomime: bibliographical checklist', *Nineteenth Century-Theatre* 18 (1990), 75–85.

158. Ryan, James G. (ed.). *Irish church records: their history, availability and use in family and local history.* Glenageary; Flyleaf; 1992. Pp 188.

159. Schreuder, M. Wilhelmina H. 'Sylvia Pankhurst's papers as a source', Bc50, 192–98.

160. Silke, John J. 'Lattice-works, open and closed: some reflections on diocesan archives', *Archivium Hibernicum* 45 (1990), 76–81.

161. Smith, M.M.; Lindsay, A. (ed.). *Index of English literary manuscripts, vol. 3: 1700–1800, part 3. Alexander Pope—Sir Richard Steele with a first line index to parts 1–3.* London; Mansell; 1992. Pp 480.

162. Spargo, P.E. 'Sotheby's Keynes and Yahuda—the 1936 sale of Newton's manuscripts', Bc87, 115–34.

163. Stacey, Stephen (ed.). *Neville Chamberlain: a bibliography.* London; Meckler; 1992. Pp 150.

164. Street, Sarah. 'Archival report: the Conservative Party archives', *Twentieth-Century British History* 3 (1992), 103–11.
165. Stuart, Denis. *Manorial records.* Chichester; Phillimore; 1992. Pp viii, 120.
166. Szarmach, Paul E. 'Cotton Tiberius A. iii, Arts. 26 and 27', Bc222, 29–42.
167. Taylor, Barry (comp.). *A bibliography of Norfolk history, vol. 2: 1974–1988.* Norwich; Centre of East Anglian Studies, University of East Anglia; 1991. [Pp].
168. Teviotdale, E.C. 'Some classified catalogues of the Cottonian library', *British Library J.* 18 (1992), 74–87.
169. Tweedale, Geoffrey; Procter, Timothy. 'Catalogue of the papers of Professor Sir William Boyd Dawkins in the John Rylands University Library of Manchester', *B. of the John Rylands University Library of Manchester* 74/2 (1992), 3–36.
170. van Selm, Bert. 'Dutch book trade catalogues printed before 1801 now in the British library', Bc133, 55–65.
171. Wark, Wesley K. 'In never-never land?: the British archives on intelligence', *Historical J.* 35 (1992), 195–203.
172. Webb, Cliff. 'The starting dates of English and Welsh parish registers', *Genealogists' Magazine* 24 (1992), 59–61.
173. Webb, Cliff (ed.). *National index of parish registers, vol. 9, part 1: Bedfordshire and Huntingdonshire.* London; Soc. of Genealogists; 1991. Pp 120.
174. Webb, Cliff (ed.). *National index of parish registers, vol. 9, part 2: Northamptonshire.* London; Soc. of Genealogists; 1991. Pp 86.
175. Wijffels, Alain. *Late sixteenth-century lists of law books at Merton college.* Cambridge; LP Publications; 1992. Pp 148.
176. Wilson, Eunice. *The records of the Royal Air Force: how to find the Few.* Birmingham; Federation of Family history Societies; 1991. Pp 66.
177. Wright, Erik Olin; Levine, Andrew; Sober, Elliott (ed.). *Reconstructing Marxism: essays on explanation and the theory of history.* London; Verso; 1992. Pp 260.
178. Wu, Duncan. 'Basil Montagu's manuscripts', *Bodleian Library Record* 14 (1992), 246–51.

(b) *Works of Reference*

1. Anon. *Science preserved: directory of scientific instruments in collections in the United Kingdom and Eire.* London; HMSO; 1992. Pp 270.
2. Anon. *The concise dictionary of national biography from earliest times to 1985.* Oxford; Oxford UP; 1992. 3 vols. Pp vi, 1090; v, 1048; v, 1193.
3. Anon. 'Register of research in progress on the history of London: a supplement', *London J.* 17 (1992), 93–97.
4. Arestis, Philip; Sawyer, Malcolm (ed.). *A biographical dictionary of dissenting economists.* Aldershot; Elgar; 1991. Pp 640.
5. Bausell, R. Barker. *Advanced research methodology: an annotated guide to sources.* Metuchen (NJ); Scarecrow; 1991. Pp viii, 903.

6. Baynton-Williams, Ashley. *Town and city maps of the British Isles, 1800–1855.* London; Studio; 1992. Pp 128.
7. Bell, David N. (ed.). *Libraries of the Cistercians, Gilbertines and Premonstratensians.* London; British Library; 1992. Pp 368.
8. Bellamy, Joyce M.; Saville, John (ed.). *Dictionary of Labour biography,* vol. 9. London; Macmillan; 1992. Pp 304.
9. Beresford, Maurice. '"The spade might soon determine it" : the representation of deserted medieval villages on Ordnance Survey plans, 1849–1910', *Agricultural History R.* 40 (1992), 64–70.
10. Cameron, Kenneth (ed.). *The place-names of Lincolnshire, part 2: the place-names of Yarborough wapentake.* Nottingham; English Place-name Soc.; 1991. Pp xix, 326.
11. Cameron, Kenneth (ed.). *The place-names of Lincolnshire, part 3: the place-names of Walshcroft wapentake.* Nottingham; English Place-name Soc.; 1992. Pp 248.
12. Carlsson, Stig. *Studies on middle English local bynames in East Anglia.* Lund; Lund UP; 1989. Pp 193.
13. Charles, B.G. *Place-names of Pembrokeshire.* Aberystwyth; National Library of Wales; 1992. 2 vols.
14. Coates, Richard. *The ancient and modern names of the Channel Islands: a linguistic history.* Stamford; Watkins; 1991. Pp xiv, 160.
15. Cole, Ann. 'Distribution and use of the Old English place-name *meretun*', *English Place-Name Soc. J.* 24 (1992), 30–41.
16. Colwell, Stella. *Genealogical dictionary of sources in the Public Record Office.* London; Weidenfeld & Nicolson; 1992. Pp xvii, 206.
17. Cross, Paul J. 'The private case: a history', Bc61, 201–40.
18. Cruft, Catherine H. 'The state of garden archaeology in Scotland', Bc64, 175–89.
19. David, Andrew (ed.). *Charts and coastal views of Captain Cook's voyages, vol. 2: the voyage of the 'Resolution' and 'Adventure'.* London; Hakluyt Soc. extra series, vol. 44; 1992. Pp 448.
20. Davies, John Whittaker. *Montgomeryshire pleadings in the court of Chancery, 1558–1714.* Aberystwyth; National Library of Wales; 1991. Pp xix, 182.
21. Derdak, Thomas (ed.). *International directory of company histories, 4.* London; St James; 1991. Pp 700.
22. Dodgson, J. McN.; Palmer, J.J.N. (ed.). *Index to Domesday Book, part 1: index of places.* Chichester; Phillimore; 1992. Pp ix, 327.
23. Dodgson, J. McN.; Palmer, J.J.N. (ed.). *Index to Domesday Book, part 2: index of persons.* Chichester; Phillimore; 1992. Pp xii, 405.
24. Fallon, John P. *Marks of London goldsmiths and silversmiths, 1837–1914.* London; Barrie & Jenkins; 1992. Pp 352.
25. Foy, J.D. (ed.). *Index to Domesday Book, part 3: subjects.* Chichester; Phillimore; 1992. Pp 318.
26. Goring, Rosemary (ed.). *Chambers' Scottish biographical dictionary.* Edinburgh; Chambers; 1992. Pp 512.
27. Gourvish, Terry. 'Writing British Rail's history', *Business Archives* 62 (1991), 1–9.

28. Green, Miranda J. *Dictionary of Celtic myth and legend*. London; Thames & Hudson; 1992. Pp 250.
29. Guy, Susanna (comp.). *English local studies handbook: a guide to resources for each county including libraries, record offices, societies, journals and museums*. Exeter; Exeter UP; 1992. Pp xiv, 343.
30. Harris, P.R. 'The move of printed books from Montagu House, 1838–1842', Bc61, 75–102.
31. Hast, Adele (ed.). *International directory of company histories, vol. 3: health and personal care products*. London; St James; 1991. Pp 857.
32. Hast, Adele (ed.). *International directory of company histories, vol. 4: mining and metals*. London; St James; 1991. Pp 847.
33. Hast, Adele (ed.). *International directory of company histories, vol. 5: retail and wholesale*. London; St James; 1992. Pp 884.
34. Henderson, Alan C. *Hop tokens of Kent and Sussex and their issuers*. London; Spink; 1990. Pp 128.
35. Hill, F.J. 'The shelving and classification of printed books', Bc61, 1–74.
36. Jones, Gwenith (ed.). *The descent of dissent: guide to the nonconformist records at the Leicestershire Record Office*. Leicester; Leicestershire Museums; 1989. Pp 75.
37. Kamaryc, R.M. (comp.). *Recipients of the Military Cross book 2: 1960 to 1991*. Harlow; Kamaryc; 1991. Pp 50.
38. Kepos, Paula (ed.). *International directory of company histories, vol. 6: service industries.*. London; St James; 1992. Pp 748.
39. King, Alec Hyatt. 'Quodlibet: some memoirs of the British Museum and its music room, 1934–1976', Bc61, 241–98.
40. King, Alec Hyatt. 'The traditional maintenance of the general catalogue of printed books', Bc61, 165–99.
41. Manchester, A.H. 'An introduction to iconographical studies of legal history, I: England and Wales', Bc195, 85–94.
42. Manley, Keith A. 'The book wolf bites a Bohn: Panizzi, Henry Bohn and legal deposit, 1850–1853', Bc61, 145–63.
43. Mills, A.D. *A dictionary of English place-names*. Oxford; Oxford UP; 1991. Pp xxxiii, 388.
44. Moody, David. *Scottish towns: a guide for local historians*. London; Batsford; 1992. Pp 224.
45. Munn, Charles W. 'Writing a business history: the Clydesdale bank', *Business Archives* 62 (1991), 10–16.
46. Murphy, Michael. *Newspapers and local history*. Chichester; Phillimore; 1991. Pp 24.
47. Nixon, Howard Millar; Foot, Mirjam Michaela (ed.). *The history of decorated bookbinding in England: Lyell lectures in bibliography*. Oxford; Oxford UP; 1992. Pp xviii, 124.
48. Owen, Tim; Pilbeam, Elaine. *Ordnance survey: map makers to Britain since 1791*. London; HMSO; 1992. Pp 196.
49. Pickford, Christopher J. (comp.). *Bedfordshire clock and watchmakers, 1352–1880: a biographical dictionary with selected documents*. Bedford; Bedfordshire Historical Record Soc. (publications 70); 1991. Pp xii, 252.

50. Procter, M.R. (ed.). *Education on Merseyside: guide to the sources.* Liverpool; Merseyside Archives Liaison Group; 1992. Pp 306.
51. Ravenhill, William L.D.; Padel, O.J. (ed.). *A map of the county of Cornwall, 1699, by Joel Gascoyne: reprinted in facsimile.* Exeter; Devon and Cornwall Record Soc. ns 34; 1991. Pp 38.
52. Richardson, William A.R. 'Lyonesse and the Wolf: a case study in place-name corruption', *English Place-Name Soc. J.* 24 (1992), 4–29.
53. Risse, Guenter B.; Warner, John Harley. 'Reconstructing clinical activities: patient records in medical history', *Social History of Medicine* 5 (1992), 183–205.
54. Ritchie, L.A. (ed.). *Shipbuilding industry: a guide to historical records.* Manchester; Manchester UP; 1992. Pp 304.
55. Robinson, Mark. 'Environmental archaeology of the river gravels: past achievements and future directions', Bc192, 47–62.
56. Salisbury, Joyce E. *Medieval sexuality: a research guide.* London; Garland; 1990. Pp 210.
57. Smallbone, Linda (comp.). *Employers' and trade associations' history.* Coventry; University of Warwick, Modern Records Centre (Sources booklets 4); 1992. Pp 39.
58. Smart, Veronica (ed.). *Sylloge of coins of the British Isles, vol. 41: cumulative index of volumes 21–40.* Oxford; Oxford UP for the British Academy; 1992. Pp 141.
59. Sternberg, Ilse. 'The acquisitions policies and funding of the department of printed books, 1837–1959', Bc61, 103–43.
60. Teed, Peter. *Dictionary of twentieth-century history, 1914–1990.* Oxford; Oxford UP; 1992. Pp 520.
61. Tiller, Kate. *English local history: an introduction.* Stroud; Sutton; 1992. Pp viii, 247.
62. Torvell, David. 'The significance of *Here-ford*', *English Place-Name Soc. J.* 24 (1992), 42–50.
63. Whimster, Rowan. 'Aerial photography and the British gravels: an agenda for the 1990s', Bc192, 1–14.
64. Wilson, David M. *Awful ends: British Museum book of epitaphs.* London; British Museum; 1992. Pp 96.

(c) *Historiography*

1. Abbattista, Guido. *Commercio, colonie e impero alla vigilia della rivoluzione americana: John Campbell pubblicista e storico nell' Inghilterra del secolo XVIII* [Commerce, colonies and empire on the eve of the American revolution: John Campbell publicist and historian of eighteenth century England]. florence; Olschki; 1989. Pp 464.
2. Aers, David. 'A whisper in the ear of early modernists; or, reflections on literary critics writing the "history of the subject"', Bc94, 177–202.
3. Aers, David. 'Introduction' [culture and history], Bc94, 1–6.
4. Alston, Robin C. 'Library history: a place in the education of librarians?', *Library History* 9 (1991), 37–51.
5. Anderson, Perry. *English questions.* London; Verso; 1992. Pp 368.

6. Bargielowska, Ina-Maria Zweiniger. 'The road from 1945: an interview with Paul Addison', *Contemporary Record* 6 (1992), 186–93.

7. Barker, T.C. *'The transport revolution from 1770* in retrospect', Bc117, 1–8.

8. Beresford, Maurice. 'Professor W.G. Hoskins: a memoir', *Agricultural History R.* 40 (1992), 164–67.

9. Berg, Maxine. 'The first women economic historians', *Economic History R.* 2nd ser. 45 (1992), 308–29.

10. Berridge, Virginia. 'Health and medicine in the twentieth century: contemporary history and health policy', *Social History of Medicine* 5 (1992), 307–16.

11. Beynon, Huw. 'Class and historical explanation', Bc43, 230–49.

12. Black, Jeremy. 'Can Scottish history be left to the academics?', *Scottish Literary J.* 36 (1992), 1–5.

13. Black, Jeremy. 'Maps and chaps: the historical atlas: a perspective from 1992', *Storia della Storiografia* 21 (1992), 91–114.

14. Black, Jeremy. 'The historical atlas: teaching tool or coffee-table book?', *History Teacher* 25 (1992), 489–512.

15. Blackburn, Richard. *The vampire of reason: an essay in the philosophy of history*. London; Verso; 1990. Pp 232.

16. Bock, Gisela. 'Challenging dichotomies: perspectives on women's history', Bc8, 1–23.

17. Botrel, J.F. 'A contribution to an "historical history" of literature', *J. of European Studies* 21 (1991), 55–66.

18. Boucher, David. 'Politics in a different mode: an appreciation of Michael Oakeshott, 1901–1990', *History of Political Thought* 12 (1991), 717–29.

19. Breuilly, John J. 'Introduction: making comparisons in history', Bc219, 1–25.

20. Brock, Michael. 'Michael Howard's contribution to historical studies', Bc150, 295–98.

21. Brooke, Christopher N.L. 'Cambridge and the antiquaries, 1500–1840', *Cambridge Antiquarian Soc. P.* 79 (1992 for 1990), 1–12.

22. Burke, Peter. *History and social theory*. Oxford; Polity; 1992. Pp 180.

23. Bush, M.L. 'An anatomy of nobility', Bc43, 26–46.

24. Cannadine, David. *Trevelyan: a life in history*. London; HarperCollins; 1992. Pp xvi, 288.

25. Canovan, Margaret. 'Reason recycled: the Enlightenment today', *Enlightenment & Dissent* 9 (1990), 3–13.

26. Cantor, David. 'Contracting cancer?: the politics of commissioned histories', *Social History of Medicine* 5 (1992), 131–42.

27. Carling, Alan. 'Bread and butter ethics', Bc9, 231–42.

28. Carroll, Peter N. *Keeping time: memory, nostalgia, and the art of history*. Athens (Ga); Georgia UP; 1990. Pp 224.

29. Carter, Alan. 'Functional explanation and the state', Bc9, 205–28.

30. Chibnall, Marjorie; Markus, Robert; Hill, Rosalind. 'Christopher Brooke at Cambridge, Liverpool and London', Bc70, xvii–xxiv.

31. Cipolla, Carlo M. *Between history and economics: an introduction to economic history*. Oxford; Blackwell; 1991. Pp 256.

32. Clark, Samuel. 'Historical anthropology, historical sociology, and the making of modern Europe', Bc124, 324.
33. Cowan, Ian B. 'Thomas Ian Rae, 1926–1989', Bc83, 1–6.
34. Cowling, Mark; Manners, Jon. 'Pre-history: the debate before Cohen', Bc9, 9–29.
35. Croad, Stephen. 'The National Buildings Record: the early years', *T. of the Ancient Monuments Soc.* 36 (1992), 79–98.
36. Davidoff, Leonore. 'Gender, class and nation', *History Today* 42/2 (1992), 49–53.
37. Davis, R.H.C. 'Geoffrey Barraclough and the lure of charters', Bc81, 23–34.
38. Diederiks, Herman; Hohenberg, Paul M. 'The visible hand and the fortune of cities: a historiographic introduction', Bc42, 1–16.
39. Doyle, William. 'Myths of order and ordering myths', Bc43, 218–29.
40. Dunbar, John G. 'The Royal Commission on the Ancient and Historical Monuments of Scotland: the first eighty years', *T. of the Ancient Monuments Soc.* 36 (1992), 13–77.
41. Duncan, A.A.M. 'Ian Borthwick Cowan, 1932–1990: an appreciation', *Scottish Historical R.* 70 (1991), 210–14.
42. Dunne, Tom. 'Oliver MacDonagh', Bc143, 1–13.
43. English, Barbara; Palmer, J.J.N. 'The making of the Royal Historical Society's *Annual Bibliography*', *History & Computing* 4 (1992), 110–14.
44. Forman, Paul. 'Independence, not transcendance, for the historian of science', *Isis* 82 (1991), 71–86.
45. Fowler, Peter J. *The past in contemporary society*. London; Routledge; 1992. Pp 288.
46. Frank, Robert. 'Les negationnistes britanniques', *Relations Internationales* 65 (1991), 39–47.
47. Gelling, Margaret. 'The present state of English place-names studies', *Local Historian* 22 (1992), 114–27.
48. Grafsky, V.G. 'Paul Vinogradoff as legal historian', Bc195, 211–21.
49. Grant, Don. 'The genesis of Australian studies', Bc121, 1–09.
50. Gruffudd, Ceris. 'The Welsh in Australia: a bibliography', Bc121, 144–73.
51. Hakfoort, C. 'The missing syntheses in the historiography of science', *History of Science* 29 (1991), 207–16.
52. Halfpenny, Peter. 'A refutation of historical materialism?', Bc9, 133–58.
53. Harris, Tim. 'From rage of party to age of oligarchy? Rethinking the later Stuart and early Hanoverian period', *J. of Modern History* 64 (1992), 700–20.
54. Haughton, Joseph P. 'John Harwood Andrews: an appreciation', Bc118, 1–04.
55. Hendry, Joy. 'Snug in the asylum of taciturnity: women's history in Scotland', Bc6, 125–42.
56. Hussey, John. 'John Fortescue, James Edmonds, and the history of the great war: a case of "ritual murder"', *J. of the Soc. for Army Historical Research* 70 (1992), 101–13.
57. Jenkins, K. *Re-thinking history*. London; Routledge; 1991. Pp 112.

58. Kavanagh, Dennis. 'Why political science needs history', *Political Studies* 39 (1991), 479–95.
59. Kemp, Anthony. *The estrangement of the past: study in the origins of modern historical consciousness.* New York; Oxford UP; 1991. Pp xi, 228.
60. Kent, Christopher A. 'The establishment of British social history?', *Canadian J. of History* 26 (1991), 267–75 [review article].
61. Kenyon, John. 'Revisionism and post-revisionism in early Stuart history', *J. of Modern History* 64 (1992), 686–99.
62. Knight, Alan. 'Revisionism and revolution: Mexico compared to England and France', *Past & Present* 134 (1992), 159–99.
63. Lawrence, Christopher. 'Democratic, divine and heroic: the history and historiography of surgery', Bc24, 1–47.
64. le Patourel, Jean. 'John G. Hurst—a potted biography', Bc191, 1–6.
65. MacCurtain, Margaret; O'Dowd, Mary; Luddy, Maria. 'An agenda for women's history in Ireland, 1500–1900', *Irish Historical Studies* 28 (1992), 1–37.
66. MacDonagh, Oliver. '*Apologia Pro Vita Sua*: in vita mea vademecum', Bc143, 224–40.
67. Macdonald, Graham. 'Philosophical foundations for functional sociology', Bc9, 67–105.
68. Mayfield, David; Thorne, Susan. 'Social history and its discontents: Gareth Stedman Jones and the politics of language', *Social History* 17 (1992), 165–88.
69. Mellar, Harvey. 'Historical explanation and educational computing', *History & Computing* 3 (1991), 178–82.
70. Middleton, Roger; Wardley, Peter; *et al.* 'Review of information technology, 1990', *Economic History R.* 2nd ser. 44 (1991), 343–72.
71. Morris, R.J. 'Nominal record linkage: into the 1990s', *History & Computing* 4 (1992), iii-vii.
72. Murray, Hugh. *The Yorkshire Architectural and York Archaeological Society, 1842–1992: a sesquicentenary retrospect.* York; Yorkshire Architectural & York Archaeological Society; 1992. Pp ii, 66.
73. Offen, Karen; Pierson, Ruth Roach; Rendall, Jane. 'Introduction' [writing women's history], Bc8, xix-xli.
74. O'Neill, Robert. 'In appreciation of Michael Howard', Bc150, 299–302.
75. Osterhammel, Jürgen. 'Nation und Zivilisation in der britischen Historiographie von Hume bis Macaulay' [Nation and civilisation in British historiography from Hume to Macaulay], *Historische Zeitschrift* 254 (1992), 281–340.
76. Ouen-John, Henry. 'From antiquarianism to archaeology' [in Swansea], Bc212, 244–59.
77. Palmer, Bryan. 'Is there now, or has there ever been, a working class?', *History Today* 42/3 (1992), 51–54.
78. Persson, Karl Gunnar. 'Was feudalism inevitable?', *Scandinavian Economic History R.* 39/1 (1991), 68–76.
79. Pettegree, Andrew. 'Re-writing the English Reformation', *Nederlands Archief voor Kerkgeschiedenis* 72 (1992), 37–58.

80. Phythian-Adams, Charles. 'Hoskins's England: a local historian of genius and the realisation of a theme', *Leicestershire Arch. & Historical Soc. T.* 66 (1992), 143–59.

81. Pocock, J.G.A. 'History and sovereignty: the historiographical response to Europeanization in two British cultures', *J. of British Studies* 31 (1992), 358–89.

82. Porter, Roy. 'Introduction' [myths of the English], Bc190, 1–11.

83. Reddy, William M. 'The concept of class', Bc43, 13–25.

84. Reeder, David A. 'Schooling in the city: educational history and the urban variable', *Urban History* 19 (1992), 23–38.

85. Rendall, Jane. 'Uneven developments: women's history, feminist history, and gender history in Great Britain', Bc8, 45–57.

86. Reynolds, Susan. 'The writing of medieval urban history in England', *Theoretische Geschiedenis* 19/1 (1992), 43–57.

87. Robbins, Michael. 'The progress of transport history', *J. of Transport History* 3rd ser. 12 (1991), 74–87.

88. Rodger, Richard. '*Urban History*: prospect and retrospect', *Urban History* 19 (1992), 1–22.

89. Rogers, Nicholas. 'The anthropological turn in social history', Bc124, 325–70.

90. Ross, Seamus. 'The historian and software engineering considerations', *History & Computing* 3 (1991), 141–50.

91. Sanders, Richard. 'Contemporary history in the media', *Contemporary Record* 6 (1992), 220–24.

92. Savage, Hugh; Challinor, Ray; Heffer, Eric S. 'Remembering Harry' [McShane], Bc175, 10–24.

93. Schürer, Kevin. 'The future for local history: boom or recession?', *Local Historian* 21 (1991), 99–108.

94. Schürer, Kevin; Anderson, S.J.; Duncan, J.A. (ed.). *A guide to historical datafiles held in machine-readable form.* London; Association for History and Computing; 1992. Pp 340.

95. Shapiro, Alan E. 'Beyond the dating game: watermark clusters and the composition of Newton's *Opticks*', Bc87, 181–227.

96. Shaw, Gareth. 'The study of retail development', Bc35, 1–14.

97. Silverman, Marilyn; Gulliver, P.H. 'Historical anthropology and the ethnographic tradition: a personal, historical, and intellectual account', Bc124, 3–74.

98. Simonton, Dean Keith. *Psychology, science, and history: an introduction to historiometry.* London; Yale UP; 1991. Pp 304.

99. Smith, K.E. 'Beyond public and private spheres: another look at women in Baptist history and historiography', *Baptist Q.* 34 (1991), 79–87.

100. Stevenson, John. 'The Cambridge social history of England' [review article], *History* 76 (1991), 418–32.

101. Stone, Lawrence; Spiegel, Gabriele M. 'History and post-modernism', *Past & Present* 135 (1992), 189–208.

102. Sweet, P.R. 'Der Versuch amtlicher Einflussnahme auf die Edition der *Documents on German foreign policy, 1933–1941*: ein Fall aus den

fünfziger Jahren', *Vierteljahrshefte für Zeitgeschichte* 39 (1991), 265–303.

103. Thompson, Dorothy. '19th century hidden agendas', *History Today* 42/2 (1992), 45–48.

104. Tomaselli, Sylvana. 'Reflections on the history of the science of woman', *History of Science* 29 (1991), 185–205.

105. Trela, D.J. *Thomas Carlyle's writing of 'Oliver Cromwell's letters and speeches'*. Lampeter; Mellen; 1992. Pp 220.

106. Wahrman, Dror. 'National society, communal culture: an argument about the recent historiography of eighteenth-century Britain', *Social History* 17 (1992), 43–72.

107. Waite, Peter B. 'Invading privacies: biography as history', *Dalhousie R.* 69 (1990), 479–95.

108. Wallwork, Janet. 'Irish treasures of the John Rylands Library, Manchester', *History Workshop J.* 31 (1991), 136–44.

109. Wetherly, Paul. 'Introduction' [Marx's theory of history], Bc9, 1–6.

110. Wetherly, Paul. 'Mechanisms, methodological individualism and marxism: a response to Elster', Bc9, 107–31.

111. Wetherly, Paul; Carling, Alan. 'An analytical outline of historical materialism', Bc9, 31–64.

112. Whaples, R. 'A quantitative history of the *Journal of Economic History* and the cliometric revolution', *J. of Economic History* 51 (1991), 289–301.

113. Williams, Glanmor. 'Romantic and realist: Theophilus Evans and Theophilus Jones', *Archaeologia Cambrensis* 140 (1991), 21–27.

114. Williamson, Jeffrey G. *Inequality, poverty and history*. Oxford; Blackwell; 1991. Pp 144.

115. Yeo, R. 'Reading encyclopedias: science and the organization of knowledge in British dictionaries of arts and sciences, 1730–1850', *Isis* 82 (1991), 24–49.

B. GENERAL

(a) *Long Periods—National*

1. Alderman, Geoffrey. *Modern British Jewry*. Oxford; Oxford UP; 1992. Pp xiii, 397.

2. Allen, Robert C. 'Agrarian fundamentalism and English agricultural development', *Rivista di Storia Economica* 7 (1990), 153–61.

3. Anderson, J. Stuart. *Lawyers and the making of English land law, 1832–1940*. Oxford; Oxford UP; 1992. Pp viii, 360.

4. Anderson, R.D. *Universities and elites in Britain since 1800*. London; Macmillan; 1992. Pp 88.

5. Archer, Ian. *The history of the Haberdashers' Company*. Chichester; Phillimore; 1991. Pp xv, 329.

6. Ashley, Raymond E. 'The search for longitude', *American Neptune* 51 (1991), 252–66.

7. Bailey, Douglass W. 'The living house: signifying continuity', Bc38, 19–48.

8. Bailey, Richard W. *Images of English: a cultural history of the language.* Cambridge; Cambridge UP; 1992. Pp 329.
9. Barry, Jonathan; Jones, Colin. 'Introduction' [medicine and charity], Bc26, 1–13.
10. Barry, Jonathan; Melling, Joseph. 'Introduction' [culture in history], Bc217, 3–27.
11. Bassett, Steven; Dyer, Christopher C.; Holt, Richard. 'Introduction' [death in towns], Bc216, 1–7.
12. Beckett, John V. 'Life and leisure in early modern England' [review article], *J. of British Studies* 31 (1992), 191–98.
13. Beckford, J.A. 'Politics and religion in England and Wales', *Daedalus* 120/2 (1991), 179–201.
14. Beedell, Ann. *Decline of the English musician, 1788–1888: a family of English musicians in Ireland, England, Mauritius and Australia.* Oxford; Oxford UP; 1992. Pp xv, 329.
15. Begg, Paul; Skinner, Keith. *Scotland Yard files: 150 years of the CID, 1842–1992.* London; Headline; 1992. Pp 320.
16. Bence-Jones, Mark. *The Catholic families.* London; Constable; 1992. Pp 336.
17. Bennett, Judith M. 'Conviviality and charity in medieval and early modern England', *Past & Present* 134 (1992), 19–41.
18. Black, Jeremy. *Convergence or divergence?: Britain and the continent, 1660–1992.* Durham; Durham university (Centre for European Studies no. 1); 1992. Pp 24.
19. Black, Michael. *Short history of Cambridge university press.* Cambridge; Cambridge UP; 1992. Pp 115.
20. Bonfield, Lloyd. 'Canon law in colonial America: some evidence of the transmission of English ecclesiastical court law and practice to the American colonies', Bc125, 253–71.
21. Brailsford, Dennis. *British sport: a social history.* London; Lutterworth; 1992. Pp 176.
22. Brand, Paul. *Origins of the English legal profession.* Oxford; Blackwell; 1992. Pp ix, 236.
23. Brooke, Robert. *A history of the county cricket championship.* Enfield; Guiness; 1991. Pp 192.
24. Brown, Desmond H. 'Abortive attempts to codify English criminal law', *Parliamentary History* 11 (1992), 1–39.
25. Brown, Peter. *Keeping of Christmas: England's festive tradition, 1760–1840.* York; York Civic Trust; 1992. Pp 32.
26. Buchet, C. 'La 'royal navy' et les levées d'hommes aux Antilles (1689–1763): difficultés rencontrées et modalités évolutives', *Histoire, Economie et Société* 9 (1990), 521–43.
27. Bush, Helen; Zvelebil, Marek. 'Pathology and health in past societies: an introduction', Bc211, 3–10.
28. Bynum, William F.; Lock, Stephen; Porter, Roy. 'Introduction' [medical journals and medical knowledge], Bc28, 1–5.
29. Caldwell, John. *The Oxford history of English music, vol. 1: from the beginnings to c.1715.* Oxford; Oxford UP; 1991. Pp xvii, 691.

30. Campbell, W.A.; Morris, P.J.T. 'Introduction: milestones in the history of the chemical industry', Bc36, 3–13.
31. Caple, C. 'The detection and definition of an industry: the English medieval and post-medieval pin industry', *Arch. J.* 148 (1991), 241–55.
32. Carey, John. 'A British myth of origins?', *Historical Religion* 31 (1991), 24–38.
33. Castiglione, Dario. 'Excess, frugality and the spirit of capitalism: readings of Mandeville on commercial society', Bc217, 155–79.
34. Cavallo, Sandra. 'The motivations of benefactors: an overview of approaches to the study of charity', Bc26, 46–62.
35. Challis, Christopher E. 'Appendix 1: mint output, 1220–1985', Bc226, 673–98.
36. Challis, Christopher E. 'Appendix 2: mint contracts, 1279–1817', Bc226, 699–758.
37. Chapman, Stanley D. *Merchant enterprise in Britain: from the industrial revolution to world war I.* Cambridge; Cambridge UP; 1992. Pp xv, 339.
38. Charsley, Simon R. *Wedding cakes and cultural history.* London; Routledge; 1992. Pp 162.
39. Clapson, Mark. *A bit of a flutter: popular gambling in England, c.1820–1961.* Manchester; Manchester UP; 1992. Pp 240.
40. Clark, Gregory. 'The economics of exhaustion, the Postan thesis, and the agricultural revolution', *J. of Economic History* 52 (1992), 61–84.
41. Coleman, D.C. 'New business history for old' [review article], *Historical J.* 35 (1992), 239–44.
42. Colley, Linda. 'Britishness and otherness: an argument', *J. of British Studies* 31 (1992), 309–29.
43. Colpi, Terri. *The Italian factor: the Italian community of Great Britain.* London; Mainstream; 1991. Pp 256.
44. Corfield, Penny. 'The democratic history of the English gentleman', *History Today* 42/12 (1992), 40–47.
45. Corrigan, Philip. 'Setting up the seen', Bc217, 217–46.
46. Cressy, David. 'The fifth of November remembered', Bc190, 68–90.
47. Crouzet, François M. 'Angleterre-Brésil, 1697–1850: un siècle et demi d'échanges commerciaux', *Histoire, Economie et Société* 9 (1990), 286–317.
48. Crowther, M. Anne. 'The tramp', Bc190, 91–113.
49. Currie, Christopher K. 'The early history of the carp and its economic significance in England', *Agricultural History R.* 39 (1991), 97–107.
50. Davenport-Hines, R.P.T. *Macmillans.* London; Heinemann; 1992. Pp 365.
51. Davies, Andrew. *To build a new Jerusalem: the labour movement from the 1890s to the 1990s.* London; Joseph; 1992. Pp vi, 344.
52. Davies, David. 'The birth of the imperial navy? Aspects of maritime strategy, c.1650–1850', Bc173, 14–38.
53. Dintenfass, Michael. *Decline of industrial Britain, 1870–1980.* London; Routledge; 1992. Pp 96.
54. Donnelly, J.F. 'Science, technology and industrial work in Britain, 1860–1930: towards a new synthesis', *Social History* 16 (1991), 191–202.

55. Dosi, Giovanni; Giannetti, Renato; Toninelli, Pier Angelo. 'Introduction' [technology and business], Bc134, 1–26.
56. Duffy, Michael. 'Introduction' [naval power], Bc173, 1–13.
57. Dyer, G.P.; Gaspar, P.P. 'Reform, the new technology and Tower hill, 1700–1966', Bc226, 398–606.
58. Eley, Geoff. 'Culture, Britain, and Europe', *J. of British Studies* 31 (1992), 390–414.
59. Elton, Sir Geoffrey R. *The English.* Oxford; Blackwell; 1992. Pp 256.
60. Emmerson, Robin. *British teapots and tea drinking, 1700–1850.* London; HMSO; 1992. Pp xii, 330.
61. Emsley, Clive. 'The English bobby: an indulgent tradition', Bc190, 114–35.
62. Everson, Paul L. 'Field survey and garden earthworks', Bc64, 6–19.
63. Ferguson, Moira. *Subject to others: British women writers and colonial slavery, 1670–1834.* London; Routledge; 1992. Pp 432.
64. Fielding, Steven. *Class and ethnicity: Irish Catholics in England, 1880–1939.* Buckingham; Open UP; 1992. Pp 192.
65. Fradenburg, Louise Olga. 'Introduction: rethinking queenship', Bc91, 1–13.
66. Freyer, Tony Allan. *Regulating big business: anti-trust in Great Britain and America, 1880–1990.* Cambridge; Cambridge UP; 1992. Pp xiii, 399.
67. Fyfe, Janet. *Books behind bars: role of books, reading and libraries in British prison reform, 1701–1911.* London; Greenwood; 1992. Pp 289.
68. Gibson, William. '"Withered branches and weighty symbols": surname substitution in England, 1660–1880', *British J. for Eighteenth-Century Studies* 15 (1992), 17–33.
69. Giddings, Robert. 'Delusive seduction: pride, pomp, circumstance and military music', Bc10, 25–49.
70. Ginter, Donald E. *A measure of wealth: the English land tax in historical analysis.* London; Hambledon; 1992. Pp xxvi, 711.
71. Gittings, Clare. 'Urban funerals in late medieval and Reformation England', Bc216, 170–83.
72. Glasscock, Robin E. 'Introduction' [historic landscapes], Bc181, 9–14; 239.
73. Goodings, R.F.; Dunford, J.E. 'Her Majesty's Inspectorate of Schools, 1839–1989: the question of independence', *J. of Educational Administration & History* 22/1 (1990), 1–8.
74. Grant, Brian. *The deaf advance: a history of the British Deaf Association, 1890–1990.* Edinburgh; Pentland; 1990. Pp xiv, 163.
75. Hampsher-Monk, Iain. 'Prices as descriptions: reasons as explanations', Bc217, 47–74.
76. Hansford-Miller, Frank. *A history and geography of English religion, vol. 1: the provinces and the dioceses.* Canterbury; Abcado; 1990. Pp 90.
77. Hansford-Miller, Frank. *A history and geography of English religion, vol. 3: rectories, vicarages, clergy, chapels, peculiars, gilds and chantries.* Canterbury; Abcado; 1990. Pp 100.

78. Hareven, T.K. 'The history of the family and the complexity of social change', *American Historical R*. 96 (1991), 95–124.
79. Harrison, D.F. 'Bridges and economic development, 1300–1800', *Economic History R*. 2nd ser. 45 (1992), 240–61.
80. Hart, Jenifer. *Proportional representation: critics of the British electoral system, 1820–1945*. Oxford; Oxford UP; 1992. Pp viii, 312.
81. Harvey, A.D. *Collision of empires: Britain in three world wars, 1793–1945*. London; Hambledon; 1992. Pp 784.
82. Head-König, Anne-Lise. 'Demographic history and its perception of women from the seventeenth to the nineteenth century', Bc8, 25–44.
83. Healey, Edna. *Coutts and Co., 1692–1992: the portrait of a private bank*. London; Hodder & Stoughton; 1992. Pp 480.
84. Helmholz, Richard H. 'Canon law in post-Reformation England', Bc125, 203–22.
85. Hennessy, Alistair. 'Argentines, Anglo-Argentines and others', Bc136, 9–48.
86. Hennessy, Alistair. 'Introduction' [Argentina and Britain], Bc136, 1–6.
87. Holmes, Ken. *Two centuries of shoemaking: Start-rite, 1792–1992*. Norwich; Start-rite Shoes; 1992. Pp 112.
88. Honeyman, Katrina; Goodman, Jordan. 'Women's work, gender conflict, and labour markets in Europe, 1500–1900', *Economic History R*. 2nd ser. 44 (1991), 608–28.
89. Horwitz, Henry. 'The disadvantaged in early modern Britain' [review article], *Historical J*. 35 (1992), 233–38.
90. Howkins, Alun. 'Social history and agricultural history' [review article], *Agricultural History R*. 40 (1992), 160–63.
91. Hughes, Geoffrey. *Swearing: a social history of foul language, oaths and profanity in English*. Oxford; Blackwell; 1991. Pp 304.
92. Hughes, Thomas P. 'The dynamics of technological change: salients, critical problems, and industrial revolutions', Bc134, 97–118.
93. Ingram, Edward. *Britain's Persian connection, 1798–1828: prelude to the great game in Asia*. Oxford; Oxford UP; 1992. Pp xx, 351.
94. Kaelble, Hartmut; Thomas, Mark. 'Introduction' [income distribution], Bc34, 1–56.
95. Kearney, Hugh. 'The Irish and their history', *History Workshop J*. 31 (1991), 149–55.
96. Kerridge, Eric. *Common fields of England*. Manchester; Manchester UP; 1992. Pp viii, 216.
97. King, John. 'The influence of British culture in Argentina', Bc136, 159–72.
98. Kuklick, Henrika. *The savage within: the social history of British anthropology, 1885–1945*. Stevenage; Peregrinus; 1991. Pp ix, 325.
99. Lacy, Brian. *Siege city: the story of Derry and Londonderry*. Belfast; Blackstaff; 1991. Pp ix, 293.
100. Laing, Stuart. 'Images of the rural in popular culture, 1750–1990', Bc54, 133–51.
101. Lamb, H.H. 'Climate and the history of medieval Europe and its off-lying seas', Bc207, 1–26.

102. Lamoine, Georges (ed.). *Charges to the grand jury, 1689–1803*. London; Royal Historical Soc. (Camden 4th ser. 43); 1992. Pp vi, 642.

103. Landes, David S. 'Introduction: on technology and growth', Bc32, 1–32.

104. Lasdun, Susan. *The English park: royal, private and public*. London; Deutsch; 1991. Pp 256.

105. Lawrence, Ghislaine. 'The ambiguous artifact: surgical instruments and the surgical past', Bc24, 295–314.

106. Lawton, Richard; Pooley, Colin G. *Britain, 1740–1950: an historical geography*. London; Arnold; 1992. Pp xiv, 328.

107. Laybourn, Keith. *History of British trade unionism, c.1770–1990*. Stroud; Sutton; 1992. Pp vi, 238.

108. Lazonick, William. 'Organizations and markets in capitalist development', Bc142, 253–301.

109. Leach, Douglas Edward. *Roots of conflict: British armed forces and colonial America, 1677–1763*. Chapel Hill (NC); North Carolina UP; 1989. Pp 232.

110. Lee, C.H. 'Regional inequalities in infant mortality in Britain, 1861–1971: patterns and hypotheses', *Population Studies* 45 (1991), 55–65.

111. Lemire, Beverly. *Fashion's favourite: the cotton trade and the consumer in Britain, 1660–1800*. Oxford; Oxford UP; 1991. Pp viii, 244.

112. Lindert, Peter H. 'Toward a comparative history of income and wealth inequality', Bc34, 212–31.

113. Livingstone, D.N. 'Of design and dining clubs: geography in America and Britain, 1770–1860', *History of Science* 29 (1991), 153–83.

114. Loschky, David. 'New perspectives on seven centuries of real wages', *J. of European Economic History* 21 (1992), 169–82.

115. Loudon, Irvine. *Death in childbirth: an international study of maternal care and maternal mortality, 1800–1950*. Oxford; Oxford UP; 1992. Pp 650.

116. Loudon, Irvine. 'On maternal and infant mortality, 1900–1960', *Social History of Medicine* 4 (1991), 29–73.

117. Lowe, Lisa. *Critical terrains: French and British orientalisms*. Ithaca (NY); Cornell UP; 1992. Pp 216.

118. Lowenthal, D. 'British national identity and the English landscape', *Rural History* 2 (1991), 205–30.

119. Lowerson, John. 'The mystical geography of the English', Bc54, 152–74.

120. MacKenzie, John M. 'Heroic myths of empire', Bc10, 109–38.

121. Manchester, Keith. 'The palaeopathology of urban infections', Bc216, 8–14.

122. Mandler, Peter. 'Taking the state out again: the social history of modern Britain' [review article], *J. of Interdisciplinary History* 23 (1992), 465–76.

123. Marglin, Stephen A. 'Understanding capitalism: control versus efficiency', Bc142, 225–52.

124. Marshall, Oliver. 'Peasants or planters? British pioneers on Argentina's tropical frontier', Bc136, 143–56.

125. Martel, Gordon. 'The meaning of power: rethinking the decline and fall of Great Britain', *International History R.* 13 (1991), 662–94.

126. McLaren, A. *A history of contraception: from antiquity to the present day.* Oxford; Blackwell; 1992. Pp 256.

127. McLeod, C. 'The paradoxes of patenting: invention and its diffusion in eighteenth- and nineteenth-century Britain, France, and North America', *Technology & Culture* 32 (1991), 885–910.

128. Mead, Helen. *Wyving and thryving: the making of the English gentlewoman, 1550–1750.* Oxford; Copeman; 1992. Pp 104.

129. Melman, Billie. *Women's orients: English women and the middle east, 1718–1918.* London; Macmillan; 1992. Pp 417.

130. Menard, Russell R. 'Transport costs and long-range trade, 1300–1800: was there a European "transport revolution" in the early modern era?', Bc31, 228–75.

131. Mennell, Stephen. 'Momentum and history', Bc217, 28–46.

132. Midgley, Clare. *Women against slavery: British campaigns, 1780–1870.* London; Routledge; 1992. Pp 288.

133. Mooers, C. *The making of bourgeois Europe: absolutism, revolution and the rise of capitalism in England, France and Germany.* London; Verso; 1991. Pp 208.

134. Moorehead, Caroline. *Bertrand Russell.* London; Sinclair-Stevenson; 1992. Pp 572.

135. Murphy, Peter; Scaife, Robert G. 'The environmental archaeology of gardens', Bc64, 83–99.

136. Newbold, P.; Agiakloglou, C. 'Looking for evolving growth rates and cycles in British industrial production, 1700–1913', *J. of the Royal Statistical Soc.* 154 (1991), 341–48.

137. Nicholson, Timothy Robin. *Take the strain: the Alexandra Towing Company and the British tugboat business, 1833–1987.* Liverpool; Alexandra Towing Co.; 1990. Pp xiv, 303.

138. Noel, E.B.; Clark, J.O.M. *A history of tennis.* London; Duckworth; 1991. Pp 738.

139. North, Douglass C. 'Institutions, transaction costs, and the rise of merchant empires', Bc31, 22–40.

140. Norton, Rictor. *Mother Clap's molly house: gay subculture in England, 1700–1830.* London; Gay Men's; 1992. Pp 288.

141. Orme, Nicholas I. 'Education in England, 597–1559: the growth of policy and administration', *Historia administrativa y ciencia de la administracion comparada* ed. Manuel J. Pelaez (Barcelona; University of Malaga; 1990), 4299–4314.

142. Orton, Michael; Cleary, John. *'So long as the world shall endure': the five-hundred-year history of Bond's and Ford's hospitals.* Coventry; Coventry Church Charity; 1991. Pp 168.

143. Ottaway, Patrick J. *Archaeology in British towns: from the Emperor Claudius to the Black Death.* London; Routledge; 1992. Pp 272.

144. Outhwaite, R.B. *Dearth, public policy and social disturbance in England, 1550–1800.* London; Macmillan; 1992. Pp 96.

145. Park, Katharine. 'Medicine and society in medieval Europe, 500–1500', Bc23, 59–90.
146. Pearce, Malcolm L.; Stewart, Geoffrey. *British political history, 1867–1990: democracy and decline.* London; Routledge; 1992. Pp 608.
147. Pearson, M.N. 'Merchants and states', Bc31, 41–116.
148. Phibbs, J.L. 'The archaeology of parks—the wider perspective', Bc64, 118–22.
149. Phythian-Adams, Charles. 'Local history and national history: the quest for the peoples of England', *Rural History* 2 (1991), 1–23.
150. Pollard, Sidney. 'Cultural influences on economic action', Bc217, 137–54.
151. Porter, Roy. *Popularization of medicine, 1650–1850.* London; Routledge; 1992. Pp 288.
152. Porter, Roy. 'Addicted to modernity: nervousness in the early consumer society', Bc217, 180–94.
153. Porter, Roy. 'Madness and its institutions', Bc23, 277–301.
154. Porter, Roy. 'The rise of medical journalism in Britain to 1800', Bc28, 6–28.
155. Powell, David. *British politics and the labour question, 1868–1990.* London; Macmillan; 1992. Pp 180.
156. Powell, Geoffrey. *History of the Green Howards: the centuries of service.* London; Arms & Armour; 1992. Pp 288.
157. Rack, Henry D. 'Evangelical endings: death-beds in evangelical biography', *B. of the John Rylands University Library of Manchester* 74/1 (1992), 39–56.
158. Richards, Sandra. *The rise of the English actress.* London; Macmillan; 1992. Pp 352.
159. Robb, George. *White-collar crime in modern England: financial fraud and business morality, 1845–1929.* Cambridge; Cambridge UP; 1992. Pp 250.
160. Roberts, Richard. *Schroders: merchants and bankers.* London; Macmillan; 1992. Pp 512.
161. Roper, Michael; Tosh, John. 'Introduction: historians and the politics of masculinity', Bc223, 1–24.
162. Rose, Jonathan. 'Rereading the English common reader: a preface to a history of audiences', *J. of the History of Ideas* 53 (1992), 47–70.
163. Rosen, Ulla. 'Privategendom och jordfördelning: om agrarkapitalismens framväxt i England, Danmark och Sverige' [Private ownership and the distribution of land: the development of agrarian capitalism in England, Denmark and Sweden], *Scandia* 57 (1991), 257–81 [English summary, 353–54].
164. Ruane, Joseph. 'Colonialism and the interpretation of Irish historical development', Bc124, 293–323.
165. Ruggles, Richard I. *A country so interesting: the Hudson's Bay Company and two centuries of mapping, 1670–1870.* Montreal; McGill-Queen's UP; 1990. Pp 304.
166. Ruggles, Stephen. 'Marriage, migration and mortality: correcting sources of bias in English family reconstitutions', *Population Studies* 46 (1992), 507–22.

167. Salt, Denis. *The domestic packets between Great Britain and Ireland, 1635–1840*. Beckenham; Postal History Soc.; 1991. Pp vi, 50.

168. Samson, Ross. 'Introduction' [social archaeology of houses], Bc38, 1–18.

169. Sanger, Chesley W. '"On good fishing ground but too early for whales I think": the impact of Greenland right whale migration patterns on hunting strategies in the northern whale fishery, 1600–1900', *American Neptune* 51 (1991), 221–40.

170. Sellar, W.D.H. 'Forethocht felony, malice aforethought and the classification of homicide', Bc195, 43–60.

171. Shepherd, Robert. *Power brokers: the Tory party and its leaders*. London; Hutchinson; 1991. Pp 208.

172. Short, Brian. 'Images and realities in the English rural community: an introduction', Bc54, 1–18.

173. Short, Brian. 'The evolution of contrasting communities within rural England', Bc54, 19–43.

174. Shpayer-Makov, Haia. 'Measuring labor turnover in historical research', *Historical Methods* 24 (1991), 25–34.

175. Silverstein, Arthur M. *A history of immunology*. London; Academic; 1989. Pp 440.

176. Simpson, J. 'The local legend: a product of popular culture', *Rural History* 2 (1991), 25–36.

177. Slack, Paul. 'Introduction' [epidemics and ideas], Bc25, 1–20.

178. Smith, J.T. *English houses, 1200–1800: the Hertfordshire evidence*. London; HMSO (Royal Commission on Historical Monuments of England); 1992. Pp xx, 215.

179. Snell, Keith D.M. 'Settlement, poor law and the rural historian: new approaches and opportunities', *Rural History* 3 (1992), 145–72.

180. Stanley, Brian. *The history of the Baptist Missionary Society, 1792–1992*. Edinburgh; Clark; 1992. Pp xix, 564.

181. Stewart, Ian. 'The English and Norman mints, c.600–1158', Bc226, 1–82.

182. Stone, Lawrence. *Uncertain unions: marriage in England, 1600–1753*. Oxford; Oxford UP; 1992. Pp 281.

183. Strong, Sir Roy. 'The British obsession: an introduction to the British portrait', Bc93, 9–73.

184. Taylor, Christopher C. 'Garden archaeology: an introduction', Bc64, 1–5.

185. Taylor, Miles. 'John Bull and the iconography of public opinion in England c.1712–1929', *Past & Present* 134 (1992), 93–128.

186. Taylor, Rogan. *Football supporters and the game, 1885–1985*. London; Leicester UP; 1992. Pp 156.

187. Tracy, James D. 'Introduction' [merchant empires], Bc31, 1–21.

188. Tsoulouhas, Theofanis C. 'A new look at demographic and technological changes: England, 1550 to 1839', *Explorations in Economic History* 29 (1992), 169–203.

189. Tsushima, Jean. 'Impact—some reactions to foreign surnames: or, the art of getting it wrong', *Nomina* 14 (1992 for 1990–1), 25–40.

190. Veseth, Michael. *Mountains of debt: crisis and change in Renaissance Florence, postwar Britain and postwar America*. New York; Oxford UP; 1990. Pp ix, 246.
191. Wallwork, S.C. 'A review of the statistics of the growth of the British hosiery industry, 1844–1984', *Textile History* 22 (1991), 83–104.
192. Warner, Marina. 'Speaking with double tongue: Mother Goose and the Old Wives' tale', Bc190, 33–67.
193. Wasserstein, Bernard. *Herbert Samuel: a political life*. Oxford; Oxford UP; 1992. Pp xv, 427.
194. Wear, Andrew. 'Introduction' [medicine in society], Bc23, 1–13.
195. Werhane, Patricia H. *Adam Smith and his legacy for modern capitalism*. Oxford; Oxford UP; 1991. Pp xix, 219.
196. Wigglesworth, Neil. *A social history of English rowing*. London; Cass; 1992. Pp 232.
197. Wijnberg, Nachoem M. 'The industrial revolution and industrial economics', *J. of European Economic History* 21 (1992), 153–67.
198. Williams, E.H.D. 'Tree-ring dates: lists 42–47', *Vernacular Architecture* 23 (1992), 44–61.
199. Wilson, Arthur J. *The professionals: the Institution of Mining and Metallurgy, 1892–1992*. London; Institution of Mining and Metallurgy; 1992. Pp viii, 348.
200. Wilson, C.; Woods, Robert. 'Fertility in England: a long-term perspective', *Population Studies* 45 (1991), 399–415.
201. Wilson, D.R. 'Old gardens from the air', Bc64, 20–35.
202. Witkin, Robert. 'Bourgeois production and realist styles of art', Bc217, 195–216.
203. Woodfield, Paul. 'Early buildings in gardens in England', Bc64, 123–37.
204. Woolf, Daniel R. 'Memory and historical culture in early modern England', *J. of the Canadian Historical Association* new ser. 2 (1991), 283–308.
205. Wright, Susan. 'Image and analysis: new directions in community studies', Bc54, 195–217.

(b) *Long Periods—Local*

1. Brown, Anthony E. 'Shorter articles and notes: field survey at Grafton Regis: a village plan explained?', *Landscape History* 13 (1991), 73.
2. Abdy, Charles. *A history of Ewell*. Chichester; Phillimore; 1992. Pp xii, 116.
3. Adamson, J.L.; Hudson, L.W. (ed.). *London town miscellany, vol. 1: 1900–1939*. London; Alexius; 1992. Pp 272.
4. Addison, Paul. *Churchill on the home front, 1900–1955*. London; Cape; 1992. Pp 493.
5. Aitchison, J.; Carter, Harold. 'Rural Wales and the Welsh language', *Rural History* 2 (1991), 61–80.
6. Allen, Robert C. *Enclosure and the yeoman: the agricultural development of the south midlands, 1450–1850*. Oxford; Oxford UP; 1992. Pp xiv, 376.

7. Anon. *Argyll: an inventory of the monuments, 7. Mid Argyll and Cowal: medieval and later monuments.* Edinburgh; HMSO. 1992. Pp 592.

8. Anon. *Courage high: the history of firefighting in London.* London; HMSO; 1992. Pp 280.

9. Babbidge, Adrian. 'Problems of presentation' [Welsh industrial archaeology], Bc129, 87–90.

10. Bailey, P. 'The role of the parish constable: law and disorder in the parish of Gedling from 1665 to 1848', *Nottinghamshire Historian* 46 (1991), 16–23.

11. Baker, Nigel; Dalwood, Hal; Holt, Richard. 'From Roman to medieval Worcester: development and planning in the Anglo-Saxon city', *Antiquity* 66 (1992), 65–74.

12. Barrett, Geoffrey N. 'The great hall, Oak Street, Norwich', *Norfolk Archaeology* 41 (1991), 202–07.

13. Barry, Jonathan. 'The seventeenth and eighteenth centuries' [church in Devon and Cornwall], Bc148, 81–108; 212–16.

14. Beckett, John V. 'Lincolnshire and the east midlands', *Lincolnshire History & Archaeology* 27 (1992), 23–26.

15. Beresford, Maurice. 'The urban garden in Leeds', Bc205, 47–61.

16. Bignamini, Ilaria. 'George Vertue, art historian, and art institutions in London, 1689–1768: a study of clubs and academies', *Walpole Soc.* 54 (1991 for 1988), 1–148.

17. Bingham, Caroline. *Beyond the Highland line.* London; Constable; 1991. Pp 236.

18. Bond, C. James; Iles, R. 'Early gardens in Avon and Somerset', Bc64, 36–52.

19. Breitenbach, Esther; Gordon, Eleanor. 'Introduction' [women in Scottish society], Bc135, 1–9.

20. Briggs, C. Stephen. 'Garden archaeology in Wales', Bc64, 138–59.

21. Briggs, C. Stephen. 'Introduction: the growth of industrial archaeology in Wales', Bc129, 1–6.

22. Briggs, C. Stephen. 'Site preservation and mineral development at Parys Mountain, Anglesey', Bc129, 75–81.

23. Briggs, C. Stephen. 'The conservation of non-ferrous mines' [Wales], Bc129, 32–41.

24. Briggs, C. Stephen. 'The future of industrial archaeology in Wales', Bc129, 137–46.

25. Briggs, C. Stephen; Gerrard, S.; Kirkbride, C. 'Pembrokeshire coast', Bc129, 20–22.

26. Brighton, J. Trevor; Sprakes, Brian. 'Medieval and Georgian stained glass in Oxford and Yorkshire: the work of Thomas of Oxford (1385–1427) and William Peckitt of York (1731–95) in New college chapel, York minster and St James, High Melton', *Antiquaries J.* 70 (1990), 380–415.

27. Brown, Callum G.; Stephenson, Jayne D. '"Sprouting wings?": Women and religion in Scotland, *c.*1890–1950', Bc135, 95–120.

28. Brown, S.E.; Taylor, Christopher C. 'A relict garden at Linton, Cambridgeshire', *Cambridge Antiquarian Soc. P.* 80 (1992 for 1991), 62–67.

29. Brown, Terence W. *The making of a Yorkshire village: Thorner.* Leeds; Thorner & District History Soc.; 1991. Pp xviii, 230.

30. Bruce, Warren John. *With the Manchester Ship Canal Co., 1894–1945: recollections of Warren J. Bruce, docks manager, 1938–1945.* Manchester; Richardson; 1990. Pp 54.

31. Burns, David. *Sheriffs of Surrey.* Chichester; Phillimore; 1992. Pp 86.

32. Butler, Lawrence A.S. 'Recent archaeological work in the dioceses of Ripon and Wakefield, 1970–1990', *Yorkshire Arch. J.* 64 (1992), 203–09.

33. Cassell, Michael. *Dig it burn it sell it!: the inside story of Ibstock Johnsen, 1825–1990.* London; Pencorp; 1990. Pp 193.

34. Chainey, Graham. 'King's college chapel delineated', *Cambridge Antiquarian Soc. P.* 80 (1992 for 1991), 38–61.

35. Chainey, Graham. 'Royal visits to Cambridge: Henry VI to Henry VIII', *Cambridge Antiquarian Soc. P.* 80 (1992 for 1991), 30–37.

36. Chainey, Graham. 'The lost stained glass of Cambridge', *Cambridge Antiquarian Soc. P.* 79 (1992 for 1990), 70–81.

37. Champ, Judith F. 'The Franciscan mission in Birmingham, 1657–1824', *Recusant History* 21 (1992), 40–50.

38. Chapman, John. *Guide to the parliamentary enclosures of Wales.* Cardiff; Wales UP; 1992. Pp 200.

39. Cheape, H. *Tartan: the Highland habit.* Edinburgh; National Museums of Scotland; 1991. Pp 72.

40. Cherry, Steven. 'Change and continuity in the cottage hospitals, c.1859–1948: the experience of East Anglia', *Medical History* 36 (1992), 271–89.

41. Cleggett, David A.H. *History of Leeds castle and its families.* Leeds Castle (Kent); Leeds Castle Foundation; 1990. Pp 223.

42. Clough, Charlotte; Briggs, C. Stephen. 'Dolaucothi gold mines', Bc129, 101–03.

43. Coleman, Delphine. *Orcop: the story of a Herefordshire village from prehistory to present times.* Hanley Swan; Self-Publishing; 1992. Pp 288.

44. Collard, Jane. *Where the fat black canons dined: the history of Bristol cathedral school, 1140–1992.* Bristol; Bristol Cathedral School; 1992. Pp 152.

45. Cook, G.C. *From the Greenwich hulks to old St Pancras: history of tropical disease in London.* London; Athlone; 1992. Pp 400.

46. Cooper, N.H. (ed.). *The Exeter area: proceedings of the 136th summer meeting of the Royal Archaeological Institute 1990.* London; Royal Arch. Inst.; 1990.

47. Corpe, Stella; Oakley, Anne M. *The freemen of Canterbury, 1800–1853, compiled from Canterbury city archives.* Canterbury; Kent Record Collections; 1990. Pp ix, 298.

48. Costen, Michael D. *The origins of Somerset.* Manchester; Manchester UP; 1992. Pp 202.

49. Courtney, Paul; Gray, Madeleine. 'Tintern abbey after the Dissolution', *B. of the Board of Celtic Studies* 38 (1991), 145–58.

50. Credland, Arthur S. 'Hull and Beverley gunmakers: the story of a trade through three centuries', Bc103, 5–22.

51. Crew, Peter; Williams, Merfyn. 'Snowdonia', Bc129, 16–18.
52. Crocker, Glenys. 'The place of Godalming in the hosiery and knitwear industry: history and products', *Surrey Arch. Collections* 81 (1991–2), 41–70.
53. David, R.G.; Brambles, G.W. 'The slate quarrying industry in Westmorland, part 2', *T. of the Cumberland & Westmorland Antiquarian & Arch. Soc.* 92 (1992), 213–27.
54. Davies, Brian. 'Big Pit mining museum', Bc129, 104–06.
55. Davison, Alan. 'Great Hockham: a village which has moved?', *Norfolk Archaeology* 41 (1991), 145–61.
56. Denyer, Susan. *Traditional buildings and life in the Lake District.* London; Gollancz; 1991. Pp 224.
57. Dillistone, F.W. 'Do cathedrals have a future mission? Liverpool's two cathedrals in historical perspective', *Anglican & Episcopal History* 62 (1992), 57–69.
58. Dix, Brian. 'Towards the restoration of a period garden', Bc64, 60–72.
59. Donald, Archie. *The posts of Sevenoaks in Kent ... AD 1085 to 1985.* Tenterden; Woodvale; 1992. Pp xiii, 452.
60. Donnachie, Ian; Whatley, Christopher A. 'Introduction' [Scottish history], Bc6, 1–15.
61. Duncan, Robert; McIvor, Arthur J. 'Introduction' [militant workers], Bc175, 1–9.
62. Dyer, Christopher C. *Hanbury, Worcestershire: settlement and society in a woodland landscape.* Leicester; Leicester UP English Local History, 4th ser.: 1991. Pp 96.
63. Ellmers, Chris; Werner, Alex. *Dockland life: a pictorial history of London's docks, 1860–1970.* London; Mainstream, for the Museum of Dockland; 1991. Pp 208.
64. Elvin, Raymond; Muckle, James. 'Caroline Lockwood and the Radford Woodhouse free library', *T. of the Thoroton Soc.* 94 (1991 for 1990), 92–97.
65. Emms, Margaret. 'Education in Swanage, 1787–1902', *Dorset Natural History & Arch. Soc. P.* 113 (1992 for 1991), 5–16.
66. English, Barbara; Miller, Keith. 'The deserted village of Eske, East Yorkshire', *Landscape History* 13 (1991), 5–32.
67. English, Judie; Field, David. 'A survey of earthworks at Hammer meadow, Abinger', *Surrey Arch. Collections* 81 (1991–2), 91–95.
68. Evans, J. Daryll. *Clergy of the ancient parishes of Gwent.* Panteg; Archangel; 1991. Pp 120.
69. Evans, Neil. 'Patterns of protest and regional labour implantation in south Wales and the north-east of England, 1780–1950', *Tijdschift voor Sociale Geschiedenis* 18 (1992), 212–30.
70. Faraday, Michael. *Ludlow, 1085–1660: a social, economic and political history.* Chichester; Phillimore; 1991. Pp 272.
71. Farrant, John H.; *et al.* 'Laughton Place: a manorial and architectural history, with an account of recent restoration and excavation', *Sussex Arch. Collections* 129 (1991), 99–164.

72. Feenstra, R. 'Scottish-Dutch legal relations in the 17th and 18th centuries', Bc72, 25–45.
73. Fernie, Eric C. *Architectural history of Norwich cathedral.* Oxford; Oxford UP; 1992. Pp 228.
74. Firth, Gary. *Bradford and the industrial revolution: an economic history, 1760–1840.* Halifax; Ryburn; 1990. Pp 224.
75. Ford, Deborah. 'Pottery production and distribution in Staffordshire, from the ninth to fifteenth century', Bc209, 105–12.
76. Fraser, Constance Mary (ed.). *Durham quarter sessions rolls, 1471–1625.* Newcastle; Surtees Soc. vol. 199; 1991. Pp 422.
77. Fry, Michael. *The Dundas despotism.* Edinburgh; Edinburgh UP; 1992. Pp 480.
78. Glover, Elizabeth. *A history of the Ironmongers' company.* London; Worshipful Company of Ironmongers; 1991. Pp xiv, 197.
79. Godwin, Jeremy. 'Rickerby: an estate and its owners, part 1', *T. of the Cumberland & Westmorland Antiquarian & Arch. Soc.* 92 (1992), 229–50.
80. Gow, Ian. *The Scottish interior: Georgian and Victorian decor.* Edinburgh; Edinburgh UP; 1992. Pp 230.
81. Grenter, Stephen; Williams, Ann. 'Bersham ironworks', Bc129, 96–100.
82. Griffiths, Gwym. 'Industrial archaeology and the role of the Welsh Development Agency', Bc129, 73–74.
83. Hall, Mary B. (ed.). *The parish of Largs by David Baxter.* Largs; Largs and District Historical Soc.; 1992. Pp 180.
84. Hall, Richard A.; Kenward, H.K. *Environmental evidence from the Colonia [at York]: General Accident and Rougier Street.* York; CBA (Archaeology of York 14/6); 1990. Pp 146.
85. Hamilton, James T. *Mid Kirk of Greenock: a development and social history, 1741–1991.* Greenock; Mid Kirk of Greenock; 1991. Pp iv, 72.
86. Harborne, Leslie R.; White, Robin L.W. *The history of freemasonry in Berkshire and Buckinghamshire.* Wokingham; Provincial Grand Lodge of Berkshire and Provincial Grand Lodge of Buckinghamshire; 1990; Pp 224.
87. Harris, S.D. 'The church of St Mary the Virgin, Stone-in-Oxney, with particular reference to recent excavation of the north chapel and to the fire of 1464', *Archaeologia Cantiana* 109 (1992 for 1991), 121–38.
88. Hatchwell, Richard. 'The life and work of John Britton (1771–1857)', *Wiltshire Arch. & Natural History Magazine* 85 (1992), 101–13.
89. Hawkey, D.; Marlow, A. 'Bedford borough police, 1836–1947', *Bedfordshire Magazine* 23 (1991), 61–68.
90. Hay, Tempest. 'The ledger slabs of Canterbury cathedral, 1991', *Archaeologia Cantiana* 109 (1992 for 1991), 5–28.
91. Henderson, Diana M. 'The Scottish soldier abroad: the sociology of acclimatization', Bc21, 122–31.
92. Henrywood, R.K. *Bristol potters, 1775–1906.* Bristol; Redcliffe; 1992. Pp 112.
93. Hepple, L.W.; Doggett, Alison. *The Chilterns.* Chichester; Phillimore; 1992. Pp xvi, 272.

94. Hill, James Michael. 'The distinctiveness of Gaelic warfare, 1400–1750', *European History Q.* 22 (1992), 323–45.
95. Hill, John (ed.). *Hertfordshire militia lists: Bramfield, Sacomb and Stapleford.* Ware; Hertfordshire Family and Population History Soc.; 1992. Pp 46.
96. Hill, John (ed.). *Hertfordshire militia lists: Layston.* Ware; Hertfordshire Family & Population History Soc.; 1990. Pp 44.
97. Hinchcliffe, Tanis. *North Oxford.* London; Yale UP; 1992. Pp 272.
98. Hinde, Thomas. *Imps of promise: a history of the King's School, Canterbury.* London; James & James; 1991. Pp x, 118.
99. Hocart, R. 'Guernsey: local history in an insular context', *Local Historian* 21 (1991), 20–26.
100. Holt, Geoffrey, S. J. 'Stapehill in Dorset before the Cistercians', *Recusant History* 21 (1992), 26–39.
101. Hope, Annette. *Londoner's larder: English cuisine from Chaucer to present.* Edinburgh; Mainstream; 1990. Pp 256.
102. Horn, Joyce M. (ed.). *John le Neve: fasti ecclesiae anglicanae, 1541–1857, vol. 7: Ely, Norwich, Westminster and Worcester dioceses.* London; University of London, Inst. of Historical Research; 1992. Pp x, 142.
103. Houston, R.A. 'Mortality in early modern Scotland: the life expectancy of advocates', *Continuity & Change* 7 (1992), 47–69.
104. Howes, Lesley. 'Archaeology as an aid to restoration at Painshill Park, Surrey', Bc64, 73–82.
105. Huelin, Gordon. *Sion college and library.* London; Sion College; 1992. Pp xiv, 130.
106. Hughes, Colin. *Lime, lemon and sarsaparilla: the Italian community in south Wales, 1881–1945.* Bridgend; Seren; 1991. Pp 144.
107. Hughes, Stephen. 'Industrial archaeology and the Royal Commission on Ancient and Historical Monuments in Wales', Bc129, 49–57.
108. Hughes, Stephen. 'Panel archaeolog Ddiwydiannol Cymru—the Welsh industrial archaeology panel', Bc129, 131–36.
109. Hunt, David. *History of Preston.* London; Carnegie; 1992. Pp 288.
110. Imray, Jean. *The Mercers' Hall*, ed. Ann Saunders. London; London Topographical Soc. 143; 1991. Pp viii, 509.
111. Jackson, Gordon; Kinnear, Kate. *The trade and shipping of Dundee, 1780–1850.* Dundee; Abertay Historical Soc.; 1991. Pp viii, 125.
112. Jenkins, Geraint H. 'Industrial archaeology and the National Museum of Wales', Bc129, 91–94.
113. Jobey, George. 'Cock-fighting in Northumberland and Durham during the eighteenth and nineteenth centuries', *Archaeologia Aeliana* 5th ser. 20 (1992), 1–25.
114. Johnson, Christine. *Scottish Catholic secular clergy, 1879–1989.* Edinburgh; Donald; 1992. Pp 280.
115. Johnson, Christopher; Vince, Alan. 'The South Bail gates of Lincoln', *Lincolnshire History & Archaeology* 27 (1992), 12–16.
116. Jones, Andrew. 'Fish, fisheries and other resources in the North sea, 400–1500', Bc207, 93–98.

117. Jones, B.L. 'Place-names, signposts to the past in Anglesey', *T. of the Anglesey Antiquarian Soc. & Field Club* (1991), 23–38.
118. Jones, Donald. *History of Clifton*. Chichester; Phillimore; 1992. Pp xii, 196.
119. Jones, R. Merfyn. 'Beyond identity?: the reconstruction of the Welsh', *J. of British Studies* 31 (1992), 330–57.
120. Keen, Richard. 'Problems of protection' [Welsh industrial archaeology], Bc129, 9–11.
121. Keen, Richard. 'The National Trust and industrial archaeology' [Wales], Bc129, 95.
122. Kent, Timothy. *West country silver spoons and their makers, 1550–1750*. London; Bourdon-Smith; 1992. Pp 192.
123. Kerkham, Caroline; Briggs, C. Stephen. 'A review of the archaeological potential of the Hafod landscape, Cardiganshire', Bc64, 160–74.
124. Key, Michael. *A century of Stamford coachbuilding: a history of Henry Hayes and son, carriage and wagon builders of Stamford, Peterborough and London (1825–1924)*. Stamford; Watkins; 1990. Pp 56.
125. Kingsley, Nicholas. *The country houses of Gloucestershire, vol. 2: 1660–1830*. Chichester; Phillimore; 1992. Pp xviii, 316.
126. Kjølbye-Biddle, Birthe. 'Dispersal or concentration: the disposal of the Winchester dead over 2000 years', Bc216, 210–47.
127. Knight, Jeremy K. 'Problems of scheduling: the role of Cadw', Bc129, 12–15.
128. Knox, R. 'The origins and development of the Nayland feoffees', *P. of the Suffolk Inst. for Archaeology & History* 37 (1991), 225–37.
129. Ledbury, Chris; Briggs, C. Stephen. 'Brecon Beacons', Bc129, 19–20.
130. Leitch, Roger. '"Here chapman billies tak their stand": a pilot study of Scottish chapmen, packmen and pedlars', *Soc. of Antiquaries of Scotland P*. 120 (1990), 173–88.
131. Levine, David; Wrightson, Keith. *The making of an industrial society: Whickham, 1560–1765*. Oxford; Oxford UP; 1991. Pp xxi, 456.
132. Lowe, Michael C. 'Toll houses of the Exeter turnpike trust', *Devonshire Association Report & T*. 124 (1992), 87–99.
133. Ludlow, Neil. 'The first Kidwelly tinplate works: an archaeological and technological history', Bc101, 79–100.
134. Marsden, Peter. 'Roman and medieval shipping of south-east England', Bc207, 125–30.
135. Marshall, J.D. 'Proving ground or the creation of regional identity?: the origins and problems of regional history in Britain', Bc205, 11–26.
136. Mayer, David. 'The world on fire ...: pyrodramas at Belle Vue Gardens, Manchester, *c*.1850–1950', Bc10, 179–97.
137. McAllister, William. *History of Inverness*. Edinburgh; Donald; 1992. Pp 200.
138. Michie, Ranald, C. *The city of London: continuity and change, 1850–1990*. London; Macmillan; 1992. Pp 238.
139. Miller, George C. *Blackburn: the evolution of a cotton town*. London; THCL; 1992. Pp 456.

140. Miller, Mervyn; Gray, A. Stuart. *Hampstead garden suburb*. Chichester; Phillimore; 1992. Pp xii, 274.

141. Moore, Lindy. 'Educating for the "woman's sphere": domestic training versus intellectual discipline', Bc135, 10–41.

142. Moore, Pam. 'The industrial archaeology of regions of the British Isles, no. 3: the Isle of Wight', *Industrial Archaeology R*. 13 (1991), 172–81.

143. Moore, Susan. 'Kidwelli tinplate works', Bc129, 115–17.

144. Moore-Colyer, Richard J. 'Horse and equine improvement in the economy of modern Wales', *Agricultural History R*. 39 (1991), 126–42.

145. Moran, Peter A. 'Grisy, the Scots college farm near Paris', *Innes R*. 43 (1992), 60–64.

146. More, Charles. *Training of teachers, 1847–1947: history of the church colleges at Cheltenham*. London; Hambledon; 1992. Pp xv, 206.

147. Morgan, Kenneth O. 'Montgomeryshire's Liberal century: Rendel to Hooson, 1880–1979', *Welsh History R*. 16 (1992–3), 93–109.

148. Morgan, Roy. *Chichester: a documentary history*. Chichester; Phillimore: 1992. Pp xiv, 226.

149. Murfet, G.J. 'The Henry Ashwell legacy: the vicissitudes of a family business in bleaching, dyeing and finishing, 1855–1975', *Textile History* 22 (1991), 105–19.

150. Murphy, Michael. *Catholic poor schools in Tower Hamlets (London), 1765–1865, part 1: Wapping and Commercial Road*. North Harrow; Murphy; 1991. Pp vii, 93.

151. Nash, Gerallt D. 'Welsh corn mills—the past, present ... and future?', Bc129, 125–30.

152. Neave, David; Turnbull, Deborah. *Landscaped parks and gardens of East Yorkshire*. Beverley; Georgian Soc.; 1992. Pp 82.

153. Oxley, John. 'York, research, and the conservation of archaeological deposits', Bc214, 21–26.

154. Packer, Brian. 'Nonconformity in Tenterden: 1640–1750', *T. of the Unitarian Historical Soc*. 20 (1991–2), 81–97.

155. Palmer, Marilyn. 'Problems of recording', Bc129, 45–48.

156. Palmer, Marilyn; Neaverson, Peter. *Industrial landscapes of the east midlands*. Chichester; Phillimore; 1992. Pp xv, 208.

157. Parker, M. 'The vill of Fasham', *T. of the Hunter Arch. Soc*. 16 (1991), 1–6.

158. Parnell, Geoffrey. 'The rise and fall of the Tower of London', *History Today* 42/3 (1992), 13–19.

159. Partridge, Colin; Davenport, Trevor George. *Fortifications of Alderney: a concise history and guide to the defences of Alderney from Roman times to the second world war*. Alderney; Alderney; 1992. Pp 128.

160. Pearson, Jane. 'Merthyr Tydfil Heritage Trust', Bc129, 107–14.

161. Peaty, Ian P. *Essex brewers: and the malting and hop industries of the county*. Dartford; Brewery History Soc.; 1992. Pp 180.

162. Pounds, N.J.G. *The St Andrews area: proceedings of the 137th summer meeting of the Royal Archaeological Institute, 1991*. London; 1991 [published 1992].

163. Powell, Dilys. 'Industrial archaeology of Cwm Twrch', Bc129, 66–69.

164. Powell, Joyce. *Kenilworth at school: education and charity, 1700–1914.* Kenilworth; Odibourne; 1991. Pp 64.

165. Priddy, Deborah. 'The protection of historic gardens', Bc64, 190–91.

166. Priestley, Ursula. *The fabric of stuffs: the Norwich textile industry from 1565.* Norwich; Centre of East Anglian Studies; 1990. Pp 44.

167. Prochaska, F.K. *Philanthropy and the hospitals of London: the King's Fund, 1897–1990.* Oxford; Oxford UP; 1992. Pp xi, 308.

168. Proctor, Ray. 'The industrial heritage: British Coal' [Wales], Bc129, 82–83.

169. Reilly, John W. *Policing Birmingham: an account of 150 years of police in Birmingham.* Birmingham; West Midlands Police; 1989. Pp 240.

170. Riden, Philip. 'The charcoal iron industry in the east midlands, 1580–1780', *Derbyshire Arch. J.* 111 (1991), 64–84.

171. Roberts, Alasdair F.B. 'Aspects of highland and lowland Catholicism on Deeside', *Northern Scotland* 10 (1990), 19–30.

172. Roberts, Edward. 'Tichborne: an historical introduction', *The Hatcher R.* 4/34 (1992), 40–49.

173. Rodwell, K.A. 'Court farm, Lower Almondsbury', *T. of the Bristol & Gloucestershire Arch. Soc.* 109 (1992 for 1991), 179–93.

174. Rollison, David. *Local origins of modern society: Gloucestershire, 1500–1800.* London; Routledge; 1992. Pp 304.

175. Rose, J.K.H. *Education in the potteries, 1870–1974.* Stafford; Staffordshire Libraries, Arts and Archives; 1991. Pp 88.

176. Rowe, J. 'Bringing home the sheaves', *J. of the Royal Institution of Cornwall* 10 (1990), 385–403.

177. Rowland, A.B. *Essays on Devon history.* Dawlish; Rowland; 1991. Pp 48.

178. Royden, Michael W. *Pioneers and perseverance: a history of the Royal School for the Blind, Liverpool (1791–1991).* Liverpool; Countyvise in conjunction with the Royal School for the Blind; 1991; Pp 132.

179. Rudling, David. 'Excavations at Cliffee, Lewes, 1987 and 1988 (medieval-mid eighteenth century)', *Sussex Arch. Collections* 129 (1991), 165–81

180. Sample, Paul. *'The oldest and the best': the history of Wiltshire constabulary, 1839–1989.* Salisbury; No Limits Public Relations; 1989. Pp 48.

181. Schwarz, L.D. *London in the age of industrialization: entrepreneurs, labour force and living conditions, 1700–1850.* Cambridge; Cambridge UP; 1992. Pp 320.

182. Scola, Roger. *Feeding the Victorian city: the food supply of Manchester, 1770–1870.* Manchester; Manchester UP; 1992. Pp 384.

183. Scotland, Nigel. *Agricultural trade unionism in Gloucestershire, 1872–1950.* Cheltenham; Cheltenham and Gloucester College of Higher Education; 1991. Pp 238.

184. Scott, Miriam; Golding, Arthur; Spearman, Margaret (ed.). *Index to Dorset wills and administrations proved in the PCC, 1821–1858*, part 2. Sherborne; Somerset & Dorset Family History Soc.; 1992. Pp iv, 116.

185. Scourfield, E. 'Carmarthen craftsmen and implement makers', *Carmarthenshire Antiquarian* 27 (1991), 61–70.

186. Shaw, John P. 'Water power and rural industry in East Lothian', *T. of the East Lothian Antiquarian Soc.* 20 (1989 for 1988), 33–58.
187. Shoemaker, Robert B. 'Reforming the city: the reformation of manners campaign in London, 1690–1738', Bc82, 99–120.
188. Simm, Geoff.; Winstanley, Ian G. *Mining memories: an illustrated record of coal-mining in St Helens.* St Helens; St Helens Borough Council; 1990. Pp xxiii, 92.
189. Slater, S.C.; Dow, Derek A. (ed.). *The Victoria Infirmary of Glasgow, 1890–1990: a centenary history.* Glasgow; Victoria Infirmary Centenary Committee; 1990. Pp xvi, 325.
190. Smout, T.C. *Scotland and the sea.* Edinburgh; Donald; 1992. Pp 250.
191. Stagg, D. 'Silvicultural inclosure in the New Forest from 1780 to 1850', *P. of the Hampshire Field Club & Arch. Soc.* 46 (1991), 131–44.
192. Steppler, Glenn A. *Britons, to arms! The story of the British volunteer soldier.* Stroud; Sutton; 1992. Pp 200.
193. Stidder, Derek. *The watermills of Surrey.* Buckingham; Barracuda; 1990.
194. Stout, Geoffrey. *History of North Ormesby hospital, 1858–1948.* Stokesley; Stout; 1989. Pp viii, 195.
195. Summers, David W. 'Demographic evolution in the fishing villages of east Aberdeenshire, 1696–1880', *Scottish Geographical Magazine* 106/1 (1990), 49–53.
196. Surtees, John. *The story of St Mary's Eastbourne, 1794–1990: barracks, work-house and hospital.* Eastbourne; Eastbourne Local History Soc.; 1992. Pp 157.
197. Symons, Malcolm. 'Naming of seams in the Llanelli coalfield', Bc101, 101–20.
198. Tatton-Brown, Tim. 'Kent churches: some new architectural notes', *Archaeologia Cantiana* 109 (1992 for 1991), 111–19.
199. Taylor, Robert. 'Population explosions and housing, 1550–1850', *Vernacular Architecture* 23 (1992), 24–29.
200. Taylor, W. Robert. *Powering forward: 100 years of electricity in Birmingham.* Studley; Brewin; 1991. Pp viii, 64.
201. Thomas, D.L.B. 'The chronology of Devon's bridges', *Devonshire Association Report & T.* 124 (1992), 175–206.
202. Tilson, Barbara (ed.). *Made in Birmingham: design and industry, 1889–1989.* Studley; Brewin; 1990. Pp viii, 320.
203. Tittler, Robert. 'Harmony in the metropolis: writings on medieval and Tudor London and Westminster' [review article], *J. of British Studies* 31 (1992), 187–91.
204. Tonks, Eric Sidney. *The ironstone quarries of the midlands: history, operation and railways, part 8: south Lincolnshire.* Cheltenham; Runpast; 1990. Pp 256.
205. Turner, Barbara Carpenter. *A history of Winchester.* Chichester; Phillimore; 1992. Pp xiii, 210.
206. Venn, P. 'Exceptional Eakring: Nottinghamshire's other open field parish', *T. of the Thoroton Soc.* 94 (1991 for 1990), 69–74.
207. Weatherall, Mark; Kamminga, Harmke. *Dynamic science: biochemistry*

in Cambridge, 1898–1949. Cambridge; Wellcome Unit for the History of Medicine; 1992. Pp 87.

208. Whatley, Christopher A. (ed.). *The remaking of Juteopolis: Dundee, 1891–1991.* Dundee; Abertay Historical Soc.; 1992. Pp 104.

209. Wheeler, Geoffrey. *St Peter's [Hale] centenary, 1892–1992.* Hale; St Peter's Centenary Committee; 1992. Pp v, 80.

210. Wiliam, Eurwyn. *Welsh long-houses: four centuries of farming at Gilewent.* Cardiff; Wales UP; 1992. Pp 44.

211. Williams, David W.; *et al.* 'Two sites in Betchworth, 1986: excavations at Church Barn and in The Street', *Surrey Arch. Collections* 81 (1991–2), 103–32.

212. Williams, E.H.D. 'Church houses in Somerset', *Vernacular Architecture* 23 (1992), 15–23.

213. Williams, E.H.D. 'The building materials of Somerset's vernacular houses', *Somerset Archaeology & Natural History* 135 (1992 for 1991), 123–34.

214. Williams, Glanmor. 'Religion and belief' [in Swansea], Bc212, 17–33.

215. Williams, Glyn. 'Neither Welsh nor Argentine: the Welsh in Patagonia', Bc136, 109–22.

216. Williams, Robin. 'Melingriffith water pump', Bc129, 118–21.

217. Wilson, John (ed.). *Index of British ships registered at the port of Grimsby, 1824–1918.* Grimsby; Humberside County Archives Service; 1990. Pp 72.

218. Winstanley, Michael (ed.). *A traditional grocer: T.D. Smith's of Lancaster, 1858–1981.* Lancaster; North-West Regional Studies (Occasional paper no. 21); 1991. Pp 48.

219. Wood, Florence; Wood, Kenneth (ed.). *A Lancashire gentleman: the letters and journals of Richard Hodgkinson, 1763–1847.* Stroud; Sutton; 1992. Pp viii, 407.

220. Wood, George O. 'Roberton, the making of a parish', *Hawick Arch. Soc. T.* (1991), 17–66.

221. Yallop, H.J. *History of the Honiton lace industry.* Exeter; Exeter UP; 1992. Pp xii, 342.

222. Yates, Nigel. 'Church buildings of the Protestant establishments in Wales and Scotland: some points of comparison', *J. of Welsh Ecclesiastical History* 9 (1992), 1–19.

(c) *Collective Volumes*

1. Gunn, Steven J.; Lindley, Phillip G. (ed.). *Cardinal Wolsey: church, state and art.* Cambridge; Cambridge UP; 1991. Pp xvi, 329.

2. Bailyn, Bernard; Morgan, Philip D. (ed.). *Strangers within the realm: cultural margins of the first British empire.* Chapel Hill (NC); North Carolina UP; 1991. Pp 456.

3. Ramsay, Nigel; Sparks, Margaret; Tatton-Brown, Tim (ed.). *St Dunstan: his life, times and cult.* Woodbridge; Boydell; 1992. Pp 336.

4. Goodman, Anthony E.; Tuck, Anthony (ed.). *War and border societies in the middle ages.* London; Routledge; 1992. Pp ix, 198.

5. Anon. *Studies in European arms and armour: the C. Otto von Kienbusch collection in the Philadelphia Museum of Art.* Philadelphia (Pa); Scolar; 1992. Pp 207.

6. Donnachie, Ian; Whatley, Christopher A. (ed.). *The manufacture of Scottish history.* Edinburgh; Polygon; 1992. Pp 189.

7. Bailey, Victor (ed.). *Forged in fire: the history of the Fire Brigades Union.* London; Lawrence & Wishart; 1991. Pp xxiii, 486.

8. Offen, Karen; Pierson, Ruth Roach; Rendall, Jane (ed.). *Writing women's history: international perspectives.* London; Macmillan; 1991. Pp xli, 552.

9. Wetherly, Paul (ed.). *Marx's theory of history: the contemporary debate.* Aldershot; Avebury; 1992. Pp 242.

10. MacKenzie, John M. (ed.). *Popular imperialism and the military, 1850–1950.* Manchester; Manchester UP; 1992. Pp ix, 228.

11. Anderson, David M.; Killingray, David (ed.). *Policing and de-colonization: politics, nationalism and the police, 1917–1965.* Manchester; Manchester UP; 1992. Pp xi, 227.

12. Wood, Diana (ed.). *The Church and sovereignty, c.590–1918: essays in honour of Michael Wilks.* Oxford; Blackwell (Studies in Church History: Subsidia 9); 1991. Pp xx, 513.

13. Davis, R.W.; Helmstadter, R.J. (ed.). *Religion and irreligion in Victorian society: essays in honor of R.K. Webb.* London; Routledge; 1992. Pp viii, 205.

14. Skinner, Quentin; Tuck, Richard; Thomas, William; Singer, Peter (ed.). *Great political thinkers: Machiavelli, Hobbes, Mill, Marx* [reprints not itemised]. Oxford; Oxford UP; 1992. Pp vii, 459.

15. Dunn, John; Urmson, J.O.; Ayer, A.J. (ed.). *The British empiricists: Locke, Berkeley, Hume* [reprints not itemised]. Oxford; Oxford UP; 1992. Pp vii, 287.

16. Arneson, Richard J. (ed.). *Liberalism* [reprints not itemised]. Aldershot; Elgar; 1992. 3 vols.

17. Cortazzi, Sir Hugh; Daniels, Gordon (ed.). *Britain and Japan, 1859–1991: themes and personalities.* London; Routledge; 1991. Pp xxi, 319.

18. Jeffreys-Jones, Rhodri; Lownie, Andrew (ed.). *North American spies: new revisionist essays.* Edinburgh ; Edinburgh UP; 1991. Pp ix, 256.

19. Fraser, T.G.; Lowe, Peter (ed.). *Conflict and amity in east Asia: essays in honour of Ian Nish.* London; Macmillan; 1992. Pp xiv, 190.

20. Boog, Horst (ed.). *The conduct of the air war in the second world war: an international comparison.* Oxford; Berg; 1992. Pp xii, 763.

21. Simpson, Grant G. (ed.). *The Scottish soldier abroad, 1247–1967.* Edinburgh; Donald; 1992. Pp xii, 173.

22. Mathias, Peter; Davis, John A. (ed.). *Innovation and technology in Europe: from the eighteenth century to the present day.* Oxford; Blackwell; 1991. Pp vi, 192.

23. Wear, Andrew (ed.). *Medicine in society: historical essays.* Cambridge; Cambridge UP; 1992. Pp ix, 397.

24. Lawrence, Christopher (ed.). *Medical theory, surgical practice: studies in the history of surgery.* London; Routledge; 1992. Pp x, 331.

25. Ranger, Terence; Slack, Paul (ed.). *Epidemics and ideas: essays on the historical perception of pestilence.* Cambridge; Cambridge UP; 1992. Pp ix, 346.

26. Barry, Jonathan; Jones, Colin (ed.). *Medicine and charity before the welfare state.* London; Routledge; 1991. Pp x, 259.

27. Chibnall, Marjorie (ed.). *Anglo-Norman studies, 14.* Woodbridge; Boydell (Proceedings of the Battle conference, 1991); 1992. Pp 337.

28. Bynum, William F.; Lock, Stephen; Porter, Roy (ed.). *Medical journals and medical knowledge: historical essays.* London; Routledge; 1992. Pp xii, 279

29. Jones, Peter; Skinner, Andrew S. (ed.). *Adam Smith reviewed.* Edinburgh; Edinburgh UP; 1992. Pp xii, 252.

30. Fry, Michael (ed.). *Adam Smith's legacy: his place in the development of modern economics.* London; Routledge; 1992. Pp xv, 203

31. Tracy, James D. (ed.). *The political economy of merchant empires.* Cambridge; Cambridge UP; 1991. Pp 504.

32. Higonnet, Patrice; Landes, David S.; Rosovsky, Henry (ed.). *Favorites of fortune: technology, growth and economic development since the industrial revolution.* London; Harvard UP; 1991. Pp vii, 558.

33. Tolliday, Steven W. (ed.). *Government and business* [reprints not itemised]. Aldershot; Elgar; 1991. Pp xix, 576.

34. Brenner, Y.S.; Kaelble, Hartmut; Thomas, Mark (ed.). *Income distribution in historical perspective.* Cambridge; Cambridge UP; 1991. Pp xii, 261.

35. Benson, John; Shaw, Gareth (ed.). *The evolution of retail systems, c.1800–1914.* Leicester; Leicester UP; 1991. Pp 207.

36. Morris, P.J.T.; Campbell, W.A.; Roberts, H.L. (ed.). *Milestones in 150 years of the chemical industry.* Cambridge; Royal Society of Chemistry; 1991. Pp viii, 307.

37. Grant, Wyn; Nekkers, Jan; van Waarden, Frans (ed.). *Organising business for war: corporatist economic organisation during the second world war.* Oxford; Berg; 1991. Pp xviii, 310.

38. Samson, Ross (ed.). *The social archaeology of houses.* Edinburgh; Edinburgh UP; 1990. Pp 282.

39. Brant, Clare; Purkiss, Diane (ed.). *Women, texts and histories, 1575–1760.* London; Routledge; 1992. Pp xi, 299.

40. Birkett, Jennifer; Harvey, Elizabeth D. (ed.). *Determined women: studies in the construction of the female subject, 1900–1990.* London; Macmillan; 1991. Pp x, 213.

41. Good, G.H.; Jones, R.H.; Ponsford, M.W. (ed.). *Waterfront archaeology.* London; CBA (Research report 74); 1991. Pp xii, 201.

42. Diederiks, Herman; Hohenberg, Paul M.; Wagenaar, Michael (ed.). *Economic policy in Europe since the late middle ages: the visible fortune of cities.* Leicester; Leicester UP; 1991. Pp 229.

43. Bush, M.L. (ed.). *Social orders and social classes in Europe since 1500: studies in social stratification.* London; Longman; 1992. Pp vii, 267.

44. Bock, Gisela; Thane, Pat (ed.). *Maternity and gender policies: women and the rise of the European welfare states, 1880s-1950s.* London; Routledge; 1991. Pp 259.

45. Magdalino, Paul (ed.). *The perception of the past in twelfth-century Europe*. London; Hambledon; 1992. Pp xvi, 240.

46. Saul, Nigel (ed.). *Age of chivalry: art and society in late medieval England* [some reprints not itemised]. London; Collins & Brown; 1992. Pp 140.

47. Barker, Katherine; Kain, Roger J.P. (ed.). *Maps and history in south-west England*. Exeter; Exeter UP; 1991. Pp ix, 148.

48. Schwoerer, Lois G. (ed.). *The revolution of 1688–1689: changing perspectives*. Cambridge; Cambridge UP; 1992. Pp xxii, 288.

49. Bernard, G.W. (ed.). *The Tudor nobility*. Manchester; Manchester UP; 1992. Pp 312.

50. Bullock, Ian; Pankhurst, Richard (ed.). *Sylvia Pankhurst: from artist to anti-fascist*. London; Macmillan; 1992. Pp ix, 210.

51. Harper-Bill, Christopher (ed.). *Religious belief and ecclesiastical careers in late medieval England*. Woodbridge; Boydell; 1991. Pp xvii, 238.

52. Cairncross, Frances; Cairncross, Sir Alec (ed.). *The legacy of the golden age: the 1960s and their economic consequences*. London; Routledge; 1992. Pp viii, 204.

53. Crowfoot, Elisabeth; Pritchard, Frances; Staniland, Kay (ed.). *Textiles and clothing, c.1150-c.1450: medieval finds from excavations in London, 4* [not itemised]. London; HMSO; 1992. Pp x, 234.

54. Short, Brian (ed.). *The English rural community: image and analysis*. Cambridge; Cambridge UP; 1992. Pp vii, 239.

55. Kearns, Gerry; Withers, Charles W.J. (ed.). *Urbanising Britain: essays on class and community in the nineteenth century*. Cambridge; Cambridge UP; 1991. Pp 177.

56. Shires, Linda M. (ed.). *Rewriting the Victorians: theory, history, and the politics of gender*. London; Routledge; 1992. Pp xiv, 196.

57. Cooter, Roger (ed.). *In the name of the child: health and welfare, 1880–1940*. London; Routledge; 1992. Pp xii, 292.

58. Riddy, Felicity (ed.). *Regionalism in late medieval manuscripts and texts: essays celebrating the publication of 'A linguistic atlas of late mediaeval English'*. Cambridge; Brewer; 1991. Pp xiii, 214.

59. Cain, T.G.S.; Robinson, Ken (ed.). *Into another mould: change and continuity in English culture, 1625–1700*. London; Routledge; 1992. Pp xxvii, 196.

60. Kroll, Richard; Ashcraft, Richard; Zagorin, Perez (ed.). *Philosophy, science, and religion in England, 1640–1700*. Cambridge; Cambridge UP; 1992. Pp xv, 287.

61. Harris, P.R. (ed.). *The library of the British Museum: retrospective essays on the department of printed books*. London; British Library; 1991. Pp xii, 305.

62. Grundy, Isobel; Wiseman, Susan (ed.). *Women, writing, history, 1640–1740*. London; Batsford; 1992. Pp 239.

63. Gardner, Viv; Rutherford, Susan (ed.). *The new woman and her sisters: feminism and theatre, 1850–1914*. Hemel Hempstead; Harvester Wheatsheaf; 1992. Pp xxi, 238.

64. Brown, Anthony E. (ed.). *Garden archaeology: papers presented to a conference at Knuston Hall, Northamptonshire, April 1988.* London; CBA (Research report 78); 1991. Pp ix, 198.

65. Belchem, John (ed.). *Popular politics, riot and labour: essays in Liverpool history, 1790–1940.* Liverpool; Liverpool UP; 1992. Pp xii, 257.

66. John, Angela V. (ed.). *Our mothers' land: chapters in Welsh women's history, 1830–1939.* Cardiff; Wales UP; 1991. Pp xii, 207.

67. Tiratsoo, Nick (ed.). *The Attlee years.* London; Pinter; 1991. Pp ix, 214.

68. Cerasano, S.P.; Wynne-Davies, Marion (ed.). *Gloriana's face: women, public and private, in the English Renaissance.* Hemel Hempstead; Harvester Wheatsheaf; 1992. Pp xiv, 234.

69. Fissel, Mark Charles (ed.). *War and government in Britain, 1598–1650.* Manchester; Manchester UP; 1991. Pp ix, 293.

70. Abulafia, David; Franklin, Michael; Rubin, Miri (ed.). *Church and city, 1000–1500: essays in honour of Christopher Brooke.* Cambridge; Cambridge UP; 1992. Pp xxiv, 354.

71. Reuter, Timothy (ed.). *Warriors and churchmen in the high middle ages: essays presented to Karl Leyser.* London; Hambledon; 1992. Pp xxiii, 232.

72. de Ridder-Symoens, H.; Fletcher, J.M. (ed.). *Academic relations between the Low Countries and the British Isles, 1450–1700.* Ghent; [Publisher]; 1991. [Pp].

73. Anon. *Il secolo di ferro: mito e realità del secolo X* [The iron century: myth and reality of the tenth century]. Spoleto; Centro Italiano di Studi sull'alto Medievo; 1991. [Pp].

74. Kunze, Bonnelyn Young; Brautigam, Dwight D. (ed.). *Court, country and culture: essays in early modern British history in honor of Perez Zagorin.* Rochester (NY); Rochester UP; 1992. Pp xvii, 249.

75. Hicks, Carola (ed.). *England in the eleventh century: proceedings of the 1990 Harlaxton symposium.* Stamford; Watkins (Harlaxton Medieval Studies 2); 1992. Pp xi, 356 + plates.

76. Barrow, G.W.S. (ed.). *Scotland and its neighbours in the middle ages* [reprints not itemised]. London; Hambledon; 1992. Pp xxii, 255.

77. Abraham, Gerald (ed.). *Romanticism, 1830–1890.* Oxford; Oxford UP (New Oxford History of Music 9); 1990. Pp xx, 935.

78. Crocker, Richard; Hiley, David (ed.). *The early middle ages to 1300.* Oxford; Oxford UP (New Oxford History of Music 2); 1990. Pp xx, 795.

79. Morrill, John S. (ed.). *Revolution and Restoration: England in the 1650s.* London; Collins & Brown; 1992. Pp 160.

80. Hexter, J.H. (ed.). *Parliament and liberty from the reign of Elizabeth to the English civil war.* Stanford (Ca); Stanford UP; 1992. Pp xx, 333.

81. Thacker, Alan T. (ed.). *The earldom of Chester and its charters: a tribute to Geoffrey Barraclough.* Chester; Chester Arch. Soc. vol. 71; 1991. Pp 247.

82. Davison, Lee; Hitchcock, Tim; Keirn, Tim; Shoemaker, Robert B. (ed.). *Stilling the grumbling hive: the response to social and economic problems in England, 1689–1750.* Stroud; Sutton; 1992. Pp liv, 170.

83. Anon. *Miscellany of the Scottish History Society, XI.* Edinburgh; Scottish History Soc., 5th ser., vol. 3; 1991 for 1990. Pp 413.

84. Anon. *Thoresby Society Miscellany*. Leeds; Thoresby Soc. 2nd ser. vol. 1; 1991. Pp 96.

85. Welsh, William Jeffrey; Skaggs, David Curtis (ed.). *War on the Great Lakes: essays commemorating the 175th anniversary of the battle of Lake Erie*. Kent (Oh); Kent State UP; 1991. Pp 154.

86. Harris, John R. (ed.). *Essays in industry and technology in the eighteenth century: England and France* [reprints not itemised]. Aldershot; Variorum (Collected Studies); 1992. Pp 223.

87. Harman, P.M.; Shapiro, Alan E. (ed.). *The investigation of difficult things: essays on Newton and the history of the exact sciences in honour of D.T. Whiteside*. Cambridge; Cambridge UP; 1992. Pp xvi, 531.

88. Oates, John C.T. *Studies in English printing and libraries* [reprints not itemised]. London; Pindar; 1991. Pp 409.

89. Barrell, John. *The birth of Pandora and the division of knowledge* [mainly reprints not itemised]. London; Macmillan; 1992. Pp xvii, 263.

90. Campbell, Sheila; Hall, Bert; Klausner, David (ed.). *Health, disease and healing in medieval culture*. London; Macmillan; 1992. Pp xxiv, 204.

91. Fradenburg, Louise Olga (ed.). *Women and sovereignty*. Edinburgh; Edinburgh UP (Yearbook of the Traditional Cosmology Soc., vol. 7); 1992. Pp 344.

92. Goldberg, P.J.P. (ed.). *Woman is a worthy wight: women in English society, c.1200–1500*. Stroud; Sutton; 1992. Pp xvii, 229.

93. Charlton-Jones, Richard; *et al.* (ed.). *The British portrait, 1660–1960*. Woodbridge; Antique Collectors Club; 1991. Pp 443.

94. Aers, David (ed.). *Culture and history, 1350–1600: essays on English communities, identities and writings*. Hemel Hempstead; Harvester Wheatsheaf; 1992. Pp 213.

95. Cassis, Youssef (ed.). *Finance and financiers in European history, 1880–1960*. Cambridge; Cambridge UP; 1992. Pp 445.

96. Edwards, Nancy; Lane, Alan (ed.). *The early church in Wales and the west: recent work in early Christian archaeology and place-names*. Oxford; Oxbow (Monograph 16); 1992. Pp viii, 168.

97. Farrell, Robert T.; Neuman de Vegvar, Carol L. (ed.). *Sutton Hoo: fifty years after*. Oxford (Oh); Western Michigan UP (American Early Medieval Studies 2); 1992. Pp 198.

98. Harper-Bill, Christopher; Harvey, Ruth (ed.). *Medieval knighthood, IV: papers from the fifth Strawberry Hill conference, 1990*. Woodbridge; Boydell; 1992. Pp xiv, 240.

99. Hart, Cyril R. *The Danelaw*. London; Hambledon; 1992. Pp xviii, 702.

100. Blair, W. John; Sharpe, Richard (ed.). *Pastoral care before the parish*. Leicester; Leicester UP (Studies in the Early History of Britain); 1992. Pp ix, 298.

101. James, Heather (ed.). *Sir Gar: studies in Carmarthenshire history: essays in memory of W.H. Morris and M.C.S. Evans*. Carmarthen; Carmarthenshire Antiquarian Soc. (monograph ser. 4); 1991. Pp xi, 284.

102. Munro, John H.A. *Bullion flows and monetary policies in England and the Low Countries, 1350–1500* [reprints not itemised]. London; Variorum; 1992. Unpaginated.

103. English, Barbara (ed.). *East Yorkshire miscellany*, vol. 1. Beverley; East Yorkshire Local History Soc.; 1992. Pp 55.

104. Kelly, Francis (ed.). *Medieval art and architecture at Exeter cathedral.* Leeds; Meaney for British Archaeological Association (Conference transactions vol. 9, 1985); 1992. Pp 236.

105. Kahn, Deborah (ed.). *The Romanesque frieze and its spectator.* London; Miller for Lincoln cathedral; 1992. Pp 232.

106. Evans, D.H.; Tomlinson, D.G. (ed.). *Excavations at 33–35 Eastgate, Beverley, 1983–1986* [not itemised]. Sheffield; Collis (Sheffield Excavation Reports); 1992. Pp 340.

107. Richardson, R.C. (ed.). *Town and countryside in the English revolution.* Manchester; Manchester UP; 1992. Pp ix, 278.

108. Graham-Campbell, James A. (ed.). *Viking treasure from the north-west: the Cuerdale hoard in its context.* Liverpool; National Museums and Galleries on Merseyside (Occasional Papers 5); 1992. Pp viii, 115.

109. Williams, Ann; Martin, G.H. (ed.). *The Kent Domesday.* London; Alecto; 2 vols. 1992. Pp viii, 94 & fos. 0–15v.

110. Williams, Ann; Martin, G.H. (ed.). *The Yorkshire Domesday.* London; Alecto; 2 vols. 1992. Pp viii, 120 & fos. 297–334v, 373–82v.

111. Williams, Ann; Martin, G.H. (ed.). *The Lincolnshire Domesday.* London; Alecto; 2 vols. 1992. Pp viii, 76 & fos. 335–78v.

112. Coss, Peter R.; Lloyd, Simon D. (ed.). *Thirteenth century England, IV: proceedings of the Newcastle-upon-Tyne conference, 1991.* Woodbridge; Boydell; 1992. Pp xii, 202.

113. Carver, Martin O.H. (ed.). *The age of Sutton Hoo.* Woodbridge; Boydell; 1992. Ppxviii, 406.

114. Bridbury, A.R. (ed.). *The English economy from Bede to the Reformation* [mostly reprints]. Woodbridge; Boydell; 1992. Pp 328.

115. Smith, Lesley; Ward, Benedicta (ed.). *Intellectual life in the middle ages: essays presented to Margaret Gibson.* London; Hambledon; 1992. Pp xiv, 322.

116. Stapleton, Barry (ed.). *Conflict and community in southern England: essays in the social history of rural and urban labour from medieval to modern times.* Stroud; Sutton; 1992. Pp xvii, 253.

117. Wrigley, Chris; Shepherd, John (ed.). *On the move: essays in Labour and transport history presented to Philip Bagwell.* London; Hambledon; 1992. Pp xxiii, 261.

118. Aalen, F.H.A.; Whelan, Kevin (ed.). *Dublin city and county from prehistory to the present: studies in honour of J.H. Andrews.* Dublin; Geography Publications; 1992. Pp 450.

119. Dunning, Robert W. (ed.). *History of the county of Somerset, vol. 6: Andersfield, Cannington and north Petherton hundreds (Bridgwater and neighbouring parishes)* [not itemised]. Oxford; Oxford UP (Victoria History of the Counties of England); 1992. Pp 380.

120. Morland, S.G. (ed.). *Glastonbury, Domesday and related studies* [reprints not itemised]. Glastonbury; Glastonbury Antiquarian Soc.; 1991. Pp xii, 142.

121. Edwards, Gavin; Sumner, Graham (ed.). *The historical and cultural connections and parallels between Wales and Australia*. Lampeter; Mellen; 1992. Pp 184.

122. Hoyle, Richard W. (ed.). *The estates of the English crown, 1558–1640*. Cambridge; Cambridge UP; 1992. Pp xviii, 440.

123. Laine, Michael (ed.). *A cultivated mind: essays on J.S. Mill presented to J.M. Robson*. Toronto; Toronto UP; 1991. Pp xv, 281.

124. Silverman, Marilyn; Gulliver, P.H. (ed.). *Approaching the past: historical anthropology through Irish case studies*. New York; Columbia UP; 1992. Pp 420.

125. Helmholz, Richard H. (ed.). *Canon law in Protestant lands*. Berlin; Dunker & Humblot; 1992. Pp 271.

126. Brachmann, Hans-Jurgen; Herrmann, Joachim (ed.). *Frühgeschichte der europäischen Stadt*. Berlin; Akademie; 1991. Pp 325.

127. Johnson, J.K.; Wilson, G. (ed.). *Historical essays on upper Canada*. Ottawa; Carleton UP; 1989. Pp xiii, 368.

128. Miller, David Carey (ed.). *Comparative and historical essays in Scots law*. London; Butterworth; 1992. Pp xiv, 182.

129. Briggs, C. Stephen (ed.). *The Welsh industrial heritage: a review*. London; CBA (Research report 79); 1992. Pp 168.

130. Twaddle, Michael (ed.). *Imperialism, the state and the third world* London; British Academic; 1992. Pp vii, 292.

131. Mercer, Helen; Rollings, Neil; Tomlinson, J.D. (ed.). *Labour governments and private industry: the experience of 1945–1951*. Edinburgh; Edinburgh UP; 1992. Pp vii, 244.

132. Blakemore, Steven (ed.). *Burke and the French revolution: bicentennial essays*. Athens (Ga); Georgia UP; 1992. Pp xvi, 180.

133. Roach, Susan (ed.). *Across the narrow seas: studies in the history and bibliography of Britain and the Low Countries presented to Anna E.C. Simoni*. London; British Library; 1991. Pp xv, 223.

134. Dosi, Giovanni; Giannetti, Renato; Toninelli, Pier Angelo (ed.). *Technology and enterprise in a historical perspective*. Oxford; Oxford UP; 1992. Pp 416.

135. Breitenbach, Esther; Gordon, Eleanor (ed.). *Out of bounds: women in Scottish society, 1800–1945*. Edinburgh; Edinburgh UP; 1992. Pp 228.

136. Hennessy, Alistair; King, John (ed.). *The land that England lost: Argentina and Britain, a special relationship*. London; British Academic; 1992. Pp x, 330.

137. Woodiwiss, Simon (ed.). *Iron age and Roman salt production and the medieval town of Droitwich: excavations at the Old Bowling Green and Friar Street* [not itemised]. London; CBA (Research report 81); 1992. Pp xiv, 223.

138. Dickson, A.D.R.; Treble, John H. (ed.). *People and society in Scotland: a social history of modern Scotland, vol. 3: 1914–1990*. Edinburgh; Donald; 1992. Pp 300.

139. Jonsson, Kenneth (ed.). *Studies in late Anglo-Saxon coinage in memory of Bror Emil Hildebrand*. Stockholm; Svenska Numismatiska Foreningen (Numismatiska Meddelanden 35); 1990. Pp 520.

140. Birke, Adolf M.; Mayring, Eva A. (ed.). *Britische Besatzung in Deutschland: Aktenerschliessung und Forschungsfelder* [The British occupation of Germany: documentary publications and areas of research]. London; Deutsches Historisches Institut; 1992. Pp xi, 196.

141. Strickland, Matthew (ed.). *Anglo-Norman warfare: studies in late Anglo-Saxon and Anglo-Norman military organization and warfare* [reprints not itemised]. Woodbridge; Boydell; 1992. Pp xxiii, 277.

142. Gustafsson, Bo (ed.). *Power and economic institutions: reinterpretations in economic history.* Aldershot; Elgar; 1991. Pp xiv, 344.

143. Smith, F.B. (ed.). *Ireland, England and Australia: essays in honour of Oliver MacDonagh.* Cork; Cork UP; 1990. Pp vii, 240.

144. Capie, Forrest H. (ed.). *Major inflations in history* [reprints not itemised]. Aldershot; Elgar; 1991. Pp 627.

145. Berrios, German E.; Freeman, Hugh. (ed.). *150 years of British psychiatry, 1841–1991.* London; Gaskell for the Royal College of Psychiatrists; 1991. Pp xv, 464.

146. Morton, Alan D. *Excavations at Hamwic, vol. 1: excavations, 1946–1983* [not itemised]. London; CBA (Research report 84); 1992. Pp 237.

147. Ingram, Edward (ed.). *Anglo-Ottoman encounters in the age of revolution: collected essays of Allan Cunningham* [not itemised]. London; Cass; 1992. Pp 210.

148. Orme, Nicholas I. (ed.). *Unity and variety: a history of the church in Devon and Cornwall.* Exeter; Exeter UP; 1991. Pp 242.

149. Corner, John; Harvey, Sylvia. (ed.). *Enterprise and heritage: crosscurrents of national culture.* London; Routledge; 1991. Pp 288.

150. Freedman, Lawrence; Hayes, Paul; O'Neill, Robert (ed.). *War, strategy and international politics: essays in honour of Sir Michael Howard.* Oxford; Oxford UP; 1992. Pp xi, 322.

151. Coleman, D.C. (ed.). *Myth, history and the industrial revolution* [reprints not itemised]. London; Hambledon; 1992. Pp viii, 225.

152. Gransden, Antonia (ed.). *Legends, traditions and history in medieval England* [reprints not itemised]. London; Hambledon; 1992. Pp 361.

153. Dinwiddy, J.R. (ed.). *Radicalism and reform in Britain, 1780–1850* [reprints not itemised]. London; Hambledon; 1992. Pp xx, 452.

154. Dumville, David N. (ed.). *Britons and Anglo-Saxons in the early middle ages* [reprints not itemised]. London; Variorum; 1992. Pp 352.

155. Harpham, Edward J. (ed.). *John Locke's 'Two treatises on government': new interpretations.* Lawrance (Ks); Kansas UP; 1992. Pp 239.

156. Tweddle, Dominic. (ed.). *The Anglian helmet from 16–22 Coppergate* [not itemised]. York; CBA (Archaeology of York 17/8); 1992. 2 parts. Pp 350.

157. Ottaway, Patrick J. (ed.). *Anglo-Scandinavian ironwork from 16–22 Coppergate* [not itemised]. York; CBA (Archaeology of York 17/6); 1992. Pp 454–730.

158. Rahtz, Philip; Meeson, Robert (ed.). *An Anglo-Saxon watermill at Tamworth: excavations in the Bolebridge street area of Tamworth, Staffordshire, in 1971 and 1978* [not itemised]. London; CBA; 1992. Pp xv, 160.

159. Roesdahl, Else; Wilson, David M. (ed.). *From viking to crusader: the Scandinavians in Europe, 800–1200* [not itemised]. Udevalla (Sweden); Bohusläningens Boktryckeri; 1992. Pp 436.

160. Sherlock, Stephen J.; Welch, Martin G. (ed.). *An Anglo-Saxon cemetery at Norton, Cleveland* [not itemised]. London; CBA (Research report 82); 1992. Pp x, 225.

161. Crummy, Philip (ed.). *Excavations at Culver street, the Gilberd school and other sites in Colchester, 1971–1985* [not itemised]. Colchester; Colchester Arch. Trust; 1992. 2 vols. Pp xx, 426 + microfiches.

162. Wickenden, N.P. (ed.). *The temple and other sites in the north-eastern sector of Caesaromagus* [not itemised]. London; CBA (Research report 75) for Chelmsford Arch. Trust (Report 9); 1992. Pp 150.

163. Hinchcliffe, John; Williams, John H.; Williams, Frances (ed.). *Roman Warrington: excavations at Wilderspool, 1966–1969 and 1976* [not itemised]. Manchester; Manchester Arch. Unit (Brigantia monograph 2); 1992. Pp 188.

164. Rahtz, Philip (ed.). *Cadbury Congresbury, 1968–1973* [not itemised]. London; BAR (British ser. 223); 1992. Pp 261.

165. Steedman, Ken; Dyson, Tony; Schofield, John (ed.). *Aspects of Anglo-Norman London, III: the bridgehead and Billingsgate to 1200* [not itemised]. London; London & Middlesex Arch. Soc. (Special paper 14); 1992. Pp 216.

166. Milne, Gustav (ed.). *Timber building techniques in London, c.900–1400: an archaeological study of waterfront installations and related material* [not itemised]. London; London & Middlesex Arch. Soc. (Special paper 15); 1992. Pp 152.

167. Milne, Gustav (ed.). *From Roman basilica to medieval market* [not itemised]. London; HMSO; 1992. Pp 143.

168. Armit, Ian (ed.). *The later prehistory of the western isles of Scotland* [not itemised]. London; BAR (British ser. 221); 1992. Pp 185.

169. Hayes, P.P.; Lane, T.W. (ed.). *Lincolnshire survey: the south-west fens* [not itemised]. [Norwich]; Fenland Project (East Anglian Arch. report 55); 1992. Pp 267.

170. Hall, David (ed.). *The south-western Cambridgeshire fenlands* [not itemised]. [Norwich]; Fenland Project (East Anglian Arch. report 56); 1992. Pp 118.

171. Davenport, Peter (ed.). *Archaeology in Bath, 1976–1985: excavations at Orange grove (Swallow street), the Crystal palace (Abbey street)* [not itemised]. Pp 170.

172. Milne, Gustav; Richards, Julian D. (ed.). *Two Anglo-Saxon buildings and associated finds* [not itemised]. York; University of York Department of Archaeology (Wharram Percy 7); 1992. Pp iv, 114.

173. Duffy, Michael (ed.). *Parameters of British naval power, 1650–1850.* Exeter; Exeter UP (Maritime studies 7); 1992. Pp vi, 144.

174. Jorgensen, Lise Bender (ed.). *North European textiles until AD 1000.* Aarhus (Denmark); Aarhus UP; 1992. Pp 258.

175. Duncan, Robert; McIvor, Arthur J. (ed.). *Militant workers—labour and*

class conflict on the Clyde, 1900–1950: essays in honour of Harry McShane. Edinburgh; Donald; 1992. Pp 197.

176. Breuilly, John J.; Niedhard, Goltfried; Taylor, Tony (ed.). *British labour politics from chartism to the reform league.* Oxford; Berg; 1992. Pp 368.

177. Rosenthal, Joel T. (ed.). *Medieval women and the sources of medieval history.* Athens (Ga); Georgia UP; 1990. Pp 384.

178. Anon. *Camden miscellany,* vol. 31. London; Royal Historical Soc. (Camden 4th ser. 44); 1992. Pp 411.

179. Cohen, Michael J.; Kolinsky, Martin (ed.). *Britain and the middle east in the 1930s: security problems, 1935–1939.* London; Macmillan; 1992. Pp xvii, 231.

180. Godden, Malcolm R.; Lapidge, Michael (ed.). *The Cambridge companion to Old English literature.* Cambridge; Cambridge UP; 1991. Pp 298.

181. Glasscock, Robin E. (ed.). *Historic landscapes of Britain from the air.* Cambridge; Cambridge UP; 1992. Pp 256.

182. O'Brien, John; Travers, Pauric (ed.). *The Irish emigrant experience in Australia.* Dublin; Poolbeg; 1991. Pp 279.

183. Kendall, Calvin B.; Wells, Peter S. (ed.). *Voyage to the other world: the legacy of Sutton Hoo.* Minneapolis ; Minnesota UP (Medieval studies 5); 1992. Pp xx, 222.

184. Jones, Richard F.J. (ed.). *Britain in the Roman period: recent trends.* Sheffield; Collis; 1991. Pp viii, 120.

185. Dickinson, Tania M.; James, Edward (ed.). *Medieval Europe, 1992: vol. 4: death and burial.* York; Medieval Europe (Conference on medieval archaeology in Europe); 1992. Pp 210.

186. Evans, D.R.; Metcalf, Vivienne M. (ed.). *Roman gates, Caerleon: the 'Roman gates' site in the fortress of the second Augustan legion at Caerleon, Gwent. The excavations of the Roman buildings and the evidence for early medieval activity* [not itemised]. Oxford; Oxbow for Glamorgan-Gwent Arch. Trust; 1992. Pp 216.

187. Reece, Bob (ed.). *Exiles from Erin: convict lives in Ireland and Australia.* London; Macmillan; 1991. Pp xv, 336.

188. Noakes, Jeremy (ed.). *The civilian in war: the home front in Europe, Japan and the USA in world war II.* Exeter; Exeter UP; 1992. Pp 216.

189. Wood, Diana (ed.). *Christianity and Judaism.* Oxford; Blackwell (Studies in Church history 29); 1992. Pp xvii, 493.

190. Porter, Roy (ed.). *Englishness.* Oxford; Polity; 1992. Pp 280.

191. Gaimster, David; Redknap, Mark (ed.). *Everyday and exotic pottery from Europe: studies in honour of John G. Hurst.* Oxford; Oxbow (Monograph 23); 1992. Pp 382.

192. Fulford, Michael G.; Nichols, Elizabeth (ed.). *Developing landscapes of lowland Britain: the archaeology of the British gravels—a review.* London; Soc. of Antiquaries (Occasional paper 14); 1992. Pp 145.

193. Filmer-Sankey, William; Hawkes, Sonia Chadwick; Campbell, James; Brown, David (ed.). *Anglo-Saxon studies in archaeology and history,* vol. 5. Oxford; Oxford University Committee for Archaeology; 1992. Pp 134.

194. Wood, Diana (ed.). *The Church and the arts*. Oxford; Blackwell (Studies in Church History 28); 1992. Pp 496.
195. Gordon, W.M. (ed.). *Legal history in the making: Proceedings of the ninth British legal history conference*. London; Hambledon; 1991. Pp 270.
196. Smout, T.C. (ed.). *Victorian values*. Oxford; Oxford UP (Proceedings of the British Academy 78); 1992. Pp 240.
197. Hall, Catherine (ed.). *White, male and middle class: explorations in feminism and history* [reprints not itemised]. Cambridge; Polity; 1992. Pp xii, 307.
198. Greaves, Richard L. (ed.). *John Bunyan and English nonconformity* [reprints not itemised]. London; Hambledon; 1992. Pp 230.
199. Brand, Paul. *The making of the common law* [reprints not itemised]. London; Hambledon; 1992. Pp 490.
200. Jones, Bill; Robins, Lynton J. (ed.). *Two decades in British politics: essays to mark twenty-one years of the Politics Association, 1969–1990*. Manchester; Manchester UP; 1992. Pp xi, 347.
201. George, Stephen (ed.). *Britain and the European Community: the politics of semi-detachment*. Oxford; Oxford UP; 1992. Pp xii, 214.
202. Smith, David M. (ed.). *Studies in clergy and ministry in medieval England*. York; Borthwick Inst. (Studies in History 1); 1991. Pp ix, 163.
203. Holstun, James (ed.). *Pamphlet wars: prose in the English revolution*. London; Cass; 1992. Pp 230.
204. Leslie, Michael; Raylor, Timothy (ed.). *Culture and cultivation in early modern England: writing and the land*. Leicester; Leicester UP; 1992. Pp xv, 240.
205. Swan, Philip; Foster, David (ed.). *Essays in regional and local history: essays in honour of Eric M. Sigsworth*. Cherry Burton; Hutton; 1992. Pp 174.
206. Grenville, Jane; Carver, Martin O.H.; James, Edward (ed.). *Medieval Europe, vol. 6: religion and belief*. York; Medieval Europe (Conference on medieval archaeology in Europe); 1992. Pp 210.
207. Carver, Martin O.H.; Heal, Veryan; Sutcliffe, Ray (ed.). *Medieval Europe, vol. 2: maritime studies, ports and ships*. York; Medieval Europe (Conference on medieval archaeology in Europe); 1992. Pp 160.
208. Jennings, Sarah; Vince, Alan (ed.). *Medieval Europe, vol. 3: religion and belief*. York; Medieval Europe (Conference on medieval archaeology in Europe); 1992. Pp 240.
209. Hall, Richard A.; Hodges, Richard M.; Clarke, Helen (ed.). *Medieval Europe, vol. 5: exchange and trade*. York; Medieval Europe (Conference on medieval archaeology in Europe); 1992. Pp 210.
210. Aberg, Alan; Mytum, Harold (ed.). *Medieval Europe, vol. 8: religion and belief*. York; Medieval Europe (Conference on medieval archaeology in Europe); 1992. Pp 240.
211. Bush, Helen; Zvelebil, Marek (ed.). *Health in past societies: biocultural interpretations of human skeletal remains in archaeological contexts*. London; BAR (International ser. 567); 1991. Pp vii, 145.

212. Griffiths, Ralph A. (ed.). *The city of Swansea: challenges and change.* Stroud; Sutton; 1990. Pp x, 346.
213. Nelson, Janet L. (ed.). *Richard Coeur de Lion in history and myth.* London; King's College Centre for Late Antique and Medieval Studies (Medieval studies 7); 1992. Pp xiii, 165.
214. Addyman, Peter; Roskams, Steve P. (ed.). *Medieval Europe, vol. 1: urbanism.* York; Medieval Europe (Conference on medieval archaeology in Europe); 1992. Pp 244.
215. Holbrook, Neil; Bidwell, Paul T. (ed.). *Roman finds from Exeter* [not itemised]. Exeter; Exeter UP (Arch. reports 4); 1991. Pp xvi, 313.
216. Bassett, Steven (ed.). *Death in towns: urban responses to the dying and the dead, 100–1600.* Leicester; Leicester UP; 1992. Pp vi, 258.
217. Melling, Joseph; Barry, Jonathan (ed.). *Culture in history: production, consumption and values in historical perspective.* Exeter; Exeter UP; 1992. Pp 246.
218. Gray, Todd; Rowe, Margaret; Erskine, Audrey (ed.). *Tudor and Stuart Devon: the common estate and government.* Exeter; Exeter UP; 1992. Pp 250.
219. Breuilly, John J. (ed.). *Labour and Liberalism in 19th-century Europe: essays in comparative history.* Manchester; Manchester UP; 1992. Pp viii, 316.
220. Beal, Peter; Griffiths, Jeremy (ed.). *English manuscript studies, 1100–1700,* vol. 3. London; British Library; 1992. Pp 300.
221. Porter, Roy; Teich, Mikuláš (ed.). *The Renaissance in national context.* Cambridge; Cambridge UP; 1992. Pp ix, 239.
222. Korhammer, Michael (ed.). *Words, texts and manuscripts: studies in Anglo-Saxon culture.* Cambridge; Brewer; 1992. Pp 498.
223. Roper, Michael; Tosh, John (ed.). *Manful assertions: masculinities in Britain since 1800.* London; Routledge; 1991. Pp x, 221.
224. Hill, Myrtle; Barber, Sarah (ed.). *Aspects of Irish studies.* Belfast; Institute of Irish Studies; 1990. Pp 153.
225. Boos, Florence S. (ed.). *History and community: essays in Victorian medievalism.* London; Garland; 1992. Pp xv, 289.
226. Challis, Christopher E. (ed.). *A new history of the Royal Mint.* Cambridge; Cambridge UP; 1992. Pp 816.
227. Hogg, Richard M. (ed.). *The Cambridge history of the English language, vol. 1: the beginnings to 1066.* Cambridge; Cambridge UP; 1992. Pp 680.
228. Blake, Norman F. (ed.). *The Cambridge history of the English language, vol. 2: 1066–1476.* Cambridge; Cambridge UP; 1992. Pp 600.
229. Ward, Benedicta. *Signs and wonders: saints, miracles and prayer from the fourth century to the fourteenth* [reprints not itemised]. Aldershot; Variorum; 1992. Pp xviii, 300.

(d) Genealogy and Heraldry

1. Anon. *Bangor-on-Dee parish registers, vol. 4: baptisms, 1797–1812.* Clwyd; Clwyd Family History Soc.; 1990. Pp 16.

2. Anon. *Caerwys parish registers, vol. 3: baptisms and burials, 1790–1812; marriages, 1754–1812.* Clwyd; Clwyd Family History Soc.; 1991. Pp viii, 48.

3. Anon. *Gwaenysgor parish registers, 1538–1812.* Clwyd; Clwyd Family History Soc.; 1992. Pp xv, 79.

4. Anon. *Halkyn parish registers, vol. 4: baptisms, 1765–1803; burials, 1764–1803.* Clwyd; Clwyd Family History Soc.; 1990. Pp xiii, 58.

5. Anon. *Hope parish registers, vol. 3: marriages, 1754–1812.* Clwyd; Clwyd Family History Soc.; 1991. Pp xiii, 72.

6. Anon. *Llanelidan parish registers: baptisms, 1755–1812; marriages, 1696–1812; burials, 1755–1812.* Wrexham; Clwyd Family History Soc.; 1992. Pp ix, 86.

7. Anon. *Llanfair D.C. parish registers, vol. 1: baptisms and burials, 1680–1782; marriages, 1691–1754.* Clwyd; Clwyd Family History Soc.; 1991. Pp x, 140.

8. Anon. *Mold parish registers, vol. 1: baptisms, 1612–1612; marriages, 1604–1613; burials, 1611–1623; vol. 2: baptisms, 1624–1659; marriages, 1624–1644; burials, 1624–1647; vol. 5: marriages and burials, 1674–1721; vol. 6: baptisms, 1722–1772; vol. 8: marriages, 1754–1812.* Clwyd; Clwyd Family History Soc., 1990–1. 5 vols.

9. Anon. *Monumental inscriptions: St John's Boxmoor, Hertfordshire.* Ware; Hertfordshire Family & Population History Soc.; 1990. Pp 23.

10. Anon. *Northop parish registers, vol 2: marriages, 1656–1698; burials, 1656–1698; vol. 3: baptisms, 1698–1749; vol. 4: marriages, 1698–1754; burials, 1698–1749; vol. 6: marriages, 1754–1812; burials, 1750–1812.* Clwyd; Clwyd Family History Soc.; 1990–1. 4 vols.

11. Anon. *Registers of the Church of St Bartholomew, Edgbaston: baptisms, 1813–1837; late baptisms, 1838–1850; burials, 1813–1868.* Birmingham; Birmingham & Midland Soc. for Genealogy and Heraldry; 1992. Pp 142.

12. Anon. *Ruabon parish registers, vol. 2: 1683–1731.* Wrexham; Clwyd Family History Soc.; 1990. Pp xxxiii, 143.

13. Anon. *Sproatley monumental inscriptions.* Cottingham; East Yorkshire Family History Soc.; 1991. Pp v, 32.

14. Anon. *The bishop's transcripts of the registers of St Michael's Coventry, Warwickshire: baptisms, marriages and burials, 1640 and 1662–1691.* Birmingham; Birmingham & Midland Soc. for Genealogy and Heraldry; 1992. Pp 171.

15. Anon. *Tremeirchion parish registers, vol. 2: 1695–1812.* Clwyd; Clwyd Family History Soc.; 1990. Pp xiv, 81.

16. Anon. *Welton monumental inscriptions.* Cottingham; East Yorkshire Family History Soc.; 1991. Pp viii, 47.

17. Anon. *Whitford parish registers, vol. 1 [part 2]: 1643–1742, marriages.* Wrexham; Clwyd Family History Soc.; 1991. Pp vi, 40.

18. Barnett, R.D. (ed.). *The circumcision register of Isaac and Abraham de Paiba, 1715–1775.* London; Jewish Historical Soc. of England (Bevis Marks records vol. 4); 1991. Pp viii, 149.

19. Bayliss, Mary. 'Poor journeymen: Prévosts and Provosts in London and Leek', *Huguenot Soc. P.* 25 (1992), 356–70.

20. Beattie, Alastair G.; Beattie, Margaret H. (ed.). *Pre-1855 gravestone inscriptions in Lochaber and Skye: a summary of, and index to, pre-1855 gravestone inscriptions found in burial grounds in the western part of the mainland portion of Invernessshire, the small isles, Skye and associated islands*. Edinburgh; Scottish Genealogy Soc.; 1990. Pp 95.
21. Bettey, J.H. 'From Quaker traders to Anglican gentry: the rise of a Somerset dynasty', *Somerset Archaeology & Natural History* 135 (1992 for 1991), 1–9.
22. Bonthrone, Mark A. (ed.). *Auchtermuchty deaths, 1701–1851: extracts from the kirk session account books and the birth, marriage and death registers of Auchtermuchty*. London; Tron; 1992. Pp 84.
23. Carson, G.M. 'The Wisemans—loyal servants to the crown', *Family History* 16 (1991), 33–40.
24. Crosby, Alan (ed.). *The family records of Benjamin Shaw, mechanic of Dent, Dolphinholme and Preston, 1772–1841*. Manchester; Record Soc. of Lancashire and Cheshire vol. 130; 1991. Pp lxxx, 152.
25. Dewar, Peter Beauclerk. *The house of Dewar, 1296–1991*. Chichester; Phillimore; 1991. Pp 137.
26. Eldred, Nelson B.; Eldred, Trevor. *The Saxon house of Eldred*. Chichester; Phillimore; 1992. Pp x, 158.
27. Farrington, Susan Maria. *Peshawar: monumental inscriptions, II: northwest frontier province, Pakistan*. London; BACSA; 1991. Pp vi, 160.
28. Friar, Stephen. *Heraldry for the local historian and genealogist*. Stroud; Sutton; 1992. Pp xi, 271.
29. Goodall, John. 'A fifteenth-century Anglo-French-Burgundian heraldic collection (ET)', *Antiquaries J.* 70 (1990), 424–38.
30. Goring, Jeremy. 'A Sussex dissenting family: the Ridges of Westgate chapel, Lewes', *Sussex Arch. Collections* 129 (1991), 195–215.
31. Hayward, Valerie (comp.). *Monumental inscriptions: Holy Trinity Leverstock Green (Hertfordshire)*. Ware; Hertfordshire Family & Population History Soc.; 1991. Pp 37.
32. Hood, Christine (comp.). *Norwich archdeaconry marriage licence bonds, 1813–1837 (held at the Norfolk Record Office)*. Norwich; Norfolk and Norwich Genealogical Soc. (Norfolk genealogy; vol. 23); 1991. Pp vi, 280.
33. Hudleston, C. Roy; Cockerill, Timothy. 'Millom families, part 1. Askew, Latus and Thwaites; Cragg and Lewthwaite', *T. of the Cumberland & Westmorland Antiquarian & Arch. Soc.* 92 (1992), 91–95.
34. Ince, Laurence. *The Knight family and the British iron industry, 1695–1902*. Solihull; Ferric; 1991. Pp viii, 132.
35. Landon, Theodore Luke Giffard. 'The Landons, the first two hundred years: arrival, Spitalfields and onwards', *Huguenot Soc. P.* 25 (1992), 327–39.
36. Laslett, Peter. 'Masham of Oates—the rise and fall of an English family', *History Today* 41/9 (1991), 42–48.
37. O'Brien, O. 'An embryo university press', *Factotum* 35 (1992), 12–18.
38. Price, Jacob M. *Perry of London: a family and a firm on the seaborne frontier, 1615–1753*. Cambridge (Ma); Harvard UP; 1992. Pp 206.

39. Rice, Dorothy. 'Wootton hall and the Harris family', *Northamptonshire Past & Present* 8 (1992), 287–92.
40. Simons, Janice (comp.). *Marriage and obituary notices, Lynn Advertiser, Norfolk, for 1848, 1880, 1881, 1890, and 1900.* King's Lynn; Simons; 1990–2. 5 vols. Pp x, 90; x, 66; x, 72; x, 80; x, 82.
41. van der Kiste, John. *George V's children.* Stroud; Sutton; 1991. Pp xi, 193.
42. Wagner, Sir Anthony; Rowse, A.L. *John Anstis: Garter king of arms.* London; HMSO; 1992. Pp 80.
43. Warneford, Francis E. *An English family through eight centuries: the Warnefords.* Henfield; Warneford; 1991. Pp 385.

C. ROMAN BRITAIN

See also Bb137,162–163,167,184,186,215,222,227–228; Da8,f54,69

(a) *Archaeology*

1. Allen, J.R.L.; Fulford, Michael G. 'Romano-British wetland reclamations at Longney, Gloucestershire, and evidence for the early settlement of the inner Severn estuary', *Antiquaries J.* 70 (1990), 288–326.
2. Andrews, J.S.; Andrews, J.A. 'A Roman road from Kendal to Ambleside: part 2, Broadgate to Ambleside—a field survey', *T. of the Cumberland & Westmorland Antiquarian & Arch. Soc.* 92 (1992), 57–65.
3. Barber, A.J.; Walker, G.T.; Paddock, J.; Henig, M. 'A bust of Mars or a hero from Cirencester', *Britannia* 23 (1992), 217–18.
4. Benfield, Stephen; Garrod, Simon. 'Two recently-discovered Roman buildings in Colchester', *Essex Archaeology & History* 3rd ser. 23 (1992), 25–38.
5. Black, E.W. 'Newstead: the buildings in the western annexe', *Soc. of Antiquaries of Scotland P.* 121 (1991), 215–22.
6. Boyle, S.D. 'Excavations at Hen Waliau, Caernarfon, 1952–1985', *B. of the Board of Celtic Studies* 38 (1991), 191–212.
7. Bradley, Richard. 'Roman salt production in Chichester harbour: rescue excavations at Chidham, West Sussex', *Britannia* 23 (1992), 27–44.
8. Butterworth, Christine A.; *et al.* 'Excavations at Norton Bavant borrow pit, Wiltshire, 1987', *Wiltshire Arch. & Natural History Magazine* 85 (1992), 1–26.
9. Butterworth, Christine A.; Lobb, Susan J. *Excavations in the Burghfield area, Berkshire: developments in the bronze age and Saxon landscapes.* Salisbury; Trust for Wessex Archaeology (Report 1); 1992. Pp 182.
10. Caruana, I.D. 'Carlisle: excavation of a section of the annexe ditch of the first Flavian fort, 1990', *Britannia* 23 (1992), 45–109.

11. Collingwood, R.G.; Wright, Richard P. *The Roman inscriptions of Britain*, vol. 2, fasc. 4 ed. Sheppard S. Frere and R.S.O. Tomlin. Stroud; Sutton; 1992. Pp xvii, 206 + plates.

12. Crowley, N.; Betts, I.M. 'Three *classis Britannica* stamps from London', *Britannia* 23 (1992), 218–22.

13. Currie, Christopher K. 'Excavations and surveys at the Roman kiln site, Brinkworth, 1986', *Wiltshire Arch. & Natural History Magazine* 85 (1992), 27–50.

14. Dark, Ken R. 'A sub-Roman re-defence of Hadrian's wall?', *Britannia* 23 (1992), 111–20.

15. de Micheli, C. 'A bronze bowl of Irchester type from Stainfield, Lincolnshire', *Britannia* 23 (1992), 238–41.

16. Elliot, Walter. 'Animal footprints on Roman bricks from Newstead', *Soc. of Antiquaries of Scotland P.* 121 (1991), 223–26.

17. Ferris, I.M.; Jones, Richard F.J. 'Binchester—a northern fort and vicus', Bc184, 103–09.

18. Field, F.N.; Palmer-Brown, C.P.H. 'New evidence for a Romano-British greyware pottery industry in the Trent valley', *Lincolnshire History & Archaeology* 26 (1991), 40–56.

19. Frere, Sheppard S. 'Sites explored', *Britannia* 23 (1992), 255–308.

20. Fulford, Michael G. 'Iron age to Roman: a period of radical change on the gravels', Bc192, 23–38.

21. Fulford, Michael G.; Allen, J.R.L. 'Iron-making at the Chesters villa, Woolaston, Gloucestershire: survey and excavation, 1987–1991', *Britannia* 23 (1992), 159–214.

22. Gaffney, V.; Tingle, M. 'Field survey in Roman Britain: the experience of Maddle farm', Bc184, 81–84.

23. Gethyn-Jones, Eric. 'Roman Dymock: a personal record', *T. of the Bristol & Gloucestershire Arch. Soc.* 109 (1992 for 1991), 91–98.

24. Goodburn, Damian M. 'A Roman timber framed building tradition', *Arch. J.* 148 (1991), 182–204.

25. Graham-Campbell, James A. 'Norrie's Law, Fife: on the nature and dating of the silver hoard', *Soc. of Antiquaries of Scotland P.* 121 (1991), 241–60.

26. Gregory, Tony. 'Metal-detecting on a scheduled ancient monument', *Norfolk Archaeology* 41 (1991), 186–96.

27. Grew, Francis; Griffiths, Nick. 'The pre-Flavian military belt: the evidence from Britain', *Archaeologia* 109 (1991), 47–84.

28. Halkon, Peter. 'Romano-British face pots from Holme on Spalding moor and Shiptonthorpe, East Yorkshire', *Britannia* 23 (1992), 222–28.

29. Harrison, A.C. 'Excavation of a Belgic and Roman site at 50–54 High street, Rochester', *Archaeologia Cantiana* 109 (1992 for 1991), 41–50.

30. Hassall, M.W.C.; Tomlin, R.S.O. 'Inscriptions', *Britannia* 23 (1992), 309–23.

31. Heywood, Brenda; Jarvis, P.; Webster, P.V. 'The Roman fort at Penydarren, Glamorgan', *B. of the Board of Celtic Studies* 38 (1991), 167–91.

32. Jarvis, P.; Webster, P.V. 'Roman and other pottery from the Uskmouth area of Gwent', *B. of the Board of Celtic Studies* 38 (1991), 213–19.

33. Jones, Dilwyn; Whitwell, John Benjamin. 'Survey of the Roman fort and multi-period settlement complex at Kirmington on the Lincolnshire wolds: a non-destructive approach', *Lincolnshire History & Archaeology* 26 (1991), 57–62.

34. Jones, G.D.B. 'Old police house camp, Bowness-on-Solway', *Britannia* 23 (1992), 230–31.

35. Keevill, Graham D. 'A frying pan from Great Lea, Binfield, Berkshire', *Britannia* 23 (1992), 231–33.

36. King, Anthony; Soffe, Grahame. 'Hayling island', Bc184, 111–13.

37. Knight, Jeremy K. 'The early Christian Latin inscriptions of Britain and Gaul: chronology and context', Bc96, 45–50.

38. Lambrick, George. 'The development of late prehistoric and Roman farming on the Thames gravels', Bc192, 78–105.

39. Ling, Roger. 'A collapsed building facade from Carsington, Derbyshire', *Britannia* 23 (1992), 233–36.

40. Lucas, Ronald Norman. 'The Halstock mosaic found in 1817', *Dorset Natural History & Arch. Soc. P.* 113 (1992 for 1991), 133–38.

41. Mackreth, D.F. 'Roman brooches from Gastard, Corsham, Wiltshire', *Wiltshire Arch. & Natural History Magazine* 85 (1992), 51–62.

42. Mann, John C. 'A note on *RIB* 2054', *Britannia* 23 (1992), 236–38.

43. Philp, Brian J. *The Roman villa site at Keston, Kent: first report, excavations, 1968–1978.* Dover; Kent Arch. Rescue Unit (Kent monograph ser. 6); 1991. Pp 313.

44. Rawes, Bernard. 'A prehistoric and Romano-British settlement at Vineyards farm, Charlton Kings, Gloucestershire', *T. of the Bristol & Gloucestershire Arch. Soc.* 109 (1992 for 1991), 25–89.

45. Smith, Victor T.C. 'The Roman road (R2) at Springhead', *Archaeologia Cantiana* 109 (1992 for 1991), 332–33.

46. Smithson, Peter; Branigan, Keith. 'Poole's cavern, Buxton: investigation of a Romano-British working environment', *Derbyshire Arch. J.* 111 (1991), 40–45.

47. Spearman, R.M.; Wilthew, P. 'The Helmsdale bowls, a re-assessment', *Soc. of Antiquaries of Scotland P.* 120 (1990), 63–77.

48. Stuart, Robert. *Caledonia Romana: descriptive account of the Roman antiquities of Scotland.* Stevenage; Strong Oak; 1992. Pp 380.

49. Whitworth, A.M. 'The cutting of the turf wall at Appletree', *T. of the Cumberland & Westmorland Antiquarian & Arch. Soc.* 92 (1992), 49–55.

50. Wise, P. 'A Roman gold ring from Kinwarton near Alcester, Warwickshire', *Britannia* 23 (1992), 254.

51. Woolliscroft, D.J.; Swain, S.A.M.; Lockett, N.J. 'Barcombe B.: a second Roman "signal" tower on Barcombe hill', *Archaeologia Aeliana* 5th ser. 20 (1992), 57–62.

52. Zeepvat, R.J. 'Roman gardens in Britain', Bc64, 53–59.

(b) *History*

1. Adams, J.N. 'British Latin: the text, interpretation and language of the Bath curse tables', *Britannia* 23 (1992), 1–26.

2. Alcock, Joan P. 'The Bassingbourn Diana: a comparison with other bronze figurines of Diana found in Britain', *Cambridge Antiquarian Soc. P.* 79 (1992 for 1990), 39–44.

3. Anderson, James D. *Roman military supply in north-east England: an analysis of and an alternative to the Piercebridge formula.* London; BAR (British ser. 224); 1992. Pp vi, 196.

4. Barrett, Gillian. 'The prehistoric and Roman periods', Bc181, 15–42; 239–40.

5. Blagg, T.F.C. 'Buildings' [in Roman Britain], Bc184, 3–14.

6. Bland, Roger (ed.). *The Chalfont hoard and other Roman coin hoards.* London; British Museum; 1992. Pp 408.

7. Branigan, Keith; Dearne, M.J. *Romano-British cavemen.* Oxford; Oxbow (Monograph 19); 1992. Pp 120.

8. Breeze, David J. 'Agricola in the Highlands?', *Soc. of Antiquaries of Scotland P.* 120 (1990), 55–60.

9. Breeze, David J. 'Q. Lollius Urbicus and A. Claudius Charax, Antonine commanders in Britain', *Soc. of Antiquaries of Scotland P.* 121 (1991), 227–30.

10. Cleary, Simon Esmonde. 'Town and country in Roman Britain?', Bc216, 28–42.

11. Davies, Glenys. 'Roman cineraria in "Monumenta Mattheiana" and the collection of Henry Blundell at Ince', *Antiquaries J.* 70 (1990), 34–39

12. de la Bédoyère, Guy. *Roman towns in Britain.* London; Batsford; 1992. Pp 143.

13. Evans, Jeremy. 'Pottery in the later Roman north: a case study', Bc184, 49–52.

14. Frend, W.H.C. 'Pagans, Christians, and "the barbarian conspiracy" of AD 367 in Roman Britain', *Britannia* 23 (1992), 121–31.

15. Fulford, Michael G. 'Britain and the Roman empire: the evidence for regional and long distance trade', Bc184, 35–47.

16. Haley, E. 'The Roman bronze coinage in Britain and monetary history from AD 293 to 350', *American J. of Numismatics* 1 (1989), 89–116.

17. Hanson, W.S.; Macinnes, L. 'Soldiers and settlement in Wales and Scotland', Bc184, 85–92.

18. Harries, Jill. 'Death and the dead in the late Roman west', Bc216, 56–67.

19. Higham, Nicholas J. *Rome, Britain and the Anglo-Saxons.* London; Seaby; 1992. Pp 263.

20. Higham, Nicholas J. 'Gildas, Roman walls and British dykes', *Cambridge Medieval Celtic Studies* 22 (1991), 1–14.

21. Higham, Nicholas J. 'Soldiers and settlement in northern England', Bc184, 93–101.

22. Hill, P.R.; Dobson, Brian. 'The design of Hadrian's wall and its implications', *Archaeologia Aeliana* 5th ser. 20 (1992), 27–52.

23. Hingley, Richard. 'Domestic organisation and gender relations in iron age and Romano-British households', Bc38, 125–47.

24. Hingley, Richard. 'The Romano-British countryside: the significance of rural settlement forms', Bc184, 75–80.

25. James, Heather. 'The Roman roads of Carmarthenshire', Bc101, 53–77.
26. Jermy, K.E. '*Longford* and *Langford* as significant names in establishing lines of Roman roads', *Britannia* 23 (1992), 228–29.
27. Jones, Martin J. 'Food production and consumption—plants', Bc184, 21–27.
28. Jones, Martin J. 'Lincoln', Bc184, 69–73.
29. Jones, Richard F.J. 'Cultural change in Roman Britain', Bc184, 115–20.
30. Jones, Richard F.J. 'The urbanisation of Roman Britain', Bc184, 53–65.
31. King, Anthony. 'Food production and consumption—meat', Bc184, 15–20.
32. Lewit, Tamara. *Agricultural production in the Roman economy, AD 200–400.* London; BAR (International ser. 568); 1991. Pp 261.
33. Mann, John C. 'Loca', *Archaeologia Aeliana* 5th ser. 20 (1992), 53–55.
34. Mann, John C. 'The "turning" of Scotland', *Soc. of Antiquaries of Scotland P.* 120 (1990), 61–62.
35. Marsden, Peter; West, Barbara. 'Population change in Roman London', *Britannia* 23 (1992), 133–40.
36. Molleson, Theya. 'Mortality patterns in the Romano-British cemetery at Poundbury camp, Dorchester', Bc216, 43–55.
37. Moorwood, R.D.; Hodgson, N. 'Roman bridges on the Devil's Causeway?', *Britannia* 23 (1992), 241–45.
38. Peterson, J.W.M. 'Roman cadastres in Britain, part 2, eastern area: signs of a large system in the northern English home counties', *Dialogues d'Histoire Ancienne* 16 (1990), 233–72.
39. Potter, Timothy W.; Johns, Catherine. *Roman Britain.* London; British Museum; 1992. Pp 248.
40. Reece, Richard. *Roman coins from 140 sites in Britain.* Dorchester; Cotswold (Studies 4); 1991. Pp 107.
41. Reece, Richard. 'Money in Roman Britain: a review', Bc184, 29–34.
42. Roskams, Steve P. 'London—new understanding of the Roman city', Bc184, 67–68.
43. Samson, Ross. 'Comment on Eleanor Scott's "Romano-British villas and the social construction of space"', Bc38, 173–80.
44. Scott, Eleanor. 'Romano-British villas and the social construction of space', Bc38, 149–72.
45. Stuart-Macadam, Patty. 'Anaemia in Roman Britain: Poundbury camp', Bc211, 101–14.
46. Toft, L.A. 'Roman quays and tide levels', *Britannia* 23 (1992), 249–54.
47. Tomlin, R.S.O. 'The twentieth legion at Wroxeter and Carlisle in the first century: the epigraphic evidence', *Britannia* 23 (1992), 141–58.
48. Ward-Perkins, Bryan. 'Roman "continuity"', Bc214, 33–36.
49. Whitwell, John Benjamin. *Roman Lincolnshire.* Lincoln; History of Lincolnshire Committee; new edn. 1992. Pp 155.

ENGLAND 450–1066

See also Aa66,90,b10–15;Bc97,99–100,108,113,139,154,156–160,172,180, 183,185,193,229; Ea11–12,c3,11–15,e41,58,k42; Je1

(a) *General*

1. Alcock, Leslie. 'Message from the dark side of the moon: western and northern Britain in the age of Sutton Hoo', Bc113, 205–15.
2. Barlow, Frank (ed.). *The life of King Edward who rests at Westminster.* Oxford; Oxford UP; rev. edn. 1992. Pp lxxxii, 172.
3. Bately, Janet. 'John Joscelyn and the laws of the Anglo-Saxon kings', Bc222, 435–66.
4. Campbell, James. 'The impact of the Sutton Hoo discovery on the study of Anglo-Saxon history', Bc183, 79–101.
5. Colgrave, Bertram; Mynors, R.A.B. (ed.). *Bede's ecclesiastical history of the English people.* Oxford; Oxford UP; rev. edn. 1992. Pp lxxvi, 618.
6. Costen, Michael D. 'Dunstan, Glastonbury and the economy of Somerset in the tenth century', Bc3, 25–44.
7. Dumville, David N. *Wessex and England from Alfred to Edgar: six essays on political, cultural and ecclesiastical revival.* Woodbridge; Boydell; 1992. Pp xiii, 234.
8. Evans, Jeremy. 'From the end of Roman Britain to the "Celtic west"', *Oxford J. of Archaeology* 9 (1990), 91–103.
9. Farrell, Robert T. 'Anglo-Saxon literary studies and archaeology: a nuts and bolts approach', Bc97, 13–20.
10. Foster, S.M. 'The state of Pictland in the age of Sutton Hoo', Bc113, 217–34.
11. Frantzen, Allen J. 'Literature, archaeology, and Anglo-Saxon studies: reconstruction and deconstruction', Bc97, 21–30.
12. Halsall, Guy. 'Playing by whose rules? A further look at viking atrocity in the ninth century', *Medieval History* 2/2 (1992), 2–10.
13. Hart, Cyril R. 'The earliest *Life* of St Neot', Bc99, 605–11.
14. Kendall, Calvin B.; Wells, Peter S. 'Introduction: Sutton Hoo and early medieval northern Europe', Bc183, ix-xix.
15. O'Sullivan, Deirdre. 'Changing views of the viking age', *Medieval History* 2/1 (1992), 3–13.
16. Rands, Susan. 'West Pennard's Saxon charter', *Notes & Queries for Somerset & Dorset* 33 (1992), 117–21.
17. Santoro, Verio. 'Sul concetto di *Britannia* tra antichità e medioevo' [on the concept of 'Britannia' in antiquity and the middle ages], *Romanobarbarica* 11 (1991), 321–34.
18. Seebold, Elmar. 'Kentish—and Old English texts from Kent', Bc222, 409–34.
19. Taylor, C.M. 'Elmet: boundaries and Celtic survival in the post-Roman period', *Medieval History* 2/1 (1992), 111–29.

20. van Houts, Elisabeth M.C. 'Women and the writing of history in the early middle ages: the case of the abbess Matilda of Essen and Æthelweard', *Early Medieval Europe* 1 (1992), 53–68.
21. Welch, Martin G. *Anglo-Saxon England*. London; Batsford; 1992. Pp 152.
22. Werner, Joachim. 'A review of *The Sutton Hoo ship burial* volume 3: some remarks, thoughts and proposals', Bc193, 1–24.
23. Whitehouse, David. 'The Mediterranean perspective', Bc97, 117–28.
24. Wieland, Gernot. 'England in the German legends of Anglo-Saxon saints', Bc222, 193–212.
25. Wilson, David M. 'Sutton Hoo—pros and cons', Bc97, 5–12.
26. Wood, I.N. 'Frankish hegemony in England', Bc113, 235–41.

(b) *Politics and Institutions*

1. Baines, Arnold H.J. 'Wynflæd v. Leofwine: a Datchet lawsuit of 990', *Records of Buckinghamshire* 32 (1991 for 1990), 63–75.
2. Biddle, Martin; Kjølbye-Biddle, Birthe. 'Repton and the vikings', *Antiquity* 66 (1992), 36–51.
3. Brooks, Nicholas P. 'Church, crown and community: public work and seigneurial responsibilities at Rochester bridge', Bc71, 1–20.
4. Bullough, D.A. *Friends, neighbours and fellow-drinkers: aspects of community and conflict in the early medieval west.* Cambridge; Chadwick lecture no. 7; 1990. Pp 27.
5. Gamby, Erik. 'Den angelsachsiska kronikans Anlaf: Olaf Tryggvesson eller Olof Skotkonung?' ['Anlaf' in the *Anglo-Saxon Chronicle*: Olaf Tryggvasson or Olaf Skotkonung?], *Scandia* 57 (1991), 119.
6. Hart, Cyril R. 'A charter of King Edgar for Brafield-on-the-Green', Bc99, 487–94.
7. Hart, Cyril R. 'Athelstan "half king" and his family', Bc99, 569–604.
8. Hart, Cyril R. 'Danelaw and Mercian charters of the mid-tenth century', Bc99, 431–53.
9. Hart, Cyril R. 'Eadnoth I of Ramsey and Dorchester', Bc99, 613–23.
10. Hart, Cyril R. 'Oundle: its province and the eight hundreds', Bc99, 141–76.
11. Hart, Cyril R. 'The battle of Maldon', Bc99, 533–51.
12. Hart, Cyril R. 'The battles of *the Holme, Brunanburh* and Ringmere', Bc99, 511–32.
13. Hart, Cyril R. 'The church of St Mary, Huntingdon', Bc99, 221–28.
14. Hart, Cyril R. 'The ealdordom of Essex', Bc99, 115–40.
15. Hart, Cyril R. 'The earliest Suffolk charter', Bc99, 467–85.
16. Hart, Cyril R. 'The eastern Danelaw', Bc99, 25–113.
17. Hart, Cyril R. 'The Mersea charter of Edward the Confessor', Bc99, 495–508.
18. Hart, Cyril R. 'The origins of Lincolnshire', Bc99, 177–203.
19. Hart, Cyril R. 'The St Paul's estates in Essex', Bc99, 205–20.
20. Hart, Cyril R. 'The will of Ælfgifu', Bc99, 455–65.
21. Hart, Cyril R. 'What is the Danelaw?', Bc99, 3–24.
22. Higham, Nicholas J. 'King Cearl, the battle of Chester and the origins of the Mercian 'overkingship', *Midland History* 17 (1992), 1–15.

23. Higham, Nicholas J. 'Northumbria, Mercia and the Irish sea Norse, 893–926', Bc108, 21–30.
24. Hooke, Della. 'Early units of government in Herefordshire and Shropshire', Bc193, 47–64.
25. Hudson, Benjamin T. 'Cnut and the Scottish kings', *English Historical R.* 107 (1992), 350–60.
26. Johnson-South, Ted. 'Competition for King Alfred's aura in the last century of Anglo-Saxon England', *Albion* 23 (1991), 613–26.
27. Keynes, Simon. 'Rædwald the bretwalda', Bc183, 103–23.
28. Keynes, Simon. 'The Fonthill letter', Bc222, 53–98.
29. Lawson, M.K. 'Archbishop Wulfstan and the homiletic element in the laws of Æthelred II and Cnut', *English Historical R.* 107 (1992), 565–86.
30. Loyn, Henry R. 'Kings, gesiths and thegns', Bc113, 75–79.
31. Moberg, Ove. 'Den angelsachsiska kronikans Anlaf: Olaf Tryggvesson eller Olof Skotkonung?' ['Anlaf' in the *Anglo-Saxon Chronicle*: Olaf Tryggvasson or Olaf Skotkonung?], *Scandia* 57 (1991), 115–18.
32. Newton, Sam. 'Beowulf and the East Anglian royal pedigree', Bc113, 65–74.
33. Parker, M.S. 'The province of Hatfield', *Northern History* 28 (1992), 42–69.
34. Raw, Barbara C. 'Royal power and the royal symbols in *Beowulf*', Bc113, 167–74.
35. Reynolds, Susan. 'Bookland, folkland and fiefs', Bc27, 211–27.
36. Scull, Christopher J. 'Before Sutton Hoo: structures of power and society in early East Anglia', Bc113, 3–23.
37. Taylor, Pamela. 'The endowment and military obligations of the see of London: a reassessment of three sources', Bc27, 287–312.
38. Walker, Simon. 'A context for "Brunanburh"?', Bc71, 21–39.
39. Watkin, Thomas Glyn. 'Saints, seaways and dispute settlements', Bc195, 1–10.
40. Williams, Ann. 'A bell-house and a burh-geat: lordly residences in England before the Norman conquest', Bc98, 221–40.

(c) *Religion*

1. Baltrusch-Schneider, Dagmar Beate. 'Die angelsachsischen Doppelkloster' [The Anglo-Saxon double monastery], *Doppelkloster und andere Formen der Symbiose männlicher und weiblicher Religiösen im Mittelalter* ed. Kaspar Elm and Michel Parisse (Berlin; Duncker & Humblot; 1992), 57–80.
2. Bassett, Steven. 'Church and diocese in the west midlands: the transition from British to Anglo-Saxon control', Bc100, 13–40.
3. Bassett, Steven. 'Medieval ecclesiastical organisation in the vicinity of Wroxeter and its British antecedents', *J. of the British Arch. Association* 145 (1992), 1–28.
4. Blair, W. John. 'Anglo-Saxon minsters: a topographical review', Bc100, 226–66.
5. Blair, W. John. 'The making of the English parish', *Medieval History* 2/2 (1992), 13–19.

6. Blair, W. John; Sharpe, Richard. 'Introduction' [pastoral care], Bc100, 1–10.

7. Brooks, Nicholas P. 'The career of St Dunstan', Bc3, 1–23.

8. Cameron, M.L. 'The visions of Saints Anthony and Guthlac', Bc90, 152–58.

9. Corrêa, Alicia (ed.). *The Durham Collectar*. Woodbridge; Boydell; Henry Bradshaw Soc. vol. 107; 1992. Pp 352.

10. Crawford, Barbara. 'The cult of St Clement of the Danes in England and Scotland', Bc206, 1–4.

11. Cubitt, Catherine. 'Pastoral care and conciliar canons: the provisions of the 747 council of *Clofesho*', Bc100, 193–211.

12. Dales, D.J. 'The spirit of the *Regularis concordia* and the hand of St Dunstan', Bc3, 45–56.

13. Daniels, R. 'The Anglo-Saxon monastery at Hartlepool, England', Bc206, 171–76.

14. Doyle, A.I. 'A fragment of an eighth-century Northumbrian Office book', Bc222, 11–28.

15. Foot, Sarah. 'Anglo-Saxon minsters: a review of terminology', Bc100, 212–25.

16. Foot, Sarah. '"By water in the spirit": the administration of baptism in early Anglo-Saxon England', Bc100, 171–92.

17. Gatch, Milton McC. 'Piety and liturgy in the Old English *Vision of Leofric*', Bc222, 159–80.

18. Gerchow, Jan 'Prayers for King Cnut: the liturgical commemoration of a conqueror', Bc75, 219–38.

19. Gretsch, Mechthild. 'The Benedictine rule in Old English: a document of Bishop Æthelwold's reform politics', Bc222, 131–58.

20. Hill, Joyce. 'Monastic reform and the secular church: Ælfric's pastoral letters in context', Bc75, 103–17.

21. Hollis, Stephanie. *Anglo-Saxon women and the church: sharing a common fate*. Woodbridge; Boydell; 1992. Pp x, 310.

22. Jackson, Peter. 'The *vitae patrum* in eleventh century Worcester', Bc75, 119–34.

23. Lapidge, Michael. 'Abbot Germanus, Winchcombe, Ramsey and the Cambridge psalter', Bc222, 99–130.

24. Lapidge, Michael. 'The saintly life in Anglo-Saxon England', Bc180, 243–63.

25. Lapidge, Michael (ed.). *Anglo-Saxon litanies of the saints*. Woodbridge; Boydell for the Henry Bradshaw Soc.; 1991. Pp xii, 328.

26. Levy, Kenneth. 'Latin chant outside the Roman tradition', Bc78, 69–101.

27. Meaney, Audrey L. 'Anglo-Saxon idolators and ecclesiasts from Theodore to Alcuin: a source study', Bc193, 103–26.

28. Murphy, Elinor. 'Anglo-Saxon abbey, Shaftesbury: Bectun's base or Alfred's foundation', *Dorset Natural History & Arch. Soc. P.* 113 (1992 for 1991), 23–32.

29. Orme, Nicholas I. 'From the beginnings to 1050' [church in Devon and Cornwall], Bc148, 1–22; 203–05.

30. Orme, Nicholas I. 'Saint Breage: a medieval virgin saint of Cornwall', *Analecta Bollandiana* 110 (1992), 341–52.

31. Ortenberg, Veronica. *The English church and the continent in the tenth and eleventh centuries: cultural, spiritual and artistic exchanges.* Oxford; Oxford UP; 1992. Pp 360.

32. Paxton, Frederick. 'Anointing the sick and the dying in Christian antiquity and the early medieval west', Bc90, 93–102.

33. Ramsay, Nigel. 'The cult of St Dunstan at Christ Church, Canterbury', Bc3, 311–23.

34. Rollason, David. 'The concept of sanctity in the early lives of St Dunstan', Bc3, 261–72.

35. Rosser, Gervase. 'The cure of souls in English towns before 1000', Bc100, 267–84.

36. Scharer, Anton. 'Gesellschaftliche Zustande im Spiegel des Heiligenlebens: einige Folgerungen aus den Lebensbeschreibungen des heiligen Cuthberht' [Social conditions as seen in the lives of saints: some conclusions from the Life of St Cuthbert], *Mitteilungen des Instituts für Österreichische Geschichtsforschung* 100 (1992), 103–16.

37. Scragg, Donald G. 'An Old English homilist of Archbishop Dunstan's day', Bc222, 181–92.

38. Sharpe, Richard. 'The date of St Mildreth's translation from Minster-in-Thanet to Canterbury', *Mediaeval Studies* 53 (1991), 349–54.

39. Smith, Julia M.H. 'Review article: early medieval hagiography in the late twentieth century', *Early Medieval Europe* 1 (1992), 69–76.

40. Stevenson, Jane. 'Christianity in sixth- and seventh-century Southumbria', Bc113, 175–83.

41. Thacker, Alan T. 'Cults at Canterbury: relics and reform under Dunstan and his successors', Bc3, 221–45.

42. Thacker, Alan T. 'Monks, preaching and pastoral care in early Anglo-Saxon England', Bc100, 137–70.

43. Wenisch, Franz. '*Nu bidde we eow for Godes lufon*: a hitherto unpublished Old English homiletic text in CCCC 162', Bc222, 43–52.

(d) *Economic Affairs and Numismatics*

1. Archibald, M.M. 'Dating Cuerdale: the evidence of the coins', Bc108, 15–20.

2. Balzaretti, Ross; Nelson, Janet L. 'Trade, industry and the wealth of King Alfred', *Past & Present* 135 (1992), 142–63.

3. Blackburn, Mark. 'Hiberno-Norse coins of the *Helmet* type', Bc139, 9–24.

4. Blunt, C.E.; Lyon, C.S.C. 'Some notes on the mints of Wilton and Salisbury', Bc139, 25–34.

5. Crabtree, Pam. 'Urban provisioning and rural surplus extraction: a comparison of West Stow and Brandon', Bc214, 121–26.

6. Graham-Campbell, James A. 'Anglo-Scandinavian equestrian equipment in eleventh-century England', Bc27, 77–89.

7. Graham-Campbell, James A. 'The Cuerdale hoard: a viking and Victorian treasure', Bc108, 1–14.

8. Graham-Campbell, James A. 'The Cuerdale hoard: comparisons and context', Bc108, 107–15.
9. Griffiths, David. 'Territories and exchange in the Irish sea region, 400–1100', Bc209, 9–12.
10. Griffiths, David. 'The coastal trading ports of the Irish sea', Bc108, 63–72.
11. Hodges, Richard M. 'Society, power and the first English industrial revolution', Bc73, 125–50.
12. Hodges, Richard M. 'The eighth-century pottery industry at La Londe, near Rouen and its implications for cross-channel trade with Hamwic, Anglo-Saxon Southampton', *Antiquity* 65 (1991), 882–87.
13. Jansen, Henrik M. 'The archaeology of Danish commercial centers', Bc183, 171–81.
14. Jonsson, Kenneth. 'Bror Emil Hildebrand and the Borup hoard', Bc139, 35–45.
15. Jonsson, Kenneth; van der Meer, Gay. 'Mints and moneyers, c.973–1066', Bc139, 47–136.
16. Kelly, Susan. 'Trading privileges from eighth-century England', *Early Medieval Europe* 1 (1992), 3–28.
17. Kluge, Bernd. 'Das älteste Exemplar von Agnus Dei-Typ' [The oldest example of the Agnus Dei type], Bc139, 137–56.
18. Kruse, S.E. 'Metallurgical evidence of silver sources in the Irish sea province', Bc108, 73–88.
19. Leimus, Ivar. 'A fourteenth *Agnus Dei* penny of Æthelred II', Bc139, 157–63.
20. Maddicott, J.R. 'Trade, industry and the wealth of King Alfred: a reply' [to R. Balzaretti and J.L. Nelson], *Past & Present* 135 (1992), 164–88.
21. Mainman, Ailsa J. 'Pottery production in the late Anglo-Saxon period (ninth-eleventh century)—a case study from York', Bc208, 205–10.
22. Metcalf, D.M. 'Can we believe the very large figure of £72,000 for the geld levied by Cnut in 1018?', Bc139, 165–76.
23. Metcalf, D.M. 'The monetary economy of the Irish sea province', Bc108, 89–106.
24. Mortimer, Cath. 'Patterns of non-ferrous metal use during the early medieval period', Bc208, 97–102.
25. Ottaway, Patrick J. 'Invention and innovation: the blacksmith's craft in eighth-eleventh century England', Bc208, 117–24.
26. Pagan, Hugh. 'The coinage of Harold II', Bc139, 177–205.
27. Petersson, H. Bertil A. 'Coins and weights: late Anglo-Saxon pennies and mints, c.973–1066', Bc139, 207–433.
28. Schoenfeld, Edward; Schulman, Jane. 'Sutton Hoo: an economic assessment', Bc183, 15–27.
29. Smart, Veronica. 'Osulf Thein and others: double moneyers' names on the late Anglo-Saxon coinage', Bc139, 435–53.
30. Stahl, Alan M. 'The nature of the Sutton Hoo coin parcel', Bc183, 3–14.
31. Stahl, Alan M.; Oddy, W.A. 'The date of the Sutton Hoo coins', Bc97, 129–48.
32. Stewart, Ian. 'Coinage and recoinage after Edgar's reform', Bc139, 455–85.

33. Talvio, Tuuka. 'The design of Edward the Confessor's coins', Bc139, 487–99.

(e) *Intellectual and Cultural*

1. Anderson, James E. 'Exeter Latin riddle 90: a liturgical vision', *Viator* 23 (1992), 73–93.
2. Bailey, Richard N. 'Sutton Hoo and seventh-century art', Bc97, 31–42.
3. Bammesberger, Alfred. 'Five *Beowulf* notes', Bc222, 239–56.
4. Bammesberger, Alfred. 'The place of English in Germanic and Indo-European', Bc228, 26–66.
5. Bately, Janet. 'The nature of Old English prose', Bc180, 71–87.
6. Budny, Mildred. '"St Dunstan's classbook" and its frontispiece: Dunstan's portrait and autograph', Bc3, 103–42.
7. Carver, Martin O.H. 'Ideology and allegiance in East Anglia', Bc97, 173–82.
8. Clark, Cecily. 'Onomastics', Bc228, 452–89.
9. Clemoes, Peter. 'King Alfred's debt to vernacular poetry: the evidence of *ellen* and *cræft*', Bc222, 213–38.
10. Cramp, Rosemary. *Studies in Anglo-Saxon sculpture*. London; Pindar; 1992. Pp 380.
11. Creed, Robert Payson. 'Sutton Hoo and the recording of *Beowulf*', Bc183, 65–75.
12. Derolez, Réné. 'Language problems in Anglo-Saxon England: *barbara loquella* and *barbarismus*', Bc222, 285–92.
13. Dumville, David N. 'On the dating of some late Anglo-Saxon liturgical manuscripts', *T. of the Cambridge Bibliographical Soc.* 10 (1991), 40–57.
14. Fell, Christine. 'Perceptions of transience', Bc180, 172–89.
15. Frank, Roberta. '*Beowulf* and Sutton Hoo: the odd couple', Bc183, 47–64.
16. Frank, Roberta. 'Germanic legend in Old English literature', Bc180, 88–106.
17. Frank, Roberta. 'Old English *æræt* – "too much" or "too soon"?', Bc222, 293–304.
18. Freeman, Eric F. 'The identity of Ælfgyva in the Bayeux tapestry', *Annales de Normandie* 41 (1991), 117–34.
19. Gameson, Richard. 'Ælfric and the perception of script and picture in Anglo-Saxon England', Bc193, 85–102.
20. Gameson, Richard. 'Manuscript art at Christ Church, Canterbury, in the generation after St Dunstan', Bc3, 187–220.
21. Gameson, Richard. 'The fabric of the Tanner Bede', *Bodleian Library Record* 14 (1992), 176–206.
22. Gatch, Milton McC. 'Perceptions of eternity', Bc180, 190–205.
23. Gem, Richard D.H. 'Tenth-century architecture in England', Bc73, 803–36.
24. Gneuss, Helmut. 'The Old English language', Bc180, 23–54.

25. Godden, Malcolm R. 'Biblical literature: the Old Testament', Bc180, 206–26.
26. Godden, Malcolm R. 'Literary language', Bc228, 490–535.
27. Greis, Gloria Polizzotte; Geselowitz, Michael N. 'Sutton Hoo art: two millennia of history', Bc183, 29–44.
28. Henderson, George. 'The idiosyncrasy of late Anglo-Saxon religious imagery', Bc75, 239–49.
29. Hofstetter, Walter. 'The Old English adjectival suffix -cund', Bc222, 325–48.
30. Hogg, Richard M. 'Introduction' [language], Bc228, 1–25.
31. Hogg, Richard M. 'Phonology and morphology' [Old English], Bc228, 67–167.
32. Jones, Alex I. 'Ælfric's life of St Cuthbert', Parergon 10 (1992), 35–42.
33. Kastovsky, Dieter. 'Semantics and vocabulary', Bc228, 290–408.
34. Korhammer, Michael. 'Old English bolca and Maegða land - two problems, one solved', Bc222, 305–24.
35. Lapidge, Michael. 'Æthelwold and the Vita Sancti Eustachii', Scire litteras: Forschungen zum mittelalterlichen Geistesleben, ed. Sigrid Kramer and Michael Bernhard (Munich; Bayerische Akademie der Wissenschaft; 1988), 255–65.
36. Lapidge, Michael. 'B. and the Vita S. Dunstani', Bc3, 247–59.
37. Lendinara, Patrizia. 'The world of Anglo-Saxon learning', Bc180, 264–81.
38. Locherbie-Cameron, M.A.L. 'Friend or foe? The portrayal of enemies in Anglo-Saxon literature', Medieval History 2/1 (1992), 34–44.
39. Okasha, Elisabeth. 'The English language in the eleventh century: the evidence from inscriptions', Bc75, 333–45.
40. O'Keefee, Katherine O'Brien. 'Heroic values and Christian ethics', Bc180, 107–25.
41. O'Reilly, Jennifer. 'St John as a figure of the contemplative life: text and image in the art of the Anglo-Saxon Benedictine reform', Bc3, 165–85.
42. Raw, Barbara C. 'Biblical literature: the New Testament', Bc180, 227–42.
43. Raw, Barbara C. 'What do we mean by the source of a picture?', Bc75, 285–300.
44. Reichl, Karl. 'Old English giedd, Middle English yedding as genre terms', Bc222, 349–70.
45. Richards, Julian D. 'Anglo-Saxon symbolism', Bc113, 131–47.
46. Roberts, J. 'Anglo-Saxon vocabulary as a reflection of material culture', Bc113, 185–202.
47. Robinson, Fred C. 'Beowulf', Bc180, 142–59.
48. Rosenthal, Jane. 'The pontifical of St Dunstan', Bc3, 143–63.
49. Rosenthal, Joel T. 'Anglo-Saxon attitudes: men's sources, women's history', Bc177, 259–84.
50. Sauer, Hans. 'Towards a linguistic description and classification of the Old English plant names', Bc222, 381–408.
51. Schulenburg, Jane Tibbetts. 'Saints' lives as a source for the history of women, 500–1100', Bc177, 285–320.

52. Scragg, Donald G. 'Spelling variations in eleventh-century English', Bc75, 347–54.
53. Scragg, Donald G. 'The nature of Old English verse', Bc180, 55–70.
54. Sims-Williams, Patrick. 'The emergence of Old Welsh, Cornish and Breton orthography, 600–800: the evidence of archaic Old Welsh', *B. of the Board of Celtic Studies* 38 (1991), 20–86.
55. Stevens, Wesley M. 'Sidereal time in Anglo-Saxon England', Bc183, 125–52.
56. Teviotdale, E.C. 'The making of the Cotton troper', Bc75, 301–16.
57. Toon, Thomas E. 'Old English dialects', Bc228, 409–51.
58. Trahern, Joseph B., Jr. 'Fatalism and the millennium', Bc180, 160–71.
59. Traugott, Elizabeth Closs. 'Syntax' [Old English], Bc228, 168–289.
60. Wenzel, Siegfried. 'The Middle English lexicon: help from the pulpit', Bc222, 467–76.
61. Wickham-Crowley, Kelley. 'The birds on the Sutton Hoo instrument', Bc97, 43–62.
62. Wormald, Patrick. 'Anglo-Saxon society and its literature', Bc180, 1–22.
63. Wright, Rosemary Muir. 'The Virgin in the sun and in the tree', Bc91, 36–59.

(f) *Society and Archaeology*

1. Bassett, Steven. 'Anglo-Saxon Shrewsbury and its churches', *Midland History* 16 (1991), 1–23.
2. Bayley, Justine. *Anglo-Scandinavian non-ferrous metalworking from 16–22 Coppergate*. York; CBA (Archaeology of York 17/7); 1992. Pp 116.
3. Bayley, Justine. 'Anglo-Saxon non-ferrous metalworking: a survey', *World Archaeology* 17 (1991), 115–30.
4. Bayley, Justine. 'Viking age metalworking: the British Isles and Scandinavia compared', Bc208, 91–96.
5. Bettess, F. 'The Anglo-Saxon foot: a computerized assessment', *Medieval Archaeology* 35 (1991), 44–50.
6. Boddington, Andy. 'Pagans, Christians and agnostics: patterns in Anglo-Saxon burial practice', Bc185, 99–104.
7. Bourdillon, Jennifer. 'Changes in animal provisioning from middle Saxon Southampton', Bc214, 127–32.
8. Bridbury, A.R. 'Seventh-century England in Bede and the early laws', Bc114, 56–85.
9. Brooke, Daphne. 'The Northumbrian settlements in Galloway and Carrick: an historical assessment', *Soc. of Antiquaries of Scotland P.* 121 (1991), 295–328.
10. Buchet, Luc. 'La recherche des structures sociales et des conditions de vie par l'étude des squelettes', Bc185, 1–6.
11. Capelle, Torsten. *Archäologie der Angelsachsen: eigenständigkeit und kontinentale bindung vom 5 bis 9 jahrhundert* [The archaeology of the Anglo-Saxons: independence and continental influence from the fifth to

the ninth century]. Darmstadt; Wissenschaftliche Buchgesellschaft; 1990. Pp viii, 158.

12. Carver, Martin O.H. 'Conclusion: the future of Sutton Hoo', Bc183, 183–200.

13. Carver, Martin O.H. 'The Anglo-Saxon cemetery at Sutton Hoo: an interim report', Bc113, 343–71.

14. Coates, Richard. 'Bonchurch: in defence of the man on the Vectis omnibus', *Nomina* 14 (1992 for 1990–1), 41–46.

15. Coggins, D. 'Some aspects of early medieval upland settlement in northern England', Bc210, 95–100.

16. Costen, Michael D. 'Huish and Worth: Old English survivals in a later landscape', Bc193, 65–84.

17. Dalwood, Hal. 'Continuity and change in the urban fabric of Worcester, England', Bc214, 69–74.

18. Davidson, Hilda Ellis. 'Human sacrifice in the late pagan period in north-western Europe', Bc113, 331–40.

19. Davidson, Hilda Ellis. 'Royal graves as religious symbols', Bc193, 25–32.

20. de Vegvar, Carol Neuman. 'The Sutton Hoo horns as regalia', Bc97, 63–74.

21. Dickinson, Tania M.; Härke, Heinrich G. *Early Anglo-Saxon shields*. London; Soc. of Antiquaries (*Archaeologia* vol. 110); 1992. Pp 94.

22. Dickinson, Tania M.; Speake, G. 'The seventh-century cremation burial in Asthall Barrow, Oxfordshire: a reassessment', Bc113, 95–130.

23. Edwards, B.J.N. 'The vikings in north-west England: the archaeological evidence', Bc108, 43–62.

24. Everson, Paul L. 'Three case studies of ridge and furrow: Offa's dyke at Dudston in Chirbury, Shropshire—a pre-Offan field system?', *Landscape History* 13 (1991), 53–63.

25. Fernie, Eric C. 'Anglo-Saxon lengths and the evidence of the buildings', *Medieval Archaeology* 35 (1991), 1–5.

26. Filmer-Sankey, William. 'Snape Anglo-Saxon cemetery: the current state of knowledge', Bc113, 39–51.

27. Geake, H. 'Burial practice in seventh- and eighth-century England', Bc113, 83–94.

28. Gem, Richard D.H. 'Reconstructions of St Augustine's abbey, Canterbury, in the Anglo-Saxon period', Bc3, 57–73.

29. Gifford, E.W.H. 'The sailing characteristics of the Anglo-Saxon ship finds', Bc207, 99–106.

30. Gittos, Brian; Gittos, Moira. 'The surviving Anglo-Saxon fabric of East Coker church', *Somerset Archaeology & Natural History* 135 (1992 for 1991), 107–11.

31. Gough, Harold. 'Eadred's charter of AD 949 and the extent of the monastic estate of Reculver, Kent', Bc3, 89–102.

32. Haldenby, David. 'An Anglian site on the Yorkshire wolds—continued', *Yorkshire Arch. J.* 64 (1992), 25–39.

33. Hamerow, Helena. 'Settlement on the gravels in the Anglo-Saxon period', Bc192, 39–46.

34. Härke, Heinrich G. 'Changing symbols in a changing society: the Anglo-Saxon weapon burial rite in the seventh century', Bc113, 149–65.

35. Hart, Cyril R. 'The site of *Assandun*', Bc99, 553–65.
36. Haslam, Jeremy. '*Dommoc* and Dunwich: a reappraisal', Bc193, 41–46.
37. Hawkes, Sonia Chadwick. 'Valley-bottom settlement in the early Anglo-Saxon period: the alternative model', Bc210, 7–8.
38. Hines, J. 'The Scandinavian character of Anglian England: an update', Bc113, 315–29.
39. Horden, Peregrine. 'Disease, dragons and saints: the management of epidemics in the dark ages', Bc25, 45–76.
40. Horsey, I.P.; Winder, J.M. 'Late Saxon and Conquest-period oyster middens at Poole, Dorset', Bc41, 102–04.
41. Huggins, P.J. 'Anglo-Saxon timber building measurements: recent results', *Medieval Archaeology* 35 (1991), 6–28.
42. John, Eric. 'The point of Woden', Bc193, 127–34.
43. Jones, Glanville. 'The multiple estates of (Holy) Islandshire, Hovingham and Kirkbymoorside', Bc210, 79–84.
44. Keevill, Graham D. 'The rediscovery of Eynsham abbey, Oxfordshire: archaeological investigations, 1989–1992', Bc206, 195–200.
45. Klempere, Bill. 'Burial at Hulton abbey', Bc185, 85–92.
46. Knüsel, Christopher. 'The potential of osteological indicators to identify occupations in medieval skeletal populations', Bc185, 7–14.
47. Lilley, Jane. 'The medieval cemetery of Jewbury, York', Bc185, 61–66.
48. Mainman, Ailsa J. 'Ipswich ware from York: a survey of recent evidence', Bc191, 16–20.
49. Malim, Tim. 'Barrington Anglo-Saxon cemetery, 1989', *Cambridge Antiquarian Soc. P.* 79 (1992 for 1990), 45–62.
50. Marshall, Anne; Marshall, Gary. 'A survey and analysis of the buildings of early and middle Anglo-Saxon England', *Medieval Archaeology* 35 (1991), 29–43.
51. Meaney, Audrey L. 'The Anglo-Saxon view of the causes of disease', Bc90, 12–33.
52. Morton, Alan D. 'Burial in middle Saxon Southampton', Bc216, 68–77.
53. Morton, Alan D. 'The topography of mid-Saxon Southampton (Hamwic)', Bc214, 185–88.
54. Newman, J. 'The late Roman and Anglo-Saxon settlement pattern in the Sandlings of Suffolk', Bc113, 25–38.
55. Niles, John D. 'Pagan survivals and popular belief', Bc180, 126–41.
56. Ortenberg, Veronica. 'An unknown late Anglo-Saxon text about old St Peter's in Rome', *Antiquaries J.* 70 (1990), 115–17.
57. Palliser, David M. 'Town and village planning in early medieval England', Bc214, 177–84.
58. Perkins, D.R.J. 'The Jutish cemetery at Sarre revisited: a rescue evaluation', *Archaeologia Cantiana* 109 (1992 for 1991), 139–66.
59. Preston-Jones, Ann. 'Decoding Cornish churchyards', Bc96, 104–24.
60. Rahtz, Philip. 'Late Roman and post-Roman cemeteries in the west of Britain revisited', Bc185, 179–84.
61. Robinson, Paul. 'Some late Saxon mounts from Wiltshire', *Wiltshire Arch. & Natural History Magazine* 85 (1992), 63–69.

62. Roesdahl, Else. 'Princely burial in Scandinavia at the time of the conversion', Bc183, 155–70.
63. Russell, Pamela B. 'Place-name evidence for the survival of British settlements in the West Derby hundred (Lancashire), after the Anglian invasions', *Northern History* 28 (1992), 25–41.
64. Sandred, K.I. 'Det anglosaxiska London i ny belysning' [Anglo-Saxon London in a new light], *Ortnamnssallskapets i Uppsala Arsskrift* (1990), 63–69.
65. Scull, Christopher J. 'Post-Roman phase 1 at Yeavering: a re-construction', *Medieval Archaeology* 35 (1991), 51–63.
66. Smith, Ian M. 'Sprouston, Roxburghshire: an early Anglian centre of the eastern Tweed basin', *Soc. of Antiquaries of Scotland P.* 121 (1991), 261–94.
67. Tatton-Brown, Tim. 'The city and diocese of Canterbury in St Dunstan's time', Bc3, 75–87.
68. Thompson, Pauline. 'The disease that we call cancer', Bc90, 1–11.
69. Unwin, Tim. 'From Roman Britain to the Norman conquest', Bc181, 43–72; 240–42.
70. Webster, Leslie. 'Death's diplomacy: Sutton Hoo in the light of other male princely burials', Bc97, 75–82.
71. White, R.H. 'The Badley (Needham Market) bowl', Bc193, 33–40.
72. Wicker, Nancy L. Hatch. 'Swedish-Anglian contacts antedating Sutton Hoo: the testimony of the Scandinavian gold bracteates', Bc97, 149–72.
73. Wickham, Chris. 'Problems of comparing rural societies in early medieval western Europe', *T. of the Royal Historical Soc.* 6th ser. 2 (1992), 221–46.
74. Williams, Ann. 'The battle of Maldon and *The battle of Maldon*: history, poetry and propaganda', *Medieval History* 2/2 (1992), 35–44.
75. Wilson, David M. *Anglo-Saxon paganism*. London; Routledge; 1992. Pp 208.
76. Yorke, Barbara. 'Anglo-Saxon royal burial, *c*.600–1066: the documentary evidence', Bc185, 41–46.

ENGLAND 1066–1500

See also Aa7,81,105,129,151,b22–23,25; Bb26,c3–4,27,45–46,51,53,58, 70–71,73,75,78,81,90,92,98,102,104–105,109–112,115,141,152,165– 167,177,202,206–210,213–214,d29; Dc43,d9; Fa17,b34; Kc14

(a) *General*

1. Allmand, C.T. *Henry V*. London; Methuen; 1992. Pp xvi, 480.
2. Bates, David. 'The Conqueror's charters', Bc75, 1–15.
3. Bevan, Bryan. *Edward III: monarch of chivalry*. London; Rubicon; 1992. Pp vi, 162.

4. Bridbury, A.R. 'Domesday Book: a re-interpretation', Bc114, 86–110.

5. Brossard-Dandré, Michèle; Besson, Gisèle. *Richard Coeur de Lion: histoire et légende*. Paris; UGF; 1989. Pp 448.

6. Bursey, Peter. 'William of Falaise', *Notes & Queries for Somerset & Dorset* 33 (1992), 146–47.

7. Carpenter, Christine. *Locality and polity: a study of Warwickshire landed society, 1401–1499*. Cambridge; Cambridge UP; 1992. Pp xviii, 793.

8. Childs, Wendy R.; Taylor, John (ed.). *The Anonimalle chronicle, 1307 to 1334, from Brotherton Collection MS. 29*. Leeds; Yorkshire Arch. Soc. (Record ser. 147); 1991. Pp xii, 185.

9. Church, S.D. 'The knights of the household of King John: a question of numbers', Bc112, 151–65.

10. Crook, David. 'The last days of Eleanor of Castile: the death of a queen in Nottinghamshire, November 1290', *T. of the Thoroton Soc.* 94 (1991 for 1990), 17–28.

11. Crouch, David. *The image of aristocracy in Britain, 1000–1300*. London; Routledge; 1992. Pp xiv, 392.

12. Crouch, David. 'Defining an aristocracy, 1000–1300', *History Today* 42/9 (1992), 28–34.

13. Dalton, Paul. 'Aiming at the impossible: Ranulf II, earl of Chester, and Lincolnshire in the reign of King Stephen', Bc81, 109–34.

14. Dalton, Paul '*In neutro latere*: the armed neutrality of Ranulf II earl of Chester in King Stephen's reign', Bc27, 39–59.

15. Davies, C.S.L. 'Richard III, Henry VII and the island of Jersey', *The Ricardian* 9/119 (1992), 334–42.

16. Gaudefoy, Ghislain. 'D'où venaient ces chevaliers qui ont suivi Guillaume lors de la conquête d'Angleterre?', *Connaissance de Dieppe* 59 (1989), 14–19; 60 (1989), 4–9.

17. Gelling, Margaret. *The west midlands in the early middle ages*. London; Leicester UP; 1992. Pp x, 221.

18. Gênet, Jean-Philippe. 'L'Angleterre médiévale', *L'histoire médiévale en France: bilan et perspectives*, ed. Michel Balard (Paris; Seuil; 1991), 441–53.

19. Gillingham, John. 'Conquering kings: some twelfth-century reflections on Henry II and Richard I', Bc71, 163–78.

20. Goodchild, John. *Aspects of medieval Wakefield and its legacy*. Wakefield; Wakefield Historical Publications; 1991. Pp 135.

21. Goodman, Anthony E. *John of Gaunt: the exercise of princely power in fourteenth-century Europe*. London; Longman; 1992. Pp xvi, 421.

22. Goodman, Anthony E. 'Introduction' [war and border societies], Bc4, 1–29.

23. Green, Judith A. 'Earl Ranulf II and Lancashire', Bc81, 97–108.

24. Green, Judith A. 'Financing Stephen's war', Bc27, 91–114.

25. Hardwick, Kit. 'Edward IV—master of men, servant of women?', *Medieval History* 2/2 (1992), 92–103.

26. Hart, Cyril R. 'Hereward "the Wake" and his companions', Bc99, 625–48.

27. Holt, J.C. *Magna Carta*. Cambridge; Cambridge UP; 2nd edn. 1992. Pp xxii, 553.
28. Howell, Margaret. 'The children of King Henry III and Eleanor of Provence', Bc112, 57–72.
29. Jones, Michael K.; Underwood, Malcolm G. *The king's mother: Lady Margaret Beaufort, countess of Richmond and Derby*. Cambridge; Cambridge UP; 1992. Pp xv, 332.
30. Keats-Rohan, K.S.B. 'The Bretons and Normans of England, 1066–1154: the family, the fief and the feudal monarchy', *Nottingham Medieval Studies* 36 (1992), 42–78.
31. Lewis, C.P. 'The formation of the honor of Chester, 1066–1100', Bc81, 37–68.
32. Martindale, Jane. 'Eleanor of Aquitaine', Bc213, 17–50.
33. Morris, Christopher J. *Marriage and murder in eleventh-century Northumbria: a study of "De obsessione Dunelmi"*. York; the University (Borthwick paper 82); 1992 Pp iii, 31.
34. Neillands, Robin. *The Wars of the Roses*. London; Cassell; 1992. Pp 223.
35. Patterson, Robert B. 'The ducal and royal *acta* of Henry fitz Empress in Berkeley castle', *T. of the Bristol & Gloucestershire Arch. Soc.* 109 (1992 for 1991), 117–37.
36. Pollard, A.J. 'New monarchy renovated: England, 1461–1509', *Medieval History* 2/1 (1992), 78–82.
37. Potts, Cassandra. 'The early Norman charters: a new perspective on an old debate', Bc75, 25–40.
38. Prestwich, Michael. 'Gilbert de Middleton and the attack on the cardinals, 1317', Bc71, 179–94.
39. Richmond, Colin. 'A letter of 19 April 1483 from John Gigur to William Wainfleet', *Historical Research* 65 (1992), 112–16.
40. Thacker, Alan T. 'Introduction: the earls and their earldom', Bc81, 7–22.
41. van Houts, Elisabeth M.C. (ed.). *William of Jumièges: Gesta Normannorum ducum, vol. 1: introduction and books 1–4*. Oxford; Oxford UP (Oxford Medieval Texts); 1992. Pp cxxxiii, 156.
42. Ward, Jennifer C. *English noblewomen in the later middle ages*. London; Longman; 1992. Pp xii, 190.
43. Webster, Norman William. *Blanche of Lancaster*. Driffield; Halstead; 1990. Pp 96.
44. Wood, Charles T. 'The first two Queens Elizabeth, 1464–1502', Bc91, 121–31.

(b) *Politics*

1. Carpenter, D.A. 'English peasants in politics, 1258–1267', *Past & Present* 136 (1992), 3–42.
2. Carpenter, D.A. 'King Henry III's "statute" against aliens: July, 1263', *English Historical R.* 107 (1992), 925–44.
3. Horrox, Rosemary. 'Local and national politics in fifteenth-century England' [review article], *J. of Medieval History* 18 (1992), 391–403.

4. McHardy, Alison K. 'Richard II', *Medieval History* 2/1 (1992), 45–57.
5. Richmond, Colin. 'An English mafia?' [review article], *Nottingham Medieval Studies* 36 (1992), 235–43.
6. Richmond, Colin. 'Propaganda in the Wars of the Roses', *History Today* 42/7 (1992), 12–18.
7. Storey, R.L. 'Simon Islip, archbishop of Canterbury (1349–66): church, crown and parliament', *Ecclesia militans: Studien zur Konzilien- und Reformationsgeschichte*, ed. W. Brandmüller (Paderborn; Schöningh; 1988), vol. 1, 129–55.
8. Vigier, Anil de Silva. *This most highe prince ... John of Gaunt, 1340–1399*. Edinburgh; Pentland; 1992. Pp xxii, 376.
9. Vincent, Nicholas C. 'Jews, Poitevins and the bishop of Winchester, 1231–1234', Bc189, 119–32.
10. Vincent, Nicholas C. 'Simon de Montfort's first quarrel with King Henry III', Bc112, 167–77.

(c) *Constitution, Administration and Law*

1. Aberth, John. 'Crime and justice under Edward III: the case of Thomas de Lisle', *English Historical R*. 107 (1992), 283–301.
2. Attreed, Lorraine. 'Arbitration and the growth of urban liberties in late medieval England', *J. of British Studies* 31 (1992), 205–35.
3. Bailey, Keith. 'The hidation of Buckinghamshire', *Records of Buckinghamshire* 32 (1991 for 1990), 1–34.
4. Bateson, Mark; Denton, Jeffrey. 'Usury and comital disinheritance: the case of Ferrers versus Lancaster, St Paul's, London, 1301', *J. of Ecclesiastical History* 43 (1992), 60–96.
5. Beckwith, Sarah. 'Ritual, church and theatre: medieval dramas of the Sacramental Body', Bc94, 65–90.
6. Biancalana, J. 'Widows at common law: the development of common law dower', *Irish Jurist* 23 (1988), 255–329.
7. Burns, J.H. *Lordship, kingship and empire: the idea of monarchy, 1400–1525*. Oxford; Oxford UP (Carlyle lectures 1980); 1992. Pp xi, 178.
8. Crouch, David. 'The administration of the Norman earldom', Bc81, 69–95.
9. DeWindt, Anne Reiber. 'Local government in a small town: a medieval leet jury and its constituents', *Albion* 23 (1991), 627–54.
10. Donahue, Charles, Jnr. '"Clandestine" marriage in the later middle ages: a reply', *Law & History R*. 10 (1992), 315–22.
11. Hart, Cyril R. 'The carucation of Lindsey', Bc99, 337–85.
12. Hart, Cyril R. 'The carucation of Nottinghamshire', Bc99, 387–427.
13. Hart, Cyril R. 'The hide, the carucate and the ploughland', Bc99, 289–335.
14. Hart, Cyril R. 'The sokes of the Danelaw', Bc99, 231–79.
15. Hart, Cyril R. 'Wapentakes and hundreds', Bc99, 281–88.
16. Haskett, Timothy S. 'The presentation of cases in medieval Chancery bills', Bc195, 11–28.

17. Hudson, John G.H. 'Administration, family and perceptions of the past in late twelfth-century England: Richard FitzNigel and the *Dialogue of the Exchequer*', Bc45, 75–98.
18. Jewell, Helen M. 'The value of *Fleta* as evidence about parliament', *English Historical R.* 107 (1992), 90–94.
19. Kerr, Margaret H.; Forsyth, Richard K.; Plyley, Michael J. 'Cold water and hot iron: trial by ordeal in England', *J. of Interdisciplinary History* 22 (1992), 573–95.
20. King, Edmund. 'Dispute settlement in Anglo-Norman England', Bc27, 115–30.
21. Loengard, Janet Senderowitz. '"Legal history and the medieval English-woman" revisited: some new directions', Bc177, 210–36.
22. Loyn, Henry R. '*De iure domini regis*: a comment on royal authority in eleventh-century England', Bc75, 17–24.
23. Mason, Emma. '"The site of king-making and consecration": Westminster abbey and the crown in the eleventh and twelfth centuries', Bc12, 57–76.
24. Parsons, John Carmi. 'Ritual and symbol in the English medieval queenship to 1500', Bc91, 60–77.
25. Richardson, Malcolm. 'Early equity judges: keepers of the rolls of Chancery, 1415–1447', *American J. of Legal History* 36 (1992), 441–65.
26. Spinosa, Charles D. 'The legal reasoning behind the common, collusive recovery: Taltarum's case (1472)', *American J. of Legal History* 36 (1992), 70–102.
27. Swanson, Keith. ' "God woll have a stroke": judicial combat in the *Morte d'Arthur*', *B. of the John Rylands University Library of Manchester* 74/1 (1992), 155–73.
28. van Caenegem, Raoul C. (ed.). *English lawsuits from William I to Richard I: vol. 2, Henry II and Richard I.* London; Selden Soc. vol. 107; 1991.
29. Verduyn, Anthony. 'The revocation of urban peace commissions in 1381: the Lincoln petition', *Historical Research* 65 (1992), 108–11.
30. Vincent, Nicholas C. 'Hugh de Neville and his prisoners', *Archives* 20 (1992), 190–97.
31. Watt, J.A. 'The Jews, the law, and the Church: the concept of Jewish serfdom in thirteenth-century England', Bc12, 153–72.
32. Wilkinson, S. 'Derbyshire justices of the peace, 1388–1414', *Derbyshire Miscellany* 12/3 (1990), 70–82.
33. Wormald, Patrick. 'Domesday lawsuits: a provisional list and preliminary comment', Bc75, 61–102.

(d) *External Affairs*

1. Bolton, Brenda M. 'Philip Augustus and John: two sons in Innocent III's vineyard?', Bc12, 113–34.
2. Evans, Michael R. 'Brigandage and resistance in Lancastrian Normandy: a study of the remission evidence', *Reading Medieval Studies* 18 (1992), 103–34.

3. Gross, Anthony. 'Lancastrians abroad, 1461–1471', *History Today* 42/8 (1992), 31–37.

4. Harvey, John H. 'Political and cultural exchanges between England and the Iberian peninsula in the middle ages', *Literature, culture and society of the middle ages*, ed. Miguel Martinez Lopez (Barcelona; Promociones Publicaciones Universitarios; 1989), 2629–37.

5. Harvey, Margaret M. 'Eugenius IV, cardinal Kemp, and archbishop Chichele: a reconsideration of the role of Antonio Caffarelli', Bc12, 329–44.

6. Harvey, Margaret M. 'Martin V and the English, 1422–1431', Bc51, 59–86.

7. Jenks, Stuart. *England, die Hanse und Preussen: Handel und Diplomatie, 1377–1474* [England, the Hanse, and the Prussians: trade and diplomacy]. Cologne; Böhlau (Quellen und Darstellungen zur Hansischen Geschichte, neue Folge 38); 1992. 3 vols. Pp xxxii, 1265.

8. Jones, Michael. 'Guillaume, sire de Latimer, et la Bretagne: un nouveau témoignage (1365)'. *Charpiana. Mélanges offerts par ses amis à Jacques Charpy* (Bannalec; Fédération des sociétés savantes de Bretagne; 1991), 257–65.

9. Kicklighter, Joseph. 'Appeal, negotiation, and conflict: the evolution of the Anglo-French legal relationship before the Hundred Years War', *P. of the Annual Meeting of the Western Soc. for French History* 18 (1991), 45–59.

10. Lucíč, Josip. 'On the earliest contacts between Dubrovnik and England', *J. of Medieval History* 18 (1992), 373–89.

11. Quinn, David B. 'Columbus and the north: England, Iceland, and Ireland', *William & Mary Q.* 3rd ser. 49 (1992), 278–97.

12. Tanner, Heather J. 'The expansion of the power and influence of the counts of Boulogne under Eustace II', Bc27, 251–86.

13. Visser-Fuchs, Livia. 'Edward IV's "memoir on paper" to Charles, duke of Burgundy: the so-called "Short version of the Arrivall"', *Nottingham Medieval Studies* 36 (1992), 167–227.

(e) *Religion*

1. Abulafia, Anna Sapir. 'Theology and the commercial revolution: Guibert of Nogent, St Anselm and the Jews of northern France', Bc70, 23–40.

2. Aird, W.M. 'The making of a medieval miracle collection: the *Liber de translationibus et miraculis Sancti Cuthberti*', *Northern History* 28 (1992), 1–24.

3. Barrell, A.D.M. 'The effect of papal provisions on Yorkshire parishes, 1342–1370', *Northern History* 28 (1992), 92–109.

4. Barrow, Julia. 'How the twelfth-century monks of Worcester perceived their past', Bc45, 53–74.

5. Bates, David. *Bishop Remigius of Lincoln, 1067–1092*. Lincoln; Honywood; 1992. Pp 48.

6. Beckerman, John S. 'Procedural innovation and institutional change in medieval English manorial courts', *Law & History R.* 10 (1992), 197–252.

7. Beer, Frances F. *Women and mystical experience in the middle ages.* Woodbridge; Boydell; 1992. Pp vi, 186.

8. Birch, Debra J. 'Selling the saints: competition among pilgrimage centres in the twelfth century', *Medieval History* 2/2 (1992), 20–34.

9. Brink, Daphne H. *The parish of St Edward, king and martyr, Cambridge: survival of a late medieval appropriation.* Cambridge; Cambridge Antiquarian Soc.; 1992. Pp 96.

10. Brooke, Christopher N.L. 'Chaucer's parson and Edmund Gonville: contrasting roles of fourteenth century incumbents', Bc202, 1–19.

11. Brundage, James A. 'The bar of the Ely consistory court in the fourteenth century: advocates, proctors and others', *J. of Ecclesiastical History* 43 (1992), 541–60.

12. Burgess, Clive. 'Strategies for eternity: perpetual chantry foundation in late medieval Bristol', Bc51, 1–32.

13. Burgess, Clive. 'The benefactions of mortality: the lay response in the late medieval urban parish', Bc202, 65–86.

14. Cherry, John. 'The ring of bishop Grandisson', Bc104, 205–09.

15. Dalton, John P. *The archiepiscopal and deputed seals of York, 1114–1500.* York; Borthwick Inst. (Text and Calendar no. 17); 1992. Pp xviii, 81.

16. Denton, Jeffrey. 'From the foundation of Vale Royal abbey to the statute of Carlisle: Edward I and ecclesiastical patronage', Bc112, 123–37.

17. Dobson, Barrie. 'Citizens and chantries in late medieval York', Bc70, 311–32.

18. Dohar, William J. 'Medieval ordination lists: the origins of a record', *Archives* 20 (1992), 17–35.

19. Duffy, Eamon. *The stripping of the altars: traditional religion in England, c.1400-c.1580.* New Haven (Ct); Yale UP; 1992. Pp xii, 654.

20. Dunning, Robert W. 'The west-country Carthusians', Bc51, 33–42.

21. Evans, A.K.B. 'Cirencester abbey: the first hundred years', *T. of the Bristol & Gloucestershire Arch. Soc.* 109 (1992 for 1991), 99–116.

22. Flannery, S. *Los agustinos recoletos en Inglaterra.* Madrid; Editorial Augustinus; 1989. Pp 61.

23. Fletcher, Alan J. '"Magnus predicator et deuotus": a profile of the life, work, and influence of the fifteenth-century Oxford preacher, John Felton', *Mediaeval Studies* 53 (1991), 125–75.

24. Foster, Meryl R. 'Thomas of Westoe: a monastic book-buyer at Oxford about 1300', *Viator* 23 (1992), 189–99.

25. Frankis, John. 'St Zita, St Sythe and St Osyth', *Nottingham Medieval Studies* 36 (1992), 148–50.

26. Franklin, Michael. 'The cathedral as parish church: the case of southern England', Bc70, 173–98.

27. Gilchrist, Roberta. '"Blessed art thou among women": the archaeology of female piety', Bc92, 212–26.

28. Goering, Joseph; Taylor, Daniel S. 'The *summulae* of Bishops Walter de Cantilupe (1240) and Peter Quinel (1287)', *Speculum* 67 (1992), 576–94.

29. Greatrex, Joan. 'Episcopal relations with monastic chapters in England as

reflected in 14th-century visitation records', *Regula Benedicti Studia* 14–15 (1988 for 1985/6), 309–22.

30. Greatrex, Joan. 'Monastic charity for Jewish converts: the requisition of corrodies by Henry III', Bc189, 133–43.
31. Greene, J. Patrick. *Medieval monasteries.* Leicester; Leicester UP (Archaeology of Medieval Britain); 1992. Pp 288.
32. Hagerty, R.F. 'The puzzle of the priestly effigy in Ivinghoe church: early rectors of Ivinghoe and the true founder of Ravenstone priory', *Records of Buckinghamshire* 32 (1991 for 1990), 105–19.
33. Hansford-Miller, Frank. *A history and geography of English religion, vol. 6: monasteries and monastic orders, part 3: the minor monastic orders and the monasteries as a whole.* Canterbury; Abcado; 1992. Pp 100.
34. Haren, Michael J. 'Social ideas in the pastoral literature of fourteenth-century England', Bc51, 43–57.
35. Harper-Bill, Christopher. 'Who wanted the English Reformation?', *Medieval History* 2/1 (1992), 68–77.
36. Harvey, Margaret M. 'England, the council of Florence and the end of the council of Basle', *Christian unity: the council of Ferrara-Florence, 1438/1439–1989,* ed. G. Alberigo (Louvain; Louvain UP; 1991), 203–25.
37. Heslop, T.A. 'Twelfth-century forgeries as evidence for earlier seals: the case of St Dunstan', Bc3, 299–310.
38. Hicks, Michael. 'Four studies in conventional piety', *Southern History* 13 (1991), 1–21.
39. Hockey, S.F. (ed.). *The charters of Quarr abbey.* Newport; Isle of Wight County Record Office (Record ser. vol. 3); 1991. Pp xviii, 160.
40. Holdsworth, Christopher J. *The piper and the tune: medieval patrons and monks.* Reading; University of Reading (Stenton lecture, 1990); 1991. Pp 27.
41. Holdsworth, Christopher J. 'From 1050 to 1307' [church in Devon and Cornwall], Bc148, 23–51; 205–09.
42. Holdsworth, Christopher J. 'Royal Cistercians: Beaulieu, her daughters and Rewley', Bc112, 139–50.
43. Hughes, Jill. 'Walter Langton, bishop of Coventry and Lichfield 1296–1321: his family background', *Nottingham Medieval Studies* 35 (1991), 70–76.
44. Hughes, Jonathan. 'The administration of confession in the diocese of York in the fourteenth century', Bc202, 87–163.
45. Isserlin, Raphael. 'Aspects of English and continental Jewries', Bc206, 37–42.
46. Lepine, David N. 'The origins and careers of the canons of Exeter cathedral, 1300–1455', Bc51, 87–120.
47. Maccioni, P.A. 'The "Egmond"-account of the Thomas Becket story', *Heiligenlevens, annalen en kronieken,* ed. G.N.M. Vis *et al.* (Hilversum; Verloren; 1990), 93–114.
48. Marcombe, David. 'The preceptory of the knights of St Lazarus at Locke', *Derbyshire Arch. J.* 111 (1991), 51–63.
49. Martindale, Jane. 'Monasteries and castles: the priories of St-Florent de Saumur in England after 1066', Bc75, 135–56.

50. Martyn, Isolde. 'How posterity beheaded Morton: the case of the missing head', *The Ricardian* 9/118 (1992), 311–14.
51. McHardy, Alison K. 'Some patterns of ecclesiastical patronage in the later middle ages', Bc202, 20–37.
52. Milis, Ludo J.R. *Angelic monks and earthly men: monasticism and its meaning to medieval society.* Woodbridge; Boydell; 1992. Pp 192.
53. Morrissey, Thomas E. 'Surge, illuminare: a lost address by Richard Fleming at the council of Constance', *Annuarium Historiae Conciliorum* 22 (1990), 86–130.
54. Munster, M.B. zu. 'Das Priorät zur heiligen Mildred in Minster/England' [The priorate of St Mildred in Minster, England], *Studien und Mitteilungen zur Geschichte des Benediktiner-Ordens und seiner Zweige* 99 (1988), 268–73.
55. Orme, Nicholas I. 'Bishop Grandisson and popular religion', *Devonshire Association Report & T.* 124 (1992), 107–18.
56. Orme, Nicholas I. 'The later middle ages and the Reformation' [church in Devon and Cornwall], Bc148, 52–80; 209–12.
57. Pfaff, Richard W. 'Eadui Basan: scriptorum princeps?', Bc75, 267–83.
58. Pfaff, Richard W. 'Lanfranc's supposed purge of the Anglo-Saxon calendar', Bc71, 95–108.
59. Pfaff, Richard W. 'Prescription and reality in the rubrics of Sarum rite service books', Bc115, 197–205.
60. Pierrepont, J.R. 'Edward I versus the dean and chapter of Lichfield *re* the advowson of the church of Bakewell', *Bakewell & District Historical Soc. J.* 17 (1990), 23–29.
61. Richmond, Colin. 'The English gentry and religion, *c.*1500', Bc51, 121–50.
62. Robson, M. 'Notice about Bishop Moorman's index of Franciscans in England, 1124–1539', *Antonianum* 66 (1991), 420–35.
63. Rogers, Nicholas. 'The Waltham abbey relic-list', Bc75, 157–81.
64. Rollason, David. 'Symeon of Durham and the community of Durham in the eleventh century', Bc75, 183–98.
65. Rubin, Miri. 'Religious culture in town and country: reflections on a great divide', Bc70, 3–22.
66. Rubin, Miri. 'The eucharist and the construction of medieval identities', Bc94, 43–64.
67. Rubin, Miri. 'What did the eucharist mean to thirteenth-century villagers?', Bc112, 47–55.
68. Sayers, Jane E. 'William of Drogheda and the English canonists', *P. of the VIIth International Congress of Medieval Canon Law*, ed. P. Linehan (Vatican; Vatican Library (Monumenta iuris canonici ser. C vol. 8); 1988), 205–22.
69. Soden, Iain. 'The Carthusians of Coventry: exploding myths', Bc206, 77–82.
70. Stacey, Robert C. 'The conversion of the Jews to Christianity in thirteenth-century England', *Speculum* 67 (1992), 263–83.
71. Storey, R.L. 'Papal provisions to English monasteries', *Nottingham Medieval Studies* 35 (1991), 77–91.

72. Storey, R.L. (ed.). *The register of John Kirkby, bishop of Carlisle, 1332–1352, and the register of John Ross, bishop of Carlisle, 1325–1332*, vol. 1. Twickenham; Canterbury & York Soc. vol. 79; 1992. Pp xx, 204.

73. Summerson, Henry. 'The king's *clericulus*: the life and career of Silvester de Everdon, bishop of Carlisle, 1247–1254', *Northern History* 28 (1992), 70–91.

74. Swanson, Robert N. 'Medieval liturgy as theatre: the props', Bc194, 239–53.

75. Thompson, Benjamin. '*Habendum et tenendum*: lay and ecclesiastical attitudes to the property of the Church', Bc51, 197–238.

76. Townley, Simon. 'Unbeneficed clergy in the thirteenth century: two English dioceses', Bc202, 38–64.

77. Tsurushima, H. 'The fraternity of Rochester cathedral priory about 1100', Bc27, 313–37.

78. Turville-Petre, Thorlac. 'A Middle English Life of St Zita', *Nottingham Medieval Studies* 35 (1991), 102–05.

79. van Houts, Elisabeth M.C. 'Nuns and goldsmiths: the foundation and early benefactors of St Radegund's priory at Cambridge', Bc70, 59–79.

80. Ward, Benedicta. 'Two letters relating to relics of St Thomas of Canterbury', Bc115, 175–78.

81. Wilcox, Jonathan. 'The dissemination of Wulfstan's homilies: the Wulfstan tradition in eleventh-century vernacular preaching', Bc75, 199–217.

82. Winstead, Karen A. 'Piety, politics and social commitment in Capgrave's *Life of St Katherine*', *Medievalia et Humanistica* ns 17 (1991), 59–80.

83. Wogan-Browne, Jocelyn. 'Queens, virgins and mothers: hagiographic representations of the abbess and her powers in twelfth- and thirteenth-century Britain', Bc91, 14–35.

84. Wood, Jason. 'Furness abbey: an integrated and multi-disciplinary approach to the survey, recording, analysis and interpretation of a monastic building', Bc206, 163–70.

(f) *Economic Affairs*

1. Archer, Rowena E. '"How ladies ... who live on their manors ought to manage their households and estates": women as landholders and administrators in the later middle ages', Bc92, 149–81.

2. Atkin, Malcolm A. 'The medieval exploitation and division of Malham moor', *Nomina* 14 (1992 for 1990–1), 61–71.

3. Ayers, Brian S. 'From cloth to creel—riverside industries in Norwich', Bc41, 1–8

4. Bailey, Mark. 'Coastal fishing off south-east Suffolk in the century after the Black Death', *P. of the Suffolk Inst. of Archaeology & History* 37 (1990), 102–14.

5. Bailey, Mark; Allnutt, Richard (ed.). *The bailiffs' minute book of Dunwich, 1404–1430*. Woodbridge; Boydell; Suffolk Record Series, vol. 34; 1992. Pp 153.

6. Birrell, Jean. 'Deer and deer farming in medieval England', *Agricultural History R.* 40 (1992), 112–26.

7. Bolton, J.L. 'Inflation, economics and politics in thirteenth-century England', Bc112, 1–14.

8. Bonney, Margaret. 'The English medieval wool and cloth trade: new approaches for the local historian', *Local Historian* 22 (1992), 18–40.

9. Brand, Paul. 'The rise and fall of the hereditary steward in English ecclesiastical institutions, 1066–1300', Bc71, 145–62.

10. Brewster, T.C.M.; Hayfield, Colin. 'The medieval pottery industries at Staxton and Potter Brompton, East Yorkshre', *Yorkshire Arch. J.* 64 (1992), 49–82.

11. Bridbury, A.R. 'Introduction' [the English economy], Bc114, 1–42.

12. Bridbury, A.R. 'The Domesday valuation of manorial income', Bc114, 111–32.

13. Brown, Vivien (ed.). *Eye priory cartulary and charters*, part 1. Woodbridge; Boydell; 1992. Pp xvi, 240.

14. Campbell, Bruce M.S.; Galloway, James A.; Murphy, Margaret. 'Rural land-use in the metropolitan hinterland, 1270–1339: the evidence of *inquisitiones post mortem*', *Agricultural History R.* 40 (1992), 1–22.

15. Childs, Wendy R. 'Anglo-Portuguese trade in the fifteenth century', *T. of the Royal Historical Soc.* 6th ser. 2 (1992), 195–219.

16. Christensen, Arne-Emil. 'Medieval shipping in the North sea', Bc207, 87–92.

17. Cosgel, Metin M. 'Risk sharing in medieval agriculture', *J. of European Economic History* 21 (1992), 99–110.

18. Crawford, Anne (ed.). *Household books of John Howard, duke of Norfolk*. Stroud; Sutton; new edn. 1992. Pp 336.

19. Currie, Christopher K. 'Construction methods used in medieval dams and water control', Bc208, 85–90.

20. Currie, Christopher K. 'The role of freshwater fish in medieval society', Bc206, 89–94.

21. Daniels, R. 'Medieval Hartlepool: evidence of and from the waterfront', Bc41, 43–50.

22. Dyer, Alan. *Decline and growth in English towns, 1400–1600*. London; Macmillan; 1992. Pp 96.

23. Dyer, Christopher C. 'Small-town conflict in the later middle ages: events at Shipston-on-Stour', *Urban History* 19 (1992), 183–210.

24. Eaglen, R.J. 'The evolution of coinage in thirteenth-century England', Bc112, 15–24.

25. Egan, Geoffrey. 'Industry and economics on the medieval and later London waterfront', Bc41, 9–18.

26. Egan, Geoffrey; Blades, Nigel; Brown, Dana Goodburn. 'Evidence for the mass production of copper alloy dress accessories and other items in the late medieval city of London', Bc208, 111–16.

27. Farmer, David L. 'Millstones for medieval manors', *Agricultural History R.* 40 (1992), 97–111.

28. Fenoaltea, Stefano. 'Transaction costs, Whig history and the common fields', Bc142, 107–69.

29. Goldberg, P.J.P. *Women, work and life cycle in a medieval economy: women in York and Yorkshire: c.1300–1520*. Oxford; Oxford UP; 1992. Pp 420.

30. Goldberg, P.J.P. '"For better, for worse": marriage and economic opportunity for women in town and country', Bc92, 108–25.

31. Goodburn, Damian M. 'An archaeology of medieval boat-building practice', Bc207, 131–38.

32. Goodburn, Damian M. 'New light on early ship- and boat-building in the London area', Bc41, 105–15.

33. Graham, Helena. '"A woman's work ...": labour and gender in the late medieval countryside', Bc92, 126–48.

34. Hilton, Rodney H. 'Why was there so little champart rent in medieval England?', *J. of Peasant Studies* 17 (1990), 510–19.

35. Hodges, Richard M.; Wildgoose, Martin. 'Roystone grange: excavations of the Cistercian grange, 1980–1987', *Derbyshire Arch. J.* 111 (1991), 46–50.

36. Holt, Richard. 'Mills in medieval England', Bc208, 15–20.

37. Horsey, I.P. 'Poole: the medieval waterfront and its usage', Bc41, 51–54.

38. Jenkins, H.J.K. 'Medieval barge traffic and the building of Peterborough cathedral', *Northamptonshire Past & Present* 8/4 (1992–93), 255–61.

39. Jennings, Sarah. 'A ceramic fish trap', Bc191, 66–69.

40. Jones, R.H. 'Industry and environment in medieval Bristol', Bc41, 19–26.

41. Masschaele, James. 'Market rights in thirteenth-century England', *English Historical R.* 107 (1992), 78–89.

42. Mayhew, Nicholas J. 'From regional to central minting, 1158–1464', Bc226, 83–178.

43. Mayhew, Nicholas J. 'The mathematics of minting: the assaying of silver', Bc208, 103–10.

44. McDonnell, John. 'Pressures on Yorkshire woodland in the later middle ages', *Northern History* 28 (1992), 110–25.

45. Newman, Caron. 'Small town trade: evidence from Romsey, Hampshire', Bc226, 99–104.

46. O'Brien, C. 'Newcastle-upon-Tyne and its North sea trade', Bc41, 36–42.

47. Oggins, Virginia Darrow; Oggins, Robin S. 'Hawkers and falconers along the Ouse: a geographic principle of location in some serjeanty and related holdings', *Cambridge Antiquarian Soc. P.* 80 (1992 for 1991), 7–20.

48. Penn, Simon A.C. 'A hidden workforce: building workers in fourteenth-century Bristol', *T. of the Bristol & Gloucestershire Arch. Soc.* 109 (1992 for 1991), 171–78.

49. Piper, Alan J. 'Evidence of accounting and local estate services at Durham, *c.*1240', *Archives* 20 (1992), 36–39.

50. Postles, David A. 'Brewing and the peasant economy: some manors in late medieval Devon', *Rural History* 3 (1992), 133–44.

51. Postles, David A. 'Gifts in frankalmoign, warranty of land, and feudal society', *Cambridge Law J.* 50 (1991), 330–46.

52. Postles, David A. 'Tenure in frankalmoign and knight service in twelfth-century England: interpretation of the charters', *J. of the Soc. of Archivists* 13 (1992), 18–28.

53. Power, John P.; Campbell, Bruce M.S. 'Cluster analysis and the classification of medieval demesne-farming systems', *T. of the Institute of British Geographers* ns. 17 (1992), 227–45.
54. Roffe, David. 'The *Descriptio terrarum* of Peterborough abbey', *Historical Research* 65 (1992), 1–16.
55. Russett, V.E.J. 'Hythes and bows: aspects of river transport in Somerset', Bc41, 60–66.
56. Saaler, Mary. 'The manor of Tillingdown: the changing economy of the demesne, 1325–1371', *Surrey Arch. Collections* 81 (1991–2), 19–40.
57. Sakata, Toshio. *English medieval towns* [in Japanese, with appendices etc in English]. Tokyo; Yuhikatu; 1991. Pp xi, 300.
58. Salisbury, C.R. 'Primitive British fishweirs', Bc41, 76–87.
59. Smart, Veronica. 'The influence of the Norman conquest on the moneyers' names of York', *Festskrift till Lars O. Lagerkvist* (Stockholm; Svenska Numismatiska Foreningen (Numismatiska Meddelanden 3); 1989), 387–91.
60. Stahl, Alan M. 'Coinage in the name of medieval women', Bc177, 321–41.
61. Steane, J.M.; Foreman, M. 'The archaeology of medieval fishing tackle', Bc41, 88–101.
62. Swanson, Robert N. 'Standards of livings: parochial revenues in pre-Reformation England', Bc51, 151–96.
63. Thornton, Christopher. 'Efficiency in medieval livestock farming: the fertility and mortality of herds and flocks at Rimpton, Somerset, 1208–1349', Bc112, 25–46.
64. Turner, R.C. 'Shop, stall and undercroft in "the rows" of Chester', Bc214, 161–66.
65. Witney, K.P. 'Kentish land measurements of the thirteenth century', *Archaeologia Cantiana* 109 (1992 for 1991), 29–39.
66. Woolgar, C.M. *Household accounts from medieval England, part 1: introduction, glossary, diet accounts.* Oxford; Oxford UP for the British Academy (Records of Social and Economic History, ns 17); 1992. Pp xvi, 430.

(g) Social Structure and Population

1. Acheson, Eric. *A gentry community: Leicestershire in the fifteenth century, c.1442–1485.* Cambridge; Cambridge UP; 1992. Pp 290.
2. Alexander, Philip S. 'Madame Eglentyne, Geoffrey Chaucer and the problem of medieval anti-semitism', *B. of the John Rylands University Library of Manchester* 74/1 (1992), 109–20.
3. Bennett, Judith M. 'Medieval women, modern women: across the great divide', Bc94, 147–76.
4. Biller, P.P.A. 'Marriage patterns and women's lives: a sketch of a pastoral geography', Bc92, 60–107.
5. Colbourne, Amanda. 'Sir William Martyn: the true history of the builder of Athelhampton hall', *Dorset Natural History & Arch. Soc. P.* 113 (1992 for 1991), 192–3.

6. Coulton, Barbara. 'The wives of Sir William Stanley', *The Ricardian* 9/118 (1992), 315–18.
7. Cullum, P.H. *Cremetts and corrodies: care of the poor and sick at St Leonard's Hospital, York, in the middle ages.* York; the University (Borthwick paper 79); 1991. Pp 35.
8. Cullum, P.H. '"And hir name was Charitie": charitable giving by and for women in late medieval Yorkshire', Bc92, 182–211.
9. Dobson, Barrie. 'The role of Jewish women in medieval England', Bc189, 145–68.
10. Dyer, Christopher C. 'Current studies of medieval rural settlements in England', Bc210, 227–40.
11. Foréville, Raymonde. 'Du Domesday Book à la grande charte: guildes, franchises et chartes urbaines', *Les origines des libertés urbaines: actes du XVIè Congrès des Historiens Médiévistes de l'Enseignement Supérieur* (Mont-Saint-Aignan; University of Rouen; 1990), 163–74.
12. Getz, Faye Marie. 'Black Death and the silver lining: meaning, continuity and revolutionary change in histories of medieval plague', *J. of Historical Biology* 24 (1991), 265–90.
13. Gidman, Jean M. 'The wives and children of Sir William Stanley of Holt', *The Ricardian* 9/116 (1992), 206–10.
14. Gilchrist, Roberta. 'Christian bodies and souls: the archaeology of life and death in later medieval hospitals', Bc216, 101–18.
15. Goldberg, P.J.P. 'Marriage, migration, and servanthood: the York cause paper evidence', Bc92, 1–15.
16. Gordon, Eleanora C. 'Accidents among medieval children as seen from the miracles of six English saints and martyrs', *Medical History* 35 (1991), 145–63.
17. Grauer, Anne. 'Patterns of life and death: the palaeodemography of medieval York', Bc211, 67–80.
18. Hanawalt, Barbara A.; McRee, Ben R. 'The guilds of *homo prudens* in late medieval England', *Continuity & Change* 7 (1992), 163–79.
19. Hare, J.N. 'The lords and their tenants: conflict and stability in fifteenth-century Wiltshire', Bc116, 16–34.
20. Herlihy, David. *Opera muliebria: women and work in medieval Europe.* Maidenhead; McGraw-Hill; 1989. Pp 300.
21. Hilton, Rodney H. *English and French towns in feudal society.* Cambridge; Cambridge UP; 1992. Pp 200.
22. Hudson, Hazel; Neale, Frances. 'A busy day at Wedmore church, 1350', *Notes & Queries for Somerset & Dorset* 33 (1992), 171–73.
23. Hughes, Jonathan. 'Stephen Scrope and the circle of Sir John Fastolf: moral and intellectual outlooks', Bc98, 109–46.
24. Jones, E.D. 'The medieval leyrwite: a historical note on female fornication', *English Historical R.* 107 (1992), 945–53.
25. Jones, E.D. 'Villein mobility in the later middle ages: the case of Spalding priory', *Nottingham Medieval Studies* 36 (1992), 151–66.
26. Lepine, David N. 'The Courtenays and Exeter cathedral in the later middle ages', *Devonshire Association Report & T.* 124 (1992), 41–58.

27. Lewis, Carenza. 'Medieval rural settlement in south Wiltshire', Bc210, 181–86.
28. Maddern, Philippa C. *Violence and social order: East Anglia, 1422–1442*. Oxford; Oxford UP; 1992. Pp ix, 270.
29. Mate, Mavis. 'The economic and social roots of medieval popular rebellion: Sussex in 1450–1451', *Economic History R.* 2nd ser. 45 (1992), 661–76.
30. McHardy, Alison K. (ed.). *Clerical poll-taxes of the diocese of Lincoln, 1377–1381*. Lincoln; Lincoln Record Soc. vol. 81; 1992. Pp xliv, 252.
31. McRee, Ben R. 'Religious gilds and civic order: the case of Norwich in the late middle ages', *Speculum* 67 (1992), 69–97.
32. Moore, John S. 'The Anglo-Norman family: size and structure', Bc27, 153–96.
33. Moreton, C.E. *The Townshends and their world: gentry, law and the land in Norfolk, c.1450–1551*. Oxford; Oxford UP; 1992. Pp xiv, 279.
34. Olson, Sherri. 'Family linkages and the structure of the local elite in the medieval and early modern village', *Medieval Prosopography* 13 (1992), 53–82.
35. Owen, A.E.B. 'Castle Carlton: the origins of a medieval "new town"', *Lincolnshire History & Archaeology* 27 (1992), 17–22.
36. Parfitt, Keith. 'St John's Hospital reredorter, Canterbury', *Archaeologia Cantiana* 109 (1992 for 1991), 298–308.
37. Payling, S.J. 'Social mobility, demographic change, and landed society in late medieval England', *Economic History R.* 2nd ser. 45 (1992), 51–73.
38. Postles, David A. 'Demographic change in Kibworth Harcourt, Leicestershire, in the later middle ages', *Local Population Studies* 48 (1992), 41–48.
39. Postles, David A. 'The baptismal name in thirteenth-century England: processes and patterns', *Medieval Prosopography* 13 (1992), 1–52.
40. Postles, David A. 'The pattern of rural migration in a midlands county: Leicestershire, c.1270–1350', *Continuity & Change* 7 (1992), 139–61.
41. Prescott, Elizabeth. *The English medieval hospital, 1050–1640*. London; Seaby; 1992. Pp 288.
42. Reeves, A. Compton. 'Histories of English families published in the 1980s', *Medieval Prosopography* 13/2 (1992), 83–120.
43. Richmond, Colin. 'What a difference a manuscript makes: John Wyndham of Felbrigg, Norfolk (d. 1475)', Bc58, 129–41.
44. Ridgeway, Huw. 'William de Valence and his *familiares*, 1247–1272', *Historical Research* 65 (1992), 239–57.
45. Rösener, Werner. *Peasants in the middle ages*. Oxford; Blackwell; 1992. Pp 333.
46. Rosenthal, Joel T. 'Introduction' [medieval women], Bc177, vii-xvii.
47. Schofield, John. 'London im frühen Mittelalter', Bc126, 80–84.
48. Smith, Richard M. 'Geographical diversity in the resort to marriage in late medieval Europe: work, reputation, and unmarried females in the household formation systems of northern and southern Europe', Bc92, 16–59.
49. Stafford, Pauline. 'Women in Domesday', *Reading Medieval Studies* 15 (1989), 75–94.

50. Tuck, Anthony. 'The Percies and the community of Northumberland in the later fourteenth century', Bc4, 178–95.
51. Watts, D.G. 'Popular disorder in southern England, 1250–1450', Bc116, 1–15.

(h) *Naval and Military*

1. Ailes, Adrian. 'The knight, heraldry and armour: the role of recognition and the origins of heraldry', Bc98, 1–21.
2. Arthurson, Ian. 'Espionage and intelligence from the Wars of the Roses to the Reformation', *Nottingham Medieval Studies* 35 (1991), 134–54.
3. Ayton, Andrew. 'Military service and the development of the Robin Hood legend in the fourteenth century', *Nottingham Medieval Studies* 36 (1992), 126–47.
4. Ayton, Andrew. 'War and the English gentry under Edward III', *History Today* 42/3 (1992), 34–40.
5. Bishop, M.C. 'The White Wall, Berwick-upon-Tweed', *Archaeologia Aeliana* 5th ser. 20 (1992), 117–19.
6. Blair, Claude. 'Crediton: the story of two helmets', Bc5, 152–77.
7. Bliese, John R.E. 'The courage of the Normans: a comparative study of battle rhetoric', *Nottingham Medieval Studies* 35 (1991), 1–26.
8. Bradbury, J. *The medieval siege*. Woodbridge; Boydell; 1992. Pp xvi, 362.
9. Curry, Anne. 'The nationality of men-at-arms serving in English armies in Normandy and the *pays de conquête*, 1415–1450: a preliminary survey', *Reading Medieval Studies* 18 (1992), 135–63.
10. Dobson, Barrie. 'The church of Durham and the Scottish borders, 1378–1388', Bc4, 124–54.
11. Dockray, Keith. 'The battle of Wakefield and the Wars of the Roses', *The Ricardian* 9/117 (1992), 238–56.
12. Grant, Alexander. 'The Otterburn war from the Scottish point of view', Bc4, 30–64.
13. Knowles, Richard. 'The battle of Wakefield: the topography', *The Ricardian* 9/117 (1992), 257–65.
14. Lloyd, Simon D. 'William Longespee II: the making of an English crusading hero' [part 1], *Nottingham Medieval Studies* 35 (1991), 41–69.
15. Lloyd, Simon D.; Hunt, Tony. 'William Longespee II: the making of an English crusading hero, part II', *Nottingham Medieval Studies* 36 (1992), 79–125.
16. McNamee, Colm. 'Buying off Robert Bruce: an account of monies paid to the Scots by Cumberland communities in 1313–1314', *T. of the Cumberland & Westmorland Antiquarian & Arch. Soc.* 92 (1992), 77–89.
17. Neville, C.J. 'A plea roll of Edward I's army in Scotland, 1296', Bc83, 7–133.
18. Oakeshott, Ewart. *Records of the medieval sword*. Woodbridge; Boydell; 1991. Pp xii, 276.

19. Plaisse, André. *La delivrance de Cherbourg et du clos du Cotentin à la fin de la Guerre de Cent Ans.* Cherbourg; Presse de la Manche; 1989. Pp 200.

20. Prestwich, J.O. 'Richard Coeur de Lion: *rex bellicosus*', Bc213, 1–16.

21. Prestwich, Michael. 'Why did Englishmen fight in the Hundred Years War?', *Medieval History* 2/1 (1992), 58–65.

22. Ralegh Radford, C.A. 'Appendix: comments on the helmet in Crediton church', Bc5, 178–83.

23. Ryder, Peter F. 'The Cow Port at Berwick-upon-Tweed', *Archaeologia Aeliana* 5th ser. 20 (1992), 99–116.

24. Strickland, Matthew. 'Arms and the men: war, loyalty and lordship in Jordan Fantosme's chronicle', Bc98, 187–220.

25. Strickland, Matthew. 'Slaughter, slavery or ransom: the impact of the Conquest on conduct in warfare', Bc75, 41–59.

26. Summerson, Henry. 'Responses to war: Carlisle and the west march in the later fourteenth century', Bc4, 155–77.

27. Tyson, Colin. 'The battle of Otterburn: when and where was it fought?', Bc4, 65–93.

28. Wright, Nicholas. 'Ransoms of non-combatants during the Hundred Years War', *J. of Medieval History* 17 (1991), 323–32.

(i) *Intellectual and Cultural*

1. Anderson, T. 'An example of meningiomatous hyperostosis from medieval Rochester', *Medical History* 36 (1992), 207–13.

2. Beadle, Richard. 'Prolegomena to a literary geography of later medieval Norfolk', Bc58, 89–108.

3. Bell, David N. 'The English Cistercians and the practice of medicine', *Cîteaux* 40 (1989), 139–74.

4. Benskin, Michael. 'The "fit"-technique explained', Bc58, 9–26.

5. Black, Maggie. *The medieval cookbook.* London; British Museum; 1992. Pp 143.

6. Blake, Norman F. 'Introduction' [language], Bc227, 1–22.

7. Blake, Norman F. 'The literary language', Bc227, 500–41.

8. Blamires, David. 'Folktales and fairytales in the middle ages', *B. of the John Rylands University Library of Manchester* 74/1 (1992), 97–107.

9. Boffey, Julia; Meale, Carol M. 'Selecting the text: Rawlinson C. 86 and some other books for London readers', Bc58, 143–69.

10. Bowers, Roger. 'Fixed points in the chronology of English fourteenth-century polyphony', *Music and Letters* 71 (1990), 113–35.

11. Brett, Martin. 'The annals of Bermondsey, Southwark and Merton', Bc70, 279–310.

12. Brett, Martin. 'The *Collectio Lanfranci* and its competitors', Bc115, 157–71.

13. Brewer, D. 'Feasts in England and English literature in the fourteenth century', *Feste und Feiern im Mittelalter* ed. Detle Altenburg, Jorg Jarnut and Hans H. Steinhoff Altenburg (Sigmaringen; J. Thorbecke; 1991), 13–26.

14. Brooke, Christopher N.L. 'Aspects of John of Salisbury's *Historia pontificalis*', Bc115, 185–95.
15. Brown, S.F. 'The eternity of the world discussion at early Oxford', *Mensch und Natur im Mittelalter* part 1, ed. Albert Zimmermann and Andreas Speer (Berlin; de Gruyter; 1991), 89–104.
16. Burnley, David. 'Lexis and semantics', Bc227, 409–99.
17. Carling, Alan. 'Marx, Cohen and Brenner: functional explanation versus rational choice in the transition from feudalism to capitalism', Bc9, 161–79.
18. Carter, John Marshall. *Medieval games: sports and recreations in feudal society*. London; Greenwood; 1992. Pp 299.
19. Clanchy, M.T. *From memory to written record: England, 1066–1307*. Oxford; Blackwell; 2nd edn. 1992. Pp 320.
20. Clark, Cecily. 'Domesday Book—a great red-herring: thoughts on some late eleventh-century orthographies', Bc75, 317–31
21. Clark, Cecily. 'Onomastics', Bc227, 542–606.
22. Cobban, Alan B. 'Pembroke college: its educational significance in late medieval Cambridge', *T. of the Cambridge Bibliographical Soc.* 10 (1991), 1–16.
23. Cobban, Alan B. 'Reflections on the role of medieval universities in contemporary society', Bc115, 227–41.
24. Copeland, Rita. 'Lydgate, Hawes and the science of rhetoric in the late middle ages', *Modern Language Q.* 53 (1992), 57–82.
25. Crick, Julia. 'Geoffrey of Monmouth: prophecy and history', *J. of Medieval History* 18 (1992), 357–71.
26. Crocker, Richard. 'Medieval chant', Bc78, 225–309.
27. Crocker, Richard. 'Polyphony in England in the thirteenth century', Bc78, 679–720.
28. de Medeiros, Marie-Thérèse. 'Voyage et lieux de mémoire: le retour de Froissart en Angleterre', *Le Moyen Age* 98 (1992), 419–28.
29. Edwards, John. 'The cult of "St" Thomas of Lancaster and its iconography', *Yorkshire Arch. J.* 64 (1992), 103–22.
30. Fischer, Olga. 'Syntax' [Middle English], Bc227, 207–408.
31. French, Katherine I. 'The legend of Lady Godiva and the image of the female body', *J. of Medieval History* 18 (1992), 3–19.
32. Friedman, John B. 'Cluster analysis and the manuscript chronology of William du Stiphel, a fourteenth-century scribe at Durham', *History & Computing* 4 (1992), 75–97.
33. Fuller, Sarah. 'Early polyphony', Bc78, 485–556.
34. Getz, Faye Marie. 'Introduction' [health and disease], Bc90, xiii-xx.
35. Getz, Faye Marie. 'Medical practitioners in medieval England', *Social History of Medicine* 3 (1990), 245–83.
36. Getz, Faye Marie. 'To prolong life and promote health: Baconian alchemy and pharmacy in the English learned tradition', Bc90, 141–51.
37. Gillingham, John. 'Some legends of Richard the Lionheart: their development and their influence', Bc213, 51–69.
38. Goering, Joseph; Mantello, F.A.C. 'Two *opuscula* of Robert Grosseteste: *De vniversi complecione* and *Exposicio canonis misse*', *Mediaeval Studies* 53 (1991), 89–123.

39. Harvey, Margaret M. 'The diffusion of the *Doctrinale* of Thomas Netter in the fifteenth and sixteenth centuries', Bc115, 281–94.
40. Harvey, P.D.A. 'Matthew Paris's maps of Britain', Bc112, 109–21.
41. Hook, David. 'The figure of Richard I in medieval Spanish literature', Bc213, 117–40.
42. Hudson, Anne. 'The king and erring clergy: a Wycliffite contribution', Bc12, 269–78.
43. Hudson, John G.H. 'Diplomatic and legal aspects of the charters', Bc81, 153–78.
44. Hunt, Tony. *Medieval surgery*. Woodbridge; Boydell; 1992. Pp 128.
45. Jones, Martin H. 'Richard the Lionheart in German literature of the middle ages', Bc213, 70–116.
46. Laborderie, Olivier de. 'Du souvenir à la réincarnation: l'image de Richard Coeur de Lion dans *La vie et mort du roi Jean* de William Shakespeare', Bc213, 141–65.
47. Laing, Margaret. 'Anchor texts and literary manuscripts in early middle English', Bc58, 27–52.
48. Lang, S.J. 'John Bradmore and his book *Philomena*', *Social History of Medicine* 5 (1992), 121–30.
49. Lass, Roger. 'Phonology and morphology' [Middle English], Bc227, 23–156.
50. Lefferts, Peter M. (ed.). *Robertus de Handlo, 'regule' ... and Johannes Hanboys, 'summa': Greek and Latin music theory*. Lincoln (Ne); Nebraska UP; 1991. Pp 400.
51. Lewis, Neil. 'The first recension of Robert Grosseteste's *De libero arbitrio*', *Mediaeval Studies* 53 (1991), 1–88.
52. Lodge, R.A. 'Language attitudes and linguistic norms in France and England in the thirteenth century', Bc112, 73–83.
53. Lord, Mary Louise. 'Virgil's *Eclogues*, Nicholas Trevet, and the harmony of the spheres', *Mediaeval Studies* 54 (1992), 186–273.
54. Lovatt, Roger. 'The library of John Blacman and contemporary Carthusian spirituality', *J. of Ecclesiastical History* 43 (1992), 195–230.
55. McGrade, Arthur Stephen. 'Somersaulting sovereignty: a note on reciprocal lordship and servitude in Wyclif', Bc12, 261–68.
56. McKinley, Richard. 'Medieval Latin translations of English personal bynames: their value for surname history', *Nomina* 14 (1992 for 1990–1), 1–6.
57. McLaren, Mary-Rose. 'The textual transmission of the London chronicles', Bc220, 38–72.
58. Means, Laurel. '"Ffor as moche as yche man may not have Œþe astrolabe": popular Middle English variations on the computus', *Speculum* 67 (1992), 595–623.
59. Mentgen, G. 'Richard of Devizes und die Juden: ein Beitrag zur Interpretation seiner *Gesta Richardi*' [Richard of Devizes and the Jews: a contribution to the interpretation of his 'Gesta Richardi'], *Kairos* 30/31 (1988/9), 95–104.
60. Meredith, Peter. 'Manuscript, scribe and performance: further looks at the N.town manuscript', Bc58, 109–28.

61. Middleton-Stewart, J. 'Singing for souls in Suffolk, 1300–1548', *Suffolk R.* 16 (1991), 1–19.
62. Milroy, James. 'Middle English dialectology', Bc227, 156–206.
63. Nederman, Cary J. 'Kings, peers and parliament: virtue and co-rulership in Walter Burley's *Commentarius in VIII libros politicorum Aristotelis*', *Albion* 24 (1992), 391–407.
64. Nicholson, Helen. 'Steamy Syrian scandals: Matthew Paris on the Templars and Hospitallers', *Medieval History* 2/2 (1992), 68–85.
65. Noble, Peter. 'Perversion of an ideal', Bc98, 177–86.
66. Noone, Timothy B. 'Evidence for the use of Adam of Buckfield's writings at Paris: a note on New Haven, Yale University, Historical-Medical Library 12', *Mediaeval Studies* 54 (1992), 308–16.
67. Patterson, Lee. 'Court politics and the invention of literature: the case of Sir John Clanvowe', Bc94, 7–42.
68. Patterson, Robert B. 'The author of the "Margam annals": early thirteenth-century Margam abbey's compleat scribe', Bc27, 197–210.
69. Pearsall, Derek. *The life of Geoffrey Chaucer: a critical biography.* Oxford; Blackwell; 1992. Pp xii, 365.
70. Pearsall, Derek. 'Lydgate as innovator', *Modern Language Q.* 53 (1992), 5–22.
71. Rankin, Susan. 'Liturgical drama', Bc78, 310–56.
72. Reed, James. 'The ballad and the source: some literary reflections on the *Battle of Otterburn*', Bc4, 94–123.
73. Richards, M.P. 'Texts and their traditions in the medieval library of Rochester cathedral priory', *T. of the American Philosophical Soc.* 78/3 (1988), i-xii, 1–120.
74. Rothwell, W. 'Chaucer and Stratford atte Bowe', *B. of the John Rylands University Library of Manchester* 74/1 (1992), 3–28.
75. Rubin, Miri. 'Imagining medieval hospitals: considerations on the cultural meaning of institutional change', Bc26, 14–25.
76. Samuels, Michael L. 'Scribes and manuscript traditions', Bc58, 1–7.
77. Saul, Nigel. 'Culture and society in the age of chivalry', Bc46, 8–23.
78. Saul, Nigel. 'The pre-history of an Oxford college; Hart hall and its neighbours in the middle ages', *Oxoniensia* 54 (1989), 327–44.
79. Scase, Wendy. 'Reginald Pecock, John Carpenter and John Colop's "common-profit" books: aspects of book ownership and circulation in fifteenth-century London', *Medium Aevum* 61 (1992), 261–74.
80. Sekules, Veronica. 'Women's piety and patronage', Bc46, 120–31.
81. Shirland, Ann. 'Paget's disease (osteitis deformans): a classic case?', *International J. of Osteoarchaeology* 1 (1991), 173–77.
82. Shirland, Ann. 'Pre-Columbian trepanematosis in medieval Britain', *International J. of Osteoarchaeology* 1 (1991), 39–47.
83. Short, Ian. 'Patrons and polyglots: French literature in twelfth-century England', Bc27, 229–49.
84. Smith, Jeremy J. 'Tradition and innovation in south-west midland middle English', Bc58, 53–65.
85. Smith, Lesley. 'Lending books: the growth of a medieval question from Langton to Bonaventure', Bc115, 265–79.

86. Stevens, John. 'Medieval song', Bc78, 357–444.

87. Stevens, John. '*Samson dux fortissime*: an international Latin song', *Plainsong & Medieval Music* 1 (1992), 1–40.

88. Sutton, Anne F. 'Merchants, music and social harmony: the London Puy and its French and London contexts, circa 1300', *London J.* 17 (1992), 1–17.

89. Sutton, Anne F.; Visser-Fuchs, Livia. 'Richard III's books: ancestry and "true nobility"', *The Ricardian* 9/119 (1992), 343–58.

90. Sutton, Anne F.; Visser-Fuchs, Livia. 'Richard III's books: mistaken attributions', *The Ricardian* 9/118 (1992), 303–10.

91. Sutton, Anne F.; Visser-Fuchs, Livia. 'Richard III's books, XIII: chivalric ideals and reality', *The Ricardian* 9/116 (1992), 190–205.

92. Sutton, Anne F.; Visser-Fuchs, Livia. '"Richard liveth yet": an old myth', *The Ricardian* 9/117 (1992), 266–69.

93. Todisci, O. 'Lo spirito critico da G. Duns Scoto a G. d'Occam: dalla ragione nella fede alla fede senza ragione' [The critical spirit from Duns Scotus to Occam: from reason in faith to faith without reason], *Miscellanea Franciscana* 89 (1989), 39–108.

94. Virgoe, Roger. 'Hugh atte Fenne and books at Cambridge', *T. of the Cambridge Bibliographical Soc.* 10 (1991), 92–98.

95. Voigts, Linda E.; Hudson, Robert P. '"A drynke Œþat men callen dwale to make a man to slepe whyle men kerven hem": a surgical anesthetic from late medieval England', Bc90, 34–56.

96. Waldron, Ronald. 'Dialect aspects of manuscripts of Trevisa's translation of the *Polychronicon*', Bc58, 67–87.

97. Wathey, Andrew. 'The marriage of Edward III and the transmission of French motets to England', *J. of the American Musicological Soc.* 45 (1992), 1–29.

98. Wathey, Andrew. 'The peace of 1360–1369 and Anglo-French musical relations', *Early Music History* 9 (1990), 129–74.

99. Webber, M.T.J. 'The scribes and handwriting of the original charters', Bc81, 137–51.

100. Webber, Teresa. *Scribes and scholars at Salisbury cathedral, c.1075-c.1125*. Oxford; Oxford UP; 1992. Pp xii, 220.

101. Webber, Teresa. 'Building a library in early Norman England: the manuscripts of the first canons of Salisbury cathedral', *The Hatcher R.* 4/33 (1992), 4–9.

102. Wood, Diana. 'Introduction: "straight answers to the problem of sovereignty?"', Bc12, 1–5.

103. Zier, Mark. 'The healing power of the Hebrew tongue: an example from late thirteenth-century England', Bc90, 103–18.

(j) *Visual Arts*

1. Allan, John. 'A note on the building stones of the cathedral', Bc104, 10–18.

2. Allan, John; Blaylock, Stuart. 'The structural history of the west front' [of Exeter cathedral], Bc104, 94–115.

3. Binski, Paul. 'The murals in the nave of St Albans abbey', Bc70, 249–78.
4. Bizzarro, Tina Waldeier. *Romanesque architectural criticism: a prehistory.* Cambridge; Cambridge UP; 1992. Pp 253.
5. Brosnahan, Leger. 'The pendant in the Chaucer portraits', *The Chaucer R.* 26 (1992), 424–431.
6. Brownsword, R.; Homer, R.F. 'The Norwich castle flagon', *J. of the Pewter Soc.* 8/2 (1992), 62–65.
7. Cahn, Walter. 'Romanesque sculpture and the spectator', Bc105, 45–60; 194–96.
8. Cherry, Bridget. 'Flying angels and bishops' tombs, a fifteenth-century conundrum', Bc104, 199–204.
9. Cherry, John. 'The use of seals as decoration on medieval pottery and bronze vessels', Bc191, 59–65.
10. Cocke, Thomas. 'Lincoln cathedral—the west front and the Romanesque reliefs: post-medieval perceptions', Bc105, 163–76; 208.
11. de Hamel, Christopher. 'A contemporary miniature of Thomas Becket', Bc115, 179–84.
12. Draper, Peter. 'The architectural setting of Gothic art', Bc46, 60–75.
13. Egan, Geoffrey. 'Marks on butterpots', Bc191, 97–100.
14. Egan, Geoffrey; Pritchard, Frances. *Dress accessories: medieval finds from excavations in London, c.1150-c.1450.* London; HMSO; 1991. Pp xi, 410.
15. Ford, Judy Ann. 'Art and identity in the parish communities of late medieval Kent', Bc194, 225–37.
16. French, T.W. 'The dating of York minster choir', *Yorkshire Arch. J.* 64 (1992), 123–33.
17. Hayfield, Colin. 'Humberware: the development of a late medieval pottery tradition', Bc191, 38–44.
18. Henry, Avril. 'The iconography of the west front' [of Exeter cathedral], Bc104, 134–46.
19. Heslop, T.A. 'The seals of the twelfth-century earls of Chester', Bc81, 179–97.
20. Hicks, Carola. 'The borders of the Bayeux tapestry', Bc75, 251–65.
21. Hillewaert, Bieke. 'An English lady in Flanders: reflections on a head in Scarborough ware', Bc191, 76–82.
22. Hoey, Laurence. 'The design of Romanesque clerestories with wall passages in Normandy and England', *Gesta* (1989), 78–101.
23. Hulbert, Anna. 'An examination of the polychromy of Exeter cathedral roof-bosses, and its documentation', Bc104, 188–98.
24. James, Susan E. 'Parr memorials in Kendal parish church', *T. of the Cumberland & Westmorland Antiquarian & Arch. Soc.* 92 (1992), 99–103.
25. Jansen, Virginia. 'The design and building sequence of the eastern area of Exeter cathedral, c.1270–1310: a qualified study', Bc104, 35–56.
26. Jennings, Sarah. *Medieval pottery in the Yorkshire museum.* York; Yorkshire Museum; 1992. Pp 56.
27. Jones, Malcolm; Tracy, Charles. 'A medieval choirstall desk-end at Haddon Hall: the fox-bishop and the geese-hangmen', *J. of the British Arch. Association* 144 (1991), 107–15.

28. Kahn, Deborah. 'Anglo-Saxon and early Romanesque frieze sculpture in England', Bc105, 61–74; 196–97.
29. Kahn, Deborah. 'Editor's introduction' [Lincoln frieze], Bc105, 9–13.
30. Kalinowski, Lech. 'The "frieze" at Malmesbury', Bc105, 85–96; 198–201.
31. Larson, John. 'An outline proposal for the conservation of the Romanesque frieze at Lincoln cathedral', Bc105, 183–92; 209.
32. Lasko, Peter. 'The principles of restoration' [sculpture], Bc105, 143–62; 207.
33. Lüdtke, Hartwig. 'English pottery in Hedeby, Schleswig and Hollingstedt', Bc191, 70–75.
34. Macpherson-Grant, Nigel. '"... many reduced wares have not been recognised"', Bc191, 83–96.
35. Martindale, Andrew. 'Patrons and minders: the intrusion of the secular into sacred spaces in the late middle ages', Studies in Church History 28 (1992), 143–78.
36. McAleer, J. Philip. 'Le problème du transept occidental en Grande-Bretagne', Cahiers de Civilisation Médiévale 34 (1991), 349–56.
37. McAleer, J. Philip. 'The problem(s) of the St Edmund's chapel, Exeter cathedral', Bc104, 147–61.
38. McCarthy, M.R.; Brooks, C.M. 'The establishment of a medieval pottery sequence in Cumbria, England', Bc191, 21–37.
39. McNulty, J. Bard. The narrative art of the Bayeux tapestry master. New York; AMS; 1989. Pp 151.
40. Moore, B. 'Wild men in the misericords of St Mary's church, Beverley', Yorkshire Arch. J. 64 (1992), 135–43.
41. Morris, Richard K. 'Thomas of Witney at Exeter, Winchester and Wells', Bc104, 57–84.
42. Morrison, Kathryn; Baxter, Ronald. 'Fragments of 12th-century sculpture in Bosham church', Sussex Arch. Collections 129 (1991), 33–38.
43. Murray, Hugh. 'The Scrope tapestries', Yorkshire Arch. J. 64 (1992), 145–56.
44. Murray, Mary Charles. 'The Christian zodiac on a font at Hook Norton: theology, church and art', Bc194, 87–97.
45. Musset, Lucien. La tapisserie de Bayeux: oeuvre d'art et document historique. St Léger-Vauban; Zodiaque; 1989. Pp 309.
46. Nenk, Beverley. 'A medieval sgraffito-decorated jug from Mill Green', Essex Archaeology & History 3rd ser. 23 (1992), 51–56.
47. Norris, Malcolm. 'Later medieval monumental brasses: an urban funerary industry and its representation of death', Bc216, 184–209; 248–51.
48. Orme, Nicholas I. 'The charnel chapel of Exeter cathedral', Bc104, 162–71.
49. Parsons, David. 'The cruciform arrangement of urban churches: some possible English examples', Bc206, 117–22.
50. Price, Clifford. 'The conservation of architectural sculpture', Bc105, 177–82; 209.

51. Ramsay, Nigel. 'Artists and craftsmen', Bc46, 48–59.
52. Ray, Anthony. 'Spanish lustreware imported into England—three new 14th-century examples', Bc191, 198–201.
53. Russell, Georgina. 'Some aspects of the Decorated tracery of Exeter cathedral', Bc104, 85–93.
54. Ryder, Peter F. *Medieval cross slab grave covers in west Yorkshire.* Wakefield; West Yorkshire Archaeology Service; 1991. Pp 72.
55. Sauerländer, Willibald. 'Romanesque sculpture in its architectural context', Bc105, 17–43; 193–94.
56. Sekules, Veronica. 'The liturgical furnishings of the choir of Exeter cathedral', Bc104, 172–79.
57. Sinclair, Eddie. 'The west front polychromy' [of Exeter cathedral], Bc104, 116–33.
58. Thomson, Robert G.; Brown, Duncan H. 'Archaic Pisan Maiolica and related Italian wares in Southampton', Bc191, 177–85.
59. Thurlby, Malcolm. 'The Romanesque cathedral of St Mary and St Peter at Exeter', Bc104, 19–34.
60. Thurmer, John. 'Some observations on views of the interior of Exeter cathedral in the nineteenth and twentieth centuries', Bc104, 210–15.
61. Tracy, Charles. 'Dating the misericords from the thirteenth-century choir-stalls at Exeter cathedral', Bc104, 180–87.
62. Tracy, Charles. 'The St Albans abbey watching chamber: a re-assessment', *J. of the British Arch. Association* 145 (1992), 104–11.
63. Warren, John. 'Greater and lesser gothic roofs: a study of the crown-post roof and its antecedents', *Vernacular Architecture* 23 (1992), 1–9.
64. Woodcock, Sally. 'The building history of St Mary de Haura, New Shoreham', *J. of the British Arch. Association* 145 (1992), 89–103.
65. Wright, Sylvia. 'The author portraits in the Bedford psalter-hours: Gower, Chaucer and Hoccleve', *British Library J.* 18 (1992), 190–201.

(k) *Topography*

1. Aldred, David; Dyer, Christopher C. 'A medieval Cotswold village: Roel, Gloucestershire', *T. of the Bristol & Gloucestershire Arch. Soc.* 109 (1992 for 1991), 139–70.
2. Ayers, Brian S. 'The influence of minor streams on urban development: Norwich, a case study', Bc214, 173–76.
3. Baker, Nigel. 'Approaches to the structure of medieval towns', Bc214, 167–72.
4. Barrow, Julia. 'Urban cemetery location in the high middle ages', Bc216, 78–100.
5. Binney, M. 'Bolton castle, Yorkshire', *Country Life* 185/21 (1992), 60–63.
6. Birrell, Jean. 'The forest and the chase in medieval Staffordshire', *Staffordshire Studies* 3 (1991), 23–50.
7. Blood, N. Keith; Bowden, Mark. 'The Comby hills at North Charlton, Northumberland', *Landscape History* 13 (1991), 65–67.

8. Blood, N. Keith; Taylor, Christopher C. 'Cawood: an archiepiscopal landscape', *Yorkshire Arch. J.* 64 (1992), 83–102.

9. Brooks, Howard. 'Two rural medieval sites in Chignall St James: excavations 1989', *Essex Arch. & History* 23 (1992), 39–50.

10. Brown, Anthony E.; Taylor, Christopher C. *Moated sites in northern Bedfordshire: some surveys and wider implications.* Leicester; University of Leicester, Dept. of Adult Education; 1991. Pp 51.

11. Chitwood, P. 'Lincoln's ancient docklands: the search continues', Bc41, 169–76.

12. Coulson, Charles. 'Some analysis of the castle of Bodiam, east Sussex', Bc98, 51–107.

13. Counihan, J. 'Mottes, Norman or not!', *Fortress* 11 (1991), 53–60.

14. Croom, Jane N. 'The topographical analysis of medieval town plans: the examples of Much Wenlock and Bridgnorth', *Midland History* 17 (1992), 16–38.

15. Dinn, Robert. 'Death and rebirth in late medieval Bury St Edmunds', Bc216, 151–69.

16. Dixon, Philip. 'From hall to tower: the change in seigneurial houses on the Anglo-Scottish border after *c.*1250', Bc112, 85–107.

17. Dodd, A.; Moss, P. 'The history of Brimpsfield castle and the Giffard family', *Glevensis* 25 (1991), 34–37.

18. Emerick, K.; Szymanski, J.E. 'Recent archaeological discoveries at the medieval site of Fountains abbey', Bc206, 47–54.

19. Falvey, Heather. 'The More: Archbishop George Neville's palace in Rickmansworth, Hertfordshire', *The Ricardian* 9/118 (1992), 290–302.

20. Farley, Michael; Lawson, Jo. 'A fifteenth-century pottery and tile kiln at Leyhill, Latimer, Buckinghamshire', *Records of Buckinghamshire* 32 (1991 for 1990), 35–62.

21. Gardiner, Mark; Jones, Gwen; Martin, David. 'The excavation of a medieval aisled hall at Park farm, Salehurst, East Sussex', *Sussex Arch. Collections* 129 (1991), 81–97.

22. Gauthiez, Bernard. 'Hypothèses sur la fortification de Rouen au onzième siècle: le donjon, la tour de Richard II et l'enceinte de Guillaume', Bc27, 61–76.

23. Gilchrist, Roberta. 'Knight clubs: archaeology of the military orders', Bc206, 65–70.

24. Glasscock, Robin E. 'The early middle ages: 1066–1348', Bc181, 73–102; 242.

25. Glasscock, Robin E. 'The later middle ages: 1348- *c.*1540', Bc181, 103–34; 242–43.

26. Hall, Richard A. 'The waterfronts of York', Bc41, 177–84.

27. Harding, Vanessa. 'Burial choice and burial location in later medieval London', Bc216, 119–35.

28. Heslop, T.A. 'Orford castle, nostalgia and sophisticated living', *Architectural History* 34 (1991), 36–58.

29. Higham, Robert; Barker, Philip. *Timber castles.* London; Batsford; 1992. Pp 390.

30. Higham, Robert; Hamlin, A. 'Bampton castle, Devon: history and archaeology', *P. of the Devon Arch. Soc.* 48 (1990), 101–10.
31. Hislop, M.J.B. 'The castle of Ralph, fourth baron Neville, at Raby', *Archaeologia Aeliana* 5th ser. 20 (1992), 91–97.
32. Hooke, Della. 'The relationship between ridge and furrow and mapped strip holdings', *Landscape History* 13 (1991), 69–71.
33. Leppard, M.J. 'East Grinstead before the town', *Sussex Arch. Collections* 129 (1991), 29–32.
34. McNeill, Tom E. *Castles*. London; Batsford; 1992. Pp 143.
35. Metcalf, Vivienne M. 'The Wood hall moated manor project', Bc210, 201–06.
36. Milne, Gustav. 'Waterfront archaeology and vernacular architecture: a London study', Bc41, 116–20.
37. Moorhouse, Stephen. 'Ceramics in the medieval garden', Bc64, 100–17.
38. Moran, Madge. 'A terrace of crucks at Much Wenlock, Shropshire', *Vernacular Architecture* 23 (1992), 10–14.
39. Potter, Geoff. 'The medieval bridge and waterfront at Kingston-on-Thames', Bc41, 137–49.
40. Potter, Geoff. 'The medieval bridge and waterfront, Kingston-on-Thames, England', Bc208, 1–8.
41. Rady, Jonathan; Tatton-Brown, Tim; Bowen, John Atherton. 'The archbishop's palace, Canterbury: excavations and building recording works from 1981 to 1986', *J. of the British Arch. Association* 144 (1991), 1–60.
42. Roffe, David. 'Place-naming in Domesday Book: settlements, estates, and communities', *Nomina* 14 (1992 for 1990–1), 47–60.
43. Ryder, Peter F. 'The gatehouse of Morpeth castle, Northumberland', *Archaeologia Aeliana* 5th ser. 20 (1992), 63–77.
44. Saunders, Andrew. 'Administrative buildings and prisons in the earldom of Cornwall', Bc71, 195–216.
45. Saunders, Tom. 'The feudal construction of space: power and domination in the nucleated village', Bc38, 181–96.
46. Sherlock, David. 'Wisbech Barton's farm buildings in 1412/13', *Cambridge Antiquarian Soc. P.* 80 (1992 for 1991), 21–29.
47. Sherlock, Stephen J. 'Excavations at Castle Hill, Castleton, North Yorkshire', *Yorkshire Arch. J.* 64 (1992), 41–47.
48. Slocombe, Pamela M. *Medieval houses of Wiltshire*. Stroud; Sutton; 1992. Pp 112.
49. Smith, Lance. 'Sydenhams moat: a 13th-century moated manor-house in the Warwickshire Arden', *Birmingham & Warwickshire Arch. Soc. T.* 96 (1991 for 1989–90), 27–68.
50. Steane, Kate. 'Excavations at Ratley Castle, 1968–1973', *Birmingham & Warwickshire Arch. Soc. T.* 96 (1991 for 1989–90), 5–26.
51. Stevens, Lawrence; Stevens, Patricia. 'Excavations on the south lawn, Michelham priory, Sussex 1971–1976', *Sussex Arch. Collections* 129 (1991), 45–79.
52. Stevenson, Janet H. 'The castles of Marlborough and Ludgershall in the middle ages', *Wiltshire Arch. & Natural History Magazine* 85 (1992), 70–79.

53. Studd, Robin. 'Medieval Newcastle-under-Lyme: a hidden Domesday borough?', *Staffordshire Studies* 3 (1991), 1–22.
54. Szymanski, J.E.; Campbell, T. 'Non-destructive site diagnosis at medieval abbey sites in the UK', Bc206, 201–06.
55. Tatton-Brown, Tim. 'Medieval building stone at the Tower of London', *London Archaeologist* 6/13 (1991), 361–66.
56. Tatton-Brown, Tim. 'The buildings and topography of St Augustine's abbey, Canterbury', *J. of the British Arch. Association* 144 (1991), 61–91.
57. Taylor, Arnold J. '"Belrem"', Bc27, 1–23.
58. Thompson, M.W. 'Keep or country house? Thin-walled Norman "proto-keeps"', *Fortress* 12 (1992), 13–22.
59. Tolley, R.J. 'Stokesay castle, Shropshire: the repair of a major monument', *T. of the Association for Studies in the Conservation of Historic Buildings* 15 (1990), 3–24.
60. Turnbull, Percival; Walsh, Deborah. 'Monastic remains at Ravenstonedale', *T. of the Cumberland & Westmorland Antiquarian & Arch. Soc.* 92 (1992), 67–76.
61. Walker, David. 'Gloucestershire castles', *T. of the Bristol & Gloucestershire Arch. Soc.* 109 (1992 for 1991), 5–23.
62. Warnicke, Retha M. 'More's *Richard III* and the mystery plays', *Historical J.* 35 (1992), 761–78.
63. Watson, B. 'The excavation of a Norman fortress on Ludgate hill', *London Archaeologist* 6/14 (1992), 371–77.
64. Weale, C. 'Priest and parish: the town of Hertford, 1209–1393', *Hertfordshire Past & Present* 28 (1990), 1–5.
65. Williams, John H.; Gidney, L.; Howard-Davis, C.; Moore, D.T.; McCarthy, M.R.; Wild, F. 'Excavations at Brougham castle, 1987', *T. of the Cumberland & Westmorland Antiquarian & Arch. Soc.* 92 (1992), 105–34.
66. Wilson, Barbara. 'Medieval York, 1320?', *Interim: Archaeology in York* 16/4 (winter 1991), 25–29.
67. Wilson-North, W.R.; Dunn, C.J. '"The Rings", Loddiswell: a new survey', *P. of the Devon Arch. Soc.* 48 (1990), 87–100.
68. Worsley, G. 'Bamburgh castle, Northumberland', *Country Life* 186/35 (1992), 46–49.

F. ENGLAND AND WALES 1500–1714

See also Aa2,50,52,98,175,b51,61,95; Bc1,48–49,59–60,62,68–69,74,79–80,87,107,122,155,203,218,221,d14; Ee35,56,61–62,f22,25,g33–34,41,i39; Ga18,30,e24; Hg46

(a) *General*

1. Bailyn, Bernard; Morgan, Philip D. 'Introduction' [cultural margins of the first British empire], Bc2, 1–31.

2. Bidwell, William B.; Jansson, Maija (ed.). *Proceedings in parliament, 1626, vols. 2–3: House of Commons*. London; Yale UP; 1992. 2 vols.
3. Brautigam, Dwight D. 'The *Court and the country* revisited', Bc74, 55–64.
4. Dickinson, J.R. 'The earl of Derby and the Isle of Man, 1643–1651', *T. of the Historic Soc. of Lancashire & Cheshire* 141 (1991), 39–76.
5. Dray, William H. 'Causes, individuals and ideas in Christopher Hill's interpretation of the English revolution', Bc74, 21–40.
6. Elton, Sir Geoffrey R. *Studies in Tudor and Stuart politics and government, vol. 4: papers and reviews, 1982–1990*. Cambridge; Cambridge UP; 1992. Pp x, 321.
7. Hardin, Richard F. 'Geoffrey among the lawyers: *Britannica* (1607) by John Ross of the Inner Temple', *Sixteenth Century J.* 23 (1992), 235–49.
8. Hexter, J.H. 'Introduction' [parliament and liberty], Bc80, 1–19.
9. Keeble, N.H. 'Rewriting the Restoration' [review article], *Historical J.* 35 (1992), 223–25.
10. Palliser, David M. *The age of Elizabeth: England under the later Tudors, 1547–1603*. London; Longman; 2nd edn. 1992. Pp xxvi, 516.
11. Smith, Nigel. 'Literature as history' [review article], *Historical J.* 35 (1992), 213–21.
12. Starkey, David. 'England' [Renaissance], Bc221, 146–63.
13. Stevenson, David. 'Solomon and son, British style' [review article], *Historical J.* 35 (1992), 205–11.
14. Teich, Mikuláš; Porter, Roy. 'Introduction' [Renaissance], Bc221, 1–5.
15. Thomas, P.W. 'Conclusion: another pattern. Seventeenth-century Britain revisited', Bc59, 151–86.
16. Thornes, Robin; Leach, John. 'Buxton old hall: the earl of Shrewsbury's tower house re-discovered', *Arch. J.* 148 (1991), 256–68.
17. Westcott, Margaret. 'Katherine Courtenay, countess of Devon, 1479–1527', Bc218, 13–38.

(b) *Politics*

1. Adams, Simon. 'The Dudley clientele, 1553–1563', Bc49, 241–65.
2. Anglo, Sydney. *Images of Tudor kingship*. London; Seaby; 1992. Pp 148.
3. Ashley, Maurice. *The battle of Naseby and the fall of King Charles I*. Stroud; Sutton; 1992. Pp xii, 172.
4. Barber, Sarah. 'Irish undercurrents of the politics of April 1653', *Historical Research* 65 (1992), 315–35.
5. Barnes, Thomas Garden. 'Deputies not principals, lieutenants not captains: the institutional failure of lieutenancy in the 1620s', Bc69, 58–86.
6. Barrell, Rex A. *Anthony Ashley Cooper, earl of Shaftesbury (1671–1713) and 'le refuge français' correspondence*. Lampeter; Mellen; 1989. Pp 264.
7. Bennett, Martyn. 'Between Scylla and Charybdis: the creation of rival administrations at the beginning of the English civil war', *Local Historian* 22 (1992), 191–202.

8. Bennett, Ronan. 'War and disorder: policing the soldiery in civil war Yorkshire', Bc69, 248–73.
9. Bergeron, David M. *Royal family, royal lovers: King James of England and Scotland*. London; Missouri UP; 1991. Pp 232.
10. Bernard, G.W. 'Introduction: the Tudor nobility in perspective', Bc49, 1–48.
11. Bernard, G.W. 'The downfall of Sir Thomas Seymour', Bc49, 212–40.
12. Bernard, G.W. 'The fall of Anne Boleyn: a rejoinder', *English Historical R.* 107 (1992), 665–74.
13. Bush, M.L. 'Captain Poverty and the Pilgrimage of Grace', *Historical Research* 65 (1992), 17–36.
14. Cogswell, Thomas. 'War and the liberties of the subject', Bc80, 225–51.
15. Croft, Pauline. 'The parliamentary installation of Henry, prince of Wales', *Historical Research* 65 (1992), 177–93.
16. Cust, Richard. 'Anti-Puritanism and urban politics: Charles I and Great Yarmouth', *Historical J.* 35 (1992), 1–26.
17. Cust, Richard. 'Charles I, the Privy Council and the parliament of 1628', *T. of the Royal Historical Soc.* 6th ser. 2 (1992), 25–50.
18. Cust, Richard. 'Parliamentary elections in the 1620s: the case of Great Yarmouth', *Parliamentary History* 11 (1992), 179–91.
19. de Krey, Gary S. 'Revolution *redivivus*: 1688–1689 and the radical tradition in seventeenth-century London politics', Bc48, 198–217.
20. Dean, David. 'Locality and parliament: the legislative activities of Devon's MPs during the reign of Elizabeth', Bc218, 75–95.
21. Dean, David. 'Pressure groups and lobbies in the Elizabethan and early Jacobean parliaments', *Parliaments, Estates & Representation* 11 (1991), 139–52.
22. Dickinson, H.T. 'The letters of Bolingbroke to the earl of Orrery, 1712–1713', Bc178, 349–71.
23. Dunlop, David. 'The "masked comedian"?: Perkin Warbeck's adventures in Scotland and England from 1495 to 1497', *Scottish Historical R.* 70 (1991), 97–128.
24. Ellis, Steven G. 'A border baron and the Tudor state: the rise and fall of lord Dacre of the north', *Historical J.* 35 (1992), 253–77.
25. Evans, Robert Rees. *Pantheisticon: the career of John Toland*. New York; Lang; 1991. Pp xi, 232.
26. Ferris, John. 'Before Hansard: records of debate in the seventeenth-century House of Commons', *Archives* 20 (1992), 198–207.
27. Fissel, Mark Charles. 'Scottish war and English money: the Short parliament of 1640', Bc69, 193–223.
28. Fradenburg, Louise Olga. 'Sovereign love: the wedding of Margaret Tudor and James IV of Scotland', Bc91, 78–100.
29. Fraser, Antonia. *Six wives of Henry VIII*. London; Weidenfeld & Nicolson; 1992. Pp 479.
30. Frye, Susan. 'The myth of Elizabeth at Tilbury', *Sixteenth Century J.* 23 (1992), 95–114.
31. Gabbey, Alan. 'Cudworth, More, and the mechanical analogy', Bc60, 109–27.

32. Gray, Madeleine. 'Power, patronage and politics: office-holding and administration on the crown's estates in Wales', Bc122, 137–62.

33. Gunn, Steven J. 'Early Tudor dates for the death of Edward V', *Northern History* 28 (1992), 213–16.

34. Gunn, Steven J. 'Henry Bourchier, earl of Essex (1472–1540)', Bc49, 134–79.

35. Gunn, Steven J.; Lindley, Phillip G. 'Introduction' [Wolsey], Bc1, 1–53.

36. Guy, John. 'The "imperial crown" and the liberty of the subject: the English constitution from Magna Carta to the Bill of Rights', Bc74, 65–88.

37. Guy, John. 'Wolsey and the Tudor polity', Bc1, 54–75.

38. Hammer, Paul E.J. 'An Elizabethan spy who came in from the cold: the return of Anthony Standen to England in 1593', *Historical Research* 65 (1992), 277–95.

39. Hartley, T.E. *Elizabeth's parliaments: queen, lords and commons, 1559–1601.* Manchester; Manchester UP; 1992. Pp 184.

40. Hexter, J.H. 'Parliament, liberty, and freedom of elections', Bc80, 21–55.

41. Hirst, Derek M. 'Freedom, revolution and beyond', Bc80, 252–74.

42. Holmes, Clive. 'Parliament, liberty, taxation, and property', Bc80, 122–54.

43. Hoyle, Richard W. 'Henry Percy, sixth earl of Northumberland, and the fall of the house of Percy, 1527–1537', Bc49, 180–211.

44. Hoyle, Richard W. 'Letters of the Cliffords, lords Clifford and earls of Cumberland, *c*.1500- *c*.1565', Bc178, 1–189.

45. Israel, Jonathan I. 'Propaganda in the making of the Glorious revolution', Bc133, 167–77.

46. Ives, E.W. 'Henry VIII's will: a forensic conundrum', *Historical J.* 35 (1992), 779–804.

47. Ives, E.W. 'The fall of Anne Boleyn reconsidered', *English Historical R.* 107 (1992), 651–64.

48. Ives, E.W. 'The fall of Wolsey', Bc1, 286–315.

49. Keirn, Tim; Davison, Lee. 'The reactive state: English governance and society, 1689–1750', Bc82, xi-liv.

50. Kershaw, Stephen E. 'Power and duty in the Elizabethan aristocracy: George, earl of Shrewsbury, the Glossopdale dispute and the council', Bc49, 266–95.

51. Knowlden, Patricia E. 'West Wickham and north-west Kent in the civil war', *Local Historian* 22 (1992), 138–43.

52. Ligou, Daniel. 'Voltaire et la 'Glorieuse révolution' anglaise de 1688', *Parliaments, Estates & Representation* 11 (1991), 153–62.

53. Loades, D.M. *The mid-Tudor crisis, 1545–1565.* London; Macmillan; 1992. Pp 215.

54. MacCaffrey, Wallace Trevithie. *Elizabeth I: war and politics, 1588–1603.* Oxford; Princeton UP; 1992. Pp xv, 592.

55. Malcolm, Joyce Lee. 'Charles II and the reconstruction of royal power', *Historical J.* 35 (1992), 307–30.

56. Manning, Brian. *1649: the crisis of the English revolution.* London; Bookmarks; 1992. Pp 288.

57. Nenner, Howard. 'Pretense and pragmatism: the response to uncertainty in the succession crisis of 1689', Bc48, 83–94.
58. Patterson, Catherine F. 'Leicester and Lord Huntingdon: urban patronage in early modern England', *Midland History* 16 (1991), 45–62.
59. Pincus, Steven C.A. 'Popery, trade and universal monarchy: the ideological context of the outbreak of the second Anglo-Dutch war', *English Historical R.* 107 (1992), 1–29.
60. Pocock, J.G.A. 'The fourth English civil war: dissolution, desertion and alternative histories in the Glorious revolution', Bc48, 52–64.
61. Prest, Wilfred. 'Predicting civil war allegiances: the lawyer's case considered', *Albion* 24 (1992), 225–36.
62. Pugh, T.B. 'Henry VII and the English nobility', Bc49, 49–110.
63. Robbins, Christopher A. *The earl of Wharton and Whig party politics, 1679–1715.* Lampeter; Mellen; 1992. Pp 484.
64. Robinson, W.R.B. 'Royal service in North Wales under the early Tudors: the career of John Puleston (d. 1524) of Berse and Hafod-y-Wern', *T. of the Denbighshire Historical Soc.* 40 (1991), 29–42.
65. Roots, Ivan. 'English politics, 1625–1700', Bc59, 18–52.
66. Rumbold, Valerie. 'The Jacobite vision of Mary Caesar', Bc62, 178–98.
67. Sacks, David Harris. 'Parliament, liberty and the commonweal', Bc80, 85–121.
68. Sargent, Mark L. 'Thomas Hutchinson, Ezra Stiles, and the legend of the regicides', *William & Mary Q.* 3rd ser. 49 (1992), 431–48.
69. Schwoerer, Lois G. 'Introduction' [revolution of 1688], Bc48, 1–20.
70. Schwoerer, Lois G. 'The coronation of William and Mary, April 11, 1689', Bc48, 107–30.
71. Scott, Andrew Murray. 'Letters of John Graham of Claverhouse', Bc83, 135–268.
72. Sharpe, Kevin. *Personal rule of Charles I.* London; Yale UP; 1992. Pp xxiv, 983.
73. Shephard, Robert. 'Court factions in early modern England', *J. of Modern History* 64 (1992), 721–45.
74. Sherwood, Roy. *Civil war in the midlands, 1642–1651.* Stroud; Sutton; new edn. 1992. Pp xii, 209.
75. Skerpan, Elizabeth. *Rhetoric of politics in the English revolution, 1642–1660.* London; Missouri UP; 1992. Pp 288.
76. Smith, David L. 'The 4th earl of Dorset and the politics of the sixteen-twenties', *Historical Research* 65 (1992), 37–53.
77. Snow, Vernon F.; Young, Anne Steel (ed.). *Private journals of the Long Parliament, 2 June to 17 September 1642.* London; Yale UP; 1992. Pp xxxviii, 515.
78. Sommerville, Johann P. 'Parliament, privilege and the liberties of the subject', Bc80, 56–84.
79. Speck, William A. 'William—and Mary?', Bc48, 131–46.
80. Stater, Victor. 'War and the structure of politics: lieutenancy and the campaign of 1628', Bc69, 87–109.
81. Stern, Virginia F. *Sir Stephen Powle of court and country: memorabilia of*

a government agent for Queen Elizabeth I, chancery official and English country gentleman. Selinsgrove (Pa); Susquehanna UP; 1992. Pp 245.

82. Szechi, Daniel. 'The diary and speeches of Sir Arthur Kaye, 1710–1721', Bc178, 321–48.

83. Teague, Frances. 'Queen Elizabeth in her speeches', Bc68, 63–78.

84. Walker, Greg. 'John Skelton, cardinal Wolsey and the English nobility', Bc49, 111–33.

85. Ward, Philip. 'The politics of religion: Thomas Cromwell and the Reformation in Calais, 1534–1540', *J. of Religious History* 17 (1992), 152–71.

86. Weil, Rachel J. 'The politics of legitimacy: women and the warming-pan scandal', Bc48, 65–82.

87. Zaller, Robert. 'Parliament and the crisis of European liberty', Bc80, 201–24.

88. Zook, Melinda. 'History's Mary: the propagation of Queen Mary II, 1689–1694', Bc91, 170–91.

(c) *Constitution, Administration and Law*

1. Beattie, John M. 'London crime and the making of the "bloody code" 1689–1718', Bc82, 49–76.

2. Beattie, John M. 'The cabinet and the management of death at Tyburn after the revolution of 1688–1689', Bc48, 218–33.

3. Carter, Patrick. 'An internecine administrative feud of the Commonwealth: Thomason tract 669.f.20 (18)', *British Library J.* 18 (1992), 205–07.

4. Croft, Pauline. 'Sir John Doddridge, King James I, and the antiquity of parliament', *Parliaments, Estates & Representation* 12 (1992), 95–107.

5. Cromartie, Alan. 'The rule of law', Bc79, 55–69.

6. Gray, Charles M. 'Parliament, liberty and the law', Bc80, 155–200.

7. Gray, Madeleine. 'Exchequer officials and the market in crown property, 1558–1640', Bc122, 112–36.

8. Greene, Robert A. 'Whichcote, the candle of the Lord, and synderesis', *J. of the History of Ideas* 52 (1991), 617–44.

9. Haslam, Graham. 'Jacobean Phoenix: the duchy of Cornwall in the principates of Henry Frederick and Charles Graham Haslam', Bc122, 263–96.

10. Haslam, Graham. 'The Elizabethan duchy of Cornwall, an estate in stasis', Bc122, 88–111.

11. Haynes, Alan. *Invisible power: the Elizabethan secret services, 1570–1603.* Stroud; Sutton; 1992. Pp xxi, 179.

12. Hey, D. 'The establishment of a legal profession in Sheffield, 1660–1740', *T. of the Hunter Arch. Soc.* 16 (1991), 16–23.

13. Hoyle, Richard W. 'Customary tenure on the Elizabethan estates', Bc122, 191–203.

14. Hoyle, Richard W. 'Introduction: aspects of the crown's estate, c.1558–1640', Bc122, 1–57.

15. Hoyle, Richard W. 'Reflections on the history of the crown lands, 1558–1640', Bc122, 418–32.

16. Hoyle, Richard W. '"Shearing the hog": the reform of the estates, c.1598–1640', Bc122, 204–62.
17. Hoyle, Richard W. 'Tenure on the Elizabethan estates', Bc122, 163–68.
18. Keirn, Tim. 'Parliament, legislation and the regulation of English textile industries, 1689–1714', Bc82, 1–24.
19. Manning, Roger B. 'Sir Robert Cotton, antiquarianism and estate administration: a Chancery decree of 1627', British Library J. 18 (1992), 88–96.
20. Mendle, Michael. 'The great council of parliament and the first ordinances: the constitutional theory of the civil war', J. of British Studies 31 (1992), 133–62.
21. Patterson, Annabel. 'The egalitarian grant: representations of justice in history/literature', J. of British Studies 31 (1992), 97–132.
22. Robertson, Una A. 'An Edinburgh lawyer and his bees', Book of the Old Edinburgh Club ns 1 (1991), 79–81.
23. Seipp, David J. 'The structure of English common law in the seventeenth-century', Bc195, 61–83.
24. Smith, David L. 'The struggle for new constitutional and institutional forms', Bc79, 15–34.
25. Sokol, B.J. 'The Merchant of Venice and the Law Merchant', Renaissance Studies 6 (1992), 60–67.
26. Stacey, William R. 'Impeachment, attainder, and the "revival" of parliamentary judicature under the early Stuarts', Parliamentary History 11 (1992), 40–56.
27. Wormald, Jenny. 'The creation of Britain: multiple kingdoms or core and colonies?', T. of the Royal Historical Soc. 6th ser. 2 (1992), 175–94.

(d) External Affairs

1. Allen, D.F. 'James II and the court of Rome: John Caryll's contributions', Durham University J. 84 (1992), 21–27.
2. Black, Jeremy. 'A parliamentary foreign policy?: the 'Glorious revolution' and the conduct of British foreign policy', Parliaments, Estates & Representation 11 (1991), 69–80.
3. Breslaw, E.G. '"Price's—his deposition": kidnapping Amerindians in Guyana, 1674', J. of the Barbados Museum & Historical Soc. 39 (1991), 47–51.
4. Campbell, P.F. 'Two generations of Walronds' [part 2], J. of the Barbados Museum & Historical Soc. 39 (1991), 1–23.
5. Craton, Michael. 'Reluctant creoles: the planters' world in the British West Indies', Bc2, 314–62.
6. Dailey, Barbara Ritter. 'The early Quaker mission and the settlement of meetings in Barbados, 1655–1700' [part 1], J. of the Barbados Museum & Historical Soc. 39 (1991), 24–46.
7. de Divitiis, G. Pagano. Mercanti inglesi nell'Italia dei seicento [English merchants in Italy in the seventeenth century]. Venice; Marsilio; 1990. Pp 223.

8. Gray, Todd. 'Turks, Moors and the Cornish fishermen: piracy in the early seventeenth century', *J. of the Royal Institution of Cornwall* 10 (1990), 457–75.
9. Greene, Jack P. 'The Glorious revolution and the British empire, 1688–1783', Bc48, 260–71.
10. Gunn, Steven J. 'Wolsey's foreign policy and the domestic crisis of 1527–1528', Bc1, 149–77.
11. Haley, K.H.D. 'The Dutch, the invasion of England, and the alliance of 1689', Bc48, 21–34.
12. Hart, Marjolein 't. '"The devil or the Dutch": Holland's impact on the financial revolution in England, 1643–1694', *Parliaments, Estates & Representation* 11 (1991), 39–52.
13. Hasan, F. 'Conflict and co-operation in Anglo-Mughal trade relations during the reign of Aurangzeb', *J. of the Economic & Social History of the Orient* 34 (1991), 351–60.
14. Hobelt, Lothar. 'Imperial diplomacy and the "Glorious revolution"', *Parliaments, Estates & Representation* 11 (1991), 61–67.
15. Hunt, William. 'A view from the Vistula on the English revolution', Bc74, 41–54.
16. Jones, Maldwyn A. 'The Scotch-Irish in British America', Bc2, 284–313.
17. Kirk-Smith, Harold. *William Brewster, the father of New England: his life and times, 1567–1644.* Boston; Kay; 1992. Pp 382.
18. Knapp, Jeremy. *An empire nowhere: England, America and literature from 'Utopia' to 'The Tempest'.* Berkeley (Ca); California UP; 1992. Pp 387.
19. Lemay, J.A. Leo. *Did Pocahontas save Captain John Smith?* Athens (Ga); Georgia UP; 1992. Pp 160.
20. Merrell, James H. '"The customes of our countrey": Indians and colonists in early America', Bc2, 117–56.
21. Morgan, Philip D. 'British encounters with Africans and African-Americans, c.1600–1780, Bc2, 156–219.
22. Pennell, C.R. *Piracy and diplomacy in seventeenth-century north Africa: the journal of Thomas Baker, English consul in Tripoli, 1677–1685.* Rutherford (NJ); Fairleigh Dickinson UP; 1989. Pp 261.
23. Pincus, Steven C.A. 'England and the world in the 1650s', Bc79, 129–47.
24. Poussou, J.-P. *Les îles britanniques, les Provinces Unies, la guerre et la paix au 17è siècle.* Paris; Economica; 1991. Pp 241.
25. Rhodes, Dennis E. 'Richard White of Basingstoke: the erudite exile', Bc133, 23–30.
26. Roeber, A.G. '"The origin of whatever is not English among us": the Dutch-speaking and the German-speaking peoples of colonial British America', Bc2, 220–83.
27. Rule, John C. 'France caught between two balances: the dilemma of 1688', Bc48, 35–51.
28. Seed, Patricia. 'Taking possession and reading texts: establishing the authority of overseas empires', *William & Mary Q.* 3rd ser. 49 (1991), 183–209.
29. Sparks, Carol. 'England and the Columbian discoveries: the attempt to

legitimize English voyages to the New World', *Terrae Incognitae* 22 (1990), 1–12.

30. Stobbart, Lorainne G. *Utopia—fact or fiction?: the evidence from the Americas.* Stroud; Sutton; 1992. Pp xiv, 143.

31. Storrs, Christopher. 'Machiavelli dethroned: Victor Amadeus II and the making of the Anglo-Savoyard alliance of 1690', *European History Q.* 22 (1992), 347–82.

32. van Strien, C.D. *British travellers in Holland during the Stuart period: Edward Browne and John Locke as tourists in the United Provinces.* Leiden; Brill (Publications of the Sir Thomas Browne Inst. ns 13); 1992. [Pp 270].

33. Williams, Michael E. 'Alarms and excursions in Lisbon under Castilian domination: the case of Captain Richard Butler', *Portuguese Studies* 6 (1990), 94–114.

(e) *Religion*

1. Alblas, Jacques B.H. 'Richard Allestree's *The whole duty of man* (1658) in Holland: the denominational and generic transformations of an Anglican classic', *Nederlands Archief voor Kerkgeschiedenis* 71 (1991), 92–104.

2. Arnoult, Sharon L. 'The sovereignties of body and soul: women's political and religious actions in the English civil war', Bc91, 228–49.

3. Ashcraft, Richard. 'Latitudinarianism and toleration: historical myth versus political history', Bc60, 151–77.

4. Aston, Margaret. 'The *Bishops' Bible* illustrations', Bc194, 267–85.

5. Barnard, Leslie W. 'Bishop George Bull of St David's: scholar and defender of the faith', *J. of Welsh Ecclesiastical History* 9 (1992), 37–52.

6. Barrell, Rex A. (ed.). *Correspondence of Abel Boyer, Huguenot refugee, 1667–1729.* Lampeter; Mellen; 1992. Pp 240.

7. Bartlett, Kenneth. 'Papal policy and the English crown, 1563–1565: the Bertano correspondence', *Sixteenth Century J.* 23 (1992), 643–59.

8. Baskerville, E.J. 'A religious disturbance in Canterbury, June 1561: John Bale's unpublished account', *Historical Research* 65 (1992), 340–48.

9. Bettey, J.H. 'The suppression of the Benedictine nunnery at Shaftesbury in 1539', *The Hatcher R.* 4/34 (1992), 3–11.

10. Bozeman, Theodore Dwight. 'Federal theology and the "national covenant": an Elizabethan Presbyterian case study', *Church History* 61 (1992), 394–407.

11. Brown, Keith M. 'Wolsey and ecclesiastical order: the case of the Franciscan Observants', Bc1, 219–38.

12. Carleton, Kenneth W.T. 'John Marbeck and *The Booke of Common Praier noted*', Bc194, 255–65.

13. Carlson, Eric Josef. 'Clerical marriage and the English Reformation', *J. of British Studies* 31 (1992), 1–31.

14. Carlson, Eric Josef. 'The marriage of William Turner', *Historical Research* 65 (1992), 336–39.

15. Champion, J.A.I. *The pillars of priestcraft shaken: the Church of England and its enemies, 1660–1730*. Cambridge; Cambridge UP; 1992. Pp 268.

16. Cottret, Bernard. *The Huguenots in England, immigration and settlement*. Cambridge; Cambridge UP; 1992. Pp 310.

17. Coudert, Allison P. 'Henry More, the Kabbalah, and the Quakers', Bc60, 31–67.

18. Cross, Claire. 'Monks, friars, and the royal supremacy in sixteenth-century Yorkshire', Bc12, 437–56.

19. Davie, Donald. 'Baroque in the hymn-book', Bc194, 329–42.

20. Davies, Julian. *Caroline captivity of the church: Charles I and the remoulding of Anglicanism, 1625–1641*. Oxford; Oxford UP; 1992. Pp xviii, 400.

21. Davis, J. Colin. 'Religion and the struggle for freedom in the English Revolution', *Historical J*. 35 (1992), 507–30.

22. Dever, Mark E. 'Moderation and deprivation: reappraisal of Richard Sibbes', *J. of Ecclesiastical History* 43 (1992), 396–413.

23. Dorman, Marianne (ed.). *Sermons of Lancelot Andrewes, vol. 1: Nativity, Lenten and Passion*. Soham; Pentland; 1992. Pp xxi, 224.

24. d'Uzer, Vincenette. 'The Jews in the sixteenth-century homilies', Bc189, 265–77.

25. Eales, Jacqueline. 'Iconoclasm, iconography, and the altar in the English civil war', Bc194, 313–27.

26. Eisenach, Eldon. 'Religion and Locke's *Two treatises of government*', Bc155, 50–81.

27. Evans, R.H. '"The truth sprang up first in Leicestershire": George Fox, 1624–1691, and the origins of Quakerism', *Leicestershire Arch. & Historical Soc. T*. 66 (1992), 121–35.

28. George, C.H. 'Parnassus restored, saints confounded: the secular challenge to the age of the godly, 1560–1660', *Albion* 23 (1991), 409–37.

29. Gilliam, Elizabeth; Tighe, W.J. 'To "run with the time": Archbishop Whitgift, the Lambeth articles, and the politics of theological ambiguity in late Elizabethan England', *Sixteenth Century J*. 23 (1992), 325–40.

30. Gowers, Ian. 'The clergy in Devon, 1641–1662', Bc218, 200–26.

31. Greaves, Richard L. 'Radicals, rights and revolution: British nonconformity and roots of the American experience', *Church History* 61 (1992), 151–68.

32. Greaves, Richard L. 'Shattered expectations?: George Fox, the Quakers, and the Restoration state, 1660–1685', *Albion* 24 (1992), 237–59.

33. Hansford-Miller, Frank. *A history and geography of English religion, vol. 7: diocesan changes of King Henry VIII and the friars and the lollards*. Canterbury; Abcado; 1992. Pp 100.

34. Hansford-Miller, Frank. *A history and geography of English religion, vol. 8: the diocesan and parish system of England and Wales in 1550*. Canterbury; Abcado; 1992. Pp 100.

35. Hansford-Miller, Frank. *A history and geography of English religion, vol. 9: heresy and Marian Catholicism*. Canterbury; Abcado; 1992. Pp 100.

36. Hansford-Miller, Frank. *A history and geography of English religion, vol. 10: the Elizabethan settlement*. Canterbury; Abcado; 1992. Pp 100.

37. Hansford-Miller, Frank. *A history and geography of English religion, vol. 11: Elizabethan Puritanism*. Canterbury; Abcado; 1992. Pp 100.
38. Hill, Christopher. 'Quakers and the English revolution', *J. of the Friends' Historical Soc.* 56 (1992), 165–79
39. Horowitz, Elliott. '"A different mode of civility": Lancelot Addison on the Jews of Barbary', Bc189, 309–25.
40. Hoyle, Richard W. 'Advancing the Reformation in the north: orders from York High Commission, 1583 and 1592', *Northern History* 28 (1992), 217–27.
41. Hudson, Elizabeth K. 'The Catholic challenge to Puritan piety, 1580–1620', *Catholic Historical R.* 77 (1991), 1–20.
42. Hughes, Ann. 'The frustrations of the godly', Bc79, 70–90.
43. Hunter, Michael. 'Latitudinarianism and the "ideology" of the early Royal Society: Thomas Sprat's *History of the Royal Society*. (1667) reconsidered', Bc60, 199–229.
44. Hutton, Sarah. 'Edward Stillingfleet, Henry More, and the decline of *Moses Atticus*: a note on seventeenth-century Anglican apologetics', Bc60, 68–84.
45. Ingle, H. Larry. 'George Fox, millenarian', *Albion* 24 (1992), 261–78.
46. Ingle, H. Larry. 'Richard Hubberthorne and history: the crisis of 1659', *J. of the Friends' Historical Soc.* 56 (1992), 189–200.
47. Jenkins, Geraint H. *Protestant dissenters in Wales, 1639–1689*. Cardiff; Wales UP; 1992. Pp x, 122.
48. Katz, David S. 'The phenomenon of philo-Semitism', Bc189, 327–61.
49. Kroll, Richard. 'Introduction' [philosophy, science and religion], Bc60, 1–28.
50. Kunze, Bonnelyn Young. '"Vesells fitt for the masters us[e]": a transatlantic community of religious women, the Quakers 1675–1753', Bc74, 177–98.
51. Kusunoki, Akiko. '"Their testament at their apron-strings": the representation of Puritan women in early-seventeenth-century England', Bc68, 185–204.
52. Lake, Peter. 'The Laudians and the argument from authority', Bc74, 149–76.
53. Lee, Colin. '"Fanatic magistrates": religious and political conflict in three Kent boroughs, 1680–1684', *Historical J.* 35 (1992), 43–61.
54. Levine, Joseph M. 'Latitudinarians, neoplatonists, and the ancient wisdom', Bc60, 85–108.
55. Llewellyn-Edwards, Tam. 'Richard Farnworth of Tickhill', *J. of the Friends' Historical Soc.* 56 (1992), 201–09.
56. Loades, D.M. *Revolution in religion: the English Reformation, 1530–1570*. Cardiff; Wales UP; 1992. Pp viii, 134.
57. Louis, Cameron. 'The wayward vicar of Wollaton', *T. of the Thoroton Soc.* 94 (1991 for 1990), 29–34.
58. MacDonald, Michael. '*The fearefull estate of Francis Spira*: narrative, identity, and emotion in early modern England', *J. of British Studies* 31 (1992), 32–61.
59. Marshall, John. 'John Locke and latitudinarianism', Bc60, 253–82.

60. Mayor, S.H. 'James II and the dissenters', *Baptist Q.* 34 (1991), 180–90.
61. McCoog, Thomas M. 'Apostasy and knavery in Restoration England: the checkered career of John Travers', *Catholic Historical R.* 78 (1992), 395–412.
62. McNair, Philip. 'Bernardino Ochino in Inghilterra', *Rivista di Storia Italiana* 103 (1991), 231–42.
63. Mortimer, Jean E. 'Thoresby's "poor deluded Quakers": the sufferings of Leeds Friends in the seventeenth century', Bc84, 35–57.
64. Murdoch, Tessa. 'The dukes of Montagu as patrons of the Huguenots', *Huguenot Soc. P.* 25 (1992), 340–55.
65. Murphy, Martin. 'The Cadiz letters of William Johnson, *vere* Purnell, SJ', *Recusant History* 21 (1992), 1–10.
66. Newman, Keith A. 'Holiness in beauty?: Roman Catholics, Arminians, and the aesthetics of religion in early Caroline England', Bc194, 303–12.
67. Nuttall, Geoffrey F. 'A parcel of books for Morgan Llwyd', *J. of the Friends' Historical Soc.* 56 (1992), 180–88.
68. O'Day, Rosemary. 'Hugh Latimer: prophet of the kingdom', *Historical Research* 65 (1992), 258–76.
69. Parish, Debra L. 'The power of female pietism: women as spiritual authorities and religious role models in seventeenth-century England', *J. of Religious History* 17 (1992), 33–46.
70. Penny, D. Andrew. *Freewill or predestination: the battle over saving grace in mid-Tudor England.* Woodbridge; Boydell for the Royal Historical Soc.; 1991. Pp x, 246.
71. Purkiss, Diane. 'Producing the voice, consuming the body: women prophets of the seventeenth century', Bc62, 139–58.
72. Ramsbottom, John D. 'Presbyterians and 'partial conformity' in the Restoration church of England', *J. of Ecclesiastical History* 43 (1992), 249–70.
73. Reedy, Gerard. *Robert South, 1634–1716: introduction to his life and sermons.* Cambridge; Cambridge UP; 1992. Pp xiii, 171.
74. Rex, Richard. *The theology of John Fisher.* Cambridge; Cambridge UP; 1992. Pp 293.
75. Rex, Richard. 'Cardinal Wolsey' [review article], *Catholic Historical R.* 78 (1992), 607–14.
76. Ridgway, Maurice H. 'The early church plate at Wrexham parish church, Clwyd', *Archaeologia Cambrensis* 140 (1991), 148–54.
77. Roberts, Stephen. 'The Quakers in Evesham, 1655–1660: a study in religion, politics and culture', *Midland History* 16 (1991), 63–85.
78. Sacks, David Harris. 'Bristol's wars of religion', Bc106, 100–29.
79. Schochet, Gordon J. 'John Locke and religious toleration', Bc48, 147–64.
80. Scott, David. *Quakerism in York, 1650–1720.* York; the University (Borthwick paper 80); 1991. Pp iii, 36.
81. Sell, Alan P.F. 'Robert Barclay (1648–1690), the fathers and the inward, universal saving light: a tercentenary reappraisal', *J. of the Friends' Historical Soc.* 56 (1992), 210–26.
82. Siegenthaler, David. 'Popular devotion and the English Reformation: the case of *Ave Maria*', *Anglican & Episcopal History* 62 (1992), 1–11.

83. Sommerville, C. John. *The secularization of early modern England: from religious culture to religious faith.* Oxford; Oxford UP; 1992. Pp 227.
84. Stocker, Margarita. 'From faith to faith in reason? Religious thought in the seventeenth century', Bc59, 53–85.
85. Todd, Margo. 'Puritan self-fashioning: the diary of Samuel War', *J. of British Studies* 31 (1992), 236–64.
86. Trevett, Christine. *Women and Quakerism in the 17th century.* York; Sessions; 1992. Pp xvi, 171.
87. Underdown, David. *Fire from heaven: life in an English town in the seventeenth century.* London; HarperCollins; 1992. Pp xii, 308.
88. Usher, Brett. 'The Jew that Shakespeare drew', Bc189, 279–98.
89. Vanhulst, Henri. 'Thomas Harding, Joannes Bogardus et "An answere to maister Iuelles chalenge": le contrat de 1563', *Quaerendo* 22 (1991), 20–27.
90. Vidal, Daniel. 'Mystique abstraite et intrigue financière: Benoît de Canfield et la raison comptable au XVIIè siècle', *Revue Historique* 287 (1982), 33–60.
91. Walker, Greg. 'Cardinal Wolsey and the satirists: the case of *Godly Queen Hester* re-opened', Bc1, 239–60.
92. White, Peter. *Predestination, policy and polemic: conflict and consensus in the English church from the Reformation to the civil war.* Cambridge; Cambridge UP; 1992. Pp xiv, 336.
93. Winchester, Angus. 'Travellers in grey: Quaker journals as a source for local history', *Local Historian* 21 (1991), 70–76.
94. Wykes, David L. 'James II's religious indulgence of 1687 and the early organization of dissent: the building of the first nonconformist meeting-house in Birmingham', *Midland History* 16 (1991), 86–102.
95. Yardley, Bruce. 'George Villiers, second duke of Buckingham, and the politics of toleration', *Huntington Library Q.* 55 (1992), 317–37.
96. Zakai, Avihu. *Exile and kingdom: history and apocalypse in the Puritan migration to America.* Cambridge; Cambridge UP; 1992. Pp x, 264.
97. Zakai, Avihu. 'Orthodoxy in England and New England: Puritans and the issue of religious toleration, 1640–1650', *P. of the American Philosophical Soc.* 135 (1991), 401–41.

(f) *Economic Affairs*

1. Bowler, D.P. 'The post-medieval harbour, Tay Street, Perth', Bc41, 55–59.
2. Brunelle, Gayle K. 'Early modern international trade and merchant empires:' [review article], *Sixteenth Century J.* 23 (1992), 791–95.
3. Challis, Christopher E. 'Lord Hastings to the great silver recoinage, 1464–1699', Bc226, 179–397.
4. Crouzet, François M. 'The Huguenots and the English financial revolution', Bc32, 221–66.
5. Fairclough, K.R. 'Mills and ferries along the lower Lea', *Essex Archaeology & History* 3rd ser. 23 (1992), 57–66.

6. Gittings, Clare. 'Probate accounts: a neglected source', *Local Historian* 21 (1991), 51–59.

7. Glennie, Paul. 'Late Tudor and Stuart Britain: c.1540- c.1714', Bc181, 125–53; 243–44.

8. Grant, Alison. 'Breaking the mould: north Devon maritime enterprise, 1560–1640', Bc218, 119–40.

9. Gray, Madeleine. 'An early professional group?: the auditors of land revenue in the late sixteenth and early seventeenth centuries', *Archives* 20 (1992), 45–62.

10. Gray, Todd. 'Fishing and the commercial world of early Stuart Dartmouth', Bc218, 173–99.

11. Greeves, Tom. 'Four Devon stannaries: a comparative study of tinworking in the sixteenth century', Bc218, 39–74.

12. Henderson, C.G. 'The development of Exeter quay, 1564–1701', Bc41, 124–36.

13. Irwin, D.A. 'Mercantilism as strategic trade policy: the Anglo-Dutch rivalry for the East India trade', *J. of Political Economy* 99 (1991), 1296–1314.

14. Large, Peter. 'From Swanimote to disafforestation: Feckenham forest in the early seventeenth century', Bc122, 389–417.

15. Law, Robin. *The slave coast of west Africa, 1550–1750: the impact of the Atlantic slave trade on an African society.* Oxford; Oxford UP; 1991. Pp xii, 376.

16. Leedham-Green, E.S. (ed.). *Garrett Godfrey's accounts, c.1527–1533.* Cambridge; Cambridge Bibliographical Soc. (Monograph 12); 1992. Pp xxviii, 164.

17. McRae, Andrew. 'Husbandry manuals and the language of agrarian improvement', Bc204, 35–62.

18. Nicholls, Mark. '"As happy a fortune as I desire": the pursuit of financial security by the younger brothers of Henry Percy, 9th earl of Northumberland', *Historical Research* 65 (1992), 296–314.

19. Noonkester, Myron C. 'Dissolution of the monasteries and the decline of the sheriff', *Sixteenth Century J.* 23 (1992), 677–98.

20. Ralph, Elizabeth (ed.). *Bristol apprentice book, part 3: 1552–1565.* Bristol; Bristol Record Soc. vol. 43; 1992. Pp 160.

21. Sacks, David Harris. *The widening gate: Bristol and the Atlantic economy, 1450–1700.* Berkeley (Ca); California UP; 1992. Pp 464.

22. Smail, John. 'Manufacturer or artisan?: the relationship between economic and cultural change in the early stages of the eighteenth-century industrialization', *J. of Social History* 25 (1991–2), 791–814.

23. Spence, Richard T. 'Mining and smelting in Yorkshire by the Cliffords, earls of Cumberland, in the Tudor and early Stuart period', *Yorkshire Arch. J.* 64 (1992), 157–83.

24. Stephens, W.B. 'English wine-imports, c.1603–1640, with special reference to the Devon ports', Bc218, 141–72.

25. Thirsk, Joan. 'Agrarian problems and the English revolution', Bc106, 169–97.

26. Thirsk, Joan. 'The crown as projector on its own estates, from Elizabeth I to Charles I', Bc122, 297–352.

27. Thomas, David. 'Leases of crown lands in the reign of Elizabeth I', Bc122, 169–90.
28. Thomas, David. 'The Elizabethan crown lands: their purposes and problems', Bc122, 58–87.
29. Ward, Ian. 'Rental policy on the estates of the English peerage 1649–1660', *Agricultural History R.* 40 (1992), 23–37.

(g) *Social History (General)*

1. Allan, John; Barber, James; Higgins, David. 'A seventeenth-century pottery group from the Kitto institute, Plymouth', Bc191, 225–45.
2. Anon. *The hearth tax for Agbrigg and Morley wapentakes, West Riding of Yorkshire: Lady Day, 1672.* Ripon; Ripon Historical Soc.; 1992. 2 vols. Pp 108, 88.
3. Atkinson, C.; Atkinson, J.B. 'Subordinating women: Thomas Bentley's use of biblical women in *The monument of matrones* (1582)', *Church History* 60 (1991), 289–300.
4. Barker, Katherine. 'An Elizabethan map of north-west Dorset: Sherborne, Yetminster and surrounding manors', Bc47, 29–54.
5. Beaver, Dan. '"Sown in dishonour, raised in glory": death, ritual and social organization in northern Gloucestershire, 1590–1690', *Social History* 17 (1992), 389–419.
6. Bettey, J.H. 'Manorial custom and widows' estate', *Archives* 20 (1992), 208–16.
7. Bower, Jacqueline. 'Probate accounts as a source for Kentish early modern economic and social history', *Archaeologia Cantiana* 109 (1992 for 1991), 51–62.
8. Cooper, Sheila M. 'Inter-generational social mobility in late seventeenth- and early eighteenth-century England', *Continuity & Change* 7 (1992), 283–301.
9. Crawford, Patricia. 'The challenges to patriarchalism: how did the revolution affect women?', Bc79, 112–28.
10. Dunbar, John G.; Davies, Katherine. 'Some late seventeenth-century building contracts', Bc83, 269–327.
11. Durston, Christopher G. '"Wild as colts untamed": radicalism in the Newbury area during the early-modern period', Bc116, 35–50.
12. Dyer, Alan. 'The bishops' census of 1563: its significance and accuracy', *Local Population Studies* 49 (1992), 19–37.
13. Emmison, F.G. 'Elizabethan Essex wills', *Family History* 15 (1991), 359–69.
14. Evans, Nesta. 'A scheme for re-pewing the parish church of Chesham, Buckinghamshire, in 1606', *Local Historian* 22 (1992), 203–07.
15. Faber, Sir Richard. 'Murder at Lees Court', *Archaeologia Cantiana* 109 (1992 for 1991), 167–83.
16. Fletcher, J.M.; Upton, C.A. 'Eating and drinking in Renaissance Oxford and Louvain: a comparison of food and drink purchased at Merton college and Busleyden college in the early 16th century', Bc72, 143–58.

17. Friedman, Jerome. 'The battle of the frogs and Fairford's flies: miracles and popular journalism during the English revolution', *Sixteenth Century J.* 23 (1992), 419–42.

18. Gray, Todd; Draisey, J. (ed.). 'Witchcraft in the diocese of Exeter, II: East Worlington (1558), Townstall (1558) and Moretonhampstead (1559)', *Devon & Cornwall Notes & Queries* 36 (1990), 281–87.

19. Gray, Todd; Draisey, J. (ed.). 'Witchcraft in the diocese of Exeter, III: St Thomas by Exeter (1561) and St Marychurch (1565)', *Devon & Cornwall Notes & Queries* 36 (1990), 305–14.

20. Gray, Todd; Draisey, J. (ed.). 'Witchcraft in the diocese of Exeter, IV: Whimple (1565), Chawleigh (1571) and Morwenstow (1575)', *Devon & Cornwall Notes & Queries* 36 (1990), 366–69.

21. Grubb, Farley. 'Fatherless and friendless: factors influencing the flow of English emigrant servants', *J. of Economic History* 52 (1992), 85–108.

22. Gruenfelder, John K. 'Nicholas Murford, Yarmouth salt-producer', *Norfolk Archaeology* 41 (1991), 162–70.

23. Grundy, Isobel. 'Women's history? Writings by English nuns', Bc62, 126–38.

24. Hainsworth, D.R. *Stewards, lords and people: the estate steward and his world in later Stuart England.* Cambridge; Cambridge UP; 1992. Pp xx, 278.

25. Harrison, Jennifer. 'Lord Mayor's Day in the 1590s', *History Today* 42/1 (1992), 37–43.

26. Harvey, Barbara K. 'An early seventeenth-century survey of four Wiltshire manors', *Vernacular Architecture* 23 (1992), 30–33.

27. Hebden, J.R. (ed.). *The hearth tax lists for the North Riding of Yorkshire: 1672, 1673,* parts 1–5. Ripon; Ripon Historical Soc.; 1990–2. 9 vols. Pp 87; 88; 92; 92; 96; 88; 108; 88; 104.

28. Hobby, Elaine. '"Discourse so unsavoury": women's published writings of the 1650s', Bc62, 16–32.

29. Hoyle, Richard W. 'Disafforestation and drainage: the crown as entrepreneur?', Bc122, 353–88.

30. Hoyle, Richard W. 'Some reservations on Dr Ward on the "rental policy of the English peerage, 1649–1660"', *Agricultural History R.* 40 (1992), 156–59.

31. Hughes, Pat. 'Property and prosperity: the relationship of the buildings and fortunes of Worcester, 1500–1660', *Midland History* 17 (1992), 39–58.

32. Hull, F. 'Kentish map-makers of the seventeenth century', *Archaeologia Cantiana* 109 (1992 for 1991), 63–83.

33. Jones, Clyve. 'The London life of a peer in the reign of Anne: a case study from Lord Ossulston's diary', *London J.* 16 (1991), 140–55.

34. King, Walter. 'How high is too high?: disposing of dung in seventeenth-century Prescot', *Sixteenth Century J.* 23 (1992), 443–57.

35. Morrill, John S. 'Introduction' [revolution and Restoration], Bc79, 8–14.

36. Morrill, John S. 'The impact on society' [revolution], Bc79, 91–111.

37. Newman, John. 'Cardinal Wolsey's collegiate foundations', Bc1, 103–15.

38. O'Connell, Sheila. 'Lord Shaftesbury in Naples: 1711–1713', *Walpole Soc.* 54 (1991 for 1988), 149–219.

39. O'Hara, Diana. 'The language of tokens and the making of marriage', *Rural History* 3 (1992), 1–40.

40. Orr, Bridget. 'Whores' rhetoric and the maps of love: constructing the feminine in Restoration erotica', Bc39, 195–216.

41. Peck, Linda Levy. 'Benefits, brokers and beneficiaries: the culture of exchange in seventeenth-century England', Bc74, 109–28.

42. Purkiss, Diane. 'Material girls: the seventeenth-century woman debate', Bc39, 69–101.

43. Rappaport, S.; Monfasani, J.; Musto, R.G. (ed.). 'Reconsidering apprenticeship in sixteenth-century London', *Renaissance society and culture* (New York, 1991), 239–61.

44. Robinson, Ken. 'The book of nature', Bc59, 86–106.

45. Salgado, Gamini. *The Elizabethan underworld*. Stroud; Sutton; 1992. Pp xii, 230.

46. Seaver, Paul S. 'Declining status in an aspiring age: the problem of the gentle apprentice in seventeenth-century London', Bc74, 129–48.

47. Shapin, S. '"A scholar and a gentleman": the problematic identity of the scientific practitioner in early modern England', *History of Science* 29 (1991), 279–327.

48. Sharpe, Pamela. 'Locating the "missing marriages" in Colyton, 1660–1750', *Local Population Studies* 48 (1992), 49–59.

49. Slack, Paul. 'Dearth and social policy in early modern England', *Social History of Medicine* 5 (1992), 1–17.

50. Sommerville, C. John. *The discovery of childhood in Puritan England*. Athens (Ga); Georgia UP; 1992. Pp 208.

51. Stapleton, Barry. 'Marriage, migration and mendicancy in a preindustrial community', Bc116, 51–91.

52. Tadmor, Naomi. 'Dimensions of inequality among siblings in eighteenth-century English novels: the cases of *Clarissa* and *The history of Miss Betsy Thoughtless*', *Continuity & Change* 7 (1992), 303–33.

53. Tobriner, A. 'Old age in Tudor-Stuart broadside ballads', *Folklore* 101 (1991), 149–74.

54. Vince, Alan; Bell, Robert. 'Sixteenth-century pottery from Acton Court, Avon', Bc191, 101–12.

55. Walker, R.F. 'The manor of Manorbier, Pembrokeshire, in the early seventeenth century', *National Library of Wales J.* 27 (1991–2), 131–74.

56. Wilson, Jean. 'The noble imp: the upper-class child in English Renaissance art and literature', *Antiquaries J.* 70 (1990), 360–79.

57. Wiseman, Susan. 'Gender and status in dramatic discourse: Margaret Cavendish, duchess of Newcastle', Bc62, 159–77.

58. Woolf, Daniel R. 'The mental world in Tudor and early Stuart England', *Canadian J. of History* 27 (1992), 341–52.

(h) *Social Structure and Population*

1. Addy, John. *Death, money and the vultures: inheritance and avarice, 1660–1750*. London; Routledge; 1992. Pp 240.

2. Boothman, Lyn. 'On the accuracy of a late sixteenth-century parish register', *Local Population Studies* 49 (1992), 62–67.
3. Coldham, Peter Wilson. *Emigrants in chains: a social history of forced emigration to the Americas, 1607–1776.* Stroud; Sutton; 1992. Pp iv, 188.
4. Coward, Barry. 'The experience of the gentry, 1640–1660', Bc106, 198–223.
5. Craig, J.S. 'The Bury stirs revisited: an analysis of the townsmen', *P. of the Suffolk Inst. for Archaeology & History* 37 (1991), 208–24.
6. Houston, R.A. *The population history of Britain and Ireland, 1500–1750.* London; Macmillan; 1992. Pp 96.
7. Hughes, Ann. 'Coventry and the English revolution', Bc106, 69–99.
8. Kitch, Malcolm. 'Population movement and migration in pre-industrial rural England', Bc54, 62–84.
9. Lindley, Keith. 'London's citizenry in the English revolution', Bc106, 19–45.
10. Nelson, Janet L. 'Mortality crisis in mid-Sussex, 1606–1640', *Local Population Studies* 46 (1991), 39–49.
11. Phillips, C.B. 'Landlord-tenant relationships, 1642–1660', Bc106, 224–50.
12. Pugsley, Steven. 'Landed society and the emergence of the country house in Tudor and early Stuart Devon', Bc218, 96–118.
13. Richardson, R.C. 'Town and countryside in the English revolution', Bc106, 1–18.
14. Riley, D. 'Wealth and social structure in north-western Lancashire in the later seventeenth century: a new use for probate inventories', *T. of the Historic Soc. of Lancashire & Cheshire* 141 (1991), 77–100.
15. Roy, Ian. 'The city of Oxford, 1640–1660', Bc106, 130–68.
16. Scott, David. 'Politics and government in York, 1640–1662', Bc106, 46–68.
17. Sharp, Buchanan. 'Rural discontents and the English revolution', Bc106, 251–72.
18. Sharpe, J.A. *Witchcraft in seventeenth-century Yorkshire: accusations and counter measures.* York; the University (Borthwick paper 81); 1992. Pp 28.
19. Thirsk, Joan. 'English rural communities: structures, regularities, and change in the sixteenth and seventeenth centuries', Bc54, 44–61.
20. Tittler, Robert. 'Seats of honor, seats of power: the symbolism of public seating in the English urban community, c.1560–1620', *Albion* 24 (1992), 205–23.
21. Trubowitz, Rachel. 'Female preachers and male wives: gender and authority in civil war England', Bc203, 112–33.
22. Williams, I.L. 'An Isle of Wight community in the seventeenth century: the evidence from probate inventories', *The Hatcher R.* 4/33 (1992), 10–29.

(i) *Naval and Military*

1. Andrews, Kenneth R. *Ships, money and politics: seafaring and naval*

enterprise in the reign of Charles I. Cambridge; Cambridge UP; 1991. Pp x, 240.

2. Atkin, Malcolm; Laughlin, Wayne. *Gloucester and the civil war: a city under siege.* Stroud; Sutton; 1992. Pp xviii, 206.

3. Carlton, Charles. *Going to the wars: experience of the British civil wars, 1638–1651.* London; Routledge; 1992. Pp 368.

4. Carlton, Charles. 'The face of battle in the English civil wars', Bc69, 226–47.

5. Durston, Christopher G. 'Phoney war—England, summer 1642', *History Today* 42/6 (1992), 13–19.

6. Earle, Peter. *The last fight of the Revenge.* London; Collins & Brown; 1992. Pp 192.

7. Elton, Sir Geoffrey R. 'War and the English in the reign of Henry VIII', Bc150, 1–17.

8. English, Barbara. 'Sir John Hotham and the English civil war', *Archives* 20 (1992), 217–24.

9. Fox, Frank. 'The English naval ship-building programme of 1664', *Mariner's Mirror* 78 (1992), 277–92.

10. Gentles, Ian. *The new model army in England, Ireland and Scotland, 1645–1653.* Oxford; Blackwell; 1992. Pp xii, 584.

11. Harrington, P. *Archaeology of the English civil war.* Princes Risborough; Shire; 1992.

12. Hibbard, Caroline M. 'Episcopal warriors in the British wars of religion', Bc69, 164–92.

13. Hussey, John. 'Marlborough and the loss of Anleux, 1711: accident or design?', *J. of the Soc. for Army Historical Research* 70 (1992), 5–14.

14. Lawrence, Anne. 'Women's work and the English civil war', *History Today* 42/6 (1992), 20–25.

15. Le Fevre, Peter. 'Lord Torrington's trial: a rejoinder', *Mariner's Mirror* 78 (1992), 7–15.

16. Loades, D.M. *Tudor navy: administrative, political and military history.* Aldershot; Scolar; 1992. Pp 304.

17. Longfield-Jones, G.M. 'Buccaneering doctors', *Medical History* 36 (1992), 187–206.

18. Meikle, Maureen M. 'A godly rogue: the career of Sir John Forster, an Elizabethan Border warden', *Northern History* 28 (1992), 126–63.

19. Meikle, Maureen M. 'Northumberland divided: anatomy of a sixteenth-century bloodfeud', *Archaeologia Aeliana* 5th ser. 20 (1992), 79–89.

20. Milford, Anna. *Eye and ear witnesses: English civil war from contemporary sources.* Leigh-on-Sea; Partizan; 1992. [Pp 225].

21. Nolan, John S. 'The muster of 1588', *Albion* 23 (1991), 387–407.

22. Roy, Ian. 'An English country house at war: Littlecote and the Pophams', Bc150, 19–37.

23. Spence, Richard T. (ed.). *Skipton castle in the great civil war, 1642–1645.* Skipton; Skipton Castle; 1991. Pp x, 126.

24. Stewart, Richard W. 'Arms and expeditions: the Ordnance office and the assaults on Cadiz (1625) and the Isle of Rhé (1627)', Bc69, 112–32.

25. Thrush, Andrew. 'Naval finance and the origins and development of ship money', Bc69, 133–62.
26. Vecchioni, Domenico. *La flotta Tudor, 1485–1603: nascità della potenza marinara inglese* [The Tudor fleet, 1485–1603: the birth of English maritime power]. Milan; Edizioni Italiane; 1988. Pp 145.
27. Wernham, Richard B. (ed.). *The expedition of Sir John Norris and Sir Francis Drake to Spain and Portugal, 1589*. Aldershot; Temple Smith (Navy Records Soc. vol. 127); 1988. Pp lxvi, 380.
28. West, John. *Oliver Cromwell and the battle of Gainsborough*. Boston; Kay; 1992. Pp 32.
29. Wheeler, James Scott. 'The logistics of the Cromwellian conquest of Scotland, 1650–1651', *War & Soc.* 10 (1992), 1–18.

(j) *Political Thought and History of Ideas*

1. Achinstein, Sharon. 'The politics of Babel in the English revolution', Bc203, 14–44.
2. Anngoulvent, Anne-Laure. *Hobbes ou la crise de l'état baroque*. Paris; PUF; 1992. Pp 256.
3. Armitage, David. 'The Cromwellian Protectorate and the languages of empire', *Historical J.* 35 (1992), 531–55.
4. Ashcraft, Richard. 'The politics of Locke's *Two treatises of government*', Bc155, 14–49.
5. Baxter, Stephen B. 'William III as Hercules: the political implications of court culture', Bc48, 95–106.
6. Bergeron, David M. 'Francis Bacon's *Henry VII*: commentary on King James I', *Albion* 24 (1992), 17–26.
7. Burgess, Glenn. *The politics of the ancient constitution: introduction to English political thought, 1600–1642*. London; Macmillan; 1992. Pp 304.
8. Burgess, Glenn. 'The divine right of kings reconsidered', *English Historical R.* 107 (1992), 837–61.
9. Burke, Peter. 'The language of orders in early modern Europe', Bc43, 1–12.
10. Burtt, Shelly G. *Virtue transformed: political argument in England, 1688–1740*. Cambridge; Cambridge UP; 1992. Pp x, 182.
11. Butler, John. *Lord Herbert of Chirbury (1582–1648): an intellectual biography*. Lampeter; Mellen; 1990. Pp xx, 573.
12. Cameron, James K. 'The conciliarism of John Mair: a note on *A disputation on the authority of a council*', Bc12, 429–35.
13. Coquillette, Daniel R. *Francis Bacon*. Edinburgh; Edinburgh UP; 1992. Pp 200.
14. d'Andrea, Antonio. 'Aspiring minds: a Machiavellian motif from Marlowe to Milton', Bc74, 211–22.
15. Gelderblom, Arie-Jan. 'The publisher of Hobbes's Dutch *Leviathan*', Bc133, 163–66.
16. Gobetti, Daniela. *Private and public: individuals, households and body politic in Locke and Hutcheson*. London; Routledge; 1992. Pp viii, 211.

17. Gobetti, Daniela. 'Goods of the mind, goods of the body and external goods: sources of conflict and political regulation in seventeenth-century natural law theory', *History of Political Thought* 13 (1992), 32–49.
18. Goldie, Mark. 'John Locke's circle and James II', *Historical J.* 35 (1992), 557–87.
19. Hamowy, Ronald. '*Cato's letters* John Locke, and the republican paradigm', Bc155, 148–72.
20. Hartman, Joan E. 'Restyling the king: Clarendon writes Charles I', Bc203, 45–59.
21. Hinds, Hilary. '"Who may binde where God hath loosed?": responses to sectarian women's writing in the second half of the seventeenth century', Bc68, 205–27.
22. Holstun, James. 'Introduction' [pamphlets], Bc203, 1–13.
23. Holstun, James. 'Rational hunger: Gerrard Winstanley's *Hortus inconclusus*', Bc203, 158–204.
24. Houston, Alan Craig. *Algernon Sidney and the republican heritage in England and America*. Princeton (NJ); Princeton UP; 1991. Pp xiii, 335.
25. Lund, William R. 'Hobbes on opinion, private judgment and civil war', *History of Political Thought* 13 (1992), 51–72.
26. Martinich, A.P. *Two gods of Leviathan: Thomas Hobbes on religion and politics*. Cambridge; Cambridge UP; 1992. Pp xiv, 430.
27. McEntee, Ann Marie. '"The [un]civill-sisterhood of oranges and lemons": female petitioners and demonstrators, 1642–1653', Bc203, 92–111.
28. Nelson, Byron. 'The Ranters and the limits of language', Bc203, 60–75.
29. Newman, Stephen L. 'Locke's *Two treatises* and contemporary thought: freedom, community, and the Liberal tradition', Bc155, 173–208.
30. Parry, Graham. 'Cotton's counsels: the contexts of *Cottoni Posthuma*', *British Library J.* 18 (1992), 29–43.
31. Pasqualucci, Paulo. 'Metaphysical implications in Hobbes's theory of passions', Bc74, 199–210.
32. Peltonen, Markku. 'Politics and science: Francis Bacon and the true greatness of states', *Historical J.* 35 (1992), 279–305.
33. Raylor, Timothy. 'Samuel Hartlib and the commonwealth of bees', Bc204, 91–129; 223–25.
34. Roberts, R.J. 'John Dee and the matter of Britain', *Honourable Soc. of Cymmrodorion T.* (1991), 129–43.
35. Rogers, G.A.J. 'Locke and the latitude-men: ignorance as a ground of toleration', Bc60, 230–52.
36. Rogow, Arnold A. *Thomas Hobbes: un radical au service de la réaction*. Paris; P.U.F.; 1991. Pp 361.
37. Schiavone, Guiseppe. *Winstanley: il profeta della rivoluzione inglese [Winstanley: the prophet of the English revolution]*. Bari; Dedalo; 1991. Pp 294.
38. Schochet, Gordon J. 'The English revolution in the history of political thought', Bc74, 1–20.
39. Scott, Jonathan. 'The English republican imagination', Bc79, 35–54.
40. Shapiro, B. 'Early modern intellectual life: humanism, religion and

science in seventeenth-century England', *History of Science* 29 (1991), 45–71.

41. Shephard, Amanda. 'Henry Howard and the lawful regiment of women', *History of Political Thought* 12 (1991), 589–603.

42. Simmons, A. John. *The Lockean theory of rights*. Princeton (NJ); Princeton UP; 1992. Pp x, 387.

43. Sommerville, Johann P. *Thomas Hobbes: political ideas in historical context*. London; Macmillan; 1992. Pp xiv, 234.

44. Southgate, Beverley C. '"Cauterizing the tumour of Pyrrhonism": Blackloism versus skepticism', *J. of the History of Ideas* 53 (1992), 631–45.

45. Stavely, Keith W.F. 'Roger Williams: Bible politics and Bible art', Bc203, 76–91.

46. Vaughn, Karen Iversen. 'The economic background to Locke's *Two treatises of government*', Bc155, 118–47.

47. Vickers, Brian. 'Francis Bacon and the progress of knowledge', *J. of the History of Ideas* 53 (1992), 495–518.

48. Wiseman, Susan. '"Adam, the father of all flesh": porno-political rhetoric and political theory in and after the English civil war', Bc203, 134–57.

49. Wood, Neal. '*Tabula rasa*, social environmentalism, and the "English paradigm"', *J. of the History of Ideas* 53 (1992), 647–68.

50. Zagorin, Perez. 'Cudworth and Hobbes on Is and Ought', Bc60, 128–48.

51. Zwicker, Steven N. 'Representing the revolution: politics and high culture in 1689', Bc48, 165–83.

(k) *Arts and Cultural History*

1. Anon. 'Bibliography' [Locke], Bc155, 209–30.

2. Ashbee, Andrew (ed.). *Records of English court music, vol. 4: 1603–1625*. Snodland; Ashbee; 1991. Pp 288.

3. Ashbee, Andrew (ed.). *Records of English court music, vol. 5: 1625–1714*. Aldershot; Scolar; 1991. Pp 333.

4. Ashbee, Andrew (ed.). *Records of English court music, vol. 6: 1588–1603*. Aldershot; Scolar; 1992. Pp 288.

5. Austern, Linda Phyllis. '"Art to enchant": musical magic and its practitioners in English Renaissance drama', *J. of the Royal Musical Association* 115 (1990), 191–206.

6. Backhouse, Janet. 'Sir Robert Cotton's record of a royal bookshelf', *British Library J.* 18 (1992), 44–51.

7. Ballaster, Rosalind. 'Manl(e)y forms: sex and the female satirist', Bc39, 217–41.

8. Ballaster, Rosalind. 'Seizing the means of seduction: fiction and feminine identity in Aphra Behn and Delarivier Manley', Bc62, 93–108.

9. Barash, Carol. '"The native liberty ... of the subject": configurations of gender and authority in the works of Mary Chudleigh, Sarah Fyge Egerton, and Mary Astell', Bc62, 55–69.

10. Bearman, Robert. 'The "Stratford fireplace" at Packwood house', *Birmingham & Warwickshire Arch. Soc. T.* 96 (1991 for 1989–90), 83–87.

11. Bono, Barbara J. '"The chief knot of all the discourse": the maternal subtext tying Sidney's *Arcadia* to Shakespeare's *King Lear*', Bc68, 105–27.
12. Bowers, Roger. 'The cultivation and promotion of music in the household and orbit of Thomas Wolsey', Bc1, 178–218.
13. Breight, Curt. 'Realpolitik and Elizabethan ceremony: the earl of Hertford's entertainment of Elizabeth at Elvetham, 1591', *Renaissance Q.* 45 (1992), 20–48.
14. Cain, T.G.S. 'The visual arts and architecture in Britain, 1625–1700', Bc59, 107–50.
15. Cain, T.G.S.; Robinson, Ken. 'Introduction' [English culture, 1625–1700], Bc59, 1–17.
16. Canley, James P. 'The royal library as a source for Sir Robert Cotton's collection: a preliminary list of acquisitions', *British Library J.* 18 (1992), 52–73.
17. Cerasano, S.P.; Wynne-Davies, Marion. '"From myself, my other self I turned": an introduction', Bc68, 1–24.
18. Chalmers, Hero. '"The person I am or what they made me to be": the construction of the feminine subject in the autobiographies of Mary Carleton', Bc39, 164–94.
19. Chambers, Douglas. '"Wild pastorall encounter": John Evelyn, John Beale and the renegotiation of pastoral in the mid-seventeenth century', Bc204, 173–94.
20. Charlton-Jones, Richard. 'Lely to Kneller, 1650–1723', Bc93, 74–127.
21. Clegg, Cyndia Susan. 'Which Holinshed? Holinshed's *Chronicles* at the Huntington library', *Huntington Library Q.* 55 (1992), 559–78.
22. Davies, Gareth Alban. 'The English *Lazarillo di Tormes* (1586) and its translator: David Rowland of Anglesey or Richard Rowland Verstegan?' [part 1], *Honourable Soc. of Cymmrodorion T.* (1991), 99–128.
23. Dow, Helen Jeanette. *Sculptural decoration of the Henry VII chapel, Westminster abbey.* Edinburgh; Pentland; 1992. Pp xii, 118.
24. Elliott, John R. Jr. 'The Folger manuscript of *The triumph of peace* procession', Bc220, 193–215.
25. Ezell, Margaret J. 'Elizabeth Delaval's spiritual heroine: thoughts on redefining manuscript texts by early women writers', Bc220, 216–37.
26. Fletcher, J.M.; Upton, C.A. 'John Drusius of Flanders, Thomas Bodley and the development of Hebrew studies at Merton college, Oxford', Bc72, 111–29.
27. Fowler, Alastair. 'Georgic and pastoral: laws of genre in the seventeenth century', Bc204, 81–88.
28. Glanville, Philippa. 'Cardinal Wolsey and the goldsmiths', Bc1, 131–48.
29. Hackett, Helen. '"Yet tell me some such fiction": Lady Mary Wroth's *Urania* and the "femininity" of romance' [1621], Bc39, 39–68.
30. Hamersley, Lydia. 'The Tenbury and Ellesmere part-books: new findings on manuscript compilation and exchange, and the reception of the Italian madrigal in Elizabethan England', *Music and Letters* 73 (1992), 177–221.

31. Harpham, Edward J. 'Locke's *Two treatises* in perspective', Bc155, 1–13.
32. Harvey, Elizabeth D. *Ventriloqized voices: feminist theory and English Renaissance texts*. London; Routledge; 1992. Pp 208.
33. Hellinga-Querido, Lotte. 'Reading an engraving: William Caxton's dedication to Margaret of York, duchess of Burgundy', Bc133, 1–15.
34. Henderson, Diana E. 'Elizabeth's watchful eye and George Peele's gaze: examining female power beyond the individual', Bc91, 150–69.
35. Hood, Gervase. 'A Netherlandic triumphal arch for James I', Bc133, 67–82.
36. Hopkins, John T. '"Such a twin likeness there was in the pair": an investigation into the painting of the Cholmondeley sisters', *T. of the Historic Soc. of Lancashire & Cheshire* 141 (1991), 1–37.
37. Howarth, David. 'Sir Robert Cotton and the commemoration of famous men', *British Library J.* 18 (1992), 1–28.
38. Howe, Elizabeth. *The first English actresses: women and drama, 1660–1700*. Cambridge; Cambridge UP; 1992. Pp 226.
39. Hulse, Lynn. 'The musical patronage of Robert Cecil, first earl of Salisbury (1563–1612)', *J. of the Royal Musical Association* 116 (1991), 24–40.
40. Hunter, Michael. 'The making of Christopher Wren', *London J.* 16 (1991), 101–16.
41. Hutton, Sarah; Nicholson, Margaria Hope (ed.). *The Conway letters: the correspondence of Anne, viscountess Conway, Henry More, and their friends, 1642–1684*. Oxford; Oxford UP; new edn. 1992. Pp 662.
42. Killey, Kate. 'True state within: women's elegy, 1640–1740', Bc62, 72–92.
43. King, T.J. *Casting Shakespeare's plays: London actors and their roles, 1590–1642*. Cambridge; Cambridge UP; 1992. Pp xvi, 284.
44. Klein, Benjamin. '"Between the bums and bellies of the multitude": civic pageantry and the problem of audience in late Stuart London', *London J.* 17 (1992), 18–26.
45. Knowles, James. 'Marston, Skipwith and *The entertainment at Ashby*', Bc220, 137–92.
46. Krontiris, Tina. *Oppositional voices: women as writers and translators in the English Renaissance*. London; Routledge; 1992. Pp 182.
47. Kunze, Bonnelyn Young; Brautigam, Dwight D. 'Introduction' [court, country and culture], Bc74, xi-xvii.
48. Leslie, Michael. 'The spiritual husbandry of John Beale', Bc204, 151–72; 226–31.
49. Leslie, Michael; Raylor, Timothy. 'Introduction' [culture and cultivation], Bc204, 1–12.
50. Levine, Joseph M. *The battle of the books: history and literature in the Augustan age*. Ithaca (NY); Cornell UP; 1991. Pp xiv, 428.
51. Levine, Joseph M. 'Sir Walter Ralegh and the ancient wisdom', Bc74, 89–108.
52. Lilley, Kate. 'Blazing worlds: seventeenth-century women's utopian writing', Bc39, 102–33.
53. Lindley, Phillip G. 'Playing check-mate with royal majesty? Wolsey's patronage of Italian Renaissance sculpture', Bc1, 261–85.

54. Lindley, Phillip G. (ed.). *Gainsborough Old Hall*. Lincoln; Soc. for Lincolnshire History & Archaeology (Occasional papers 8); 1991. Pp 88.
55. Louis, Cameron. 'Some alternative sources of Tudor poetry texts', Bc220, 246–257.
56. Low, Anthony. 'Agricultural reform and the love poems of Thomas Carew; with an instance from Lovelace', Bc204, 63–80.
57. Maguire, Laurie E. '"Household Kates": chez Petruchio, Percy and Plantagenet', Bc68, 129–65.
58. Maguire, Nancy Klein. *Regicide and restoration: English tragicomedy, 1660–1671*. Cambridge; Cambridge UP; 1992. Pp 280.
59. Marx, Steven. 'Shakespeare's pacifism', *Renaissance Q.* 45 (1992), 49–95.
60. McCrea, Adriana. 'Reason's muse: Andrew Marvell, R. Fletcher, and the politics of poetry in the engagement debate', *Albion* 23 (1991), 655–80.
61. Medoff, Jeslyn. 'The daughters of Behn and the problem of reputation', Bc62, 33–54.
62. Milsom, John. 'English-texted chant before Merbecke', *Plainsong & Medieval Music* 1 (1992), 77–92.
63. Morehen, John. 'The English anthem text, 1549–1660', *J. of the Royal Musical Association* 117 (1992), 62–85.
64. Page, R.I. 'Audits and replacements in the Parker Library, 1590–1650', *T. of the Cambridge Bibliographical Soc.* 10 (1991), 17–39.
65. Parr, Anthony. 'Thomas Coryat and the discovery of Europe', *Huntington Library Q.* 55 (1992), 579–602.
66. Pinto, David. 'The music of the Hattons', *Royal Musical Association Research Chronicle* 23 (1990), 79–108.
67. Potter, Lois. 'Politics and popular culture: the theatrical response to the revolution', Bc48, 184–97.
68. Ravenhill, William L.D.; Rowe, Margery. 'A decorated screen map of Exeter based on John Hooker's map', Bc218, 1–12.
69. Renfrew, Jane M.; Robbins, Michael. 'Tobias Rustat and his monument in Jesus college chapel, Cambridge', *Antiquaries J.* 70 (1990), 416–23.
70. Resnick, David. 'Rationality and the *Two treatises*', Bc155, 82–117.
71. Scattergood, John. 'The London manuscripts of John Skelton's poems', Bc58, 171–82.
72. Sharrock, Catherine. 'De-ciphering women and de-scribing authority: the writings of Mary Astell', Bc62, 109–24.
73. Shaw, Watkins. *The succession of organists of the chapel royal and the cathedrals of England and Wales*. Oxford; Oxford UP (Oxford Studies in British Church Music); 1991. Pp 476.
74. Skretkowicz, Victor. 'Greville's *Life of Sidney*: the Hertford manuscript', Bc220, 102–36.
75. Thompson, Ruby Reid. 'The "Tregian" manuscripts: a study of their compilation', *British Library J.* 18 (1992), 202–04.
76. Thurley, Simon. 'The domestic building works of Cardinal Wolsey', Bc1, 76–102.
77. Tite, Colin G. '"Lost or stolen or strayed": a survey of manuscripts formerly in the Cotton library', *British Library J.* 18 (1992), 107–47.

78. Tomlinson, Sophie. '"My brain the stage": Margaret Cavendish and the fantasy of female performance', Bc39, 134–63.
79. Trapp, Joseph Burney. *Erasmus, Colet and More: the early Tudor humanists and their books.* London; British Library; 1991. Pp viii, 144.
80. Tylden-Wright, David. *John Aubrey: a life.* London; HarperCollins; 1991. Pp xv, 270.
81. Upton, C.A. 'Speaking sorrow: the English university anthologies of 1587 on the death of Philip Sidney in the Low Countries', Bc72, 130–42.
82. Usher, Brett. 'The silent community: early puritans and patronage of the arts', Bc194, 287–302.
83. van Heertum, Cis. 'Willem Christiaens van der Boxe's translation of *The parlament of women* (1640)', Bc133, 149–62.
84. van Houts, Elisabeth M.C. 'Camden, Cotton and the chronicles of the Norman conquest of England', *British Library J.* 18 (1992), 148–62.
85. Wainwright, Jonathan P. 'George Jeffrey's copies of Italian music', *Royal Musical Association Research Chronicle* 23 (1990), 109–24.
86. Wallis, Helen. 'Intercourse with the peaceful muses', Bc133, 31–54.
87. Wanko, C. 'Mary Morein *fl. 1701*: Drury Lane actress and fair performer', *Theatre Survey* 32 (1991), 22–30.
88. Wayment, Hilary. 'Wolsey and stained glass', Bc1, 116–30.
89. Wenley, Robert. 'Robert Paston and *The Yarmouth collection*', *Norfolk Archaeology* 41 (1991), 113–44.
90. Wesselius, J.W. 'Johannes Drusius the younger's last journey to England and his Hebrew letter-book, *Lias* 16 (1989), 159–76.
91. Wilcox, Helen. 'Private writing and public function: autobiographical texts by Renaissance Englishwomen', Bc68, 47–62.
92. Womack, Peter. 'Imagining communities: theatres and the English nation in the sixteenth century', Bc94, 91–146.
93. Wynne-Davies, Marion. 'The queen's masque: Renaissance women and the seventeenth-century court masque', Bc68, 79–104.
94. Ziegler, Georgianna. 'Penelope and the politics of woman's place in the Renaissance', Bc68, 25–46.

(l) *Science and Technology*

1. Bates, Don G. 'Harvey's account of his "discovery"', *Medical History* 36 (1992), 361–78.
2. Beale, Georgia R. 'Early French members of the Linnean Society of London, 1788–1802: from the Estates General to Thermidor', *P. of the Annual Meeting of the Western Soc. for French History* 18 (1991), 272–82.
3. Bechler, Zev. 'Newton's ontology of the force of inertia', Bc87, 287–304.
4. Brackenridge, J. Bruce. 'The critical role of curvature in Newton's developing dynamics', Bc87, 231–60.
5. Burl, Aubrey. 'Review article: two early plans of Avebury', *Wiltshire Arch. & Natural History Magazine* 85 (1992), 163–67.

6. Cook, A. 'Edmund Halley and Newton's *Principia*', *Notes & Records of the Royal Soc. of London* 45 (1991), 129–38.

7. Dobbs, Betty Jo Teeter. *The Janus faces of genius: the role of alchemy in Newton's thought*. Cambridge; Cambridge UP; 1992. Pp 380.

8. Fowler, D.H. 'Newton, Cotes and $\sqrt{}$ $\sqrt{}$ 2: a footnote to Newton's theory of the resistance of fluids', Bc87, 355–68.

9. Gabbey, Alan. 'Newton's *Mathematical Principles of natural philosophy*: a treatise on "mechanics"?', Bc87, 305–22.

10. Hald, Anders. *A history of probability and statistics and their application before 1750*. New York; Wiley; 1990. Pp 586.

11. Hall, A. Rupert. *Isaac Newton: adventurer in thought*. Oxford; Blackwell; 1992. Pp 448.

12. Hall, A. Rupert. 'Newton and the absolutes: sources', Bc87, 261–85.

13. Hughes, R. Elwyn. 'A recusant contribution to medicine in Wales: Gwilym Pue, OSB and his *De Sceletyrrbe ... or A Traetice of the Scorbut*, 1675', *J. of Welsh Ecclesiastical History* 9 (1992), 20–36.

14. Hunt, John Dixon. '"Gard'ning can speak proper *English*"', Bc204, 195–224.

15. Kelly, John Thomas. *Practical astronomy during the seventeenth century: a study of almanac-makers in America and England*. London; Garland; 1991. Pp vii, 319.

16. Marmoy, C.F.A. 'The "pest house", 1681–1717: predecessor of the French hospital', *Huguenot Soc. P.* 25 (1992), 385–99.

17. McCray Beier, Lucinda. 'Seventeenth-century English surgery: the casebook of Joseph Binns', Bc24, 48–84.

18. McKitterick, David. *A history of Cambridge university press, vol. 1: printing and the book trade in Cambridge, 1534–1698*. Cambridge; Cambridge UP; 1992. Pp xxiv, 500.

19. Mendelsohn, J. Andrew. 'Alchemy and politics in England, 1649–1665', *Past & Present* 135 (1992), 30–78.

20. Miller, Leta E. 'John Birchensha and the early Royal Society: grand scales and scientific composition', *J. of the Royal Musical Association* 115 (1990), 63–79.

21. Morris, G.C.R. 'On the identity of Jacques de Moulin, FRS, 1667', *Notes & Records of the Royal Soc. of London* 45 (1991), 1–10.

22. Osler, Margaret J. 'The intellectual sources of Robert Boyle's philosophy of nature: Gassendi's voluntarism, and Boyle's physicotheological project', Bc60, 178–98.

23. Parry, Graham. 'John Evelyn as hortulan saint', Bc204, 130–50.

24. Snider, A. 'Bacon, legitimation, and the "origin" of Restoration science', *Eighteenth Century: Theory & Interpretation* 32 (1991), 119–38.

25. Thirsk, Joan. 'Making a fresh start: sixteenth-century agriculture and the classical inspiration', Bc204, 15–34.

26. Thulesius, Olav. *Nicholas Culpeper: English physician and astrologer*. London; Macmillan; 1992. Pp 190.

27. Whiteside, D.T. 'The prehistory of the *Principia* from 1664 to 1686', *Notes & Records of the Royal Soc. of London* 45 (1991), 11–62.

G. BRITAIN 1714–1815

See also Aa11,54,128,161,b106; Bb26,86,132,d1–2,4–7,18; Fd9,21,e6,15, 50,80,f22,g8,52,h1,3,6,l10; He14,73,f43,99,g46,75,83,88,h39,i7,k52

(a) *General*

1. Beckett, John V. 'Nottinghamshire in the 1790s', *T. of the Thoroton Soc.* 94 (1990), 47–61.
2. Black, Jeremy. 'Britain and the continent, 1688–1815: convergence or divergence?', *British J. for Eighteenth-Century Studies* 15 (1992), 145–49.
3. Black, Jeremy. 'Non-Walpolean manuscripts in the Lewis Walpole library', *Yale University Library Gazette* 67/1–2 (1992), 58–67.
4. Black, Jeremy. 'The eighteenth-century British press', *The encyclopedia of the British press* ed. Dennis Griffiths (London; 1992), 13–23.
5. Brown, Keith M. 'Imagining Scotland' [review article], *J. of British Studies* 31 (1992), 415–25.
6. Calder, Angus. 'The Enlightenment', Bc6, 31–50.
7. Clark, Jonathan C.D. 'Reconceptualizing eighteenth-century England', *British J. for Eighteenth-Century Studies* 15 (1992), 135–39.
8. Colley, Linda. *Britons: forging the nation, 1707–1837*. New Haven (Ct); Yale UP; 1992. Pp x, 430.
9. Colman, Pamela. *The baker's diary: life in Georgian England from the book of George Sloper, a Wiltshire baker, 1753–1810*. Trowbridge; Wiltshire County Council Library & Museum Service; 1991. Pp 96.
10. Di Sciullo, Franco M. 'La povertà nella cultura politica inglese fra Burke e Malthus' [Poverty in English political thought from Burke to Malthus], *Pensiero Politico* 23 (1990), 407–29.
11. Fissell, Mary E. 'Charity universal? Institutions and moral reform in eighteenth-century Bristol', Bc82, 121–44.
12. Gilmour, Ian. *Riot, rising and revolution: governance and violence in eighteenth century England*. London; Hutchinson; 1992. Pp 512.
13. Hallaway, H.R. *Margery Jackson, 1722–1812: the life and times of a Carlisle miser*. Carlisle; Halstead; 1991. Pp viii, 207.
14. Hitchcock, Tim. 'Paupers and preachers: the SPCK and the parochial work-house movement', Bc82, 145–66.
15. Innes of Edingight, Sir Malcolm. 'Ceremonial in Edinburgh: the heralds and the Jacobite risings', *Book of the Old Edinburgh Club* ns 1 (1991), 1–6.
16. Lenman, Bruce P. 'Some recent Jacobite studies' [review article], *Scottish Historical R.* 70 (1991), 66–74.
17. Morgan, Kenneth (ed.). *An American Quaker in the British Isles: the travel journals of Jabez Maud Fisher, 1775–1779*. Oxford; Oxford UP for the British Academy (Records of Social and Economic History, ns 16); 1991. Pp xii, 356.
18. Olson, Alison G. *Making the empire work: London and American interest groups, 1690–1790*. Cambridge (Ma); Harvard UP; 1992.

19. Olson, Alison G. 'The eighteenth century empire: the London dissenters' lobbies and the American colonies', *J. of American Studies* 26 (1992), 41–58.

20. Ousby, Ian (ed.). *James Plumptre's Britain: the journals of a tourist in the 1790s.* London; Hutchinson; 1992. Pp 240.

21. Porter, Roy. 'Georgian Britain: an ancien régime', *British J. for Eighteenth-Century Studies* 15 (1992), 141–44.

22. Poussou, J.-P. 'De l'influence de la 'Glorieuse Révolution' de 1688 sur l'interprétation anglaise de la Revolution Française', *Revue d'Histoire Diplomatique* 106 (1992), 47–61.

23. Price, Jacob M. 'Who cared about the colonies? The impact of the thirteen colonies on British society and politics, *c.*1714–1775', Bc2, 395–436.

24. Ravenhill, William L.D. 'The south-west in the eighteenth-century re-mapping of England', Bc47, 1–27.

25. Rawley, James A. 'Richard Harris, slave trader spokesman', *Albion* 23 (1991), 439–58.

26. Richards, Eric. 'Scotland and the uses of the Atlantic empire', Bc2, 67–114.

27. Rule, John C. *Albion's people: English society, 1714–1815.* London; Longman; 1992. Pp 269.

28. Rule, John C. 'Labour consciousness and industrial conflict in eighteenth-century Exeter', Bc116, 92–109.

29. Speck, William A. 'The eighteenth century: England's ancien régime', *British J. for Eighteenth-Century Studies* 15 (1992), 131–33.

30. Stevenson, John. *Popular disturbances in England, 1700–1832.* London; Longman; 2nd edn. 1992. Pp 368.

31. Vickery, Amanda. 'The neglected century: writing the history of eighteenth-century women' [review article], *Gender & History* 3 (1991), 211–19.

32. Withers, Charles W.J. 'The historical creation of the Scottish Highlands', Bc6, 143–56.

(b) *Politics*

1. Adman, Peter; Baskerville, Stephen W.; Beedham, Katherine F. 'Computer-assisted record linkage: or how best to optimize links without generating errors', *History & Computing* 4 (1992), 2–15.

2. Black, Jeremy. 'Parliament and foreign policy, 1739–1763', *Parliaments, Estates & Representation* 12 (1992), 121–42.

3. Black, Jeremy. 'Politics and the press in the eighteenth century: the proposal to launch a pro-government Scottish newspaper in 1762', *Factotum* 30 (1989), 21–23.

4. Blakemore, Steven. 'Burke and the revolution: bicentennial reflections', Bc132, 144–67.

5. Bullion, John L. 'The origins and significance of gossip about Princess Augusta and lord Bute, 1755–1756', *Studies in Eighteenth-Century Culture* 21 (1991), 245–65.

6. Eastwood, David. 'Robert Southey and the meanings of patriotism', *J. of British Studies* 31 (1992), 265–87.

7. Fedorak, Charles John. 'Catholic Emancipation and the resignation of William Pitt in 1801', *Albion* 24 (1992), 49–64.

8. Fitzpatrick, Martin. 'Richard Price and the London Revolution Society', *Enlightenment & Dissent* 10 (1991), 35–50.

9. Flavell, Julie M. 'American patriots in London and the quest for talks, 1773–1775', *J. of Imperial & Commonwealth History* 20 (1992), 335–69.

10. Flavell, Julie M. 'Lord North's conciliatory proposal and the patriots in London', *English Historical R.* 107 (1992), 302–22.

11. Graham, Jenny. 'Revolutionary philosopher: the political ideas of Joseph Priestley (1733–1804)', *Enlightenment & Dissent* 8 (1989), 43–68; 9 (1990), 14–46.

12. Harding, Richard. 'Lord Cathcart, the earl of Stair and the Scottish opposition to Sir Robert Walpole, 1732–1735', *Parliamentary History* 11 (1992), 192–217.

13. Harris, Robert. 'A Leicester House political diary', Bc178, 373–411.

14. Innes, Joanna. 'Politics, property and the middle class' [review article], *Parliamentary History* 11 (1992), 286–92.

15. Jones, Clyve. 'Jacobitism and the historian: the case of William, 1st earl Cowper', *Albion* 23 (1991), 681–96.

16. Jones, Clyve; Harris, Frances. '"A question ... carried by bishops, pensioners, place-men, idiots": Sarah, duchess of Marlborough and the lords' division over the Spanish convention, 1 March 1719', *Parliamentary History* 11 (1992), 254–77.

17. Knox, Thomas. '"Peace for ages to come": the Newcastle elections of 1780 and 1784', *Durham University J.* 84 (1992), 1–19.

18. Lowe, William C. 'George III, peerage creations and politics, 1760–1784', *Historical J.* 35 (1992), 587–609.

19. McMahon, Marie P. *The radical whigs, John Trenchard and Thomas Gordon: libertarian loyalists to the new house of Hanover.* Lanham (Md); UP of America; 1990.

20. Mitchell, L.G. *Charles James Fox.* Oxford; Oxford UP; 1992. Pp xii, 338.

21. Murdoch, Alexander J. 'Politics and the people in the burgh of Dumfries, 1758–1760', *Scottish Historical R.* 70 (1991), 151–71.

22. O'Gorman, Frank. 'An age of progress and reform' [review article], *Historical J.* 35 (1992), 689–99.

23. O'Gorman, Frank. 'Campaign rituals and ceremonies: the social meaning of elections in England, 1780–1860', *Past & Present* 135 (1992), 79–115.

24. Oldfield, J.R. 'The London Committee and mobilization of public opinion against the slave trade', *Historical J.* 35 (1992), 331–43.

25. Philp, Mark. 'Burke and Paine: texts in context', *Enlightenment & Dissent* 9 (1990), 93–105.

26. Schweizer, Karl W.; Bullion, John L. 'The vote of credit controversy, 1762', *British J. for Eighteenth-Century Studies* 15 (1992), 175–88.

27. Schweizer, Karl W.; Lawson, Philip. 'A political diary by Charles Jenkinson, 13 May-29 June 1765', *Historical Research* 65 (1992), 349–58.

28. Speck, William A. 'Northumberland elections in the eighteenth century', *Northern History* 28 (1992), 164–77.
29. Stevenson, Janet H. 'Arnold Nesbitt and the borough of Winchelsea', *Sussex Arch. Collections* 129 (1991), 183–93.
30. Taylor, Stephen. '"The fac totum in ecclesiastic affairs?" The duke of Newcastle and the crown's ecclesiastical patronage', *Albion* 24 (1992), 409–33.
31. Thomas, David Oswald (ed.). *Political writings of Richard Price.* Cambridge; Cambridge UP; 1991. Pp xxxiv, 199.
32. Thomas, Peter D.G. 'The rise of Plas Newydd: Sir Nicholas Bayly and county elections in Anglesey, 1734–1784', *Welsh History R.* 16 (1992–3), 160–76.
33. Wahrman, Dror. 'Virtual representation: parliamentary reporting and language of class in the 1790s', *Past & Present* 136 (1992), 83–113.
34. Wood, Marcus. 'Thomas Spence and modes of subversion', *Enlightenment & Dissent* 10 (1991), 51–77.
35. Woodland, Patrick. 'The house of lords, the city of London and political controversy in the mid-1760s: the opposition to the cider excise further considered', *Parliamentary History* 11 (1992), 57–87.
36. Zebrowski, Martha K. 'The corruption of politics and the dignity of human nature: the critical and constructive radicalism of James Burgh', *Enlightenment & Dissent* 10 (1991), 78–103.

(c) *Constitution, Administration and Law*

1. Baines, Paul. 'Curll at the Old Bailey: "Curll or Cull?"', *Factotum* 30 (1989), 6–10.
2. Barrell, John. 'Imaginary treason, imaginary law: the state trials of 1794', Bc89, 119–43; 230–32.
3. Cairns, John W. 'The influence of Smith's jurisprudence on legal education in Scotland', Bc29, 168–89.
4. Davison, Lee. 'Experiments in the social regulation of industry: gin legislation, 1729–1751', Bc82, 25–48.
5. Dowden, M.J. 'A disputed inheritance: the Tredegar estates in the eighteenth century', *Welsh History R.* 16 (1992–3), 36–46.
6. Gibson, A. 'Territorial continuity and the administrative division of Lochtayside, 1769', *Scottish Geographical Magazine* 106/3 (1990), 174–85.
7. King, Peter. 'Legal change, customary right, and social conflict in late eighteenth-century England: the origins of the great gleanings case of 1788', *Law & History R.* 10 (1992), 1–31.
8. Panton, F.H. 'The finances and government of the city and county of Canterbury in the eighteenth and early nineteenth centuries', *Archaeologia Cantiana* 109 (1992 for 1991), 191–246.
9. Phillipson, Nicholas T. (ed.). *The Scottish Whigs and the reform of the Court of Session, 1785–1830.* Edinburgh; Stair Soc. vol. 37; 1990. Pp ix, 216.

10. Pleuger, Gilbert. *John Howard and the prisons: an essay to mark the bicentenary of his death, January 20, 1790*. Bedford; Sempringham; 1990. Pp 66.
11. Rogers, Nicholas. 'Confronting the crime wave: the debate over social reform and regulation, 1749–1753', Bc82, 77–98.
12. Rogers, Nicholas. 'Policing the poor in eighteenth-century London: the vagrancy laws and their administration', *Histoire Sociale/Social History* 24 (1991), 127–47.
13. Stoker, David A. 'The tailor of Diss: sodomy and murder in a Norfolk market town', *Factotum* 31 (1990), 18–21.
14. Viner, David; Powell, Christopher. 'Lock-ups at Cirencester and Bisley', *T. of the Bristol & Gloucestershire Arch. Soc.* 109 (1992 for 1991), 207–17.

(d) *External Affairs*

1. Andersen, Henning Soby. 'Denmark between the wars with Britain, 1801–1807', *Scandinavian J. of History* 14 (1989), 231–38.
2. Black, Jeremy. *The British abroad: the grand tour in the eighteenth century*. Stroud; Sutton; new edn. 1992. Pp 355.
3. Black, Jeremy. 'Anglo-French relations, 1763–1775', *Francia* 18/2 (1992 for 1991), 99–114.
4. Black, Jeremy. 'Anglo-Wittelsbach relations, 1730–1742', *Zeitschrift für Bayerische Landesgeschichte* 55 (1992), 307–45.
5. Black, Jeremy. 'Archival sources for the grand tour', *Archives* 20 (1992), 86–98.
6. Black, Jeremy. 'British policy towards Austria, 1780–1793', *Mitteilungen des Österreichischen Staatsarchivs* 42 (1992), 188–228.
7. Black, Jeremy. 'British views of Russia in the early eighteenth century', *Durham University J.* 84 (1992), 29–33.
8. Black, Jeremy. 'Dr Johnson, eighteenth-century pamphleteering and the Tory view on foreign policy', *Factotum* 32 (1990), 15–21.
9. Black, Jeremy. 'Genoa in 1731', *Italian Q.* 30 (1989), 39–53.
10. Bowers, Brian; Symons, Lenore (ed.). *'Curiosity perfectly satisfyed': Faraday's travels in Europe, 1813–1815*. London; Peregrinus; 1992. Pp xvi, 168.
11. Bulley, Anne. *Free mariner: John Adolphus Pope in the East Indies, 1786–1821*. London; BACSA; 1992. Pp 250.
12. Bullion, John L. 'British ministers and American resistance to the Stamp Act, October-December 1765', *William & Mary Q.* 3rd ser. 49 (1992), 89–107.
13. Bushnell, O.A. 'Aftermath: Britons' responses to news of the death of Captain James Cook', *Hawaiian J. of History* 25 (1991), 1–20.
14. Corbett, Julian S. *England in the Seven Years' War: vol. 1, 1756–1759*. London; Greenhill; new edn. 1992. Pp 448.
15. Corbett, Julian S. *England in the Seven Years' War: vol. 2, 1759–1763*. London; Greenhill; new edn. 1992. Pp 416.

16. Corbett, Margery. 'John Michael Wright, *An Account of His Excellence Roger Earl of Castlemaine's Embassy, from His Sacred Majesty James the IId. King of England, Scotland, France and Ireland & to His Holiness Innocent IX', Antiquaries J.* 70 (1990), 117–20.

17. Das, Sudipta. 'British reactions to the French bugbear in India, 1763–1783', *European History Q.* 22 (1992), 39–65.

18. Duffy, Michael. 'Realism and tradition in eighteenth-century British foreign policy' [review article], *Historical J.* 35 (1992), 227–32.

19. Gough, Barry M. *British mercantile interests in the making of the peace of Paris, 1763: trade, war and empire.* Lampeter; Mellen; 1992. Pp 148.

20. Gregory, Desmond. *Minorca, the illusory prize: a history of the British occupations of Minorca between 1708 and 1802.* Rutherford (Ca); Farleigh Dickinson UP; 1991. Pp 295.

21. Grijp, P. van der. 'Giftenverkeer, ruilhandel en wapengekletter: eerste contacten tussen Europa en Polynesie' [Gift exchange, barter and the clash of arms: the first contacts between Europe and Polynesia], *Tijdschrift voor Zeegeschiedenis* 10 (1991), 101–15.

22. Gruner, Wolf D. '"French liberty and British slavery": das britische Frankreichbild zwischen Revolution und Rekonstruktion, 1789–1815' [British perceptions of France between Revolution and Restoration, 1789–1815], *Die Französische Revolution und Europa von 1789–1799* (Saarbrucken; 1989), 455–79.

23. Guerrero, Ana Clara. *Viajeros británicos en la España del siglo XVIII* [British travellers in Spain in the eighteenth century]. Madrid; Aguilar; 1990. Pp 487.

24. Lawson, Philip. *The imperial challenge: Quebec and Britain in the age of the American revolution.* Kingston (Ont); McGill-Queen's UP; 1990. Pp x, 192.

25. Lucas, Colin. 'Great Britain and the union of Norway and Sweden', *Scandinavian J. of History* 15 (1990), 269–78.

26. Mavor, Elizabeth (ed.). *The grand tours of Katherine Wilmot: France, 1801–1803, and Russia, 1805–1807.* London; Weidenfeld & Nicolson; 1992. Pp 208.

27. Rice, Geoffrey W. 'Archival sources for the life and career of the fourth earl of Rochford (1717–81), British diplomat and statesman', *Archives* 20 (1992), 254–68.

28. Schofield, Philip. 'British politicians and French arms: the ideological war of 1793–1795', *History* 77 (1992), 183–201.

29. Scott, H.M. 'The second "Hundred Years War", 1689–1815' [review article], *Historical J.* 35 (1992), 443–69.

30. Singer, Aubrey. *The lion and the dragon: the story of the first British embassy to the court of the emperor Quian Long in Peking, 1792–1794.* London; Barrie and Jenkins; 1992. Pp 192.

31. Stevens, Paul L. 'Wabasha visits Governor Carleton, 1776: new light on a legendary episode of Dakota-British diplomacy on the Great Lakes frontier', *Michigan Historical R.* 16 (1990), 21–48.

32. Thomas, Peter D.G. *Revolution in America: Britain and the colonies, 1763–1776.* Cardiff; Wales UP; 1992. Pp x, 101.

(e) *Religion*

1. Andrew, Donna T. 'On reading charity sermons: eighteenth-century Anglican solicitation and exhortation', *J. of Ecclesiastical History* 43 (1992), 581–91.
2. Barnard, Leslie W. 'Thomas Secker and the English parliament', *Parliaments, Estates & Representation* 12 (1992), 47–57.
3. Barrie-Curien, Viviane. *Clergé et pastorale en Angleterre au XVIIIè siècle: le diocèse de Londres*. Paris; Edition de CNRS; 1992. Pp 442.
4. Baumber, Michael. 'William Grimshaw, Patrick Brontë and the Evangelical revival', *History Today* 42/11 (1992), 25–31.
5. Biggs, Barry J. 'Saints of the soil', *P. of the Wesley Historical Soc.* 48 (1991–92), 177–90.
6. Black, Jeremy. 'An attempt to create a general European Protestant Fund in 1725', *Enlightenment & Dissent* 10 (1991), 110–12.
7. Boynton, Lindsay. 'Gillows' furnishings for Catholic chapels, 1750–1800', *Studies in Church History* 28 (1992), 363–79.
8. Bradley, James E. *Religion, revolution and English radicalism: nonconformity in eighteenth-century politics and society*. Cambridge; Cambridge UP; 1990. Pp 473.
9. Cantor, Geoffrey. 'Dissent and radicalism?: the example of the Sandemanians', *Enlightenment & Dissent* 10 (1991), 3–20.
10. Chilcote, Paul Wesley. *John Wesley and the women preachers of early methodism*. Metuchen (NJ); Scarecrow; 1992. Pp xii, 375
11. Christodoulou, Joan. 'The Glasgow Universalist church and Scottish radicalism from the French Revolution to Chartism: a theology of liberation', *J. of Ecclesiastical History* 43 (1992), 608–23.
12. Cohen, Sheldon S. 'The Mill prisoners and the Englishman who continued "in the light"', *Enlightenment & Dissent* 10 (1991), 21–34.
13. Crawford, Michael J. *Seasons of grace: colonial New England's revival tradition in its British context*. New York; Oxford UP; 1991. Pp 365.
14. Ditchfield, G.M. 'Priestley riots in historical perspective', *T. of the Unitarian Historical Soc.* 20 (1991–2), 3–16.
15. Ditchfield, G.M. 'Religion, "Enlightenment" and progress in eighteenth-century England' [review article], *Historical J.* 35 (1992), 681–87.
16. Ditchfield, G.M. 'Two unpublished letters of Theophilus Lindsey', *T. of the Unitarian Historical Soc.* 20 (1991–2), 137–42.
17. Ditchfield, G.M.; Keith-Lucas, Bryan (ed.). *A Kentish parson: selections from the private papers of the Rev. Joseph Price, vicar of Brabourne, 1767–1786*. Maidstone; Kent County Council, Arts and Libraries; 1991. Pp x, 203.
18. Dybikowksi, James; Fitzpatrick, Martin. 'David Williams, John Jebb and liturgical reform', *Enlightenment & Dissent* 9 (1990), 106–13.
19. English, John C. 'John Wesley's studies as an undergraduate', *P. of the Wesley Historical Soc.* 47 (1989–90), 29–37.
20. Fitzpatrick, Martin. 'Joseph Priestley, politics and ancient prophecy', *Enlightenment & Dissent* 10 (1991), 104–09.
21. Green, Paul G. 'Charity, morality, and social control: clerical attitudes in

the diocese of Chester, 1715–1795', *T. of the Historic Soc. of Lancashire & Cheshire* 141 (1991), 207–33.

22. Hannah, Gavin (ed.). *Diary of an Oxfordshire rector: James Newton of Nuneham Courtenay, 1736–1786.* Stroud; Sutton; 1992. Pp 192.

23. Hart, Elizabeth. 'Susanna Wesley and her editors', *P. of the Wesley Historical Soc.* 48 (1991–92), 202–09.

24. Jacob, James R.; Jacob, Margaret C. 'The saints embalmed. Scientists, latitudinarians, and society: a review essay', *Albion* 24 (1992), 435–42.

25. Jacobson, David L. 'The king's four churches: the established churches of America, England, Ireland, and Scotland in the early years of George III', *Anglican and Episcopal History* 59 (1990), 181–201.

26. Jenkins, A.P. (ed.). *The correspondence of Thomas Secker, bishop of Oxford, 1737–1758.* Oxford; Oxfordshire Record Soc. vol. 57; 1991. Pp xxxv, 336.

27. Lambert, Frank. '"Pedlar in divinity": George Whitefield and the great awakening, 1737–1745', *J. of American History* 77 (1990), 812–37.

28. Loft, L. 'Quakers, Brissot and eighteenth-century abolitionists', *J. of the Friends Historical Soc.* 55/8 (1989), 277–89.

29. Maclear, J.F. 'Isaac Watts and the idea of public religion', *J. of the History of Ideas* 53 (1992), 25–45.

30. Mather, F.C. *High church prophet: bishop Samuel Horsley (1733–1806) and the Caroline tradition in the later Georgian church.* Oxford; Oxford UP; 1992. Pp viii, 333.

31. Milburn, Geoffrey E. 'Charles Wesley in the rude populous north. The Newcastle journal, 23rd September to 3rd October 1742', *P. of the Wesley Historical Soc.* 47 (1989–90), 202–20.

32. Mills, Frederick V. 'The colonial Anglican episcopate: a historiographical review', *Anglican and Episcopal History* 61 (1992), 325–45.

33. Morgan, D. Densil. *Christmas Evans a'r Ymneillduaeth newydd.* Llandysul; Gwasg Gomer; 1991. Pp 180.

34. Morgan, D. Densil. 'Christmas Evans, 1766–1838, and the birth of nonconformist Wales', *Baptist Q.* 34 (1991), 116–24.

35. Morgan, Derec Llwyd. 'Rowland a Williams: Dau camlwyddiant marw'r ddau' [Rowland and Williams: the 2nd anniversary of their deaths], *Honourable Soc. of Cymmrodorion T.* (1991), 145–60.

36. Newhouse, Neville H. 'Seeking God's will: a monthly meeting at work in 1804', *J. of the Friends' Historical Soc.* 56 (1992), 227–43.

37. Podmore, Colin J. 'The Fetter Lane Society', *P. of the Wesley Historical Soc.* 46 (1987–8), 125–53; 47 (1989–90), 156–86.

38. Rack, Henry D. 'The providential moment: church building, Methodism and evangelical entryism in Manchester, 1788–1825', *T. of the Historic Soc. of Lancashire & Cheshire* 141 (1991), 235–60.

39. Rataboul, Louis J. *John Wesley: un anglican sans frontières, 1703–1791.* Nancy; Nancy UP; 1991. Pp 240.

40. Robins, Roger. 'Anglican prophetess: Joanna Southcott and the gospel story', *Anglican & Episcopal History* 62 (1992), 277–302.

41. Robinson, Andrew. 'Identifying the beast: Samuel Horsley and the problem of papal antichrist', *J. of Ecclesiastical History* 43 (1992), 592–607.

42. Scott, Geoffrey. *Gothic rage undone: English monks in the age of Enlightenment.* Bath; Downside Abbey; 1992. Pp 314.
43. Sefton, Henry R. 'Contemporary ecclesiastical reactions to Home's *Douglas*', *Studies in Church History* 28 (1992), 355–61.
44. Stoker, David A. (ed.). *The correspondence of the Reverend Francis Blomefield, 1705–1752.* Norwich; Norfolk Record Soc. vol. 55; 1992. Pp 278.
45. Stout, Harry S. *The divine dramatist: George Whitefield and the rise of modern evangelicalism.* Grand Rapids (Mi); Eerdmans; 1991. Pp xxiv, 301.
46. Taylor, Stephen. 'William Warburton and the alliance of church and state', *J. of Ecclesiastical History* 43 (1992), 271–86.
47. Vickers, John A. 'A new Whitefield letter', *P. of the Wesley Historical Soc.* 48 (1991–2), 119–22.
48. Vickers, John A. 'Good red herring: Methodism's relations with dissent', *P. of the Wesley Historical Soc.* 47 (1989–90), 77–93.
49. Ward, William Reginald. *The Protestant evangelical awakening.* Cambridge; Cambridge UP; 1992. Pp xviii, 370.
50. Waterman, Anthony M.C. 'A Cambridge *via media* in late Georgian Anglicanism', *J. of Ecclesiastical History* 42 (1991), 419–36.
51. Welch, Edwin. 'Lady Huntingdon's chapel at Ashby', *Leicestershire Arch. and Historical Soc. T.* 66 (1992), 136–42.
52. Williams, A.H. 'John Wesley's preferment to St Daniel's church, near Pembroke', *P. of the Wesley Historical Soc.* 48 (1991–92), 155.
53. Wykes, David L. '"The spirit of persecutors exemplified": the Priestley riots and the victims of the church and king mobs', *T. of the Unitarian Historical Soc.* 20 (1991–2), 17–39.

(f) *Economic Affairs*

1. Allais, Maurice. 'The general theory of surpluses as a formalization of the underlying theoretical thought of Adam Smith, his predecessors and his contemporaries', Bc30, 29–62.
2. Barber, Jill. '"A fair and just demand"?: tithe unrest in Cardiganshire, 1796–1823', *Welsh History R.* 16 (1992–3), 177–206.
3. Beckley, Susan. 'The Blaensawdde estate maps', Bc101, 37–42.
4. Berg, Maxine. 'On the origins of capitalist hierarchy', Bc142, 173–94.
5. Berg, Maxine. 'Revisions and revolutions: technology and productivity change in manufacture in eighteenth-century England', Bc22, 43–64.
6. Berg, Maxine; Hudson, Pat. 'Rehabilitating the industrial revolution', *Economic History R.* 2nd ser. 45 (1992), 24–50.
7. Brumhead, Derek. 'An eighteenth-century coal-mining account book for New Mills', *Manchester Region History R.* 6 (1992), 91–95.
8. Buchanan, James M. 'The supply of labour and the extent of the market', Bc30, 104–16.
9. Butlin, Robin. 'Georgian Britain: c.1714–1837', Bc181, 154–85; 244.
10. Campbell, D. 'Adam Smith, Farrar on company law and the economics of the corporation', *Anglo-American Law R.* 19 (1990), 185–208.

11. Chapman, John. 'The interpretation of enclosure maps and awards', Bc47, 73–88.
12. Clark, B.E.; Clark, M.A. 'William Clark, clockmaker of Kendal, 1716–1763', *T. of the Cumberland and Westmorland Antiquarian & Arch. Soc.* 92 (1992), 135–60.
13. Crafts, N.F.R.; Harley, C.K. 'Output growth and the British industrial revolution: a restatement of the Crafts-Harley view', *Economic History R.* 2nd ser. 45 (1992), 703–30.
14. Dalton, Roger. 'Maps of the Egginton enclosure award: reconstruction and interpretation', *Derbyshire Arch. J.* 111 (1991), 85–91.
15. d'Angelo, Michela. *Mercanti inglesi a Malta, 1800–1825* [English merchants in Malta]. Milan; Angeli; 1990. Pp 287.
16. Davey, Roger (ed.). *East Sussex land tax, 1785*. Lewes; Sussex Record Soc.; 1992. Pp xxxvi, 309.
17. Dodgshon, Robert A. 'Agriculture practice in the western Highlands and islands before crofting: a study in cultural inertia or opportunity costs?', *Rural History* 3 (1992), 173–89.
18. Duncan, J.G. 'A Scottish trading house in eighteenth century Gothenburg: Carnegy and Shepherd', *Northern Scotland* 11 (1991), 1–9.
19. Evans, Chris. 'Failure in a new technology: smelting iron with coke in south Gloucestershire in the 1770s', *T. of the Bristol & Gloucestershire Arch. Soc.* 109 (1992 for 1991), 199–206.
20. Evans, Chris. 'Manufacturing iron in the north-east during the eighteenth century: the case of Bedlington', *Northern History* 28 (1992), 178–96.
21. Feliciano Ramos, Hector R. *El contrabando ingles en el Caribe y el Golfo de Mexico, 1748–1778* [English smuggling in the Caribbean and the Gulf of Mexico, 1748–1778]. Seville; Disputacion Provincial de Sevilla; 1990. Pp 414.
22. Fletcher, David H. 'Mapping and estate management on the early nineteenth-century estate: the case of the earl of Aylesford's estate atlas', *Archaeologia Cantiana* 109 (1992 for 1991), 85–109.
23. French, Christopher J. '"Crowded with traders and a great commerce": London's domination of English overseas trade, 1700–1775', *London J.* 17 (1992), 27–35.
24. Griffiths, Trevor; Hunt, Philip A.; O'Brien, Patrick K. 'Inventive activity in the British textile industry, 1700–1800', *J. of Economic History* 52 (1992), 881–906.
25. Harris, John R. 'Introduction' [essays in industry and technology], Bc86, 1–17.
26. Haslam, Graham. 'Patronising the plotters: the advent of systematic estate mapping', Bc47, 55–72.
27. Hayfield, Colin. 'Manure factories? The post-enclosure high barns of the Yorkshire wolds', *Landscape History* 13 (1991), 33–45.
28. Hoftijzer, P.G. 'Business and pleasure; a Leiden bookseller in England in 1772', Bc133, 178–87.
29. Horrell, Sara; Humphries, Jane. 'Old questions, new data, and alternative perspectives: families' living standards in the industrial revolution', *J. of Economic History* 52 (1992), 849–80.

30. Hudson, Pat. *The industrial revolution: reading history*. London; Arnold; 1992. Pp 192.

31. Jackson, R.V. 'Rates of industrial growth during the industrial revolution', *Economic History R.* 2nd ser. 45 (1992), 1–23.

32. James, Jude. 'A glimpse of an old world: the diary of James Warne for the year 1758', *The Hatcher R.* 4/33 (1992), 30–49.

33. Jones, Michael John. 'The accounting system of Magdalen college, Oxford, in 1812', *Accounting, Business & Financial History* 1 (1991), 141–61.

34. Jones, Peter. 'Introduction' [Adam Smith], Bc29, x-xii.

35. Kaiser, Thomas E. 'Public credit: John Law's scheme and the question of *confiance*', *P. of the Annual Meeting of the Western Soc. for French History* 16 (1989), 72–81.

36. Kingsley, Nicholas. 'Boulton and Watt engines supplied to Gloucestershire: a preliminary list', *Gloucestershire Soc. for Industrial Archaeology J.* (1990), 49–59.

37. Klein, Lawrence R. 'Smith's use of data', Bc30, 15–28.

38. Lemire, Beverly. 'Peddling fashion: salesmen, pawnbrokers, tailors, thieves and the second-hand clothes trade in England, *c.*1700–1800', *Textile History* 22 (1991), 67–82.

39. Lewis, C.P. 'John Chapman's maps of Newmarket', *Cambridge Antiquarian Soc. P.* 80 (1992 for 1991), 68–78.

40. Lowe, Michael C. 'Archival sources for road improvement in eighteenth-century Devon', *Archives* 20 (1992), 99–103.

41. Magnusson, Lars. 'From *Verlag* to factory: the contest for efficient property rights', Bc142, 195–222.

42. Modigliani, Franco. 'On the *Wealth of nations*', Bc30, 86–103.

43. Moore-Colyer, Richard J. 'Coastal limekilns in south-west Wales and their conservation', Bc129, 23–29.

44. Morgan, Gerald. 'Aberystwyth herring tithes and boat names in 1730', *Maritime Wales* 14 (1991), 7–15.

45. Morgan, Kenneth. 'Bristol and the Atlantic trade in the eighteenth century', *English Historical R.* 107 (1992), 626–50.

46. Murphy, A.E. 'The evolution of John Law's monetary theories and policies, 1707–1715', *European Economic R.* 35 (1991), 1109–25.

47. Neal, Larry. *The rise of financial capitalism: international capital markets in the age of reason*. Cambridge; Cambridge UP; 1990. Pp 278.

48. Pearson, Robin O. 'Fire insurance and the British textile industries during the industrial revolution (1790–1830)', *Business History* 34/4 (1992), 1–19.

49. Plank, Frans. 'Adam Smith: grammatical economist', Bc29, 21–55.

50. Pollard, Sidney. 'The concept of the industrial revolution', Bc134, 29–62.

51. Prasch, R.E. 'The ethics of growth in Adam Smith's "Wealth of nations"', *History of Political Economy* 23 (1991), 337–52.

52. Raphael, D.D. 'Adam Smith 1790: the man recalled; the philosopher revived', Bc29, 93–118.

53. Raven, James. *Judging new wealth: popular publishing and responses to commerce in England, 1750–1800*. Oxford; Oxford UP; 1992. Pp viii, 327.

54. Richardson, Christine. *The waterways revolution: from the peaks to the Trent, 1768–1778.* London; Self Publishing Association; 1992. Pp 288.

55. Richardson, David; Schofield, M.M. 'Whitehaven and the eighteenth-century British slave trade', *T. of the Cumberland and Westmorland Antiquarian & Arch. Soc.* 92 (1992), 183–204.

56. Rothschild, Emma. 'Adam Smith and conservative economics', *Economic History R.* 2nd ser. 45 (1992), 74–96.

57. Rule, John C. *The vital century: England's developing economy, 1714–1815.* London; Longman; 1992. Pp xvii, 334.

58. Samuelson, Paul A. 'The overdue recovery of Adam Smith's reputation as an economic theorist', Bc30, 1–14.

59. Sawers, Larry. 'The Navigation acts revisited', *Economic History R.* 2nd ser. 45 (1992), 262–84.

60. Schultz, Theodore W. 'Adam Smith and human capital', Bc30, 130–40.

61. Sillitoe, Paul J. 'George Marchant's river Mersey barrage, 1768, craft industry in the countryside', *T. of the Historic Soc. of Lancashire & Cheshire* 141 (1991), 329–38.

62. Skinner, Andrew S. 'Political economy: Adam Smith and his Scottish predecessors', Bc29, 217–43.

63. Smith, Catherine. 'Image and reality: two Nottinghamshire market towns in late Georgian England', *Midland History* 17 (1992), 59–74.

64. Stone, Richard. 'Public economic policy: Adam Smith on what the state and other public institutions should and should not do', Bc30, 63–85.

65. Thornthwaite, S.E. 'On the most advantageous line: the Tyne-Solway canal', *Archaeologia Aeliana* 5th ser. 20 (1992), 121–38.

66. Thwaites, W. 'The corn market and economic change: Oxford in the eighteenth century', *Midland History* 16 (1991), 103–25.

67. Tyson, Blake. 'Murton Great Field, near Appleby: a case-study of the piecemeal enclosure of a common field in the mid-eighteenth century', *T. of the Cumberland and Westmorland Antiquarian & Arch. Soc.* 92 (1992), 161–82.

68. White, A. 'Stone masons in a Georgian town', *Local Historian* 21 (1991), 60–65.

69. Wilson, G. 'Public hire chairs in Exeter', *Devon & Cornwall Notes & Queries* 36 (1990), 265–69; (1991) 314–24.

70. Winch, Donald. 'Adam Smith: Scottish moral philosopher as political economist', *Historical J.* 35 (1992), 91–113.

(g) *Social Structure and Population*

1. Ashcroft, Loraine (ed.). *Vital statistics: the Westmorland 'census' of 1787.* [Place]; Curwen Archive Texts; 1992. Pp xii, 414.

2. Brunton, Deborah. 'Smallpox inoculation and demographic trends in eighteenth-century Scotland', *Medical History* 36 (1992), 403–29.

3. Davies, G.J. 'Literacy in Dorset, 1750–1800', *Notes & Queries for Somerset & Dorset* 33 (1991), 21–28.

4. Evans, Muriel Bowen. 'Dros y Mynydd Du i Frynaman: documentary sources for some farms near Nant Melyn', Bc101, 23–35.
5. Fogelman, Aaron. 'Migrations to the thirteen British north American colonies, 1700–1775: new estimates', *J. of Interdisciplinary History* 22 (1992), 691–709.
6. Foster, Charles F. 'The landowners and residents of four north Cheshire townships in the 1740s', *T. of the Historic Soc. of Lancashire & Cheshire* 141 (1991), 101–205.
7. Hinchliffe, G. 'The Robinsons of Newby park and Newby hall, part 2: Tancred Robinson, baronet, rear admiral of the White, 1686–1754', *Yorkshire Arch. J.* 64 (1992), 185–202.
8. Hunt, Margaret. 'Wife beating, domesticity and women's independence in eighteenth-century London', *Gender & History* 4 (1992), 10–33.
9. Jacobs, Derek. 'Seasonal variation patterns in baptisms and burials for Ruislip, Middlesex', *Local Population Studies* 48 (1992), 33–40.
10. Johnston, J.A. 'Social change in the eighteenth-century: the evidence in wills from six Lincolnshire parishes, 1661–1812', *Lincolnshire History and Archaeology* 27 (1992), 27–33.
11. King, Steve. 'Record linkage in a proto-industrial community', *History & Computing* 4 (1992), 27–33.
12. Klein, Lawrence R. 'A time of progress?', *J. of British Studies* 31 (1992), 294–300.
13. Landau, Norma. 'The eighteenth-century context of the laws of settlement', *Continuity & Change* 6 (1991), 417–39.
14. Milburn, Geoffrey E. (ed.). *A Northumbrian methodist childhood: an autobiographical account of family community and chapel life in Choppington and Berwick-upon-Tweed around the turn of the 19th century by Albert Victor Murray.* Morpeth; Northumberland County Library; 1992. Pp 140.
15. Moody, David. 'Together like a horse and carriage: some eighteenth-century examples of love, marriage and divorce', *T. of the East Lothian Antiquarian Soc.* 20 (1989 for 1988), 11–20.
16. Outhwaite, R.B. 'Sweetapple of Fledborough and clandestine marriage in eighteenth-century Nottinghamshire', *T. of the Thoroton Soc.* 94 (1990), 35–46.
17. Seed, John. 'From "middling sort" to middle class in late eighteenth- and early nineteenth-century England', Bc43, 114–35.
18. Smail, John. 'The Stansfields of Halifax: a case study of the making of the middle class', *Albion* 24 (1992), 27–47.
19. Snell, Keith D.M. 'Deferential bitterness: the social outlook of the rural proletariat in eighteenth- and nineteenth-century England and Wales', Bc43, 158–84.
20. Snell, Keith D.M. 'Pauper settlement and the right to poor relief in England and Wales', *Continuity & Change* 6 (1991), 375–415.
21. Thomas, P. 'The population of Cornwall in the eighteenth century', *J. of the Royal Institution of Cornwall* 10 (1990), 416–56.
22. Tilney, Ruth,; Mills, Dennis. 'The people of Swinderby in 1771 and

1791: a study in population mobility', *Lincolnshire History & Archaeology* 26 (1991), 7–11.
23. Weil, Rachel J. 'Gender and the historians' eighteenth century', *J. of British Studies* 31 (1992), 288–93.
24. Williams, J.D. 'The noble household as a unit of consumption: the Audley End experience, 1765–1797', *Essex Archaeology & History* 3rd ser. 23 (1992), 67–78.
25. Wyatt, Grace. 'Bastardy and prenuptial pregnancy in a Cheshire town during the eighteenth century', *Local Population Studies* 49 (1992), 38–50.

(h) *Naval and Military*

1. Altoff, Gerard T. 'The battle of lake Erie: a narrative', Bc85, 5–16; 129–30.
2. Baugh, Daniel A. 'The politics of British naval failure, 1775–1777', *American Neptune* 52 (1992) 221–46.
3. Behre, Goran. 'Jacobite refugees in Gothenburg after Culloden', *Scottish Historical R.* 70 (1991), 58–65.
4. Black, Jeremy. 'British naval power and international commitments: political and strategic problems, 1688–1770', Bc173, 39–59.
5. Black, Jeremy. 'Naval power, strategy and foreign policy, 1775–1791', Bc173, 93–120.
6. Black, Jeremy. 'Sir William Fawcett and the publication of military works in the mid-eighteenth century', *Factotum* 33 (1991), 10–13.
7. Black, Jeremy. 'The battle of Long Island: a new account', *J. of the Soc. for Army Historical Research* 69 (1991), 201–05.
8. Carter-Edwards, Dennis. 'The battle of lake Erie and its consequences: denouement of the British right division and abandonment of the western district to American troops, 1813–1815', Bc85, 41–55; 136–38.
9. Coss, Edward J. 'The British soldier in the Peninsular War: the acquisition of an unjust reputation', *P. of the Annual Meeting of the Western Soc. for French History* 18 (1991), 243–51.
10. Crewe, D.G. *Yellow Jack and the worm: British naval administration in the West Indies, 1739–1748.* Liverpool; Liverpool UP; 1992. Pp 352.
11. Dening, Greg. *Mr Bligh's bad language: passion, power and the theatre of the Bounty.* Cambridge; Cambridge UP; 1992. Pp [464].
12. Douglas, W.A.B. 'The honor of the flag had not suffered: Robert Herriot Barclay and the battle of lake Erie', Bc85, 30–40; 134–36.
13. Drake, Frederick C. 'Artillery and its influence on naval tactics: reflections on the battle of lake Erie', Bc85, 17–29; 130–34.
14. Drake, Frederick C. 'The Niagara peninsula and naval aspects of the war of 1812', *The military in the Niagara peninsula* ed. Wesley B. Turner (St Catharines (Ont); Vanwell; 1990), 15–37.
15. Duffy, Michael. 'The establishment of the Western Squadron as the linchpin of British naval strategy', Bc173, 60–81.
16. Edmunds, R. David. 'Tecumseh's native allies: warriors who fought for the crown', Bc85, 56–67; 138–41.

17. Fletcher, Ian. *Craufurd's Light Division: the life of Robert Craufurd and his command of the Light Division.* Tunbridge Wells; Spellmount; 1991. Pp 248.

18. Fletcher, Ian (ed.). *Guards officer in the Peninsula: Peninsular war letters of John Rous, Coldstream Guards, 1812–1814.* Tunbridge Wells; Spellmount; 1992. Pp 136.

19. Hall, Christopher D. *British strategy in the Napoleonic war, 1803–1815.* Manchester; Manchester UP; 1992. Pp xii, 238.

20. Harley, Basil. 'The Society of Arts' model ship trials, 1758–1763', *Newcomen Soc. T.* 63 (1992), 53–71.

21. James, Lawrence. *Iron duke: the military biography of Arthur Wellesley, duke of Wellington.* London; Weidenfeld & Nicolson; 1992. Pp x, 307.

22. Langley, Harold D. 'The quest for peace in the war of 1812', Bc85, 68–77; 141–43.

23. Lanning, George. 'Thomas Hardy and the German Hussars', *Dorset Natural History & Arch. Soc. P.* 113 (1992 for 1991), 1–4.

24. Lapierre, Laurier. *1759: the battle for Canada.* Toronto; McClelland and Stuart; 1990. Pp 336.

25. Legault, Roch. 'L'organisation militaire sous le régime britannique et le rôle assigné à la gentilhommerie canadienne (1760–1815)', *Revue d'Histoire de l'Amerique Française* 45 (1991), 229–49.

26. Lewis, James A. *The final campaign of the American revolution: the rise and fall of the Spanish Bahamas.* Columbia (SC); South Carolina UP; 1991. Pp xi, 149.

27. Luff, P.A. 'The noblemen's regiments: politics and the 'forty-five', *Historical Research* 65 (1992), 54–73.

28. MacKay, John; Coleman, Ron. *The 24-gun frigate 'Pandora'.* London; Conway; 1992. Pp 128.

29. Maissonneuve, B.; Maissonneuve, M. *Le 'Maidstone', 1747: miroire d'une mémoire maritime.* Saint-Gilles-Croix-de-Vie; ARHIMS; 1991. Pp 1889.

30. Marley, David F. 'Havana surprised: prelude to the British invasion, 1762', *Mariner's Mirror* 78 (1992), 293–305.

31. Marshall, P.J. '"Cornwallis triumphant": war in India and the British public in the late eighteenth century', Bc150, 57–74.

32. Marzo, Alessandro. 'La vicenda veneziana di James Pattison: un ufficiale britannico al servizio della Serenissima (1768–1772)' [The Venetian episode of James Pattison: a British official in the service of the Serenissima, 1768–1772], *Studi Veneziani* ns 19 (1991 for 1990), 293–311.

33. McCorry, Helen C. (ed.). '"Besides, he was very drunk at the time ..."': desertion and discipline, north Britain, 1751–1753' [part 1], *J. of the Soc. for Army Historical Research* 69 (1991), 221–32.

34. McCorry, Helen C. (ed.). '"Besides, he was very drunk at the time ..."': desertion and discipline, north Britain, 1751–1753' [part 2], *J. of the Soc. for Army Historical Research* 70 (1992), 114–17.

35. McCorry, Helen C (ed.). '"Besides, he was very drunk at the time ..."': desertion and discipline, north Britain, 1751–1753' [part 3], *J. of the Soc. for Army Historical Research* 70 (1992), 189–97.

36. Murdoch, Alexander J. 'The killing of George Wood at Tranent, 3 April 1757', *T. of the East Lothian Antiquarian Soc.* 20 (1989 for 1988), 21–32.
37. Naish, John. 'Joseph Whidbey and the buildings of the Plymouth breakwater', *Mariner's Mirror* 78 (1992), 37–56.
38. Peddie, John. 'A melancholy example of the uncertainty of human affairs: disaster for Admiral Christian, 1795', *The Hatcher R.* 4/34 (1992), 12–28.
39. Richards, R. 'The cruise of the *Kingston* and the *Elligood* in 1800 and the wreck found on King Island in 1801', *The Great Circle* 13/1 (1991), 35–53.
40. Rodger, Nicholas A.M. '"A little navy of your own making." Admiral Boscawen and the Cornish connection in the Royal Navy', Bc173, 82–92.
41. Rodger, Nicholas A.M. 'The continental commitment in the eighteenth century', Bc150, 39–55.
42. Saxby, Richard. 'The blockade of Brest in the French revolutionary war', *Mariner's Mirror* 78 (1992), 25–35.
43. Screen, J.O.E. 'The "Royal Military Academy" of Lewis Lochee', *J. of the Soc. for Army Historical Research* 70 (1992), 143–56.
44. Steele, Ian K. 'Suppressed official British report of the siege and 'massacre' at Fort William Henry, 1757', *Huntington Library Q.* 55 (1992), 339–52.
45. Turner, Wesley B. *The war that both sides won: the war of 1812–1814.* Toronto; Dundurn; 1990. Pp 150.
46. Ward, S.G.P. 'Mr Eastlake's visit to Spain in 1813', *J. of the Soc. for Army Historical Research* 70 (1992), 71–86.
47. Zulueta, Julian de. 'Health and military factors in Vernon's failure at Cartagena', *Mariner's Mirror* 78 (1992), 127–41.

(i) *Intellectual and Cultural*

1. Allen, Brian. 'The age of Hogarth, 1720–1760', Bc93, 128–83.
2. Allison, A.F. 'Leonardus Lessius of Louvain and his English translator', Bc133, 89–98.
3. Allott, Stephen. *Lindley Murray, 1745–1826: Quaker grammarian of New York and old York.* York; Sessions; 1991. Pp xiv, 86.
4. Barrell, John. 'Sportive labour: the farmworker in eighteenth-century poetry and painting', Bc54, 105–32.
5. Barrell, John. 'The birth of Pandora and the origin of painting', Bc89, 145–220; 232–40.
6. Bending, Stephen. 'Re-reading the eighteenth-century English landscape garden', *Huntington Library Q.* 55 (1992), 379–99.
7. Benson, Robert (ed.). *Memoirs of the life and writings of the Rev. Arthur Collier.* London; Thoemmes; 1990. Pp xvi, 215.
8. Borsay, A. 'Cash and convenience: financing the General Hospital at Bath, *c.*1738–1750', *Social History of Medicine* 4 (1991), 207–29.

9. Brant, Clare. 'Speaking of women: scandal and the law in the mid-eighteenth century', Bc39, 242–70.

10. Brighton, J. Trevor. 'William Peckitt's commission book, 1751–1795', *Walpole Soc.* 54 (1991 for 1988), 334–453.

11. Brown, Gordon. 'Causey clash of kirk, town and college, 1747–1748', *Book of the Old Edinburgh Club* ns 1 (1991), 82–84.

12. Brown, Iain Gordon. *Monumental reputation. Robert Adam and the emperor's palace.* Edinburgh; National Library of Scotland; 1992. Pp 52.

13. Brown, Iain Gordon. 'David Hume's tomb: a Roman mausoleum by Robert Adam', *Soc. of Antiquaries of Scotland P.* 121 (1991), 391–422.

14. Brown, Kevin L. 'Dating Adam Smith's essay *Of the external senses, J. of the History of Ideas* 53 (1992), 333–37.

15. Bruyn, Frans De. 'Theater and countertheater in Burke's *Reflections on the revolution in France*', Bc132, 28–68.

16. Bryant, Julius. *Robert Adam, 1782–1792: architect of genius.* London; English Heritage; 1992. Pp 56.

17. Bryce, J.C. 'Lectures on rhetoric and belles lettres', Bc29, 1–20.

18. Buckle, Stephen; Castiglione, Dario. 'Hume's critique of the contract theory', *History of Political Thought* 12 (1991), 457–80.

19. Bushaway, R.W. 'Rite, legitimation and community in southern England, 1700–1850: the ideology of custom', Bc116, 110–34.

20. Byrne, Maurice. 'William Bull, John Stevenson and the Harriss family', *The Galpin Soc. J.* 45 (1992), 67–77.

21. Clamp, J.D. 'Potterspury's clerical mathematician', *Northamptonshire Past & Present* 8 (1992), 275–85.

22. Clarke, George. 'The moving temples of Stowe: aesthetics of change in an English landscape over four generations', *Huntington Library Q.* 55 (1992), 501–09.

23. Cox, Jeffrey N. 'Ideology and genre in the British anti-revolutionary drama of the 1790s', *J. of English Literary History* 58 (1991), 579–610.

24. Craven, Kenneth. *Jonathan Swift and the millenium of madness: the information age in Swift's 'A tale of a tub'.* Leiden; Brill; 1992. Pp xvi, 238.

25. Dalgleish, George R. 'The "Silver Jack" trophy of the Edinburgh Society of Bowlers', *Soc. of Antiquaries of Scotland P.* 120 (1990), 189–200.

26. Dalgleish, George R. 'Two Robert Adam buildings illustrated on Edinburgh trade tokens', *Book of the Old Edinburgh Club* ns 1 (1991), 28–33.

27. Devlin-Thorp, Sheila. 'Sir William Forbes, 1739–1806, banker and philanthropist', *Book of the Old Edinburgh Club* ns 1 (1991), 74–78.

28. Downie, R.S. 'Ethics and casuistry in Adam Smith', Bc29, 119–41.

29. Dwyer, John. 'Virtue and improvement: the civic world of Adam Smith', Bc29, 190–216.

30. Furniss, Tom. 'Stripping the queen: Edmund Burke's magic lantern show', Bc132, 69–96.

31. Glaister, R.T.D. 'Rural private teachers in eighteenth-century Scotland', *J. of Educational Administration & History* 23/2 (1991), 49–61.

32. Goff, Moira. 'Court and theatre dances published in England in the early eighteenth century', *Factotum* 33 (1991), 22–27.

33. Gow, Ian. 'The Edinburgh villa', *Book of the Old Edinburgh Club* ns 1 (1991), 34–46.
34. Griffin, Dustin. 'Swift and patronage', *Studies in Eighteenth-Century Culture* 21 (1991), 197–205.
35. Griffiths, Antony. 'Sir William Musgrave and British biography', *British Library J.* 18 (1992), 171–89.
36. Groarke, Leo; Solomon, Graham. 'Some sources for Hume's account of cause', *J. of the History of Ideas* 52 (1991), 645–63.
37. Hart, Vaughan. 'Vanbrugh's travels', *History Today* 42/7 (1992), 26–32.
38. Haycock, Lorna; Thomas, James. 'The subscription list of an eighteenth century book printed in Devizes', *Wiltshire Arch. & Natural History Magazine* 85 (1992), 80–92.
39. Hill, Bridget. *The republican virago: the life and times of Catharine Macaulay, historian.* Oxford; Oxford UP; 1992. Pp 276.
40. Huber, Gerhard. 'Adam Smith: der Zusammenhang von Moral-philosophie, Ökonomie und Institutionentheorie' [Adam Smith: the connection between moral philosophy, economy and the theory of institutions], *Politische Institutionen im gesellschaftlichen Umbruch*, ed. Gerhard Gohler *et al.* (Wiesbaden; Westdeutscher Verlag Opladen; 1990), 293–309.
41. Hundert, Edward. 'Performing the Enlightenment Self: Henry Fielding and the history of identity', Bc74, 223–244.
42. Hunter, David. 'English country psalmodists and their publications, 1700–1760', *J. of the Royal Musical Association* 115 (1990), 220–39.
43. Immerwahr, John. 'Hume's revised racism', *J. of the History of Ideas* 53 (1992), 481–86.
44. Inskip, Peter. 'Discoveries, challenges, and moral dilemmas in the restoration of garden buildings at Stowe', *Huntington Library Q.* 55 (1992), 511–26.
45. Johnston, Dorothy B. 'James Beattie and his students at Marischal college, Aberdeen', *Northern Scotland* ll (1991), 11–28.
46. Jones, Peter. 'The aesthetics of Adam Smith', Bc29, 56–78.
47. Kitson, Peter J. '"Sages and patriots that being dead do yet speak to us": readings of the English revolution in the late eighteenth century', Bc203, 205–30.
48. Lippincott, Louise. 'Arthur Pond's journal of receipts and expenses, 1734–1750', *Walpole Soc.* 54 (1991 for 1988), 220–333.
49. Lomas, Michael J. 'Secular civilian amateur wind bands in southern England in the late eighteenth and early nineteenth centuries', *The Galpin Soc. J.* 45 (1992), 78–98.
50. Lomas, Michael J. 'The Wiltshire militia band, 1769- *c.*1831', *Wiltshire Arch. & Natural History Magazine* 85 (1992), 93–100.
51. McKay, Peter K. 'A patron of promise: Charles, 7th earl of Nor-thampton', *Northamptonshire Past & Present* 8 (1992), 263–73.
52. Mee, Jon. *Dangerous enthusiasm: William Blake and the culture of radicalism in the 1790s.* Oxford; Oxford UP; 1992. Pp 251.
53. Meyer, Arline. 'Sir William Musgrave's "lists" of portraits; with an

account of head-hunting in the eighteenth century', *Walpole Soc. 54* (1991 for 1988), 454–502.

54. Milhous, J.; Hume, R. 'John Rich's Covent Garden account books for 1735–1736', *Theatre Survey* 31 (1990), 200–41.

55. Minay, Priscilla. 'Eighteenth and early nineteenth century Edinburgh seedsmen and nurserymen', *Book of the Old Edinburgh Club* ns 1 (1991), 7–27.

56. Mingay, Gordon. 'Professor Symond's tour, 1790', Bc117, 9–21.

57. Morgan, Paul. 'Oxford college libraries in the eighteenth century', *Bodleian Library Record* 14 (1992), 228–36.

58. Mortimer, Jean E. 'Joseph Tatham's school, Leeds', Bc84, 1–33.

59. Philp, Mark. 'Enlightenment, toleration and liberty', *Enlightenment & Dissent* 9 (1990), 47–62.

60. Postle, Martin. 'The golden age, 1760–1790', Bc93, 184–241.

61. Reid, Christopher. 'Burke's tragic muse: Sarah Siddons and the "feminization" of the *Reflections*', Bc132, 1–27.

62. Richetti, J. 'Class struggle without class: novelists and magistrates', *Eighteenth Century: Theory & Interpretation* 32 (1991), 203–18.

63. Richey, William. '"The French revolution": Blake's epic dialogue with Edmund Burke', *J. of English Literary History* 59 (1992), 817–37.

64. Ritchie, Daniel E. 'Desire and sympathy, passion and providence: the moral imaginations of Burke and Rousseau', Bc132, 120–43.

65. Ritvo, Harriet. 'At the edge of the garden: nature and domestication in eighteenth- and nineteenth-century Britain', *Huntington Library Q. 55* (1992), 363–78.

66. Salter, John. 'Adam Smith on feudalism, commerce and slavery', *History of Political Thought* 13 (1992), 219–41.

67. Sambrook, James. *James Thomson, 1700–1748: a life.* Oxford; Oxford UP; 1991. Pp viii, 352.

68. Sanderson, Margaret H.B. *Robert Adam and Scotland: portrait of an architect.* London; HMSO; 1992. Pp 144.

69. Sapiro, Virginia. *A vindication of political virtue: the political theory of Mary Wollstonecraft.* Chicago; Chicago UP; 1992. Pp 366.

70. Schmidt, Roger. 'Roger North's Examen: a crisis in historiography', *Eighteenth-Century Studies* 26 (1992), 57–75.

71. Skinner, Andrew S. 'Adam Smith: ethics and self-love', Bc29, 142–67.

72. Stanlis, Peter J. 'Burke, Rousseau, and the French revolution', Bc132, 97–119.

73. Straub, K. 'Men from boys: Cibber, Pope and the schoolboy', *Eighteenth Century: Theory & Interpretation* 32 (1991), 219–39.

74. Taylor, Barbara. 'Mary Wollstonecraft and the wild wish of early feminism', *History Workshop* 33 (1992), 197–219.

75. Thijssen, J.M.M.H. 'David Hume and John Keill and the structure of continua', *J. of the History of Ideas* 53 (1992), 271–86.

76. Tomaselli, Sylvana. 'Remembering Mary Wollstonecraft on the bicentenary of the publication of *A Vindication of the Rights of Women*' [review article], *British J. for Eighteenth-Century Studies* 15 (1992), 125–30.

77. Tressider, George A. 'Coronation day celebrations in English towns,

1685–1821: élite hegemony and local relations on a ceremonial occasion', *British J. for Eighteenth-Century Studies* 15 (1992), 1–16.
78. Turner, Cheryl. *Living by the pen: women writers in the eighteenth century.* London; Routledge; 1992. Pp 261.
79. Wagner, Hans-Peter. 'Eroticism in graphic art: the case of William Hogarth', *Studies in Eighteenth-Century Culture* 21 (1991), 53–74.
80. Walker, Ralph S. (ed.). *A selection of Thomas Twining's letters, 1734–1804: a record of a tranquil life.* Lampeter; Mellen; 1991. Pp 845.
81. Weber, William. *The rise of musical classics in 18th century England: a study in canon, ritual and ideology.* Oxford; Oxford UP; 1992. Pp 274.
82. Wilson, John. 'The romantics, 1790–1830', Bc93, 242–297.
83. Womersley, David. 'Gibbon's unfinished History: the French revolution and English political vocabularies', *Historical J.* 35 (1992), 63–89.
84. Wood, L. 'Furniture for lord Delaval: metropolitan and provincial', *Furniture History* 26 (1990), 198–234.
85. Woolley, James (ed.). *Jonathan Swift and Thomas Sheridan: the 'Intelligencer'.* Oxford; Oxford UP; 1992. Pp xv, 363.
86. Wright, H. Bunker; Wright, Deborah Kempf. 'An autobiographical ballad by Matthew Prior', *British Library J.* 18 (1992), 163–70.
87. Zachs, William. *'Without regard to good manners': a biography of Gilbert Stuart, 1743–1786.* Edinburgh; Edinburgh UP; 1992. Pp xvi, 226.

(j) *Science*

1. Andrew, Donna T. 'Two medical charities in eighteenth-century London: the Lock hospital and the lying-in charity for married women', Bc26, 82–97.
2. Barnard, John E. 'John Barnard the younger, ship-builder of Ipswich and Harwich, 1705–1784', *Mariner's Mirror* 78 (1992), 155–75.
3. Cody, Lisa. '"The doctor's in labour; or a new whim wham from Guildford"', *Gender & History* 4 (1992), 175–96.
4. Dean, Dennis. *James Hutton and the history of geology.* Ithaca (NY); Cornell UP; 1992. Pp 303.
5. Feigenbaum, Lenore. 'The fragmentation of the European mathematical community', Bc87, 383–97.
6. Fissell, Mary E. *Patients, power and the poor in eighteenth-century Bristol.* Cambridge; Cambridge UP; 1992. Pp 200.
7. Fraser, Kevin J. 'William Stukeley and the gout', *Medical History* 36 (1992), 160–86.
8. French, Roger. 'Surgery and scrophula', Bc24, 85–100.
9. Gethyn-Jones, Eric. 'Edward Jenner's family background', *T. of the Bristol & Gloucestershire Arch. Soc.* 109 (1992 for 1991), 195–98.
10. Golinski, Jan. *Science as public culture: chemistry and enlightenment in Britain, 1760–1820.* Cambridge; Cambridge UP; 1992. Pp 330.
11. Gowing, Ronald. 'A study of spirals: Cotes and Varignon', Bc87, 371–81.

12. Jacyna, Stephen. 'Physiological principles in the surgical writings of John Hunter', Bc24, 135–52.

13. James, Frank A.J.L. 'The tales of Benjamin Abbott: a source for the early life of Michael Faraday', *British J. for the History of Science* 25 (1992), 229–40.

14. Kollerstrom, Nick. 'Thomas Simpson and "Newton's method of approximation": an enduring myth', *British J. of the History of Science* 25 (1992), 347–54.

15. Lane, Joan. *Worcester infirmary in the eighteenth century*. Worcester; Worcestershire Historical Society (Occasional publications 6); 1992. Pp ii, 44.

16. Longuet-Higgins, H. Christopher. '"The history of astronomy": a twentieth-century view', Bc29, 79–92.

17. Loudon, Irvine. 'Medical practitioners, 1750–1850, and the period of medical reform in Britain', Bc23, 219–47.

18. Lowe, N.F. 'Hogarth, beauty spots, and sexually transmitted diseases', *British J. for Eighteenth-Century Studies* 15 (1992), 71–79.

19. McBride, W.M. '"Normal" medical science and British treatment of the sea scurvy, 1753–1775', *J. of the History of Medicine & Allied Sciences* 46 (1991), 158–77.

20. Nicolson, Malcolm. 'Giovanni Battista Morgagni and eighteenth-century physical examination', Bc24, 101–34.

21. O'Brien, Patrick K. 'The mainsprings of technological progress in western Europe, 1750–1850', Bc22, 6–17.

22. Porter, Roy. 'The patient in England, *c.*1660- *c.*1800', Bc23, 91–118.

23. Pratt, Hubert T. 'Peter Crosthwaite: John Dalton's "friend and colleague"', *Ambix* 38 (1991), 11–28.

24. Risse, Guenter B. 'Medicine in the age of Enlightenment', Bc23, 149–95.

25. Rosner, Lisa. 'Thistle on the Delaware: Edinburgh medical education and Philadelphia practice, 1800–1825', *Social History of Medicine* 5 (1992), 19–42.

26. Wear, Andrew. 'Making sense of health and the environment in early modern England', Bc23, 119–47.

27. Wilson, Philip K. '"Sacred sanctuaries for the sick": surgery at St Thomas's hospital 1725–1726', *London J.* 17 (1992), 36–53.

H. BRITAIN

See also Aa1,32,49,83,88–89,157,b30,42,c13,77,85,123,196,219,225,d11, 32,40;　　Ga8,b23,d11,e38,f2,9,15,48,i3,19,77,82,j17,21;Ia12,b51,63,c1,d26, 38,42,98,g27,h26–27

(a) *General*

1. Ambrosius, Gerold; Hubbard, William H. *A social and economic history of twentieth century Europe.* trans. W.H. Hubbard & Keith Tribe. London; Harvard UP; 1989. Pp xvi, 368.

2. Bentley-Cranch, Dana. *Edward VII: image of an era*. London; HMSO; 1992. Pp 160.

3. Breuilly, John J. 'Conclusion: national peculiarities?', Bc219, 273–95.

4. Crowther, Jan. *Beverley in mid-Victorian times*. Beverley; Hutton; 1990. Pp 136.

5. John, Angela V. 'Introduction' [Welsh women's history], Bc66, 1–16.

6. Morris, R.J. 'Victorian values in Scotland and England', Bc196, 31–48.

7. Oliver, R.C.B. (ed.). 'The diary of Captain Frederick Jones, IV: 1812–1819, V & VI, 1820–1827', *Radnorshire Soc. T.* 60 (1990), 41–65; 61 (1991), 54–70.

8. Rich, J. 'An Edwardian childhood: Sir Laurence Hartnett and the search for identity', *Australian Historical Studies* 24 (1991), 252–66.

(b) *Politics*

1. Anderson, Olive. 'The feminism of T.H. Green: a late-Victorian success story', *History of Political Thought* 12 (1991), 671–93.

2. Ashton, Owen R. *W.E. Adams: Chartist, radical and journalist (1832–1906)*. Tyne & Wear; Bewick; 1991. Pp v, 200.

3. Atkinson, Diane. *The purple, white & green: suffragettes in London, 1906–1914*. London; Museum of London; 1992. Pp 136.

4. Beales, Derek. 'The electorate before and after 1832: the right to vote, and the opportunity' [review article], *Parliamentary History* 11 (1992), 139–50.

5. Belchem, John. 'Introduction: the peculiarities of Liverpool', Bc65, 1–20.

6. Belchem, John. 'Liverpool in the year of revolution: the political and associational culture of the Irish immigrant community in 1848', Bc65, 68–97.

7. Benn, Caroline. *Keir Hardie*. London; Hutchinson; 1992. Pp 544.

8. Bentley, Michael. 'Gladstone's heir', *English Historical R.* 107 (1992), 901–24.

9. Bevir, Mark. 'The Marxism of George Bernard Shaw, 1883–1889', *History of Political Thought* 13 (1992), 299–318.

10. Biagini, Eugenio F. *Liberty, retrenchment and reform: popular Liberalism in the age of Gladstone, 1860–1880*. Cambridge; Cambridge UP; 1992. Pp xii, 476.

11. Biffen, John. 'Electing Oswestry's MP, 1901', *History Today* 42/10 (1992), 36–38.

12. Bohstedt, John. 'More than one working class: Protestant and Catholic riots in Edwardian Liverpool', Bc65, 173–216.

13. Brent, Richard. 'New Whigs in old bottles' [review article], *Parliamentary History* 11 (1992), 151–56.

14. Breuilly, John J. 'Artisan economy, ideology and politics: the artisan contribution to the mid-nineteenth-century European labour movement', Bc219, 76–114.

15. Breuilly, John J. 'Liberalism in mid-nineteenth-century Britain and Germany', Bc219, 228–72.

16. Breuilly, John J. 'Liberalism in mid-nineteenth-century Hamburg and Manchester', Bc219, 197–227.

17. Breuilly, John J. 'Liberalism or social democracy? Britain and Germany, 1850–1875', Bc219, 115–59.

18. Caine, Barbara. *Victorian feminists*. Oxford; Oxford UP; 1992. Pp 284.

19. Cook, Kay; Evans, Neil. '"The petty antics of the bell-ringing boisterous band"? The women's suffrage movement in Wales, 1890–1918', Bc66, 159–88.

20. Courtenay, Adrian. 'Cheltenham Spa and the Berkeleys, 1832–1848: pocket borough and patron?', *Midland History* 17 (1992), 93–108.

21. Cox, Gary W. 'The origin of whip votes in the House of Commons', *Parliamentary History* 11 (1992), 278–85.

22. Cunliffe-Charlesworth, Hilary. 'Sylvia Pankhurst as an art student', Bc50, 1–35.

23. Davis, R.W. 'The Whigs and religious issues, 1830–1835', Bc13, 29–50.

24. Derry, John W. *Charles, earl Grey: aristocratic reformer*. Oxford; Blackwell; 1992. Pp 297.

25. Dodds, Klaus-John. 'Much ado about nothing? Cholera, local politics and health in nineteenth-century Reading', *Local Historian* 21 (1991), 168–76.

26. Dutton, David. *'His Majesty's loyal opposition': the Unionist party in opposition, 1905–1915*. Liverpool; Liverpool UP; 1992. Pp xii, 321.

27. Eastwood, David. 'Peel and the Tory party reconsidered', *History Today* 42/3 (1992), 27–33.

28. Feuchtwanger, Edgar J. 'Electoral systems: an Anglo-German comparison, 1867–1933', *Historical Research* 65 (1992), 194–200.

29. Feuchtwanger, Edgar J. 'John Stuart Mill: Theorie und Praxis im englischen Liberalismus der 1860er Jahre' [John Stuart Mill: theory and practice in English Liberalism in the 1860s], *Liberale Demokratie in Europa und den USA*, ed. F. Gress and H. Vorlander (Frankfurt; Campus; 1990), 154–68.

30. Finn, Margot. '"A vent which has conveyed our principles": English radical patriotism in the aftermath of 1848', *J. of Modern History* 64 (1992), 637–59.

31. Flint, James. 'The attempt of the British government to influence the choice of the second archbishop of Westminster', *Catholic Historical R.* 77 (1991), 42–55.

32. Garner, Les. 'Suffragism and socialism: Sylvia Pankhurst, 1903–1914', Bc50, 58–85.

33. Gilbert, Bentley Brinkerhoff. *David Lloyd George: a political life, vol. 2: organizer of victory, 1912–1916*. London; Batsford; 1992. Pp 523.

34. Gutwein, Daniel. *The divided elite: economics, politics and Anglo-Jewry, 1882–1917*. Leiden; Brill; 1992. Pp 501.

35. Halstead, John. *'The voice of the West Riding*: promoters and supporters of a provincial unstamped newspaper, 1833–1834', Bc117, 22–57.

36. Harris, José. 'Political thought and the welfare state 1870–1940: an intellectual framework for British social policy', *Past & Present* 135 (1992), 116–41.

37. Hinde, Wendy. *Catholic emancipation: a shake to men's minds*. Oxford: Blackwell; 1992. Pp 211.

38. Hogan, John. 'Protectionists and Peelites: the Conservative Party in the House of Lords 1846 to 1852', *Parliaments, Estates & Representation* 11 (1991), 163–80.

39. Hostettler, John. *The politics of criminal law reform in the nineteenth century*. Chichester; Rose; 1992. Pp xxv, 234.

40. Hunt, Tamara L. 'Morality and monarchy in the Queen Caroline affair', *Albion* 23 (1991), 697–722.

41. Jaggard, E.K.G. 'Political continuity and change in late nineteenth-century Cornwall', *Parliamentary History* 11 (1992), 218–34.

42. Jenkins, T.A. 'Hartington, Chamberlain and the Unionist alliance, 1886–1895', *Parliamentary History* 11 (1992), 108–38.

43. Jones, J.Graham. 'Framing "the people's budget": a new perspective', *Honourable Soc. of Cymmrodorion T.* (1991), 285–98.

44. Kahan, Alan. *Aristocratic Liberalism: the social and political thought of Jacob Burckhardt, John Stuart Mill, and Alexis de Tocqueville*. Oxford; Oxford UP; 1992. Pp 228.

45. Kean, Hilda. *Deeds not words: the lives of suffragette teachers*. London; Pluto; 1990. Pp 179.

46. King, Elspeth. 'The Scottish women's suffrage movement', Bc135, 121–50.

47. Kinzer, Bruce L. *Moralist in and out of parliament: John Stuart Mill at Westminster, 1865–1868*. Toronto; Toronto UP; 1992. Pp viii, 317.

48. Kinzer, Bruce L. 'John Stuart Mill and the experience of political engagement', Bc123, 182–214.

49. Kross, Editha. 'Arbeiter und Reformbill 1832 in England' [Workers and the Reform Bill in England, 1832], *Proletariat und bürgerliche Revolution, 1830–1917*, ed. Manfred Kossok and Editha Kross (Vaduz; Topos; 1990), 81–99.

50. Lawrence, Jon. 'Popular radicalism and the socialist revival in Britain', *J. of British Studies* 31 (1992), 163–86.

51. Loughlin, James. 'Joseph Chamberlain, English nationalism and the Ulster question', *History* 77 (1992), 202–19.

52. McLean, Ian. 'Rational choice and the Victorian voter', *Political Studies* 40 (1992), 496–515.

53. Millar, Mary S.; Wiebe, M.G. '"This power so vast ... & so generally misunderstood": Disraeli and the press in the 1840s', *Victorian Periodicals R.* 25 (1992), 79–85.

54. Mitchell, Dennis J. *Cross and Tory democracy: a political biography of Richard Assheton Cross*. London; Garland; 1991. Pp 322.

55. Moore, J. 'Deconstructing Darwinism: the politics of evolution in the 1860s', *J. of Historical Biology* 24 (1991), 353–408.

56. Moore, Kevin. '"This Whig and Tory ridden town": popular politics in Liverpool in the Chartist era', Bc65, 38–67.

57. Morris, Frankie. 'The illustrated press and the republican crisis of 1871–1872', *Victorian Periodicals R.* 25 (1992), 114–26.

58. Murray, Bruce K. '"Battered and shattered": Lloyd George and the 1914 budget fiasco', *Albion* 23 (1991), 483–507.

59. Nash, David S. 'F.J. Gould and the Leicester Secular Society: a positivist commonwealth in Edwardian politics', *Midland History* 16 (1991), 126–40.

60. Oldfield, Geoffrey. 'The Nottingham Borough boundary extension of 1877', *T. of the Thoroton Soc.* 94 (1991 for 1990), 83–91.

61. Pears, Iain. 'The gentleman and the hero: Wellington and Napoleon in the nineteenth century', Bc190, 216–36.

62. Phillips, John A. *The great reform bill in the boroughs: English electoral behaviour, 1818–1841.* Oxford; Oxford UP; 1992. Pp xii, 337.

63. Phillips, John A.; Wetherell, Charles. 'The great reform bill of 1832 and the rise of partisanship', *J. of Modern History* 63 (1991), 621–46.

64. Pickering, Paul A. 'Chartism and the "trade of agitation" in early Victorian Britain', *History* 76 (1991), 221–37.

65. Powell, W.R. 'John Horace Round and Victorian Colchester: culture and politics, 1880–1895', *Essex Archaeology & History* 3rd ser. 23 (1992), 79–90.

66. Quinault, Roland. 'Asquith's Liberalism', *History* 77 (1992), 33–49.

67. Ridley, Jane. 'The Unionist opposition and the House of Lords, 1906–1910', *Parliamentary History* 11 (1992), 235–53.

68. Roberts, Stephen. 'Joseph Barker and the Radical cause, 1848–1851', Bc84, 59–73.

69. Rosen, Frederick. *Bentham, Byron, and Greece: constitutionalism, nationalism, and early liberal political thought.* Oxford; Oxford UP; 1992. Pp xii, 332.

70. Rowe, Violet. 'The Hertford borough bill of 1834', *Parliamentary History* 11 (1992), 88–107.

71. Russell, Alice. *Political stability in later Victorian England: a sociological analysis and interpretation.* Lewes; Book Guild; 1992. Pp 340.

72. Ryan, Alan. 'Sense and sensibility in Mill's political thought', Bc123, 121–38.

73. Searle, G.R. *The Liberal party: triumph and disintegration, 1886–1929.* London; Macmillan; 1992. Pp 234.

74. Seymour-Jones, Carole. *Beatrice Webb: woman of conflict.* London; Allison & Busby; 1992. Pp 369.

75. Shannon, Richard. *The age of Disraeli, 1868–1881: the rise of Tory democracy.* London; Longman; 1992. Pp 445.

76. Silagi, Michael. 'Henry George and Europe: George and his followers awakened the British conscience and started a new, freer society' trans. Susan N. Faulkner, *American J. of Economics & Sociology* 50 (1991), 243–55.

77. Smith, E.A. *Reform or revolution?: a diary of reform in England, 1830–1832.* Stroud; Sutton; 1992. Pp x, 165.

78. Smith, E.A. *The House of Lords in British politics and society, 1815–1911.* London; Longman; 1992. Pp xiv, 200.

79. Southall, Humphrey. 'Mobility, the artisan community and popular politics in early nineteenth-century England', Bc55, 103–30.

80. Summers, Anne. 'The correspondence of Havelock Ellis', *History Workshop J.* 32 (1991), 167–83.

81. Taylor, Anne. *Annie Besant: a biography*. Oxford; Oxford UP; 1992. Pp ix, 383.
82. Taylor, Michael W. *Men versus the state: Herbert Spencer and late Victorian individualism*. Oxford; Oxford UP; 1992. Pp x, 292.
83. Taylor, Tony. '"A flame more bright than lasting": the Tory Party, political reform and educational change, 1865–1870', *J. of Educational Administration & History* 23/2 (1991), 62–74.
84. Wells, Roger. 'Southern Chartism', *Rural History* 2 (1991), 37–59.
85. Wrigley, Chris. *Lloyd George*. Oxford; Blackwell; 1992. Pp 170.

(c) *Constitution, Administration and Law*

1. Ballhatchet, Joan. 'The police and the London dock strike of 1889', *History Workshop J.* 32 (1991), 54–68.
2. Bartrip, Peter. 'A "pennurth of arsenic for rat poison": the Arsenic Act, 1851 and the prevention of secret poisoning', *Medical History* 36 (1992), 53–69.
3. Clifton, Gloria C. *Professionalism, patronage and public service in Victorian London: the staff of the Metropolitan Board of Works, 1856–1889*. London; Athlone; 1992. Pp 239.
4. Dobash, Russell P.; McLaughlin, Pat. 'The punishment of women in nineteenth-century Scotland: prisons and inebriate institutions', Bc135, 65–94.
5. Doggett, Maeve E. *Marriage, wife-beating and the law in Victorian England: 'sub virga viva'*. London; Weidenfeld & Nicolson; 1992. Pp 210.
6. Donnachie, Ian. '"Utterly irreclaimable": Scottish convict women and Australia, 1787–1852', Bc205, 99–116.
7. Foster, David. 'Arson in East Yorkshire', Bc205, 62–71.
8. Goodman, A.T. 'The Saltash oysterage dispute, 1876–1882, I: the parties, 2: the action', *Devon & Cornwall Notes & Queries* 36 (1990), 243–47; 287–92.
9. Gregory, Roger; Sharman, Frank. 'The influence of Butch Cassidy on the development of English company law', Bc195, 173–86.
10. Jones, David J.V. *Crime in nineteenth-century Wales*. Cardiff; Wales UP; 1992. Pp xvi, 295.
11. Kain, Roger J.P.; Oliver, Richard; Baker, Jennifer. 'The tithe surveys of south-west England', Bc47, 89–118.
12. Leneman, Leah. 'When women were not "persons": the Scottish women graduates' case, 1906–1908', *Juridical R.* part 1 for 1991, 109–18.
13. Murdoch, Norman H. 'Salvation Army disturbances in Liverpool, England, 1879–1887', *J. of Social History* 25 (1991–2), 575–93.
14. Oliver, Richard. 'The Ordnance Survey in south-west England', Bc47, 119–43.
15. Perry, C.R. *The Victorian Post Office: the growth of a bureaucracy*. Woodbridge; Boydell; 1992. Pp viii, 308.

16. Pooley, Colin G. 'Contours of crime in Westmorland, c.1880–1910', T. of the Cumberland & Westmorland Antiquarian & Arch. Soc. 92 (1992), 251–71.

17. Porter, J.H. 'Net fishermen and the salmon laws: conflict in late-Victorian Devon', Bc116, 240–50.

18. Quinault, Roland. 'Westminster and the Victorian constitution', T. of the Royal Historical Soc. 6th ser. 2 (1992), 79–104.

19. Richer, A.F. 'Early years of policing in Bedfordshire', Bedfordshire Magazine 22 (1991), 309–16.

20. Skinner, Raymond J. 'The Woodlands, Calne, and the Charlesworth case', Wiltshire Arch. & Natural History Magazine 85 (1992), 114–20.

21. Smith, Roger. 'Legal frameworks for psychiatry', Bc145, 137–51.

22. Swift, Roger. 'The English urban magistracy and the administration of justice during the early nineteenth century: Wolverhampton, 1815–1860', Midland History 17 (1992), 75–92.

(d) *External Affairs*

1. Angerer, Thomas. 'Henry Wickham Steed, Robert William Seton-Watson und die Habsburgermonarchie: ihr Haltungswandel bis Kriegsanfang im Vergleich' [Henry Wickham Steed, Robert William Seton-Watson and the Habsburg monarchy: a comparison of their developing attitudes down to the outbreak of war], Mitteilungen des Instituts für Österreichische Geschichtsforschung 99 (1991), 435–73.

2. Barr, William. 'Emile Frederic de Bray and his role in the Royal Navy's search for Franklin, 1852–1854', Canada's missing dimension ed. C.R. Harrington (Ottawa; Musée Canadien de la Nature; 1990), vol. 2, 768–92.

3. Bertram, Marshall. Birth of Anglo-American friendship: the prime facet of the Venezuelan boundary dispute—a study of the interraction of diplomacy and public opinion. Lanham; UP of America; 1992. Pp 164.

4. Brailey, Nigel. 'Sir Ernest Satow, Japan and Asia: the trials of a diplomat in the age of high imperialism', Historical J. 35 (1992), 115–50.

5. Doering-Manteuffel, Anselm. Vom Wiener Kongress zur Pariser Konferenz: England, die deutsche Frage und das Machtesystem, 1815–1856 [From the congress of Vienna to the Paris conference: England, the German question and the power system, 1815–1856]. Göttingen; Vandenhoeck & Ruprecht; 1991. Pp 351.

6. Johnston, Henry Butler McKenzie. Missions to Mexico: a tale of British diplomacy in the 1820s. London; British Academic; 1992. Pp 300.

7. Judge, R. 'May Day and merrie England', Folklore 101 (1991), 131–48.

8. Kiesewetter, Hubert. 'Competition for wealth and power: the growing rivalry between industrial Britain and industrial Germany, 1815–1914', J. of European Economic History 20 (1991), 271–99.

9. Kornicki, P.F. 'Ernest Mason Satow (1843–1929)', Bc17, 76–85; 297.

10. Lahme, Rainer. 'Das Ende der Pax Britannica: England und die europäischen Mächte, 1890–1914' [The end of the Pax Britannica:

England and the European powers, 1890–1914], *Archiv für Kulturges-chichte* 73 (1991), 169–92.

11. Mallmann, Wilhelm E. 'Grossbritanniens Beziehungen zu Österreich in den Tagebuchern lord Stanleys' [Great Britain's relations with Austria in lord Stanley's diary], *Etudes Danubiennes* 4 (1988), 73–80.

12. Massie, Robert K. *Dreadnought: Britain, Germany, and the coming of the Great War.* London; Cape; 1992. Pp xxxi, 1007.

13. Neilson, Keith. '"Greatly exaggerated": the myth of the decline of Great Britain before 1914', *International History R.* 13 (1991), 695–725.

14. Nicholls, David. 'William Stokes (1803–1881)', *Manchester Region History R.* 6 (1992), 59–61.

15. Nish, Ian. 'Hayashi Tadasu (1850–1913)', Bc17, 147–56;; 303–04.

16. Thompson, Andrew. 'Informal empire?: an exploration in the history of Anglo-Argentine relations, 1810–1914', *J. of Latin American Studies* 14 (1992), 419–36.

17. Waddell, D.A.G. 'Anglo-Spanish relations and the recognition of Spanish American independence', *Anuario de Estudios Americanos* 48 (1991), 435–62.

(e) *Religion*

1. Arnstein, Walter L. 'Religious Victorians', *J. of British Studies* 31 (1992), 300–06.

2. Aspinwall, Bernard. 'Children of the dead end: the formation of the modern archdiocese of Glasgow, 1815–1914', *Innes R.* 43 (1992), 119–44.

3. Bamford, K.G. 'The Rev. Dr J.E.N. Molesworth, vicar of Rochdale, 1839–1877', *T. of the Historic Soc. of Lancashire & Cheshire* 141 (1991), 261–88.

4. Band, Thomas A. 'A turbulent priest at Stratford-upon-Avon, 1852–1857', *Warwickshire History* 8 (1991), 30–49.

5. Binfield, Clyde. 'Collective sovereignty? Conscience in the gathered Church, *c.*1875–1918', Bc12, 479–506.

6. Binfield, Clyde. '"I suppose you are not a Baptist or a Roman Catholic?"', Bc196, 81–108.

7. Brown, Stewart J. 'Thomas Chalmers and the communal ideal in Victorian Scotland', Bc196, 61–80.

8. Buckley, W. Kemmis. 'James Buckley—itinerant preacher, 1770–1839', Bc101, 261–73.

9. Chadwick. Roger. 'Sir Augustus Stephenson and the Prosecution of Offences Act of 1884', Bc195, 201–10.

10. Coakley, J.F. *The church of the east and the Church of England: a history of the archbishop of Canterbury's Assyrian mission.* Oxford; Oxford UP; 1992. Pp 422.

11. Coats, Jerry. 'John Henry Newman's "Tamworth Reading Room": adjusting rhetorical approaches for the periodical press', *Victorian Periodicals R.* 24 (1991), 173–80.

12. Coleman, Bruce. 'The nineteenth century: non-conformity' [church in Devon and Cornwall], Bc148, 129–55; 218.
13. Cox, Jeffrey. 'On the limits of social history: nineteenth-century evangelicalism', *J. of British Studies* 31 (1992), 198–203.
14. Darragh, James. 'Thomas Mitchell: a late eighteenth century convert', *Innes R.* 43 (1992), 70–76.
15. Davis, R.W.; Helmstadter, R.J., 'Introduction' [religion and irreligion in Victorian society], Bc13, 1–6.
16. Dekar, Paul R. 'Baptist peacemakers in nineteenth-century peace societies', *Baptist Q.* 34 (1991), 3–12.
17. Feldman, David. 'Popery, Rabbinism, and reform: Evangelicals and Jews in early Victorian England', Bc189, 379–86.
18. Foster, Stewart. '"Dismal Johnny": a companion of Newman recalled', *Recusant History* 21 (1992), 99–110.
19. Foster, Stewart. 'The English seminary at Bruges: some Scottish connections' [*c.*1860–1874], *Innes R.* 43 (1992), 156–65.
20. Franklin, R. William. *Nineteenth-century churches: the history of a new Catholicism in Württemberg, England, and France.* New York; Garland; 1988. Pp 556.
21. Franklin, R. William. 'The impact of Germany on the Anglican Catholic revival in nineteenth-century Britain', *Anglican & Episcopal History* 62 (1992), 433–48.
22. Gay, Peter. 'The manliness of Christ', Bc13, 102–16.
23. Gibson, William T. 'Disraeli's church patronage, 1868–1880', *Anglican & Episcopal History* 62 (1992), 197–210.
24. Gibson, William T. 'The professionalization of an elite: the nineteenth century episcopate', *Albion* 23 (1991), 459–82.
25. Gill, Sean. '"In a peculiar relation to Christianity": Anglican attitudes to Judaism in the era of political emancipation, 1830–1858', Bc189, 399–407.
26. Green, S.J.D. '"Spiritual science" and conversion experience in Edwardian Methodism: the example of west Yorkshire', *J. of Ecclesiastical History* 43 (1992), 428–46.
27. Guelzo, Allen C. 'A test of identity: the vestments controversy in the Reformed Episcopal Church, 1873–1897', *Anglican & Episcopal History* 62 (1992), 303–24.
28. Hamilton, Ian W.F. *The erosion of Calvinist orthodoxy: seceders and subscription in Scottish Presbyterianism.* Edinburgh; Rutherford; 1990. Pp vi, 213.
29. Hatcher, Stephen. 'The sacrament of the Lord's Supper in early primitive Methodism, with particular reference to the Hull Circuit and its branches', *P. of the Wesley Historical Soc.* 47 (1989–90), 221–31.
30. Helmstadter, R.J. 'The reverend Andrew Reed (1787–1862): evangelical pastor as entrepreneur', Bc13, 7–28.
31. Herbert, Sandra. 'Between Genesis and geology: Darwin and some contemporaries in the 1820s and 1830s', Bc13, 68–84.
32. Hinchliff, Peter. *God and history: aspects of British theology, 1875–1914.* Oxford; Oxford UP; 1992. Pp ix, 267.

33. Horridge, Glenn K. 'Invading Manchester: responses to the Salvation Army, 1878–1900', *Manchester Region History R.* 6 (1992), 16–29.

34. Howsam, Leslie. *Cheap bibles: nineteenth-century publishing and the British and Foreign Bible Society.* Cambridge; Cambridge UP; 1991. Pp xviii, 245.

35. Huffman, Joan B. 'For kirk and crown: the rebuilding of Crown Court Church, 1905–1909', *London J.* 17 (1992), 54–70.

36. Itzkowitz, David C. 'Cultural pluralism and the Board of Deputies of British Jews', Bc13, 85–101.

37. Jacob, W.M. 'Henry Styleman le Strange: Tractarian, artist, squire', Bc194, 393–403.

38. Jaki, S.L. 'Newman and evolution', *Downside R.* 109 (1991), 16–34.

39. Jolly, Margaret. '"To save the girls for brighter and better lives": presbyterian missions and women in the south of Vanuatu, 1848–1870', *J. of Pacific History* 26 (1991), 27–48.

40. Jones, D.R.L. *Richard and Mary Pendrill Llewelyn: a Victorian vicar of Llangynwyd and his wife.* Parochial Church Council of the Parish of Llangynwyd with Maesteg; 1991. Pp 36.

41. Kane, Paula M. '"The willing captive of home?": the English Catholic Women's League, 1906–1920', *Church History* 60 (1991), 331–55.

42. Klottrup, Alan Coates. *George Waddington: dean of Durham, 1840–1869.* Durham; Dean and Chapter of Durham; 1990. Pp 28.

43. Knight, Frances. 'The bishops and the Jews, 1828–1858', Bc189, 387–98.

44. Kollar, Rene. *The return of the Benedictines to London: a history of Ealing abbey from 1896 to independence.* Tunbridge Wells; Burns and Oates; 1989. Pp xi, 228.

45. Kollar, Rene. 'Monks, pilgrims, tourists and the attractions of Caldey island in the time of abbot Aelred Carlyle', *J. of Welsh Ecclesiastical History* 9 (1992), 52–63.

46. Kruppa, Patricia S. '"More sweet and liquid than any other": Victorian images of Mary Magdalene', Bc13, 117–32.

47. Mangan, J.A. 'Lamentable barbarians and pitiful sheep: rhetoric of protest and pleasure in late Victorian and Edwardian "Oxbridge"', *Victorian Studies* 34 (1990–1), 473–89.

48. Marsh, Joss Lutz. '"Bibliolatry" and "Bible smashing": G.W. Foote, George Meredith, and the heretic trope of the book', *Victorian Studies* 34 (1990–1), 315–36.

49. Maynard, W.B. 'The response of the Church of England to economic and demographic change: the archdeaconry of Durham, 1800–1851', *J. of Ecclesiastical History* 42 (1991), 437–62.

50. McCalman, I.D. 'Popular irreligion in early Victorian England: infidel preachers and radical theatricality in 1830s London', Bc13, 51–67.

51. McCann, Timothy J. '"A bumping pitch and a blinding light": Henry Manning and the other religion', *Recusant History* 21 (1992), 287–91.

52. McClelland, V. Alan. 'O felix Roma!: Henry Manning, Cutts Robinson and sacerdotal formation, 1862–1872', *Recusant History* 21 (1992), 180–217.

53. McElligott, Ignatius. 'Blessed Dominic Barberi and the Tractarians', *Recusant History* 21 (1992), 51–85.

54. McLeod, Hugh. 'Urbanization and religion in 19th century Britain', *Seelsorge und Diakonie in Berlin*, ed. Kaspar Elm and Hans D. Loock (Berlin; de Gruyter; 1990), 63–80.

55. McLeod, Hugh. 'Varieties of Victorian belief' [review article], *J. of Modern History* 64 (1992), 321–37.

56. Morris, Jeremy Noah. *Religion and urban change, Croydon, 1840–1914*. Woodbridge; Boydell for Royal Historical Soc. (Studies in History no. 65); 1992. Pp xii, 236.

57. Morris, Jeremy Noah. 'The origins and growth of primitive Methodism in east Surrey', *P. of the Wesley Historical Soc.* 48 (1991–92), 133–49.

58. Murdoch, Norman H. 'From militancy to social mission: the Salvation Army and street disturbances in Liverpool, 1879–1887', Bc65, 160–72.

59. Newsome, David. 'Cardinal Manning and his influence on the Church and nation', *Recusant History* 21 (1992), 136–51.

60. Nicholls, M.K. 'Charles Haddon Spurgeon, 1834–1892: church planter', *Baptist Q.* 34 (1991), 20–29.

61. O'Brien, Susan. 'Making Catholic spaces: women, decor, and devotion in the English Catholic church, 1840–1900', Bc194, 449–64.

62. O'Gorman, Christopher. 'A history of Henry Manning's religious opinions, 1808–1832', *Recusant History* 21 (1992), 152–66.

63. Pattison, Robert. *The great dissent: John Henry Newman and the liberal heresy*. Oxford; Oxford UP; 1991. Pp xiii, 231.

64. Pereiro, James. '"Truth before peace": Manning and infallibility', *Recusant History* 21 (1992), 218–53.

65. Quiney, Anthony. 'The church of St Augustine and its builders', *T. of the Ancient Monuments Soc.* 36 (1992), 1–12.

66. Quinn, Dermot A. 'Manning as politician', *Recusant History* 21 (1992), 267–86.

67. Roberts, Alasdair F.B. 'William McIntosh: an untypical link between east and west Highland Catholicism', *Innes R.* 42 (1991), 137–42.

68. Rowell, Geoffrey. '"Remember Lot's wife": Manning's Anglican sermons', *Recusant History* 21 (1992), 167–79.

69. Royle, Edward. 'The faces of Janus: free-thinkers, Jews, and Christianity in nineteenth-century Britain', Bc189, 409–18.

70. Ruston, Alan. 'Clementia Taylor', *T. of the Unitarian Historical Soc.* 20 (1991–2), 62–68.

71. Ruston, Alan. 'The Lawrence family: nineteenth century Unitarian Forsytes?', *T. of the Unitarian Historical Soc.* 20 (1991–2), 126–36.

72. Schwieso, Joshua J. 'The founding of the Agapemone at Spaxton, 1845–1846', *Somerset Archaeology & Natural History* 135 (1992 for 1991), 113–21.

73. Scotland, N.A.D. 'John Bird Sumner, 1780–1862: Claphamite evangelical pastor and prelate', *B. of the John Rylands University Library of Manchester* 74/1 (1992), 57–73.

74. Singleton, John. 'The Virgin Mary and religious conflict in Victorian Britain', *J. of Ecclesiastical History* 43 (1992), 16–34.

75. Soffer, Reba N. 'History and religion: J.R. Seeley and the burden of the past', Bc13, 133–50.

76. Stetz, Margaret Diane. 'Sex, lies, and printed cloth: bookselling at the Bodley Head in the eighteen-nineties', *Victorian Studies* 35 (1991–2), 71–86.

77. Stinchcombe, Owen. 'Elizabeth Malleson and Unitarianism', *T. of the Unitarian Historical Soc.* 20 (1991–2), 56–61.

78. Taylor, Brian. 'Alexander's apostasy: first steps to Jerusalem', Bc189, 363–71.

79. Thurmer, John. 'The nineteenth century: the Church of England' [Devon and Cornwall], Bc148, 109–28; 216–18.

80. Waddams, S.M. *Law, politics and the Church of England: the career of Stephen Lushington, 1782–1873.* Cambridge; Cambridge UP; 1992. Pp xxii, 370.

81. Walker, Pamela J. '"I live but not yet I for Christ liveth in me": men and masculinity in the Salvation Army, 1865–1890', Bc223, 92–112.

(f) *Economic Affairs*

1. Aldcroft, Derek H. 'Technical and structural factors in British industrial decline 1870 to the present', Bc22, 107–19.

2. Anon. 'Roots of red Clydeside: the labour unrest in west Scotland, 1910–1914', Bc175, 81–105.

3. Arena, Richard. 'De l'usage de l'histoire dans la formulation des hypothèses de la theorie économique' [The use of history in the formulation of economic theory], *Revue Economique* 42 (1991), 395–410.

4. Armstrong, John. 'Railways and coastal shipping in Britain in the later nineteenth century: co-operation and competition', Bc117, 76–103.

5. Barke, Michael. 'The middle-class journey to work in Newcastle upon Tyne, 1850–1913', *J. of Transport History* 3rd ser. 12 (1991), 107–34.

6. Berghoff, Hartmut. 'A reply to W.D. Rubinstein's response', *Business History* 34/2 (1992), 82–85.

7. Bissett, Allan. 'J.C. Moore-Stevens and the glovers of Great Torrington', *Devonshire Association Report & T.* 124 (1992), 153–66.

8. Boyce, Gordon. '64thers, syndicates, and stock promotions: information flows and fund-raising techniques of British shipowners before 1914', *J. of Economic History* 52 (1992), 181–205.

9. Boyce, Gordon. 'Corporate strategy and accounting systems: a comparison of developments at two British steel firms, 1898–1914', *Business History* 34/1 (1992), 42–65.

10. Boyns, Trevor. 'Industrialization' [in Swansea], Bc212, 34–50.

11. Brewer, Anthony. 'Economic growth and technical change: John Rae's critique of Adam Smith', *History of Political Economy* 23 (1991), 1–11.

12. Bryan, Tim. *The golden age of the Great Western Railway, 1895–1914.* Sparkford; Stephens; 1991. Pp 256.

13. Budgen, Chris. 'The Bramley and Rudgwick turnpike trust', *Surrey Arch. Collections* 81 (1991–2), 97–102.

14. Burk, Kathleen. 'Money and power: the shift from Great Britain to the United States', Bc95, 359–69.
15. Burt, Roger; Timbell, Martin. 'Multiple products and the economics of the mining industries: the case of arsenic production in south-west England, 1850–1914', *J. of European Economic History* 20 (1991), 379–406.
16. Burt, Roger; Waite, Peter B.; Burnley, Raymond. *The mines of Flintshire & Denbighshire: metalliferous and associated minerals, 1845–1913.* Exeter; Exeter UP; 1992. Pp 198.
17. Burt, Roger; Waite, Peter B.; Burnley, Raymond. *The mines of Shropshire and Montgomeryshire with Cheshire and Staffordshire: metalliferous and associated minerals, 1845–1913.* Exeter; Exeter UP; 1990. Pp xxxvii, 104.
18. Buurman, Gary B. 'A comparison of the single tax proposals of Henry George and the physiocrats', *History of Political Economy* 23 (1991), 481–96.
19. Campion, Peter. *Matches from Gloucester: the Standard Match Co. Ltd, Hempsted, Gloucester.* Gretton; Campion; 1991. Pp 42.
20. Chandler, Alfred D., Jr. 'Creating competitive capability: innovation and investment in the United States, Great Britain, and Germany from the 1870s to world war I', Bc32, 432–58.
21. Chapman, John. 'The later parliamentary enclosures of south Wales', *Agricultural History R.* 39 (1991), 116–25.
22. Checkland, Olive. 'Richard Henry Brunton and the Japan lights, 1868–1876, a brilliant and abrasive engineer', *Newcomen Soc. T.* 63 (1992), 217–28.
23. Collins, Michael. 'The Bank of England as lender of last resort, 1857–1878', *Economic History R.* 2nd ser. 45 (1992), 145–53.
24. Cottrell, P.L. 'The domestic commercial banks and the City of London, 1870–1939', Bc95, 39–62.
25. Coull, J.R. 'The development of the herring fishery in the Peterhead district of Scotland before world war I', *Sjofartshisstorisk Arsbok* (1990), 119–41.
26. Cull, Robert; Davis, Lance E. 'Un, deux, trois, quatre marchés?: l'intégration du marché du capital, Etats-Unis et Grande-Bretagne (1865–1913)', *Annales* 47 (1992), 633–74.
27. Daniels, Stephen. 'Victorian Britain: 1837- c.1900', Bc181, 186–210; 244–45.
28. Daunton, M.J. 'Finance and politics: comments', Bc95, 283–90.
29. Daunton, M.J. 'Financial élites and British society, 1880–1950', Bc95, 121–46.
30. Davis, Tracy C. *Actresses as working women: their social identity in Victorian culture.* London; Routledge; 1991. Pp xvi, 200.
31. Davis, Tracy C. 'The theatrical employees of Victorian Britain: demography of an industry', *Nineteenth-Century Theatre* 18 (1990), 5–34.
32. Dawson, Charles. 'A gratifying testimonial to a captain', *American Neptune* 52 (1992), 46–51.
33. Dewhurst, M. 'Fertiliser progress, 1841–1991: a review of the development of mineral and organic fertilisers', Bc36, 53–67.

34. Duncan, Colin A.M. 'Legal protection for the soil of England: the spurious context of nineteenth century "progress"', *Agricultural History* 66/2 (1992), 75–94.

35. Feltes, N.N. 'Misery or the production of misery: defining sweated labour in 1890', *Social History* 17 (1992), 441–52.

36. Ferns, H.S. 'The Baring crisis, revisited', *J. of Latin American Studies* 14 (1992), 241–73.

37. Fisher, J.R. 'The first Nottinghamshire Agricultural Association, 1837–1850', *T. of the Thoroton Soc.* 94 (1991 for 1990), 62–68.

38. Gibson, J.S.W. '"The immediate route from the metropolis to all parts ..."', *Cake & Cockhorse* 12 (1991), 10–24.

39. Gilbert, David. *Class, community, and collective action: social change in two British coalfields, 1850–1926.* Oxford; Oxford UP; 1992. Pp x, 293.

40. Greasley, David. 'The market for south Wales coal, 1874–1913', *J. of European Economic History* 21 (1992), 135–52.

41. Green, E.H.H. 'The influence of the City over British economic policy, c.1880–1960', Bc95, 193–218.

42. Griffin, Colin. 'An industrial revolution in the east midland coalfields between c.1850 and c.1880? The case of High Park "superpit", Nottinghamshire', *T. of the Thoroton Soc.* 94 (1991 for 1990), 75–82.

43. Griffiths, John C. *The third man: the life and times of William Murdoch, 1754–1839: the inventor of gas lighting.* London; Deutsch; 1992. Pp 373.

44. Gwyn, Julian. 'The economics of the transatlantic slave trade: a review', *Histoire Sociale/Social History* 25 (1992), 151–62.

45. Harris, José. 'Financial elites and society: comments', Bc95, 187–90.

46. Heim, Carol E.; Mirowski, Philip. 'Crowding out: a response to Black and Gilmore', *J. of Economic History* 51 (1991), 701–06.

47. Howe, A.C. 'Free trade and the City of London, c.1820–1870', *History* 77 (1992), 391–410.

48. Howell, David W. 'Railway safety and labour unrest: the Aisgill railway disaster of 1913', Bc117, 123–54.

49. Huberman, Michael. 'Industrial relations and the industrial revolution: evidence from M'Connel and Kennedy, 1810–1840', *Business History R.* 65 (1991), 345–78.

50. Hunt, W.M. 'The Lincoln "high bridge" scheme', *J. of the Railway & Canal Historical Soc.* 30 (1991), 270–82.

51. James, Harold. 'Banks and economic development: comments', Bc95, 113–18.

52. Jamieson, Alan G. 'Credit insurance and trade expansion in Britain, 1820–1900', *Accounting, Business & Financial History* 1 (1991), 163–76.

53. Jarvis, Adrian. 'G.F. Lyster and the role of the dock engineer, 1861–1897', *Mariner's Mirror* 78 (1992), 177–99.

54. Jervis, Martin R. 'The Padiham power loom weavers' strike of 1859', *Manchester Region History R.* 6 (1992), 30–41.

55. Kanth, Rajani K. 'Ricardo and laissez-faire: the hidden connection', *British R. of Economic Issues* 13/2 (1991), 47–55.

56. Lovell, John C. 'Employers and craft unionism: a programme of action for British ship-building, 1902–1905', *Business History* 34/4 (1992), 38–58.

57. Lynn, Martin. 'British business and the African trade: Richard & William King Ltd. of Bristol and West Africa, 1833–1918', *Business History* 34/4 (1992), 20–37.

58. MacLeod, Christine. 'Strategies for innovation: the diffusion of new technology in nineteenth-century British industry', *Economic History R.* 2nd ser. 45 (1992), 285–307.

59. Mathias, Peter. 'Resources and technology', Bc22, 18–42.

60. Mathias, Peter; Davis, John A. 'Introduction' [innovation and technology in Europe], Bc22, 1–5.

61. McDonald, John; Shlomowitz, Ralph. 'Passenger fares on sailing vessels to Australia in the nineteenth century', *Explorations in Economic History* 28 (1991), 192–208.

62. McMurry, Sally. 'Women's work in agriculture: divergent trends in England and America, 1800 to 1930', *Comparative Studies in Soc. & History* 34 (1992), 248–70.

63. Miller, C. 'Master and man: farmers and employees in nineteenth-century Gloucestershire', Bc116, 199–209.

64. Miller, Norman.; Miller, Margaret. 'The Carlisle to Glasgow road: an early 19th century attempt to improve and maintain Scotland's most important road', *Dumfriesshire & Galloway Natural History & Antiquarian Soc. T.* 65 (1990), 100–05.

65. Molteni de Villermont, Claude. 'Les principales compagnies anglaises sur l'Atlantique nord', *Neptunia* 184 (1991), 8–15.

66. Morris, R.J. 'The state, the elite and the market: the "visible hand" in the British industrial city system', Bc42, 177–99.

67. Moss, David J. 'The Bank of England and the establishment of a branch system, 1826–1829', *Canadian J. of History* 27 (1992), 48–65.

68. O'Brien, Patrick K.; Pigman, Geoffrey Allen. 'Free trade, British hegemony and international economic order in the nineteenth-century', *R. of International Studies* 18 (1992), 89–113.

69. O'Connor, Bernard. 'The coprolite industry in Buckinghamshire', *Records of Buckinghamshire* 32 (1991 for 1990), 76–90.

70. Parkinson, Anthony. 'Slate quarries: problems of survey, conservation and preservation' [Wales], Bc129, 30–31.

71. Pearson, Robin O. 'Collective diversification: Manchester cotton merchants and the insurance business in the early nineteenth century', *Business History R.* 65 (1991), 379–414.

72. Perren, Richard. 'The manufacture and marketing of veterinary products from 1850 to 1914', *Veterinary History* ns 6/2 (1990), 43–61.

73. Phillips, Martin. 'The evolution of markets and shops in Britain', Bc35, 53–75.

74. Pollard, Sidney; Ziegler, Dieter. 'Banking and industrialization: Rondo Cameron twenty years on', Bc95, 17–36.

75. Pollins, Harold. 'British horse tramway company accounting practices, 1870–1914', *Accounting, Business & Financial History* 1 (1991), 279–302.

76. Power, M.J. 'The growth of Liverpool', Bc65, 21–37.
77. Proudlock, Noel. *Leeds: a history of its tramways*. Leeds; Proudlock; 1991. Pp 184.
78. Purvis, Martin. 'Co-operative retailing in Britain', Bc35, 107–34.
79. Read, Donald. *The power of news: the history of Reuters, 1849–1989*. Oxford; Oxford UP; 1992. Pp xii, 431.
80. Reed, Peter. 'The British chemical industry and the indigo trade', *British J. for the History of Science* 25 (1992), 113–25.
81. Roberts, R. Frederick. 'Ships built on the river Clwyd', *Maritime Wales* 14 (1991), 16–21.
82. Robertson, Paul L.; Alston, Lee J. 'Technological choice and the organization of work in capitalist firms', *Economic History R*. 2nd ser. 45 (1992), 330–49.
83. Rodger, Richard. 'Managing the market—regulating the city: urban control in the nineteenth-century United Kingdom', Bc42, 200–19.
84. Ross, M.S. 'Brickmaking at Gillingham and Motcombe, Dorset', *Dorset Natural History & Arch. Soc. P*. 113 (1992 for 1991), 17–22.
85. Rubinstein, W.D. 'British businessmen as wealth-holders, 1870–1914: a response', *Business History* 34/2 (1992), 69–81.
86. Shaw, Gareth. 'The European scene: Britain and Germany', Bc35, 17–34.
87. Shaw, Gareth. 'The evolution and impact of large-scale retailing in Britain', Bc35, 135–65.
88. Sinclair, Robert C. *Across the Irish Sea: Belfast-Liverpool shipping since 1819*. London; Conway; 1990. Pp 192.
89. Smith, Dennis. 'Paternalism, craft and organizational rationality, 1830–1930: an exploratory model', *Urban History* 19 (1992), 211–28.
90. Smith, Dennis. 'The works of William Tierney Clark (1783–1852), civil engineer of Hammersmith', *Newcomen Soc. T*. 63 (1992), 181–207.
91. Stoneman, Paul L. 'Technological diffusion: the viewpoint of economic theory', Bc22, 162–84.
92. Sturgess, Roy. 'The dairy industry of north Staffordshire and Derbyshire, 1875–1900', *Staffordshire Studies* 2 (1990), 45–58.
93. Summers, David W. 'Differences in the response of two Aberdeenshire fishing villages to large scale changes within the herring fishery, 1880–1914', *Northern Scotland* 11 (1991), 45–54.
94. Taplin, Eric. 'False dawn of new unionism? Labour unrest in Liverpool, 1871–1873', Bc65, 135–159.
95. Turner, Michael. 'Output and prices in UK agriculture, 1867–1914, and the great agricultural depression reconsidered', *Agricultural History R*. 40 (1992), 38–51.
96. Tweedale, Geoffrey. 'The razor blade king of Sheffield: the forgotten career of Paul Kuehnrich', *T. of the Hunter Arch. Soc*. 16 (1991), 39–51.
97. Walton, Whitney. *France at the Crystal palace*. Berkeley (Ca)/Oxford; California UP; 1992. Pp xii, 240.
98. Waterman, Anthony M.C. '"The canonical classical model of political economy" in 1808, as viewed from 1825: Thomas Chalmers on the "national resources"', *History of Political Economy* 23 (1991), 221–42.

99. Waugh, Mary. *Smuggling in Devon and Cornwall, 1700–1850*. Newbury; Countryside; 1991. Pp 208.
100. Wilkins, Mira. 'Extra-European financial centres: comments', Bc95, 429–35.
101. Williams, Herbert. *Davies the ocean: railway king and coal tycoon*. Cardiff; Wales UP; 1991. Pp xii, 258.
102. Williamson, Jeffrey G. 'British inequality during the industrial revolution: accounting for the Kuznets curve', Bc34, 57–75.
103. Wilson, Ted. 'The battle for the standard: the bimetallic movement in Manchester', *Manchester Region History R*. 6 (1992), 49–58.
104. Young, Craig. 'Scottish sequestrations and the role of women in the small firm: an assessment of a new source for women's history', *J. of the Soc. of Archivists* 13 (1992), 143–51.
105. Ziegler, Dieter. 'The banking crisis of 1878: some remarks', *Economic History R*. 2nd ser. 45 (1992), 137–44.

(g) *Social Structure and Population*

1. Aldrich, Zoë. 'The adventuress: *Lady Audley's secret* as novel, play and film', Bc63, 159–74.
2. Aston, Elaine. 'The "new woman" at Manchester's Gaiety theatre', Bc63, 205–20.
3. Belchem, John. 'Image, myth and implantation: the peculiarities of Liverpool, 1800–1850', *Tijdschift voor Sociale Geschiedenis* 18 (1992), 263–74.
4. Bevir, Mark. 'The British Social Democratic Federation 1880–1885: from O'Brienism to Marxism', *International R. of Social History* 37 (1992), 207–29.
5. Blaikie, J.A.D. 'The country and the city: sexuality and social class in Victorian Scotland', Bc55, 80–102.
6. Booth, P. 'Herbert Minton: nineteenth century pottery manufacturer and philanthropist', *Staffordshire Studies* 3 (1991), 63–85.
7. Bratton, J.S. 'Irrational dress', Bc63, 77–91.
8. Breuilly, John J. 'Civil society and the labour movement, class relations and the law: a comparison between Germany and England', Bc219, 160–96.
9. Breuilly, John J. 'The labour aristocracy in Britain and Germany: a comparison', Bc219, 26–75.
10. Bryson, Anne. 'Riotous Liverpool, 1815–1860', Bc65, 98–134.
11. Carter, Harold; Lewis, C. Roy. *An urban geography of England and Wales in the nineteenth century*. London; Arnold; 1990. Pp v, 226.
12. Carter, Ian (ed.). *Rural life in Victorian Aberdeenshire* by William Alexander. Edinburgh; Mercat; 1992. Pp 176.
13. Chase, Malcolm. 'The implantation of working class organisation on Teesside, 1830–1874', *Tijdschift voor Sociale Geschiedenis* 18 (1992), 191–211.
14. Cherry, R. 'Race and gender aspects of Marxian macromodels: the case of the social structure of accumulation school, 1848–1868', *Science & Soc.* 55 (1991), 60–78.

15. Childs, Michael J. *Labour's apprentices: working-class lads in late Victorian and Edwardian England*. London; Hambledon; 1992. Pp 223.

16. Clark, Anna. 'The rhetoric of Chartist domesticity: gender, language and class in the 1830s and 1840s', *J. of British Studies* 31 (1992), 62–88.

17. Dalziel, Raewyn. 'Emigration and kinship: migrants to New Plymouth, 1840–1843', *New Zealand J. of History* 25 (1991), 112–28.

18. Davies, H. Rhodri. 'Automated record linkage of census enumerators' books and registration data: obstacles, challenges and solutions', *History & Computing* 4 (1992), 16–26.

19. Davies, Russell. '"Do not go gentle into that good night"? Women and suicide in Carmarthenshire, *c.*1860–1920', Bc66, 93–108.

20. Davis, G. 'The scum of Bath: the Victorian poor', Bc116, 183–98.

21. Davis, Jill. 'The new woman and the new life', Bc63, 17–36.

22. Davis, Jill; Davis, Tracy C. 'The people of the "People's theatre": the social demography of the Britannia theatre (Hoxton)', *Theatre Survey* 32 (1991), 137–72.

23. Davis, Richard W. '"We are all Americans now!": Anglo-American marriages in the later nineteenth century', *P. of the American Philosophical Soc.* 135 (1991), 140–99.

24. Day, Helen. 'Female daredevils', Bc63, 137–58.

25. Digby, Anne. 'Victorian values and women in public and private', Bc196, 195–216.

26. Dyck, Ian. *William Cobbett and rural popular culture*. Cambridge; Cambridge UP; 1992. Pp xvi, 312.

27. Dymkowski, Christine. 'Entertaining ideas: Edy Craig and the Pioneer Players', Bc63, 221–33.

28. Edmonds, Jill. 'Princess Hamlet', Bc63, 59–76.

29. Ferris, Lesley. 'The golden girl', Bc63, 37–55.

30. Fingard, Judith. 'Race and respectability in Victorian Halifax', *J. of Imperial & Commonwealth History* 20 (1992), 169–95.

31. Fitzsimmons, Linda. 'Typewriters enchained: the work of Elizabeth Baker', Bc63, 189–201.

32. Friedlander, Dov. 'The British depression and nuptiality: 1873–1896', *J. of Interdisciplinary History* 23 (1992), 19–37.

33. Gardner, Viv. 'Introduction' [feminism and theatre], Bc63, 1–14.

34. Garnett, E. 'Arkholme and its basketmakers', *T. of the Historic Soc. of Lancashire & Cheshire* 141 (1991), 339–49.

35. Garrigan, Kristine Ottesen (ed.). *Victorian scandals: representations of gender and class*. Athens (Oh); Ohio UP; 1992. Pp 337.

36. Girouard, Mark. 'Victorian values and the upper classes', Bc196, 49–60.

37. Grace, Frank. *The late Victorian town*. Chichester; Phillimore; 1992. Pp xiv, 72.

38. Green, Muriel M. (ed.). *Miss Lister of Shibden hall, Halifax*. Lewes; Book Guild; 1992. Pp 210.

39. Green, S.J.D. 'In search of bourgeois civilisation: institutions and ideals in nineteenth-century Britain' [review article], *Northern History* 28 (1992), 228–47.

40. Greenall, R.L. 'From soldier to shoemaker: Joseph Langhorn in Northampton', *Northamptonshire Past & Present* 8 (1992), 307–14.

41. Hardy, Sheila (ed.). *Diary of a Suffolk farmer's wife, 1854–1869: a woman of her time*. Basingstoke; Macmillan; 1992. Pp 224.

42. Haynes, Barry (ed.). *Working-class life in Victorian Leicester: the Joseph Dare reports*. Leicester; Leicestershire Library Services; 1991. Pp 105.

43. Horn, Pamela. *High society: the English social elite, 1880–1914*. Stroud; Sutton; 1992. Pp vii, 215.

44. Hosgood, Chris. 'A "brave and daring folk"? Shopkeepers and trade associational life in Victorian and Edwardian England', *J. of Social History* 26 (1992–3), 285–308.

45. Howkins, Alun. 'The English farm labourer in the nineteenth century: farm, family and community', Bc54, 85–104.

46. Hudson, Pat. 'Land, the social structure and industry in two Yorkshire townships, c.1660–1800', Bc205, 27–46.

47. Ittmann, Karl. 'Family limitation and family economy in Bradford, West Yorkshire, 1851–1881', *J. of Social History* 25 (1991–2), 547–573.

48. Jaffe, James A. 'Agency and ideology in modern British social history' [review article], *J. of British Studies* 31 (1992), 89–95.

49. Jones, Dot. 'Counting the cost of coal: women's lives in the Rhondda, 1881–1911', Bc66, 109–33.

50. Jones, Ieuan Gwynedd. *Mid-Victorian Wales: the observers and the observed*. Cardiff; Wales UP; 1992. Pp xii, 204.

51. Jordon, T.E. 'Linearity, gender and social class in economic influences on heights of Victorian youths', *Historical Methods* 24 (1991), 116–23.

52. Joyce, Patrick. 'A people and a class: industrial workers and the social order in nineteenth-century England', Bc43, 199–217.

53. Kearns, Gerry; Withers, Charles W.J. 'Introduction: class, community and the processes of urbanisation', Bc55, 1–11.

54. Laithwaite, Michael. *Victorian Ilfracombe: origins and architecture of a north Devon holiday resort*. Tiverton; Devon; 1992. Pp 80.

55. Lawson, Z. 'Shops, shopkeepers, and the working-class community: Preston, 1860–1890', *T. of the Historic Soc. of Lancashire & Cheshire* 141 (1991), 309–28.

56. Lieven, Dominic. *The aristocracy in Europe, 1815–1914*. London; Macmillan; 1992. Pp 300.

57. Lingham, B.F. *The railway comes to Didcot: a history of the town, 1839–1918*. Stroud; Sutton; 1992. Pp vii, 152.

58. Marks, Lara. '"The luckless waifs and strays of humanity": Irish and Jewish immigrant unwed mothers in London, 1870–1939', *Twentieth-Century British History* 3 (1992), 113–37.

59. McClelland, Keith. 'Masculinity and the "representative artisan" in Britain, 1850–1880', Bc223, 74–91.

60. Melling, Joseph. 'Employers, workplace culture and workers' politics: British industry and workers' welfare programmes, 1870–1920', Bc217, 109–36.

61. Morgan, Carol E. 'Women, work and consciousness in the mid nineteenth-century English cotton industry', *Social History* 17 (1992), 23–41.

62. Newton, Judith. 'Engendering history for the middle class: sex and political economy in the *Edinburgh Review*', Bc56, 1–17.

63. Nicholas, Stephen J.; Nicholas, Jacqueline M. 'Male literacy, "deskilling" and the industrial revolution', *J. of Interdisciplinary History* 23 (1992), 1–18.

64. Page, Stephen J. 'The mobility of the poor: a case study of Edwardian Leicester', *Local Historian* 21 (1991), 109–19.

65. Pam, David. *A Victorian suburb: a history of Enfield, vol II: 1837–1914*. Enfield; Enfield Preservation Soc.; 1992. Pp 368.

66. Parsons, Maggy (ed.). *Every girl's duty: the diary of a Victorian debutante*. London; Deutsch; 1992. Pp ix, 182.

67. Rawding, Charles. 'Society and place in nineteenth-century north Lincolnshire', *Rural History* 3 (1992), 59–85.

68. Reed, Michael. 'The peasantry in nineteenth-century England: a neglected class?', Bc116, 210–39.

69. Reid, Alastair. *Social history and social relations in Britain, 1850–1914*. London; Macmillan; 1992. Pp 78.

70. Reid, Alastair. *Social history and the British working class, 1850–1914*. London; Macmillan; 1992. Pp 120.

71. Renvoize, Edward. 'The Association of Medical Officers of Asylums and Hospitals for the Insane, the Medico-Psychological Association, and their presidents', Bc145, 29–78.

72. Rose, Sonya O. *Limited livelihoods: gender and class in nineteenth century England*. London; Routledge; 1992. Pp xi, 292.

73. Royle, Edward. 'Annie Besant's first public lecture', *Labour History R.* 57/3 (1992), 67–69.

74. Rubinstein, W.D. 'Cutting up rich: a reply to F.M.L. Thompson', *Economic History R.* 2nd ser. 45 (1992), 350–61.

75. Rubinstein, W.D. 'The structure of wealth-holding in Britain, 1809–1839: a preliminary anatomy', *Historical Research* 65 (1992), 74–89.

76. Rule, John C. 'A "configuration of quietism"? Attitudes towards trade unions and Chartism among the Cornish miners', *Tijdschift voor Sociale Geschiedenis* 18 (1992), 248–62.

77. Rutherford, Susan. 'The voice of freedom: images of the prima donna', Bc63, 95–113.

78. Seccombe, Wally. 'Working-class fertility decline in Britain: a reply', *Past & Present* 134 (1992), 207–11.

79. Sheppard, June A. 'Small farms in a Sussex Weald parish, 1800–1860', *Agricultural History R.* 40 (1992), 127–41.

80. Shuttleworth, Sally. 'Demonic mothers: ideologies of bourgeois motherhood in the mid-Victorian era', Bc56, 31–51.

81. Stowell, Sheila. 'Drama as a trade: Cicely Hamilton's *Diana of Dobson's*', Bc63, 177–88.

82. Sturman, Christopher. 'Lady Franklin in Lincolnshire, 1835', *Lincolnshire History & Archaeology* 26 (1991), 21–25.

83. Taylor, Peter. 'A divided middle class: Bolton, 1790–1850', *Manchester Region History R.* 6 (1992), 3–15.

84. Thompson, F.M.L. 'Stitching it together again' [reply to W.D. Rubinstein], *Economic History R.* 2nd ser. 45 (1992), 362–75.
85. Tosh, John. 'Domesticity and manliness in the Victorian middle class', Bc223, 44–73.
86. Wells, Roger (ed.). *Victorian village: the diaries of the Reverend John Coker Egerton, curate and rector of Burwash, East Sussex, 1851–1888.* Stroud; Sutton; 1992. Pp 369.
87. Whitbread, Helena (ed.). *No priest but love: excerpts from the diaries of Anne Lister, 1824–1826.* Otley; Smith Settle; 1992. Pp 227.
88. White, E.M. '"Little female lambs": women in the Methodist societies of Carmarthenshire, 1737–1750', *Carmarthenshire Antiquarian* 27 (1991), 31–36.
89. Williams, F.J. 'The emergence of supervisory elites in the nineteenth-century chemical industry in Widnes', *T. of the Historic Soc. of Lancashire & Cheshire* 141 (1991), 289–307.
90. Williamson, Jeffrey G. 'Did England's cities grow too fast during the industrial revolution?', Bc32, 359–94.
91. Withers, Charles W.J. 'Class, culture and migrant identity: Gaelic Highlanders in urban Scotland', Bc55, 55–79.
92. Woods, Robert. *The population of Britain in the nineteenth century.* London; Macmillan; 1992. Pp 88.
93. Woods, Robert. 'Working-class fertility decline in Britain', *Past & Present* 134 (1992), 200–07.
94. Wright, Elizabeth F. 'Thomas Hadden: architectural metalworker', *Soc. of Antiquaries of Scotland P.* 121 (1991), 427–36.

(h) *Social Policy*

1. Alvey, Norman. 'The great voting charities of the metropolis', *Local Historian* 21 (1991), 147–55.
2. Barclay, Jean. 'John Milson Rhodes, 1847–1909: Chorlton guardian and Didsbury doctor', *Manchester Region History R.* 6 (1992), 107–12.
3. Bragshay, Mark. 'Heathcoat's industrial housing in Tiverton, Devon', *Southern History* 13 (1991), 82–104.
4. Bramwell, Bill. 'Public space and local communities: the example of Birmingham, 1840–1880', Bc55, 31–54.
5. Bushrod, Emily. 'The diary of John Gent Brooks: a Victorian commentary on poverty (1844–1854)', *T. of the Unitarian Historical Soc.* 20 (1991–2), 98–113.
6. Cashman, Bernard. *A proper house: Bedford lunatic asylum, 1812–1860.* Bedford; North Bedfordshire Health Authority; 1992. Pp 179.
7. Chinn, Carl. *Homes for people: 100 years of council housing in Birmingham.* Exeter; Birmingham Books; 1991. Pp viii, 135.
8. Crowther, M. Anne. 'The work-house', Bc196, 183–94.
9. Enoch, D.G. 'Schools and inspection as a mode of social control in south-east Wales, 1839–1907', *J. of Educational Administration & History* 22/1 (1990), 9–17.

10. Hall, Mary. 'Poor relief in Eskdale in the early 1800s', *T. of the Cumberland & Westmorland Antiquarian & Arch. Soc.* 92 (1992), 205–12.
11. Harling, Philip. 'The power of persuasion: central authority, local bureauracy and the New Poor Law', *English Historical R.* 107 (1992), 30–53.
12. Harris, José. 'Victorian values and the founders of the welfare state', Bc196, 165–82.
13. Howell, David W. 'Labour organization among agricultural workers in Wales, 1872–1921', *Welsh History R.* 16 (1992–3), 63–92.
14. Jacobs, John. 'Setting Brighton's poor to work: the work of Brighton Distress Committee, 1905–1914', *Sussex Arch. Collections* 129 (1991), 217–37.
15. John, Angela V. 'Beyond paternalism: the ironmaster's wife in the industrial community', Bc66, 43–68.
16. Jones, Kathleen. 'Law and mental health: sticks or carrots?', Bc145, 89–102.
17. Jones, Kathleen. 'The culture of the mental hospital', Bc145, 17–28.
18. Jones, Rosemary A.N. 'Women, community and collective action: the *Ceffyl Pren* tradition', Bc66, 17–41.
19. Kearns, Gerry. 'Biology, class and the urban penalty', Bc55, 12–30.
20. Lebas, Elizabeth; Magri, Susanna; Topulov, Christian. 'Reconstruction and popular housing after the first world war: a comparative study of France, Great Britain, Italy and the United States', *Planning Perspectives* 6 (1991), 249–67.
21. Little, Alan. 'Appendix: Liverpool Chartists, subscribers to the National Land Company, 1847–1848', Bc65, 247–52.
22. Lloyd-Morgan, Ceridwen. 'From temperance to suffrage?', Bc66, 135–58.
23. Lord, Evelyn. 'Conflicting interests: public health, Lammas lands, and pressure groups in nineteenth-century Kingston-on-Thames', *Southern History* 13 (1991), 22–31.
24. Lovell, John C. 'The Northamptonshire Freehold Land Society and the origins of modern Far Cotton', *Northamptonshire Past & Present* 8 (1992), 299–305.
25. Lowe, Jeremy. 'Recording nineteenth century workers' housing: a case study' [Wales], Bc129, 58–61.
26. Mahood, Linda. 'Family ties: lady child-savers and girls of the street, 1850–1925', Bc135, 42–64.
27. Melling, Joseph. 'Welfare capitalism and the origins of welfare states: British industry, workplace welfare and social reform, *c*.1870–1914', *Social History* 17 (1992), 453–78.
28. Mohr, Peter D. 'Gilbert Curlew and the development of crippled children's societies in Victorian Manchester and Salford', *Manchester Region History R.* 6 (1992), 42–48.
29. Morton, Jane. *'Cheaper than Peabody': local authority housing from 1890 to 1919.* York; Joseph Rowntree Foundation; 1991. Pp 63.
30. Nardinelli, Clark. *Child labour and the industrial revolution.* Bloomington (Id); Indiana UP; 1990. Pp 194.

31. Norton, Wayne. 'Malcolm McNeill and the emigrationist alternative to Highland land reform, 1886–1893', *Scottish Historical R.* 70 (1991), 16–30.

32. Powell, Christopher; Fisk, Malcolm J. 'Early industrial housing in Rhondda, 1800 to 1850', *Morgannwg* 35 (1991), 50–78.

33. Shiman, Lilian Lewis. *Women and leadership in nineteenth-century England.* London; Macmillan; 1992. Pp 263.

34. Shoemaker, Susan Turnbull. *'To enlighten, not to frighten': a comparative study of the infant welfare movement in Liverpool and Philadelphia, 1890–1918.* London; Garland; 1991. Pp ix, 336.

35. Todd, John; Ashworth, Lawrence. 'The West Riding asylum and James Crichton-Browne, 1818–1876', Bc145, 389–418.

36. Walkowitz, Judith. *City of dreadful delight: narratives of sexual danger in late Victorian London.* London; Virago; 1992. Pp 353.

37. Wardle, Christopher J. 'Historical influences on services for children and adolescents before 1900', Bc145, 279–93.

38. Watson, Alan; Allan, Elizabeth. 'Depopulation by clearances and non-enforced emigration in the north-east highlands', *Northern Scotland* 10 (1990), 31–46.

39. Wells, Roger. 'Popular protest and social crime: the evidence of criminal gangs in rural southern England, 1790–1860', *Southern History* 13 (1991), 32–81.

(i) *Education*

1. Anderson, R.G.W. '"What is technology?": education through museums in the mid-nineteenth century', *British J. for the History of Science* 25 (1992), 169–84.

2. Bartle, G.F. 'The role of the British and Foreign School Society in Welsh elementary education, 1840–1876', *J. of Educational Administration & History* 22/1 (1990), 18–29.

3. Bayley, Susan N. 'Modern languages as emerging curricular subjects in England, 1864–1918', *History of Education Soc. B.* 47 (1991), 23–31.

4. Betts, Robin. 'The issue of technical education, 1867–1868', *History of Education Soc. B.* 48 (1991), 30–37.

5. Bovill, Donald G. 'The education of boys for the mercantile marine: a study of three nautical schools', *History of Education Soc. B.* 47 (1991), 11–22.

6. Bovill, Donald G. 'The proprietary schools of navigation and marine engineering in the ports of the north-east of England, 1822–1914', *History of Education Soc. B.* 44 (1989), 10–25.

7. Clamp, Peter G. 'A question of education: class, language and schooling in the Isle of Man, 1800–1833', *J. of Educational Administration & History* 23/1 (1991), 1–14.

8. Clamp, Peter G. 'Education and the resurgence of nationalism in the Isle of Man, 1900–1914', *History of Education Soc. B.* 47 (1991), 42–55.

9. Davison, Leigh M. 'Rural education in the late Victorian era: school attendance committees in the East Riding of Yorkshire, 1881–1903', *History of Education Soc. B.* 45 (1990), 7–24.

10. Dew, Barbara. 'The clergy and the village school: the role of the clergyman in Church of England schools in Oxfordshire villages, 1860–1902', *History of Education Soc. B.* 49 (1992), 28–35.

11. Evans, W. Gareth. 'Free education and the quest for popular control, unsectarianism and efficiency: Wales and the Free Elementary Education Act, 1891', *Honourable Soc. of Cymmrodorion T.* (1991), 203–31.

12. Firth, Anthony E. *Goldsmith's college: a centenary account.* London; Athlone; 1991. Pp 160.

13. Harrington, Brian. 'Alexander Bain: a reappraisal', *History of Education Soc. B.* 44 (1989), 46–51.

14. Higginson, J.H. 'The significance of Michael Saddler's pioneer work for researchers today', *History of Education Soc. B.* 45 (1990), 25–37.

15. Jenkins, E.W. 'David Forsyth and the city of Leeds school', Bc84, 75–96.

16. Kearney, Anthony. 'English versus history: the battle for identity and status, 1850–1920', *History of Education Soc. B.* 48 (1991), 22–29.

17. Kearney, Anthony. 'Leslie Stephen and the English studies debate, 1886–1887', *History of Education Soc. B.* 43 (1989), 41–47.

18. Leinster-Mackay, Donald. 'The case against Diggleism ...: some musings concerning criticisms of quasi-"Thatcherite" policies in the London School Board of the 1890s', *J. of Educational Administration & History* 22/2 (1990), 8–15.

19. Mackie, Peter. 'Inter-denominational education and the United Industrial School of Edinburgh, 1847–1900', *Innes R.* 43 (1992), 3–17.

20. Marriott, J. Stuart. 'University extension in the north of England and the "Leeds historians"', *Northern History* 28 (1992), 197–212.

21. Nannestad, Eleanor. 'Working-class pleasure excursions to and from Lincoln 1846 to 1914', *Lincolnshire History & Archaeology* 26 (1991), 12–20.

22. Nash, Gerallt D. *Victorian school-days in Wales.* Cardiff; Wales UP; 1991. Pp 38.

23. Pearce, R.D. 'The prep school and imperialism: the example of Orwell's St Cyprian's', *J. of Educational Administration & History* 23/1 (1991), 42–53.

24. Ray, Michael. 'The Victorian boarding school in a suburb of an English seaside resort', *Sussex Arch. Collections* 129 (1991), 255–58.

25. Robertson, Alex. 'Catherine I. Dodd and innovation in teacher training, 1892–1905', *History of Education Soc. B.* 47 (1991), 32–41.

26. Robinson, Wendy. 'Different and unequal: elementary school experiences in the school board era', *History of Education Soc. B.* 49 (1992), 36–44.

27. Roderick, Gordon W. 'Educating the worker: the mechanics' institute movement in south Wales', *Honourable Soc. of Cymmrodorion T.* (1991), 161–74.

28. Roderick, Gordon W. 'The Department of Science and Art and technical education in south Wales', *History of Education Soc. B.* 49 (1992), 20–27.

29. Roderick, Gordon W.; Stephens, M.D. 'Science and technical studies in Welsh higher education in the nineteenth century', *History of Education Soc. B.* 43 (1989), 30–40.

30. Russell-Gebbett, Jean P. 'High Pavement: Britain's first organised science school', *History of Education Soc. B.* 43 (1989), 17–29.

31. Soffer, Reba N. 'Authority in the university: Balliol, Newnham and the new mythology', Bc190, 192–215.

32. Stinchcombe, Owen. 'Cheltenham Working Men's College and Union Club: 1883–1899', *History of Education Soc. B.* 48 (1991), 38–47.

33. Vaughan, Sir Edgar. *Joseph Lancaster en Caracas (1824–1827) y sus relaciones con El Libertador Simón Bol   )ívar, con datos sobre las escuelas lancasterianas en Hispanoamérica en el siglo XIX* [Joseph Lancaster in Caracas and his relations with the revolutionary, Simon Bolivar; with information about Lancastrian schools in Spanish America]. Caracas (Venezuela); Ministerio de Educacion; 1987–1989. 2 vols.

34. Watson, M.I. 'Mutual improvement societies in nineteenth century Lancashire', *J. of Educational Administration & History* 21/2 (1989), 8–17.

35. Webber-Mortiboys, Christine. 'School attendance in Henley-in-Arden, 1860–1885', *History of Education Soc. B.* 44 (1989), 31–45.

(j) *Naval and Military*

1. Allen, Matthew. 'Rear Admiral Reginald Custance: director of naval intelligence, 1899–1902', *Mariner's Mirror* 78 (1992), 61–75.

2. Barnes, John. *The beginnings of the cinema in England, 1894–1901, vol. 4: filming the Boer war.* London; Bishopsgate; 1992. Pp 340.

3. Beckett, Ian F.W. *Johnnie Gough, VC: a biography of brigadier-general Sir John Edmond Gough, VC, KCB, 1871–1915.* London; Donovan; 1989. Pp xvi, 244.

4. Collin, Richard H. 'The Caribbean theater transformed: Britain, France, Germany, and the USA, 1900–1906', *American Neptune* 52 (1992), 102–12.

5. Cook, Hugh (ed.). 'Letters from South Africa, 1899–1902' [cont.], *J. of the Soc. for Army Historical Research* 69 (1991), 233–55.

6. Cook, Hugh (ed.). 'The Ballard letters: the Boer War writings of C.R. Ballard' [parts 2–4], *Q. B. of the South African Library* 46 (1991), 23–38; 69–82; 106–19.

7. Hayes, Paul. 'Britain, Germany, and the Admiralty's plans for attacking German territory, 1906–1915', Bc150, 95–116.

8. Hewitt, David. 'Soldiers in romantic fiction: Walter Scott's *Quentin Durward* and Thomas Hamilton's *The youth and manhood of Cyril Thornton*', Bc21, 111–21.

9. Mallinson, Allan. 'Charging ahead: transforming Britain's cavalry, 1902–1914', *History Today* 42/1 (1992), 29–36.

10. Mitch, David F. *The rise of popular literacy in Victorian England: the influence of private choice and public policy.* Philadelphia (Pa); Pennsylvania UP; 1992. Pp xxiii, 340.

11. Morris, Peter. *First aid to the battlefront: life and letters of Sir Vincent Kennett-Barrington, 1844–1903.* Stroud; Sutton; 1992. Pp x, 231.
12. Morriss, Roger. 'Sir George Cockburn and the management of the Royal Navy, 1841–1846', Bc173, 121–43.
13. Pardoe, Jon. 'Malcolm Kennedy (1895–1935) and Japan', Bc17, 177–86; 306–07.
14. Paris, Michael. *Winged warfare: literature and theory of aerial warfare in Britain, 1859–1917.* Manchester; Manchester UP; 1992. Pp 272.
15. Ranft, Bryan. 'Parliamentary debate, economic vulnerability, and British naval expansion, 1860–1905', Bc150, 75–93.
16. Russell, Dave. '"We carved our way to glory": the British soldier in music hall song and sketch, *c.*1880–1914', Bc10, 50–79.
17. Spiers, Edward M. *The late Victorian army, 1868–1902.* Manchester; Manchester UP; 1992. Pp xiii, 374.
18. Stearn, Roger T. 'War correspondents and colonial war, *c.*1870–1900', Bc10, 139–61.

(k) *Science and Medicine*

1. Allderidge, Patricia. 'The foundation of the Maudsley hospital', Bc145, 79–88.
2. Atkins, P.J. 'White poison?: the social consequences of milk consumption in London, 1850–1939', *Social History of Medicine* 5 (1992), 207–27.
3. Bartrip, Peter. 'The *British Medical Journal*: a retrospect', Bc28, 126–45.
4. Berrios, German E. 'Psychosurgery in Britain and elsewhere: a conceptual history', Bc145, 180–96.
5. Berrios, German E.; Freeman, Hugh. 'Introduction' [British psychiatry], Bc145, ix-xv.
6. Berry, Diana; Mackenzie, Campbell. *Richard Bright, 1789–1858: physician in an age of revolution and reform.* London; Royal Soc. of Medicine; 1992. Pp xiv, 296.
7. Boorman, W.H. 'Health and sanitation in Victorian Winchester or the triumph of the Amuckabites', *P. of the Hampshire Field Club & Arch. Soc.* 46 (1991), 161–80.
8. Bowler, Peter J. 'Darwinism and Victorian values: threat or opportunity', Bc196, 129–48.
9. Bradley, Margaret; Perrin, Fernand. 'Charles Dupin's study visits to the British Isles, 1816–1824', *Technology & Culture* 32 (1991), 47–68.
10. Brock, W.H. 'Medicine and the Victorian scientific press', Bc28, 70–89.
11. Brown, David K. 'William Froude and "the way of a ship in the sea"', *Devonshire Association Report & T.* 124 (1992), 207–31.
12. Buchwald, Jed C. 'Why Stokes never wrote a treatise on optics', Bc87, 451–76.
13. Bynum, William F. 'Medical values in a commercial age', Bc196, 149–64.
14. Bynum, William F. 'Tuke's *Dictionary* and psychiatry at the turn of the century', Bc145, 163–79.

15. Bynum, William F.; Wilson, Janice C. 'Periodical knowledge: medical journals and their editors in nineteenth-century Britain', Bc28, 29–48.
16. Cardwell, Donald S.L. *James Joule: a biography.* Manchester; Manchester UP; 1989. Pp ix, 333.
17. Chernin, Eli. 'Sir Patrick Manson: physician to the Colonial Office, 1897–1912', *Medical History* 36 (1992), 320–31.
18. Chernin, Eli. 'The early British and American journals of tropical medicine and hygiene: an informal survey', *Medical History* 36 (1992), 70–83.
19. Clark, J.F. McDiarmid. 'Eleanor Ormerod (1828–1901) as an economic entomologist: "pioneer of purity even more than of Paris Green"', *British J. for the History of Science* 25 (1992), 431–82.
20. Cooter, Roger. 'Introduction' [health and welfare], Bc57, 1–18.
21. Davies, Mansel. 'A university college of Wales Aberystwyth student', *Honourable Soc. of Cymmrodorion T.* (1991), 299–305.
22. Day, Kenneth; Jancar, Joze. 'Mental handicap and the Royal Medico-Psychological Association: a historical association, 1841–1991', Bc145, 268–78.
23. Desmond, Adrian J. *The politics of evolution: morphology, medicine, and reform in radical London.* London/Chicago; Chicago UP; 1989. Pp viii, 503.
24. Dupree, Marguerite W.; Crowther, M. Anne. 'A profile of the medical profession in Scotland in the early twentieth century: the *Medical Directory* as a historical source', *B. of the History of Medicine* 65 (1991), 209–33.
25. Evans, Richard J. 'Epidemics and revolutions: cholera in nineteenth-century Europe', Bc25, 149–73.
26. Fee, Elizabeth; Porter, Dorothy. 'Public health, preventive medicine and professionalization: England and America in the nineteenth century', Bc23, 249–75.
27. Freeman, Hugh; Tantam, Digby. 'Samuel Gaskell', Bc145, 445–51.
28. Fullinwider, S.P. 'Darwin faces Kant: a study in nineteenth-century physiology', *British J. for the History of Science* 24 (1991), 21–44.
29. Granshaw, Lindsay. 'Knowledge of bodies or bodies of knowledge? Surgeons, anatomists and rectal surgery, 1830–1985', Bc24, 232–62.
30. Granshaw, Lindsay. 'The rise of the modern hospital in Britain', Bc23, 197–218.
31. Hamlin, Christopher. 'Predisposing causes and public health in early nineteenth-century medical thought', *Social History of Medicine* 5 (1992), 43–70.
32. Harman, P.M. 'Maxwell and Saturn's rings: problems of stability and calculability', Bc87, 477–502.
33. Harris, Michael. 'Social diseases? Crime and medicine in the Victorian press', Bc28, 108–25.
34. Hendrick, Harry. 'Child labour, medical capital, and the School Medical Service, *c.*1890–1918', Bc57, 45–71.
35. Hinshelwood, R.D. 'Psychodynamic psychiatry before world war I', Bc145, 197–206.

36. Hornix, Willem J. 'From process to plant: innovation in the early artificial dye industry', *British J. for the History of Science* 25 (1992), 65–90.

37. Johnson, Mary Orr. 'The insane in 19th-century Britain: a statistical analysis of a Scottish insane asylum', *Historical Social Research* 17/3 (1992), 3–20.

38. Knight, David. *Humphrey Davy: science and power*. Oxford; Blackwell; 1992. Pp 218.

39. Latham, John. 'The National Trust archaeological survey: industrial sites at Aberglaslyn', Bc129, 62–65.

40. Lawrence, Christopher; Dixey, Richard. 'Practising on principle: Joseph Lister and the germ theories of disease', Bc24, 153–215.

41. Loudon, Jean; Loudon, Irvine. 'Medicine, politics and the medical periodical 1800–1850', Bc28, 49–69.

42. Marland, Hilary. 'Lay and medical conceptions of medical charity during the nineteenth century: the case of the Huddersfield General Dispensary and Infirmary', Bc26, 149–71.

43. Marland, Hilary; Swan, Philip. 'Medical practice in the West Riding of Yorkshire from nineteenth-century census data', Bc205, 73–98.

44. Marsden, Ben. 'Engineering science in Glasgow: economy, efficiency and measurement as prime movers in the differentiation of an academic discipline', *British J. for the History of Science* 25 (1992), 319–46.

45. Mason, Joan. 'Heartha Ayrton (1854–1923) and the admission of women to the Royal Society', *Notes & Records of the Royal Soc. of London* 45 (1991), 201–20.

46. Mayr, Ernst. *One long argument: Charles Darwin and the genesis of modern evolutionary thought*. London; Harvard UP; 1992. Pp xiv, 195.

47. Metcalfe, John F. 'Whewell's developmental psychologism: a Victorian account of scientific progress', *Studies in the History & Philosophy of Science* 22 (1991), 117–40.

48. Miles, B.E. 'Medical naturalists of Victorian Herefordshire', *T. of the Woolhope Naturalists Field Club* 46 (1989), 298–307.

49. Morus, Iwan Rhys. 'Correlation and control: William Robert Grove and the construction of a new philosophy of scientific reform', *Studies in the History & Philosophy of Science* 22 (1991), 589–621.

50. Morus, Iwan Rhys. 'Marketing the machine: the construction of electrotherapeutics as viable medicine in early Victorian England', *Medical History* 36 (1992), 34–52.

51. Pepper, Sarah. 'Allinson's staff of life: health without medicine in the 1890s', *History Today* 42/10 (1992), 30–35.

52. Pickstone, John V. 'Dearth, dirt and fever epidemics: rewriting the history of British "public health", 1780–1850', Bc25, 125–48.

53. Richardson, Ruth. '"Notorious abominations": architecture and the public health in *The Builder*, 1843–1883', Bc28, 90–107.

54. Rollin, Henry R. 'Whatever happened to Henry Maudsley?', Bc145, 351–58.

55. Sharp, Rita M. '"Foul and poisonous air": sanitation and public health in a rural community', *Local Historian* 21 (1991), 156–61.

56. Shpayer-Makov, Haia. 'Notes on the medical examination of provincial

applicants to the London Metropolitan Police on the eve of the first world war', *Histoire Sociale/Social History* 24 (1991), 169–79.

57. Simpson, A.D.C. 'An Edinburgh intrigue: Brewster's Society of Arts and the pantograph dispute', *Book of the Old Edinburgh Club* ns 1 (1991), 47–73.

58. Smith, F.B. 'The Contagious Diseases Acts reconsidered', *Social History of Medicine* 3 (1990), 197–215.

59. Spahr, F. *Die Ausbreitung der Cholera in der britischen Flotte im Schwartzen Meer während des Krimkrieges im August 1854* [The outbreak of cholera in the British fleet in the Black sea during the Crimean war in August 1854]. Frankfurt; Lang (Marburger Schriften zur Medizinische Geschichte); 1989. Pp x, 178.

60. Sturdy, Steve. 'The political economy of scientific medicine: science, education and the transformation of medical practice in Sheffield, 1890–1922', *Medical History* 36 (1992), 125–59.

61. Summers, Anne. 'The costs and benefits of caring: nursing charities, c.1830- c.1860', Bc26, 133–48.

62. Topham, Jonathan. 'Science and popular education in the 1830s: the role of the *Bridgewater treatises*', *British J. for the History of Science* 25 (1992), 397–430.

63. Travis, Anthony S. 'Engineering and politics: the Channel tunnel in the 1880s', *Technology & Culture* 32 (1991), 461–97.

64. Travis, Anthony S. 'Science's powerful companion: A.W. Hofmann's investigation of aniline red and its derivatives', *British J. for the History of Science* 25 (1992), 27–44.

65. Tritton, Paul. *The lost voice of Queen Victoria: the search for the first royal recording.* London; Academy; 1991. Pp 144.

66. Tunbridge, Paul. *Lord Kelvin: his influence on electrical measurements and units.* Stevenage; Peregrinus; 1992. Pp x, 106.

67. Turner, Trevor. '"Not worth powder and shot": the public profile of the Medico-Psychological Association, c.1851–1914', Bc145, 3–16.

68. Tweney, Ryan D.; Gooding, David (ed.). *Michael Faraday's 'chemical notes, hints, suggestions and objects of pursuit' of 1822.* London; Peregrinus for the Institution of Electrical Engineers; 1992. Pp xvii, 152.

69. Weindling, Paul 'From infectious to chronic diseases: changing patterns of sickness in the nineteenth and twentieth centuries', Bc23, 303–16.

70. Weindling, Paul. 'From isolation to therapy: children's hospitals and diphtheria in *fin de siècle* Paris, London and Berlin', Bc57, 124–45.

71. Williams, Naomi. 'Death in its season: class, environment and the mortality of infants in nineteenth-century Sheffield', *Social History of Medicine* 5 (1992), 71–94.

72. Wilson, Jason. 'Charles Darwin and W.H. Hudson', Bc136, 173–82.

73. Young, David. *The discovery of evolution.* Cambridge; Cambridge UP; 1992. Pp 256.

(l) *Intellectual and Cultural*

1. Adams, James Eli. 'Philosophical forgetfulness: John Stuart Mill's "Nature"', *J. of the History of Ideas* 53 (1992), 437–54.

2. Alter, Peter. 'Industrielles Mäzenatentum in England, 1870–1914' [Industrial patronage of the arts in England, 1870–1914], *Formen ausserstaatlicher Wissenschaftsforderung im 19. und 20. Jahrhunderten*, ed. Rudiger Vom Bruch and Rainer A. Muller (Stuttgart; Steiner; 1990), 241–58.

3. Anderson, Patricia J. '"Factory girl, apprentice and clerk": the readership of mass-market magazines, 1830–1860', *Victorian Periodicals R.* 25 (1992), 64–72.

4. Baker, William. *Early history of the London Library*. Lampeter; Mellen; 1992. Pp 168.

5. Ballhatchet, Helen. 'Baba Tatsui (1850–1888) and Victorian Britain', Bc17, 107–17; 300–01.

6. Bentley-Cranch, Dana. *Founders and followers: literary lectures on the occasion of the 150th anniversary of the foundation of the London Library*. London; Sinclair-Stevenson; 1992. Pp 182.

7. Bertie, David M. 'The Peterhead Institute, 1857–1867', *Northern Scotland* 10 (1990), 47–71.

8. Binfield, Clyde. 'A chapel and its architect: James Cubitt and Union Chapel, Islington, 1874–1889', Bc194, 417–47.

9. Birrell, T.A. '"A sentimental journey" through Holland and Flanders by John Gage', Bc133, 197–206.

10. Black, Alistair. 'Libraries for the many: the philosophical roots of the early public library movement', *Library History* 9 (1991), 27–36.

11. Blacker, Carmen. 'Two Piggotts: Sir Francis Taylor Piggott (1852–1925) and Major General F.S.G. Piggott (1883–1966)', Bc17, 118–27; 301–02.

12. Boos, Florence S. 'Alternative Victorian futures: "historicism", *Past and present*, and *A dream of John Bull*', Bc225, 3–38.

13. Bowring, Richard. 'An amused guest in all: Basil Hall Chamberlain (1850–1935)', Bc17, 128–36; 302–03.

14. Bradley, Ian. 'Changing the tune: popular music in the 1890s', *History Today* 42/7 (1992), 40–47.

15. Brake, Laurel. 'Theories of formation: the *Nineteenth Century*: volume 1, no 1, March 1877. Monthly. 2/6', *Victorian Periodicals R.* 25 (1992), 16–21.

16. Briggs, Asa. 'The imaginative response of the Victorians to new technology: the case of the railways', Bc117, 58–75.

17. Cannadine, David. 'Gilbert and Sullivan: the making and un-making of a British "tradition"', Bc190, 12–32.

18. Casteras, Susan P. 'Excluding women: the cult of the male genius in Victorian painting', Bc56, 116–46.

19. Casteras, Susan P. 'Pre-Raphaelite challenges to Victorian canons of beauty', *Huntington Library Q.* 55 (1992), 13–36.

20. Catch, J.R. 'A Buckinghamshire polymath: Edward John Payne', *Records of Buckinghamshire* 32 (1991 for 1990), 120–29.

21. Cawthon, Elisabeth A. 'Apocrypha from the Victorian workplace: occupational accidents and employee attitudes in England, 1830–1860', *Victorian Periodicals R.* 25 (1992), 56–63.

22. Charlton, Christopher. '"Bag in hand, and with a provision of papers for an emergency"—an impression of the 1891 census from the pages of some contemporary newspapers', *Local Population Studies* 47 (1991), 81–88.

23. Christie, Peter. 'The true story of the north Devon savages', *Devonshire Association Report & T.* 124 (1992), 59–85.

24. Clark, John. 'Charles Wirgman (1835–1891)', Bc17, 54–63; 295–96.

25. Clarke, D.V. 'The National Museum's stained-glass window', *Soc. of Antiquaries of Scotland P.* 120 (1990), 201–24.

26. Clarke, John Stock. '*Home*: a lost Victorian periodical', *Victorian Periodicals R.* 25 (1992), 85–88.

27. Clarke, Norma. 'Strenuous idleness: Thomas Carlyle and the man of letters as hero', Bc223, 25–43.

28. Collini, Stefan. 'From sectarian radical to national possession: John Stuart Mill in English culture, 1873–1945', Bc123, 242–72.

29. Cortazzi, Sir Hugh. 'The Japan Society: a hundred-year history', Bc17, 1–53; 294–95.

30. Cortazzi, Sir Hugh; Daniels, Gordon. 'Introduction' [Britain and Japan], Bc17, xv-xx.

31. Croll, Andy. 'From bar stool to choir stall: music and morality in late Victorian Merthyr', *Llafur* 6 (1992), 17–27.

32. Crosby, Christina. 'Reading the Gothic revival: "History" and *Hints on Household Taste*', Bc56, 101–15.

33. Cunningham, Colin; Waterhouse, Prudence. *Alfred Waterhouse, 1830–1905: biography of a practice.* Oxford; Oxford UP; 1992. Pp xviii, 319.

34. Cunningham, Valentine. 'Goodness and goods: Victorian literature and values for the middle class reader', Bc196, 109–28.

35. Dellheim, Charles. 'Interpreting Victorian medievalism', Bc225, 39–58.

36. Dooley, Allan C. *Author and printer in Victorian England.* Charlottesville; UP of Virginia; 1992. Pp 224.

37. Easby, Rebecca J. 'The myth of merrie England in Victorian painting', Bc225, 59–80.

38. Ehrlich, Cyril. *The piano: a history.* Oxford; Oxford UP; rev. edn. 1990. Pp 254.

39. Feaver, George (ed.). *The Webbs in Asia: the 1911–1912 travel diary.* London; Macmillan; 1992. Pp viii, 385.

40. Ferris, Ina. 'From trope to code: the novel and the rhetoric of gender in nineteenth-century critical discourse', Bc56, 18–30.

41. Filipiuk, Marion. 'John Stuart Mill and France', Bc123, 80–120.

42. Finn, Dallas. 'Josiah Conder (1852–1920) and Meiji architecture', Bc17, 86–93; 297–98.

43. Fisher, Trevor. 'Britain's unpermissive society', *History Today* 42/8 (1992), 38–44.

44. Fulton, Richard. '*The Spectator* in alien hands', *Victorian Periodicals R.* 24 (1991), 187–96.

45. Garside, Patricia L. 'Representing the metropolis—the changing relationship between London and the press, 1870–1939', *London J.* 16 (1991), 156–73.

46. Glendinning, Victoria. *Trollope*. London; Hutchinson; 1992. Pp 551.
47. Gregory, Alexis. *The golden age of travel, 1880–1939*. London; Cassell; 1991. Pp 220.
48. Griffin, Nicholas (ed.). *The selected letters of Bertrand Russell, vol. 1: the private years (1884–1914)*. London; Lane; 1992. Pp xxi, 553.
49. Hamburger, Joseph. 'Religion and *On Liberty*', Bc123, 139–81.
50. Harley, Basil; Harley, Jessie. *A gardener at Chatsworth: three years in the life of Robert Aughtie, 1848–1850*. London; Self Publishing Association; 1992. Pp 255.
51. Harris, Geraldine. 'Yvette Guilbert: *La femme moderne* on the British stage', Bc63, 115–33.
52. Hemingway, Andrew. *Landscape imagery and urban culture in early nineteenth-century Britain*. Cambridge; Cambridge UP; 1992. Pp xix, 363.
53. Henderson, Robert. 'Russian political emigrés and the British Museum library', *Library History* 9 (1991), 59–68.
54. Hendrick, Harry. 'Changing attitudes to children, 1800–1914', *Genealogists' Magazine* 24 (1992), 41–49.
55. Herbert, D. 'Place and society in Jane Austen's England', *Geography* 76 (1991), 193–208.
56. Hillsman, Walter. 'Organs and organ music in Victorian synagogues: Christian intrusions or symbols of cultural assimilation?', Bc189, 419–33.
57. Hillsman, Walter. 'The Victorian revival of plainsong in English: its usage under Tractarians and Ritualists', Bc194, 405–15.
58. Hobsbawm, E.J. 'Birth of a holiday: the first of May', Bc117, 104–22.
59. Holman, Nigel. 'A different kind of Cambridge antiquarian: Marshall Fisher and his Ely museum', *Cambridge Antiquarian Soc. P.* 79 (1992 for 1990), 82–92.
60. Ion, A.H. 'Mountain high and valley low: Walter Weston (1861–1940) and Japan', Bc17, 94–106; 298–300.
61. Jenkins, Ian D. *Archaeologists and aesthetes in the sculpture galleries of the British Museum, 1800–1939*. London; British Museum; 1992. Pp 264.
62. Johnson, Matthew. 'The Englishman's home and its study', Bc38, 245–57.
63. Jones, Emyr Wyn. 'Syr Henry M. Stanley: dirgelion y dyddian cynnar—"ail farn ar Stanley a'i fyd"' [Sir Henry M. Stanley: secrets of his early days], *Honourable Soc. of Cymmrodorion T.* (1991), 175–201.
64. Jones, H.S. 'John Stuart Mill as moralist', *J. of the History of Ideas* 53 (1992), 287–308.
65. Kenny, Anthony. 'Victorian values: some concluding thoughts', Bc196, 217–24.
66. Kent, Christopher A. 'Higher journalism and the promotion of Comtism', *Victorian Periodicals R.* 25 (1992), 51–56.
67. Kornicki, P.F. 'William George Aston (1841–1911)', Bc17, 64–75; 296–97.

68. Krueger, Christine L. 'The "female paternalist" as historian: Elizabeth Gaskell's *My lady Ludlow*', Bc56, 166–83.

69. Laidlar, John F. 'Edgar Prestage: Manchester's Portuguese pioneer', *B. of the John Rylands University Library of Manchester* 74/1 (1992), 75–94.

70. Larkham, Peter J. 'Facadism and a vernacular farmhouse: an example in East Yorkshire', *T. of the Ancient Monuments Soc.* 36 (1992), 119–28.

71. Law, Jules. 'Water rights and the "crossing o' breeds": chiastic exchange in *The Mill on the Floss*', Bc56, 52–69.

72. Leather, John. *Albert Strange: yacht designer and artist, 1855–1917.* Edinburgh; Pentland; 1990. Pp viii, 207.

73. Linkman, A.E. 'The workshy camera: photography and the labouring classes in the nineteenth century', *Costume* 25 (1991), 36–52.

74. Littlejohn, J.H. *The Scottish music hall, 1880–1990.* Wigtown; G.C. Book Publishers; 1990. Pp 118.

75. Lloyd, Thomas. 'Evan Andrews of Kidwelly (1803–1869): scholar, schoolmaster and playwright', Bc101, 227–38.

76. Lloyd, Trevor. 'John Stuart Mill and the East India Company', Bc123, 44–79.

77. Lord, P. 'Artisan painters in Carmarthen', *Carmarthenshire Antiquarian* 27 (1991), 47–60.

78. McConkey, Kenneth. '"Well-bred contortions", 1880–1918', Bc93, 352–86.

79. McGann, Jerome. '"A thing to mind": the materialist aesthetic of William Morris', *Huntington Library Q.* 55 (1992), 55–74.

80. Moore-Colyer, Richard J. 'Gentlemen, horses and the Turf in nineteenth-century Wales', *Welsh History R.* 16 (1992–3), 47–62.

81. Moss, David J. '*Circular to Bankers*: the role of a proto-typical trade journal in the evolution of middle-class professional consciousness', *Victorian Periodicals R.* 25 (1992), 129–36.

82. Munro, Jane A. '"This hateful letter-writing": selected correspondence of Sir Edward Burne-Jones in the Huntington library', *Huntington Library Q.* 55 (1992), 75–103.

83. Munsell, F. Darrell. *The Victorian controversy surrounding the Wellington war memorial: the archduke of Hyde Park Corner.* Lampeter; Mellen; 1992. Pp x, 113.

84. Myerly, Scott Hughes. '"The eye must entrap the mind": army spectacle and paradigm in nineteenth-century Britain', *J. of Social History* 26 (1992–3), 105–31.

85. Newall, Christopher. 'The Victorians, 1830–1880', Bc93, 298–351.

86. Nuding, Gertrude Prescott. 'Britishness and portraiture', Bc190, 237–70.

87. Nunokawa, Jeff. '*Tess*, tourism, and the spectacle of the woman', Bc56, 70–86.

88. Ovenden, K. 'Roast beef and plum pudding: Queen Victoria's golden jubilee, 1887', *Leicestershire History* 3/9 (1991), 17–26.

89. Parkinson, John A. *Victorian music publishers: an annotated list.* Warren (Ml); Harmonie Park; 1990. Pp xix, 315.

90. Parry, Ann. '*The National Review* and the Dreyfus affair: "the conscience of the civilized world"', *Victorian Periodicals R.* 25 (1992), 6–15.

91. Porta, Pier Luigi (ed.). *David Ricardo: notes on Malthus's measure of value*. Cambridge; Cambridge UP; 1992. Pp 62.

92. Porter, Thomas W. 'Ernest Jones and the Royal Literary Fund', *Labour History R.* 57/3 (1992), 84–94.

93. Powell, Brian. 'Tsubouchi Shoyo (1859–1935): Sherborne and Japan—an episode in cross-cultural relations', Bc17, 223–34; 309–10.

94. Robson, Ann P. 'Mill's second prize in the lottery of life', Bc123, 215–41.

95. Seidman, Steven. 'The power of desire and the danger of pleasure: Victorian sexuality reconsidered', *J. of Social History* 24 (1990–1), 47–67.

96. Sell, Alan P.F. 'In the wake of the Enlightenment: the adjustments of James Martineau and Alexander Campbell Fraser', *Enlightenment & Dissent* 9 (1990), 63–92.

97. Shaw, Marion. '"To tell the truth of sex": confession and abjection in late Victorian writing', Bc56, 87–100.

98. Sherrington, Emlyn. 'O.M. Edwards, culture and the industrial classes', *Llafur* 6 (1992), 28–41.

99. Shires, Linda M. 'Afterword: ideology and the subject as agent', Bc56, 184–90.

100. Shires, Linda M. 'Of maenads, mothers, and feminized males: Victorian readings of the French revolution', Bc56, 147–65.

101. Smith, Lindsay. '"The seed of the flower": photography and pre-Raphaelitism', *Huntington Library Q.* 55 (1992), 37–54.

102. Smout, T.C. 'Introduction' [Victorian values], Bc196, 1–8.

103. Stapleton, Julia. 'English pluralism as cultural definition: the social and political thought of George Unwin', *J. of the History of Ideas* 52 (1991), 665–84.

104. Steedman, Carolyn. 'Bodies, figures and physiology: Margaret McMillan and the late nineteenth-century remaking of working-class childhood', Bc57, 19–44.

105. Stephens, John Russell. *The profession of the playwright: British theatre, 1800–1900*. Cambridge; Cambridge UP; 1992. Pp xix, 254.

106. Stillinger, Jack. 'John Mill's education: fact, fiction, and myth', Bc123, 19–43.

107. Street, Sean. *The wreck of the "Deutschland"*. London; Souvenir; 1992. Pp 208.

108. Taylor, Clare (ed.). 'Correspondence relating to Millburn tower and its garden, 1804–1829', Bc83, 329–87.

109. Temperley, Nicholas. 'Opera, 1850–1890: Britain and the United States', Bc77, 479–87.

110. Temperley, Nicholas. 'Romantic opera, 1830–1850: Britain and the United States', Bc77, 228–36.

111. Temperley, Nicholas. 'Solo song: Britain and the United States', Bc77, 769–92.

112. Thistlewaite, Nicholas. *The making of the Victorian organ*. Cambridge; Cambridge UP (Cambridge Musical Texts and Monographs); 1990. Pp xxiv, 584.

113. Trela, D.J. 'Carlyle's *Shooting Niagara*: the writing and revising of an article and pamphlet', *Victorian Periodicals R*. 25 (1992), 30–34.
114. Ulin, Donald. 'A clerisy of worms in Darwin's inverted world', *Victorian Studies* 35 (1991–2), 294–308.
115. Usherwood, Paul. 'Officer material: representations of leadership in late nineteenth-century British battle painting', Bc10, 162–78.
116. Vaio, John. 'Gladstone and the early reception of Schliemann in England', *Heinrich Schliemann nach hundert Jahren*, ed. William M. Calder III and Justus Cobet (Frankfurt; Klostermann; 1990), 415–30.
117. Von Arx, Jeffrey P. 'Archbishop Manning and the *Kulturkampf*', *Recusant History* 21 (1992), 254–66.
118. Warner, Malcolm. 'The pre-Raphaelites and the National Gallery', *Huntington Library Q*. 55 (1992), 1–12.
119. Waters, Chris. 'Marxism, medievalism and popular culture', Bc225, 137–68.
120. Webb, R.K. 'A crisis of authority: early nineteenth-century British thought', *Albion* 24 (1992), 1–16.
121. Wheeler, Michael; Whiteley, Nigel. *The lamp of memory: Ruskin, tradition and architecture*. Manchester; Manchester UP; 1992. Pp xii, 238.
122. White, Norman. *Hopkins: a literary biography*. Oxford; Oxford UP; 1992. Pp 531.
123. Williams, Sian Rhiannon. 'The true "Cymraes": images of women in women's nineteenth-century Welsh periodicals', Bc66, 69–91.
124. Wright, Patricia. 'Queen Victoria and her palace', *History Today* 42/12 (1992), 48–54.
125. Young, David. 'East-end street names and British imperialism', *Local Historian* 22 (1992), 84–88.

I. BRITAIN SINCE 1914

See also Aa6,9,29,100–101,107,138,142,164,176;Bc20,37,67,131,138, 179,188,201;Ha1,b28,36,e9,41,f29,33,39,89,g19,58,60,h20,26,i16,k22, 29,45,60,69,l11,45,47,60,74

(a) *General*

1. Alban, J.R. 'The wider world' [in Swansea], Bc212, 114–129.
2. Bailey, Victor. 'Introduction' [Fire Brigades union], Bc7, xix-xxiii.
3. Bailey, Victor. 'The early history of the Fire Brigades union', Bc7, 3–97.
4. Black, Maggie. *A cause for our times: Oxfam—the first fifty years*. Oxford; Oxford UP; 1992. Pp 320.
5. Boorman, David. 'The city and the channel' [Swansea], Bc212, 92–113.
6. Bridges, E.M.; Morgan, Huw. 'Dereliction and pollution' [in Swansea], Bc212, 270–90.
7. Broadberry, S.N. 'The emergence of mass unemployment: a reply', *Economic History R*. 2nd ser. 45 (1992), 739–42.

8. Bromley, Rosemary D.F. 'Swansea at the end of the twentieth century', Bc212, 305–17.

9. Carver, Michael. 'Britain and the alliance', Bc150, 211–25.

10. Dickson, A.D.R.; Treble, James H. 'Introduction: Scotland, 1914–1990', Bc138, 1–11.

11. Ellis, E.L. *T.J.: a life of Dr Thomas Jones, CH.* Cardiff; Wales UP; 1992. Pp xviii, 553.

12. Foreman-Peck, James. 'The development and diffusion of telephone technology in Britain, 1900–1940', *Newcomen Soc. T.* 63 (1992), 165–79.

13. Freedman, Lawrence. 'Strategic studies and the problem of power', Bc150, 279–94.

14. Glynn, Sean; Booth, Alan. 'The emergence of mass unemployment: some questions of precision', *Economic History R.* 2nd ser. 45 (1992), 731–38.

15. Gorham, Deborah. '"The friendship of women": friendship, feminism and achievement in Vera Brittain's life and work in the interwar decades', *J. of Women's History* 3/3 (1992), 44–69.

16. Hart, M.W. 'The realignment of 1931', *Twentieth-Century British History* 3 (1992), 196–98.

17. Hennessy, Peter. *Never again: post-war Britain, 1946–1951.* London; Cape; 1992. Pp 544.

18. Herman, Gerald. *The pivotal conflict: a comprehensive chronology of the first world war, 1914–1919.* London; Greenwood; 1992. Pp 800.

19. Horner, John. 'Recollections of a general secretary', Bc7, 279–58.

20. Humphries, Enoch. 'Reminiscences of a president', Bc7, 359–85.

21. Inglis, K.S. 'The homecoming: the war memorial movement in Cambridge, England', *J. of Contemporary History* 27 (1992), 583–605.

22. Ireson, Tony. 'A Liberal globetrotter', *Northamptonshire Past & Present* 8 (1992), 315–20.

23. Isaac, Michael J. 'Caring for the natural environment' [in Swansea], Bc212, 291–304.

24. Jefford, Michael. 'The pilotage service at Poole in the nineteenth century', *The Hatcher R.* 4/34 (1992), 29–39.

25. Kirby, M.W. 'Institutional rigidities and economic decline: reflections on the British experience', *Economic History R.* 2nd ser. 45 (1992), 637–60.

26. Lewis, G.B. 'Swansea on the map', Bc212, 67–77.

27. Mandler, Peter. 'Politics and the English landscape since the first world war', *Huntington Library Q.* 55 (1992), 459–76.

28. Mein, Margaret. *Winston Churchill and Christian Fellowship.* Ilfracombe; Stockwell; 1992. Pp 60.

29. Morgan, David; Evans, Mary. *Battle for Britain: citizenship and ideology in the second world war.* London; Routledge; 1992. Pp 208.

30. Morris, Bernard. 'Buildings and topography' [Swansea], Bc212, 145–64.

31. Morris, Christopher J. 'September-December, 1960', *Contemporary Record* 6 (1992), 208–13.

32. Muirhead, B.W. 'The politics of food and the disintegration of the Anglo-Canadian trade relationship, 1947–1948', *J. of the Canadian Historical Association* new ser. 2 (1991), 215–30.

33. Proud, Edward Baxby. *The postal history of British air mails*. Heathfield; Proud-Bailey; 1991. Pp 576.
34. Reinharz, Jehuda. *Chaim Weizmann, vol. 2: the making of a statesman*. Oxford; Oxford UP; 1992. Pp 576.
35. Sampson, Anthony. *The essential anatomy of Britain: democracy in crisis*. London; Arnold; 1992. Pp 256.
36. Saville, John. 'Terry Parry: a profile', Bc7, 270–76.
37. Schwarz, Bill. 'Where horses shit a hundred sparrows feed: Docklands and East London during the Thatcher years', Bc149, 76–92; 244–46.
38. Stevens, Terry. 'Tourism and leisure' [in Swansea], Bc212, 260–69.
39. Thorpe, Andrew. *Britain in the 1930s*. Oxford; Blackwell; 1992. Pp 145.
40. Thorpe, Andrew. 'Britain', Bc188, 14–34.
41. Tiratsoo, Nick. 'Introduction' [the Attlee years], Bc67, 1–6.
42. Voeltz, Richard A. 'The antidote to "khaki fever"?: the expansion of the British Girl Guides during the first world war', *J. of Contemporary History* 27 (1992), 627–38.
43. Woodhouse, D.G. *Anti-German sentiment in Kingston upon Hull: the German community and the first world war*. Hull; Kingston upon Hull City Record Office; 1990. Pp ii, 109.

(b) *Politics*

1. Adams, Jad. *Tony Benn: a biography*. London; Macmillan; 1992. Pp 576.
2. Alderman, Geoffrey. 'Dr Robert Forgan's resignation from the British Union of Fascists', *Labour History R.* 57/1 (1992), 37–41.
3. Ali, Yasmin. 'Echoes of empire: towards a politics of representation', Bc149, 194–211; 256–57.
4. Ashford, Nigel. 'The political parties', Bc201, 119–48.
5. Baggott, Rob. 'Pressure groups and the British political system: change and decline?', Bc200, 37–52.
6. Ball, Stuart. *Parliament and politics in the age of Baldwin and Macdonald: diaries of Sir Cuthbert Headlam, 1924–1935*. London; Historians' Press; 1992. Pp x, 360.
7. Bennett, G.H. 'The wartime political truce and hopes for post war coalition: the West Derbyshire by-election, 1944', *Midland History* 17 (1992), 118–35.
8. Brooke, Stephen. *Labour's war: the Labour party and the second world war*. Oxford; Oxford UP; 1992. Pp xiii, 363.
9. Brotherstone, Terry. 'Does red Clydeside really matter any more?', Bc175, 52–80.
10. Buchanan, Tom. 'Divided loyalties: the impact of the Spanish civil war on Britain's civil service trade unions, 1936–1939', *Historical Research* 65 (1992), 90–107.
11. Bullock, Ian. 'Sylvia Pankhurst and the Russian revolution: the making of a "left-wing" Communist', Bc50, 121–48.
12. Burge, Alun. 'The 1926 General Strike in Cardiff', *Llafur* 6 (1992), 42–61.

13. Burness, Catriona. 'The long slow march: Scottish women MPs, 1918–1945', Bc135, 151–73.

14. Butler, David. 'Voting behaviour and the party system', Bc200, 129–38.

15. Cosgrave, Patrick. *The strange death of socialist Britain: British politics, 1945–1992.* London; Constable; 1992. Pp 320.

16. Coxall, Bill. 'The social context of British politics: class, gender and race in the two major parties, 1970–1990', Bc200, 3–21.

17. Davies, Sam. 'Class, religion and gender: Liverpool Labour party and women, 1918–1939', Bc65, 217–46.

18. Davies, Sam. 'The membership of the National Unemployed Workers' movement, 1923–1938', *Labour History R.* 57/1 (1992), 29–36.

19. Dawson, Michael. 'Money and the real impact of the fourth Reform act', *Historical J.* 35 (1992), 369–81.

20. Dean, Dennis. 'Preservation or renovation?: the dilemmas of Conservative educational policy, 1955–1960', *Twentieth-Century British History* 3 (1992), 3–31.

21. Denver, David; Hands, Gordon. 'The political socialization of young people', Bc200, 94–108.

22. Durham, Martin. 'Gender and the British Union of Fascists', *J. of Contemporary History* 27 (1992), 513–29.

23. Dutton, David. *Simon: a political biography of Sir John Simon.* London; Aurum; 1992. Pp x, 364.

24. Dutton, David. 'On the brink of oblivion: the post-war crisis of British Liberalism', *Canadian J. of History* 27 (1992), 426–50.

25. Ebersold, Bernd. *Machtverfall und Machtbewusstsein: Britische Friedens- und Konfliktlosungsstrategien, 1918–1956* [Decline and power consciousness: British peace and conflict solving strategies]. Munich; Oldenbourg; 1992. Pp xii, 447.

26. Fielding, Steven. '"Don't know and don't care": popular political attitudes in Labour's Britain, 1945–1951', Bc67, 106–25.

27. Fielding, Steven. 'Labourism in the 1940s', *Twentieth-Century British History* 3 (1992), 138–53.

28. Fielding, Steven. 'What did "the people" want?: the meaning of the 1945 general election', *Historical J.* 35 (1992), 623–39.

29. Finlay, Richard J. 'Pressure group or political party? The nationalist impact on Scottish politics, 1928–1945', *Twentieth-Century British History* 3 (1992), 274–97.

30. Greenaway, John R. 'British conservatism and bureaucracy', *History of Political Thought* 13 (1992), 129–60.

31. Harvie, Christopher. 'Scottish politics', Bc138, 241–60.

32. Hazlehurst, Cameron. 'Biographical propriety and the historical Asquith', Bc143, 115–45.

33. Hennessy, Peter. 'Epilogue: reasons to be cheerful' [British politics], Bc200, 327–31.

34. Hinton, James. 'Women and the Labour vote, 1945–1950', *Labour History R.* 57/3 (1992), 59–66.

35. Hope, John. 'British fascism and the state, 1918–1927: a re-examination of the documentary evidence', *Labour History R.* 57/3 (1992), 72–83.

36. Ingle, Stephen. 'All you never wanted to know about British political parties', Bc200, 22–36.
37. Jones, Bill. 'Broadcasters, politicians and the political interview', Bc200, 53–78.
38. Jones, J. Graham. 'The Parliament for Wales campaign, 1950–1956', *Welsh History R.* 16 (1992–3), 207–36.
39. Kadish, Sharman. *Bolsheviks and British Jews: the Anglo-Jewish community, Britain and the Russian revolution.* London; Cass; 1992. Pp 312.
40. Kavanagh, Dennis. 'Changes in the political class', Bc200, 79–93.
41. Kavanagh, Dennis. 'The postwar consensus', *Twentieth-Century British History* 3 (1992), 175–90.
42. Kennedy, Paul. 'Grand strategies and less-than-grand strategies: a twentieth-century critique', Bc150, 227–42.
43. Knotter, Ad. 'The historical geography of labour in Britain and the Netherlands: electoral support and regional implantation', *Tijdschift voor Sociale Geschiedenis* 18 (1992), 148–67.
44. Kolinsky, Martin. 'The collapse and restoration of public security', Bc179, 147–68.
45. Lenman, Bruce P. *The eclipse of parliament: appearance and reality in British politics since 1914.* London; Arnold; 1992. Pp 288.
46. Letwin, Shirley Robin. *The anatomy of Thatcherism.* London; Fontana; 1992. Pp viii, 377.
47. Leventhal, F.M. 'Leonard Woolf and Kingsley Martin: creative tension on the left', *Albion* 24 (1992), 279–94.
48. Little, Eddie. 'The Manchester peace manifesto, 1936–1937', *Manchester Region History R.* 6 (1992), 80–84.
49. Mason, Tony; Thompson, Peter. '"Reflections on a revolution"? The political mood in wartime Britain', Bc67, 54–70.
50. McDonough, Frank. '*The Times*, Norman Ebbut and the Nazis, 1927–1937', *J. of Contemporary History* 27 (1992), 407–24.
51. McIvor, Arthur J.; Paterson, Hugh. 'Combatting the Left: victimization and anti-labour activities on Clydeside, 1900–1939', Bc175, 129–54.
52. Mercer, Helen; Rollings, Neil; Tomlinson, J.D. 'Introduction' [Labour governments], Bc131, 1–11.
53. Messinger, Gary S. *British propaganda and the state in the first world war.* Manchester; Manchester UP; 1992. Pp 240.
54. Meyer, W.R. 'The seditions and blasphemous teaching bills, 1922–1933', *History of Education Soc. B.* 48 (1991), 48–58.
55. Morgan, Austen. *Harold Wilson.* London; Pluto; 1992. Pp xv, 625.
56. Morgan, Kenneth O. 'The challenges of democracy' [in Swansea], Bc212, 51–66.
57. Norris, Pippa. 'Change plus ça change' [review article], *Parliamentary History* 11 (1992), 293–99.
58. Norton, Philip. 'The House of Commons: from overlooked to overworked', Bc200, 139–54.
59. Nugent, Neill. 'British public opinion and the European Community', Bc201, 172–201.

60. Pankhurst, Richard. 'Sylvia and *New Times and Ethiopia News*', Bc50, 149–91.

61. Paulmann, Johannes. 'Arbeitsmarktpolitik in Grossbritannien von der Zwischenkriegszeit bis in die Zeit nach dem Zweiten Weltkrieg: zur Entwicklung eines Politikfeldes' [Labour-market politics in Britain from between the wars to after the second world war: towards the development of a political field], *Historische Zeitschrift* 255 (1992), 345–75.

62. Philip, Alan Butt. 'British pressure groups and the European Community', Bc201, 149–71.

63. Phillips, Gordon. *Rise of the Labour party, 1893–1931*. London; Routledge; 1992. Pp 96.

64. Pimlott, Ben. *Harold Wilson*. London; HarperCollins; 1992. Pp 811.

65. Pollard, Sidney. 'Die Krise des Sozialismus in der Zwischenkriegzeit: Grossbritannien' [The crisis of socialism between the wars: Great Britain], *Geschichte und Gesellschaft* 17 (1991), 160–81.

66. Potter, Karen. 'British McCarthyism', Bc18, 143–58.

67. Pugh, Martin. 'Figures of the Liberal diaspora', *Twentieth-Century British History* 3 (1992), 304–09.

68. Rawlinson, George. 'Mobilizing the unemployed: the national un-employed workers' movement in the west of Scotland', Bc175,

69. Ridley, Jane; Percy, Clayre (ed.). *Balfour/Elcho letters: Arthur James Balfour and Lady Elcho*. London; Hamilton; 1992. Pp 456.

70. Robbins, Keith. *Churchill*. London; Longman; 1992. Pp viii, 186.

71. Rollings, Neil, '"The Reichstag method of governing"? The Attlee governments and permanent economic controls', Bc131, 15–36.

72. Rothwell, Victor. *Anthony Eden: a political biography, 1931–1957*. Manchester; Manchester UP; 1992. Pp 298.

73. Saville, John. 'The Communist party and the FBU', Bc7, 225–28.

74. Schwarz, Bill. 'The tide of history: the reconstruction of conservatism, 1945–1951', Bc67, 147–66.

75. Self, Robert. 'Conservative reunion and the general election of 1923: a reassessment', *Twentieth-Century British History* 3 (1992), 249–73.

76. Shell, Donald R. 'The House of Lords: the best second chamber we have got?', Bc200, 155–70.

77. Shepherd, John. 'Labour and the trade unions: Lansbury, Ernest Bevin and the leadership crisis of 1935', Bc117, 204–36.

78. Smart, Nick. 'Crisis? What crisis?' [review article], *Twentieth-Century British History* 3 (1992), 298–303.

79. Smyth, James J. 'Rents, peace, votes: working-class women and political activity in the first world war', Bc135, 174–96.

80. Spence, Alistair. 'Fighting a Labour government', Bc7, 428–30.

81. Stevenson, John. *Third party politics since 1945: Liberals, Alliance and Social Democrats*. Oxford; Blackwell; 1992. Pp xi, 157.

82. Tanner, Duncan. 'Scottish Labour history', *Twentieth-Century British History* 3 (1992), 191–95.

83. Taylor, Andrew J. 'The Conservative trade union movement, 1952–1961', *Labour History R.* 57/1 (1992), 21–28.

84. Taylor, Ian. 'Labour and the impact of war, 1939–1945', Bc67, 7–28.
85. Thatcher, Ian D. 'John MacLean: Soviet versions', *History* 77 (1992), 421–29.
86. Thompson, Willie. *The good old cause: British communism, 1920–1990.* London; Pluto; 1992. Pp 256.
87. Tiratsoo, Nick. 'Labour and the reconstruction of Hull, 1945–1951', Bc67, 126–46.
88. Tomlinson, J.D. 'The Labour government and the trade unions, 1945–1951', Bc67, 90–105.
89. Walker, Graham. 'The Orange order in Scotland between the wars', *International R. of Social History* 37 (1992), 177–206.
90. White, Dan S. *Lost comrades: Socialists of the front generation, 1918–1945.* Cambridge (Ma); Harvard UP; 1992. Pp 255.
91. Whiteley, Patrick. *Labour's grass roots: the politics of party membership.* Oxford; Oxford UP; 1992. Pp 275.
92. Whiting, Richard. 'Taxation policy', Bc131, 117–34.
93. Wildy, Tom. 'The social and economic publicity and propaganda of the Labour governments of 1945–1951', *Contemporary Record* 6 (1992), 45–71.
94. Williamson, Philip. *National crisis and national government: British politics, the economy and empire, 1926–1932.* Cambridge; Cambridge UP; 1992. Pp xvii, 569.
95. Winslow, Barbara. 'Sylvia Pankhurst and the great war', Bc50, 86–120.
96. Young, James D. 'James Connolly, James Larkin and John MacLean: the Easter rising and Clydeside socialism', Bc175, 155–75.

(c) *Constitution, Administration and Law*

1. Baugh, G.C. 'Government grants in aid of the rates in England and Wales, 1889–1990', *Historical Research* 65 (1992), 215–37.
2. Bulmer, Simon. 'Britain and European integration: of sovereignty, slow adaptation, and semi-detachment', Bc201, 1–29.
3. Cronin, James E. 'Power, secrecy, and the British constitution: vetting Samuel Beer's *Treasury control*', *Twentieth-Century British History* 3 (1992), 59–75.
4. Edwards, Geoffrey. 'Central government', Bc201, 64–90.
5. George, Stephen. 'The legislative dimension', Bc201, 91–103.
6. Greenaway, John R. 'The civil service: twenty years of reform', Bc200, 173–90.
7. Preston, Jill. 'Local government and the European Community', Bc201, 104–18.
8. Rhodes, R.A.W. 'Local government', Bc200, 205–18.
9. Ridley, F.F. 'What happened to the constitution under Mrs Thatcher?', Bc200, 111–28.
10. Savage, Stephen P. 'The judiciary: justice with accountability?', Bc200, 191–204.

(d) *External Affairs*

1. Adams, R.J.Q. *British politics and foreign policy in the age of appeasement, 1935–1939.* London; Macmillan; 1992. Pp 208.
2. Allen, Louis. 'William Plomer (1905–1974) and Japan', Bc17, 235–60; 310–11.
3. Arcidiacono, Bruno. 'Dei rapporti tra diplomazia e aritmetica: lo "strano accordo" Churchill-Stalin sui Balcani (Mosca, ottobre 1944)' [The relationship between diplomacy and arithmetic: the 'strange agreement' of Churchill and Stalin on the Balkans], *Storia delle Relazioni Internazionali* 5 (1989), 245–77.
4. Balachandran, G. 'Gold and empire: Britain and India in the great depression', *J. of European Economic History* 20 (1991), 239–70.
5. Bartlett, C.J. *'The special relationship': a political history of Anglo-American relations since 1945.* London; Longman; 1992. Pp ix, 196.
6. Baylis, John. *Diplomacy of pragmatism: Britain and the formation of NATO, 1942–1949.* London; Macmillan; 1992. Pp 288.
7. Beck, Peter J. 'Argentina and Britain: the Antarctic dimension', Bc136, 257–70.
8. Blacker, Carmen. 'Marie Stopes (1907–1958) and Japan', Bc17, 157–65; 304–05.
9. Bowring, Walter. 'Great Britain, the United States and the disposition of Italian East Africa', *J. of Imperial & Commonwealth History* 20 (1992), 88–107.
10. Buckley, Roger. 'In proper perspective: Sir Esler Dening (1897–1977) and Anglo-Japanese relations', Bc17, 271–76; 312–13.
11. Burham, Peter. 'Re-evaluating the Washington loan agreement: a revisionist view of the limits of postwar American power', *R. of International Studies* 18 (1992), 241–59.
12. Cannata, Antonella. 'La Gran Bretagna e la nascità della UEO (1954)' [Great Britain and the birth of the Western European Union], *Storia delle Relazioni Internazionali* 6 (1990), 137–59.
13. Chihiro, Hosoya. 'Tanaka diplomacy and its pro-British orientation, 1927–1929', Bc19, 16–29.
14. Clavin, Patricia. 'The world economic conference, 1933: the failure of British internationalism', *J. of European Economic History* 20 (1991), 489–527.
15. Cohen, Michael J. 'British strategy in the middle east in the wake of the Abyssinian crisis, 1936–1939', Bc179, 21–40.
16. Cortazzi, Sir Hugh. 'Britain and Japan: a personal view of postwar economic relations', Bc19, 163–81.
17. Crampton, R.J. 'The journalist historian in politics: Joseph Swire, the Damian Velcher case and Anglo-Bulgarian relations', *East European Q.* 25 (1991), 257–96.
18. Cretella, Louis Anthony. *Italo-British relations in the eastern Mediterranean, 1919–1923: the view from Rome.* London; Garland; 1991. Pp xi, 445.
19. Croft, Stuart. *British security policy: the Thatcher years and the end of the cold war.* London; HarperCollins; 1991. Pp 240.

20. Daniels, Gordon. 'Sir George Sansom (1883–1965): historian and diplomat', Bc17, 277–88; 313–14.
21. Deighton, Anne. 'Say it with documents: British policy overseas, 1945–1952', *R. of International Studies* 18 (1992), 393–402.
22. Dixon, Piers. 'Eden after Suez', *Contemporary Record* 6 (1992), 178–85.
23. Dobson, Alan. *Peaceful air warfare: the United States, Britain, and the politics of international aviation.* Oxford; Oxford UP; 1991. Pp xiv, 307.
24. Erlich, Haggai. 'British internal security and Egyptian youth', Bc179, 98–112.
25. Ferris, John R. '"The greatest power on earth": Great Britain in the 1920s', *International History R.* 13 (1991), 726–50.
26. Fukuda, Haruko. 'The peaceful overture: Admiral Yamanashi Katsunoshin (1877–1967)', Bc17, 198–213; 308.
27. Futrell, Robert Frank. 'US army air forces intelligence in the second world war', Bc20, 527–52.
28. Gardiner, Juliet. *Over here: the GIs in Britain during the second world war.* London; Gardner; 1992. Pp 224.
29. Gearson, John P.S. 'British policy and the Berlin wall crisis, 1958–1961', *Contemporary Record* 6 (1992), 107–77.
30. George, James H. 'Another chance: Herbert Hoover and world war II relief', *Diplomatic History* 16 (1992), 389–407.
31. George, Stephen. 'Conclusion' [Britain and the European Community], Bc201, 202–07.
32. George, Stephen. 'The policy of British governments within the European Community', Bc201, 30–63.
33. Gilbert, Mark. 'Pacifist attitudes to Nazi Germany, 1936–1945', *J. of Contemporary History* 27 (1992), 493–511.
34. Glees, Anthony. 'The making of British policy on war crimes: history as politics in the UK', *Contemporary European History* 1 (1992), 171–97.
35. Gorst, Anthony. 'Facing facts? The Labour government and defence policy, 1945–1950', Bc67, 190–209.
36. Greenwood, Sean. *Britain and European cooperation since 1945.* Oxford; Blackwell; 1992. Pp 128.
37. Gros, Daniel; Thygesen, Niels. *European monetary integration: from the European monetary system to European monetary union.* London; Longman; 1992. Pp xiii, 494.
38. Hamilton, Valerie. 'Chronology of Anglo-Japanese relations, 1858–1990', Bc17, 289–93.
39. Heikkila, H. 'The British Foreign Office and the churches in Finland and the Soviet Union, 1939–1944', *Vuosikirja Arsskrift* 78 (1988), 123–38.
40. Hennessy, Alistair. 'Epilogue' [Argentina and Britain], Bc136, 287–99.
41. Hennessy, Peter; Anstey, Caroline. *Moneybags and brains: the Anglo-American 'special relationship' since 1945.* Glasgow; Dept. of Government, University of Strathclyde; 1990. Pp iii, 39.
42. Holmes, Colin. 'Sidney Webb (1859–1947) and Beatrice Webb (1858–1943) and Japan', Bc17, 166–76; 305–06.
43. Hölscher, Wolfgang. 'Die Länderbildung in der britischen Besat-

zungszone' [The development of the *Länder* in the British occupation zone], Bc140, 81–102.

44. Howells, Gwyn. 'The British press and the Peróns', Bc136, 227–45.
45. Hunter, Janet. 'British training for Japanese engineers: the case of Kikuchi Kyozo (d. 1942)', Bc17, 137–46; 303.
46. Hüttenberger, Peter. 'Deutschland unter britischer Besatzungersher-rschaft: Gesellschaftliche Prozesse' [Germany under British occupation: social processes], Bc140, 61–80.
47. Inglin, Oswald. *Der stille Krieg: der Wirtschaftskrieg zwischen Grossbritannien und der Schweiz im Zweiten Weltkrieg* [The silent war: the economic struggle between Britain and Switzerland in the Second World War]. Zürich; NZZ; 1991. Pp 478.
48. Jacobsen, M. 'Winston Churchill and the third front', *J. of Strategic Studies* 14 (1991), 337–62.
49. Jessula, Georges. '1943—de Gaulle à Alger: les "carnets" d'Harold Macmillan', *Revue d'Histoire Diplomatique* 105 (1991), 217–48.
50. Jones, Charles A. 'British capital in Argentine history: structures, rhetoric and change', Bc136, 63–77.
51. Kaiser-Lahme, Angela. 'Control Commission for Germany (British element): Bestandsbeschreibung und Forschungsfelder' [Situation discussions and research fields], Bc140, 149–65.
52. Kay, Anne. 'British newspaper reporting of the Yugoslav resistance, 1941–1945', *Storia delle Relazioni Internazionali* 6 (1990), 309–33.
53. Keller, Josepher. 'Zionism and the Palestine question', *The Jewish enigma: an enduring people* ed. David Englander (Halba; 1992), 197–233; 255–56.
54. Kettenacker, Lothar. 'Britische Besatzungspolitik im Spannungsverhältnis von Planung und Realität' [British occupation policies in context of tensions between plans and realities], Bc140, 17–34.
55. Kettenacker, Lothar. 'Die Deklassierung Grossbritanniens bei der Herausbildung des bipolaren Weltmachtesystems, 1939–1945' [The demotion of Great Britain as a result of the development of the bipolar system of world power, 1939–1945], *Historische Mitteilungen* 3 (1990), 57–72.
56. Kettle, Michael. *Churchill and the Archangel fiasco*. London; Routledge; 1992. Pp 560.
57. Kiyoshi, Ikeda. 'The road to Singapore: Japan's view of Britain, 1922–1941', Bc19, 30–46.
58. Kochavi, Arieh J. 'Britain and the establishment of the United Nations War Crimes Commission', *English Historical R.* 107 (1992), 323–49.
59. Kostiner, Joseph. 'Britain and the challenge of the Axis powers in Arabia: the decline of British-Saudi co-operation in the 1930s', Bc179, 128–43.
60. Kramer, Alan. *Die britische Demontagepolitik am Beispiel Hamburgs, 1945–1950* [Hamburg 1945–1950, as an example of British dismantling policy]. Hamburg; Verein für Hamburgische Geschichte; 1991. Pp 482.
61. Kunz, Diane. 'Lyndon Johnson's dollar diplomacy', *History Today* 42/4 (1992), 45–51.
62. Lane, Ann J. 'Putting Britain right with Tito: the displaced persons

question in Anglo-Yugoslav relations, 1946–1947', *European History Q.* 22 (1992), 217–46.

63. Lees, Michael. *The rape of Serbia: the British role in Tito's grab for power, 1943–1944.* San Diego; Harcourt Brace Jovanovich; 1990. Pp xvi, 384.

64. Lentin, Antony. 'Trick or treat?: the Anglo-French alliance, 1919', *History Today* 42/12 (1992), 28–32.

65. Levene, Mark. 'The Balfour Declaration: a case of mistaken identity', *English Historical R.* 107 (1992), 54–77.

66. Little, Walter. 'Falklands futures', Bc136, 271–86.

67. Lowe, Peter. 'Challenge and readjustment: Anglo-American exchanges over east Asia, 1949–1953', Bc19, 143–62.

68. Lukitz, Liora. 'Axioms reconsidered: the rethinking of British strategic policy in Iraq during the 1930s', Bc179, 113–27.

69. Macdonald, Callum A. 'End of empire: the decline of the Anglo-Argentine connection 1918–1951', Bc136, 79–92.

70. MacDougall, Ian. 'Scots in the Spanish civil war, 1936–1939', Bc21, 132–46.

71. Mager, Olaf. *Die Stationierung der britischen Rheinarmee: Grossbritanniens EVG-Alternative* [The stationing of the British Army on the Rhine: Britain's alternative to the European Defence Community]. Baden-Baden; Nomos; 1990. Pp 233.

72. Mayring, Eva A. 'Control Commission for Germany (British element): Vorgehensweise bei der Aktenerschliessung' [Procedure with regard to document publication], Bc140, 133–47.

73. McKercher, B.J.C. 'From enmity to co-operation: the second Baldwin government and the improvement of Anglo-American relations, November 1928-June 1929', *Albion* 24 (1992), 65–88.

74. McKercher, B.J.C. '"Our most dangerous enemy": Great Britain pre-eminent in the 1930s', *International History R.* 13 (1991), 751–83.

75. Meers, Sharon I. 'The British connection: how the United States covered its tracks in the 1954 coup in Guatemala', *Diplomatic History* 16 (1992), 409–28.

76. Melissen, Jan. 'The restoration of the nuclear alliance: Great Britain and atomic negotiations with the United States, 1957–1958', *Contemporary Record* 6 (1992), 72–106.

77. Metzger-Court, Sarah. 'Japanese birthday: Taisho II, G.C. Allen (1900–1982) and Japan', Bc17, 261–70; 311–12.

78. Moore, Bob. 'The western allies and food relief to the occupied Netherlands, 1944–1945', *War & Soc.* 10 (1992), 91–118.

79. Moradiellos, Enrique. 'British political strategy in the face of the military rising of 1936 in Spain', *Contemporary European History* 1 (1992), 123–37.

80. Morewood, Steven. 'Protecting the jugular vein of empire: the Suez canal in British defence strategy', *War & Soc.* 10 (1992), 81–107.

81. Morris, David S.; Haigh, R.H. *Britain, Spain and Gibraltar, 1945–1990: the eternal triangle.* London; Routledge; 1992. Pp 180.

82. Nevo, Joseph. 'Palestinian-Arab violent activity during the 1930s', Bc179, 169–89.

83. Newton, Douglas. 'The British power-elite and the German revolution of 1918–1919', *Australian J. of Politics & History* 37 (1991), 446–65.

84. Nilson, Bengt. 'No coal without iron ore: Anglo-Swedish trade relations in the shadow of the Korean war', *Scandinavian J. of History* 16 (1991), 45–72.

85. Nye, John Vincent. 'Guerre, commerce, guerre commerciale: l'économie politique des échanges franco-anglais re-examinée', *Annales* 47 (1992), 613–32.

86. Oberdorfer, L. 'Skandinavien in der britischen Aussenpolitik der dreissiger Jahre' [Scandinavia in British foreign policy of the 1930s], *Zeitschrift für Geschichtswissenschaft* 40 (1992), 260–08.

87. Omissi, David Enrico. 'The Mediterranean and the middle east in British global strategy, 1935–1939', Bc179, 3–20.

88. Owen, Nicholas. '"Responsibility without power": the Attlee governments and the end of British rule in India', Bc67, 167–89.

89. Pelly, M.E.; Yasamee, H.J.; Hamilton, K.A.; Bennett, G.H. (ed.). *Documents on British policy overseas, ser. 1, vol. 6: eastern Europe, August 1945-April 1946.* London; HMSO; 1991. Pp xlii, 395 + microfiches.

90. Petersen, Tore Tingvold. 'Anglo-American rivalry in the middle east: the struggle for the Buraimi oasis, 1952–1957', *International History R.* 14 (1992), 71–91.

91. Reinharz, Jehuda. 'His Majesty's Zionist emissary: Chaim Weizmann's mission to Gibraltar in 1917', *J. of Contemporary History* 27 (1992), 259–77.

92. Reinharz, Jehuda. 'The Balfour declaration and its maker: a reassessment', *J. of Modern History* 64 (1992), 455–99.

93. Reusch, Ulrich. 'Der Verwaltungsaufbau der britischen Kontrollbehörden in London und der Militärregierung in der britischen Besatzungszone' [The administrative development of the British control authorities in London and the military government in the British occupation zone], Bc140, 35–59.

94. Robins, Lynton J. 'Britain and the European Community: twenty years of not knowing', Bc200, 243–55.

95. Schulze, Rainer. 'Durch die britische Brille gesehen: Beispiele zum Ertrag der britischen Quellen für die (nordwest-)deutsche Ländes- und Regionalgeschichte' [Seen through British eyes: the utility of British sources for north-west German Lände and regional history], Bc140, 103–20.

96. Slattery, J.F. '"Oskar Zuversichtlich": a German's response to British radio propaganda during world war II', *Historical J. of Film, Radio & Television* 12 (1992), 69–85.

97. Smith, Charles. 'Communal conflict and insurrection in Palestine, 1936–1948', Bc11, 62–83.

98. Smith, Dennis. 'Sir Charles Eliot (1862–1931) and Japan', Bc17, 187–97; 307–08.

99. Sturm, P. 'Die Sowjetunion und ihre asiatischen Nachbarn in der Zwischenkriegzeit: das Problem "Grossbritannien"', [The Soviet Union and its Asian neighbours between the wars: the problem of Great Britain]. *Jahrbücher für Geschichte Osteuropas* 39 (1991), 1–32.

100. Sundback, Esa. 'Finland, Scandinavia and the Baltic states viewed within the framework of the border state policy of Great Britain from the autumn of 1918 to the spring of 1919', *Scandinavian J. of History* 16 (1991), 313–34.

101. Tang, James Tuck-Hong. *Britain's encounter with revolutionary China, 1949–1954*. London; Macmillan; 1992. Pp 232.

102. Tsokhas, Kosmas. '"A pound of flesh": war debts and Anglo-Australian relations, 1919–1932', *Australian J. of Politics & History* 38 (1992), 12–26.

103. Waddington, G.T. 'Hitler, Ribbentrop, die NSDAP und der Niedergang des Britischen Empire, 1935–1938' [Hitler, Ribbentrop, the NSDAP and the fall of the British empire], *Vierteljahrshefte für Zeitgeschichte* 40 (1992), 273–306.

104. Walker, John. 'British travel writing and Argentina', Bc136, 183–200.

105. Warner, Sir Fred. *Anglo-Japanese financial relations: a golden tide.* Oxford; Blackwell; 1991. Pp 320.

106. Werner, Wolfram. 'Überlieferungen zur britischen Besatzungszeit in deutschen Archiven' [Materials concerning the British occupation in German archives], Bc140, 121–32.

107. Westrate, Bruce. *The Arab bureau: British policy in the middle east, 1916–1920*. Philadelphia (Pa); Pennsylvania UP; 1992. Pp 240.

108. Xiang, Lanxin. 'The recognition controversy: Anglo-American relations in China, 1949', *J. of Contemporary History* 27 (1992), 319–43.

109. Yasamee, H.J.; Hamilton, K.A.; Warner, Isabel; Lane, Ann J. (ed.). *Documents on British policy overseas, ser. 2, vol. 4: Korea, June 1950-April 1951*. London; HMSO; 1991. Pp liii, 469 + microfiche.

(e) *Religion*

1. Brearley, Margaret F. 'Jewish and Christian concepts of time and modern anti-Judaism: ousting the God of time', Bc189, 481–93.

2. Brown, Callum G. 'Religion and secularisation', Bc138, 48–79.

3. Brown, Stewart J. '"A victory for God": the Scottish Presbyterian churches and the general strike of 1926', *J. of Ecclesiastical History* 42 (1991), 596–617.

4. Bruce, Steve. 'Out of the ghetto: the ironies of acceptance', *Innes R.* 43 (1992), 145–54.

5. Greene, Thomas R. 'Vichy France and the Catholic press in England: contrasting attitudes to a moral problem', *Recusant History* 21 (1992), 111–33.

6. Machin, G.I.T. 'British churches and the cinema in the 1930s', Bc194, 477–88.

7. Mews, Stuart. 'Music and religion in the first world war', Bc194, 465–75.

8. Orme, Nicholas I. 'The twentieth century, part 2: Devon and general', Bc148, 175–97; 221–22.

9. Price, David Trevor William. *A history of the church in Wales in the twentieth century*. Cardiff; Church in Wales; 1990. Pp 76.

10. Staples, Peter. *Relations between the Netherlands Reformed Church and the Church of England since 1945*. Lampeter; Mellen; 1992. Pp 244.
11. Tarn, John Nelson. 'Liverpool's two cathedrals', Bc194, 537–69.
12. Taylor, Brian. 'Church art and church discipline round about 1939', Bc194, 489–98.
13. Turner, Garth. '"Aesthete, impressario, and indomitable persuader": Walter Hussey at St Matthew's, Northampton, and Chichester cathedral', Bc194, 523–35.
14. Winter, J.M. 'Spiritualism and the first world war', Bc13, 185–200.
15. Winter, Michael. 'The twentieth century, part 1: Cornwall', Bc148, 156–74; 218–20.

(f) *Economic Affairs*

1. Adam, James S. (ed.). *The business diaries of Sir Alexander Grant*. Edinburgh; Donald; 1992. Pp 300.
2. Aldcroft, Derek H. *Education, training and economic performance, 1944–1990*. Manchester; Manchester UP; 1992 Pp 200.
3. Alexander, James. 'The long shadow of the strike', Bc7, 434–16.
4. Anon. *Lloyd's war losses: the first world war—casualties to shipping through enemy causes*. London; Lloyds; 1990. Pp viii, 672.
5. Anon. *Retail prices, 1914–1990*. London; HMSO; 1991. Pp 109.
6. Armstrong, John. 'The shipping depression of 1901 to 1911: the experience of freight rates in the British coastal coal trade', *Maritime Wales* 14 (1991), 89–112.
7. Arnold, A.J. '"No substitute for hard cash?": an analysis of returns on investment in the Royal Mail Steam Packet Company, 1903–1929', *Accounting, Business & Financial History* 1 (1991), 335–53.
8. Artis, M.; Cobham, D.; Wickham-Jones, M. 'Social democracy in hard times: the economic record of the Labour government, 1974–1979', *Twentieth-Century British History* 3 (1992), 32–58.
9. Ashworth, William. *The state in business: 1945 to the mid 1980s*. Basingstoke; Macmillan; 1991. Pp 256.
10. Bagguley, Paul. *From protest to acquiescence?: political movements of the unemployed*. London; Macmillan; 1991. Pp 223.
11. Bailey, Victor. 'The first national strike', Bc7, 229–69.
12. Bailey, Victor. 'The "spit and polish" demonstrations', Bc7, 158–75.
13. Balfour, Michael. *Alfred Dunhill: sought after since 1893*. London; Weidenfeld & Nicolson; 1992. Pp 240.
14. Barbezat, D. 'A price for every product, every place: the International Steel Export Cartel, 1933–1939', *Business History* 33/4 (1991), 68–86.
15. Blaug, M. 'Second thoughts on the Keynesian revolution', *History of Political Economy* 23 (1991), 171–92.
16. Bonavia, M.R. 'Stamp and Beeching: parallels and contrasts', *J. of the Railway & Canal Historical Soc.* 30 (1991), 219–25.
17. Bowden, Sue M.; Collins, Michael. 'The bank of England, industrial regeneration, and hire purchase between the wars', *Economic History R.* 2nd ser. 45 (1992), 120–36.

18. Bowden, Sue M.; Turner, P. 'Productivity and long-term growth potential in the UK economy, 1924–1968', *Applied Economics* 23 (1991), 1452–32.

19. Broadberry, S.N.; Crafts, N.F.R. 'Britain's productivity gap in the 1930s: some neglected factors', *J. of Economic History* 52 (1992), 531–58.

20. Broadberry, S.N.; Crafts, N.F.R. 'The implications of British macro-economic policy in the 1930s for long run growth performance', *Rivista di Storia Economica* 7 (1990), 1–19.

21. Bryant, R.V.; Leicester, Colin R. *The Professional Association of Teachers: the early years.* London; Buckland; 1991. Pp 196.

22. Burk, Kathleen; Cairncross, Sir Alec. *'Goodbye, Great Britain': the 1976 IMF crisis.* London; Yale UP; 1992. Pp xix, 268.

23. Cairncross, Sir Alec. *The British economy since 1945: economic policy and performance.* Oxford; Blackwell; 1991. Pp xii, 338.

24. Cairncross, Sir Alec. 'Economic policy after 1974', *Twentieth-Century British History* 3 (1992), 199–208.

25. Cairncross, Sir Alec. 'Introduction: the 1960s', Bc52, 1–13.

26. Cassis, Youssef. 'Introduction: the weight of finance in European societies', Bc95, 1–15.

27. Challis, Christopher E. 'A new beginning: Llantrisant' [Royal Mint], Bc226, 607–72.

28. Chapman, John; Seeliger, Sylvia. 'The influence of the Agricultural Executive Committees in the first world war: some evidence from West Sussex', *Southern History* 13 (1991), 105–22.

29. Chick, Martin. 'Private industrial investment', Bc131, 74–90.

30. Church, Roy A.; Outram, Q.; Smith, D.N. 'The "isolated mass" revisited: strikes in British coal-mining', *Sociological R.* 39 (1991), 55–87.

31. Clarke, Peter F. 'The Keynesian recipe' [review article], *Twentieth-Century British History* 3 (1992), 310–12.

32. Clarke, Peter F. 'The twentieth-century revolution in government: the case of the British Treasury', Bc143, 159–79.

33. Clavin, Patricia. '"The fetishes of so-called international bankers": central bank co-operation for the World Economic Conference', *Contemporary European History* 1 (1992), 281–311.

34. Craig, Bill. 'The strike in Scotland', Bc7, 424–27.

35. Croham, Lord. 'Were the instruments of control for domestic economic policy adequate?', Bc52, 81–109.

36. Curran, John. 'Manning the picket line', Bc7, 431–33.

37. Dahrendorf, Ralf. 'Did the sixties swing too far?', Bc52, 139–57.

38. Davenport-Hines, R.P.T.; Slinn, Judy. *Glaxo: history to 1962.* Cambridge; Cambridge UP; 1992. Pp 425.

39. Dell, Edmund. 'The Chrysler UK rescue', *Contemporary Record* 6 (1992), 1–45.

40. Dolan, J.E. 'Explosives in the service of man', Bc36, 210–22.

41. Downes, Richard. 'Autos over rails: how US business supplanted the British in Brazil, 1910–1928', *J. of Latin American Studies* 24 (1992), 551–83.

42. Dupree, Marguerite W. 'The cotton industry: a middle way between nationalisation and self-government?', Bc131, 137–61.

43. Edgerton, David. *England and the aeroplane: an essay on a militant and technological nation.* London; Macmillan; 1991. Pp 140.
44. Edgerton, David. 'Whatever happened to the British warfare state? The Ministry of Supply, 1945–1951', Bc131, 91–116.
45. Eichengreen, Barry. *Golden fetters: the gold standard and the great depression, 1919–1939.* Oxford; Oxford UP; 1992. Pp 448.
46. Elkin, P.W. 'Aspects of the recent development of the port of Bristol', Bc41, 27–35.
47. Fells, I. 'Fossil fuels 1850 to 2000', Bc36, 197–209.
48. Fermer, H. 'Hollingbury industrial estate, Brighton', *Sussex Industrial History* 21 (1991), 16–34.
49. Fforde, John S. *The bank of England and public policy, 1941–1958.* Cambridge; Cambridge UP; 1992. Pp 850.
50. Flockhart, Jim. 'The Glasgow strike', Bc7, 386–401.
51. Frantzen, J. *Growth and crisis in post-war capitalism.* Aldershot; Dartmouth; 1990. Pp 153.
52. Geroski, P.A.; Murfin, A. 'Entry and industry evolution: the UK car industry, 1958–1983', *Applied Economics* 23 (1991), 799–810.
53. Gerrard, Bill. 'Keynes' *General Theory*: interpreting the interpretations', *Economic J.* 101 (1991), 276–87.
54. Graham, Andrew. 'Thomas Balogh (1905–85)', *Contemporary Record* 6 (1992), 194–207.
55. Grant, Wyn. 'Continuity and change in British business associations', Bc37, 23–46.
56. Gustafsson, Bo. 'Introduction' [power and economic institutions], Bc142, 1–50.
57. Halsey, A.H. *Decline of donnish dominion: British academic professions in the twentieth century.* Oxford; Oxford UP; 1992. Pp xiii, 344.
58. Harvey, Charles; Jones, Geoffrey. 'Organisational capability and competitive advantage', *Business History* 34/1 (1992), 1–10.
59. Hendry, John. *Innovating for failure; government policy and the early British computer industry.* Cambridge (Ma); M.I.T.; 1989. Pp 240.
60. Hennessy, Elizabeth. *A domestic history of the Bank of England, 1930–1960.* Cambridge; Cambridge UP; 1992. Pp 449.
61. Hunter, James. *The claim of crofting: the Scottish highlands and islands, 1930–1990.* Edinburgh; Mainstream; 1991. Pp 224.
62. James, Harold. 'Financial flows across frontiers during the interwar depression', *Economic History R.* 2nd ser. 45 (1992), 594–613.
63. Johnman, Lewis. 'The Labour party and industrial policy, 1940–1945', Bc67, 29–53.
64. Johnman, Lewis. 'The ship-building industry', Bc131, 186–211.
65. Johnson, Graham. 'The FBU, the TUC and the Labour party', Bc7, 196–224.
66. Keeble, S.P. *The ability to manage: a study of British management, 1890–1990.* Manchester; Manchester UP; 1992 Pp 176.
67. Kindleberger, Charles. 'Why did the golden age last so long?', Bc52, 15–44.
68. Kleinman, Paul. 'The strike in London', Bc7, 404–16.

69. Knapp, James. 'RMT: history of a merger', Bc117, 252–56.
70. Lawrence, Jon; Dean, Martin; Robert, Jean-Louis. 'The outbreak of war and the urban economy: Paris, Berlin, and London in 1914', *Economic History R.* 2nd ser. 45 (1992), 564–93.
71. Lazonick, William. 'Business organization and competitive advantage: capitalist transformations in the twentieth century', Bc134, 119–63.
72. Lloyd, John. *Light and liberty: the history of the Electrical, Electronic, Telecommunications and Plumbing union.* London; Weidenfeld & Nicolson; 1990. Pp 696.
73. Lovell, John C. 'Collective bargaining and the emergence of national employer organisation in the British ship-building industry', *International R. of Social History* 36 (1991), 59–91.
74. Mackenzie, David. 'The Bermuda conference and Anglo-American aviation relations at the end of the second world war', *J. of Transport History* 3rd ser. 12 (1991), 61–73.
75. Manley, T.R. 'Some milestones in plastics', Bc36, 181–93.
76. McVeigh, C. 'The chemistry of solar energy', Bc36, 235–49.
77. Mercer, Helen. 'Anti-monopoly policy', Bc131, 55–73.
78. Mercer, Helen. 'The Labour governments of 1945–1951 and private industry', Bc67, 71–89.
79. Milner, N.E. 'Colour photography', Bc36, 253–64.
80. Mini, Peter V. 'The anti-Benthamism of J.M. Keynes: implications for the General theory', *American J. of Economics & Sociology* 50 (1991), 453–68.
81. Moggridge, Donald Edward. *Maynard Keynes: an economist's biography.* London; Routledge; 1992. Pp xxxi, 941.
82. Morris, Margaret. 'In search of the profiteer', Bc117, 185–203.
83. Mougel, François-Charles. 'Nationalisations et dénationalisations en Grande-Bretagne de 1945 à 1990: un enjeu de pouvoir?', *Revue d'Histoire Moderne et Contemporaine* (1992), 238–62.
84. Moure, Kenneth. 'The limits to central bank co-operation, 1916–1936', *Contemporary European History* 1 (1992), 259–79.
85. Napier, C.J. 'Secret accounting: the P. & O. Group in the interwar years', *Accounting, Business & Financial History* 1 (1991), 303–33.
86. Omissi, David Enrico. 'The Hendon air pageant, 1920–1937', Bc10, 198–220.
87. Peteri, Gyorgy. 'Central bank diplomacy: Montagu Norman and central Europe's monetary reconstruction after world war I', *Contemporary European History* 1 (1992), 233–58.
88. Price, B.J.; Dodds, M.G. 'The quest for new medicines', Bc36, 27–50.
89. Roberts, Brian. 'A mining town in wartime: the fears for the future', *Llafur* 6 (1992), 82–95.
90. Roberts, Richard. 'Regulatory responses to the rise of the market for corporate control in Britain in the 1950s', *Business History* 34/1 (1992), 183–200.
91. Robinson, Austin. 'Munitions output of the United Kingdom, 1939–1944: a comment', *Economic History R.* 2nd ser. 45 (1992), 376–77.

92. Rockley, Pete. 'The strike and the executive', Bc7, 417–23.
93. Rooth, Tim. 'The political economy of protectionism in Britain, 1919–1932', *J. of European Economic History* 21 (1992), 47–97.
94. Roxburgh, Bob. 'Unofficial action on Merseyside', Bc7, 402–03.
95. Shaw, John P. 'Pastures in the sky: Scottish tower silos, 1918–1939', *J. of the Historic Farm Buildings Group* 4 (1990), 73–91.
96. Sheail, John. 'Road surfacing and the threat to inland fisheries', *J. of Transport History* 3rd ser. 12 (1991), 135–47.
97. Skidelsky, Robert. *John Maynard Keynes vol. 2: the economist as saviour, 1921–1937.* London; Macmillan; 1992. Pp 634.
98. Smith, Anthony D. *International financial markets: the performance of Britain and its rivals.* Cambridge; Cambridge UP; 1992. Pp 190.
99. Solow, Robert. 'Did policy errors of the 1960s sow the seeds of trouble in the 1970s?', Bc52, 158–81.
100. Southall, Humphrey. 'The twentieth century: c.1900- c.1960', Bc181, 211–37; 245–46.
101. Sowrey, F. 'A country garage—Quick's of Handcross', *Sussex Industrial History* 21 (1991), 2–8.
102. Stait, Bruce A. *Rotol: the history of an airscrew company, 1937–1960.* Stroud; Sutton; 1990. Pp xii, 180.
103. Targetti, Ferdinando. *Nicholas Kaldor: the economics and politics of capitalism as a dynamic system.* Oxford; Oxford UP; 1992. Pp 400.
104. Thain, Colin. 'Government and the economy', Bc200, 221–42.
105. Thor, Jon Th. 'The extension of Iceland's fishing limits in 1952 and the British reaction', *Scandinavian J. of History* 17 (1992), 25–43.
106. Tinbergen, Jan. 'Economics: recent performance and future trends', Bc30, 146–65.
107. Tiratsoo, Nick. 'The motor car industry', Bc131, 162–85.
108. Tobin, James. 'The invisible hand in modern macroeconomics', Bc30, 117–29.
109. Tomlinson, J.D. 'Planning: debate and policy in the 1940s', *Twentieth-Century British History* 3 (1992), 154–74.
110. Tomlinson, J.D. 'Productivity policy', Bc131, 37–54.
111. Tomlinson, S. 'Planning for cotton, 1945–1951', *Economic History R.* 2nd ser. 44 (1991), 523–26.
112. Travis, Anthony S. 'Synthetic dyestuffs: modern colours for the modern world', Bc36, 144–57.
113. Tsokhas, Kosmas. 'Wheat in wartime: the Anglo-Australian experience', *Agricultural History* 66 (1992), 1–18.
114. Turnbull, P.; Sapsford, D. 'Why did Devlin fail?: capitalism and conflict in the docks', *British J. of Industrial Relations* 29 (1991), 237–57.
115. Tweedale, Geoffrey. 'Marketing in the second industrial revolution: a case study of the Ferranti computer group, 1949–1963', *Business History* 34/1 (1992), 96–127.
116. van Waarden, Frans. 'Introduction: crisis, corporatism and continuity', Bc37, 1–19.
117. van Waarden, Frans. 'Wartime economic mobilisation and state-business relations: a comparison of nine countries', Bc37, 271–304.

118. Waas, B. 'Streik und Streikrecht in England, 1945–1985' [Strikes and industrial relations legislation in England, 1945–1985], *Archiv für Sozialgeschichte* 31 (1991), 275–96.
119. Whipp, Richard. 'Crisis and continuity: innovation in the British automobile industry, 1896–1986', Bc22, 120–41.
120. Williamson, John. 'Could international policy co-ordination have been more effective?', Bc52, 110–38.
121. Willis, R.J. 'The contribution of chemistry in animal health', Bc36, 88–105.
122. Worswick, David. 'How was it possible to run economies at such high pressure without accelerating wage rates?', Bc52, 45–80.
123. Zweiniger-Bargielowskan, Ina. 'Colliery managers and nationalisation: the experience in south Wales', *Business History* 34/4 (1992), 59–78.

(g) *Social Structure and Population*

1. Alban, J.R. 'The second world war' [in Swansea], Bc212, 130–144.
2. Ambrose, Peter. 'The rural/urban fringe as battleground', Bc54, 175–94.
3. Anderson, M. 'Population and family life', Bc138, 12–47.
4. Beddoe, Deirdre. 'Munitionettes, maids and mams: women in Wales, 1914–1939', Bc66, 189–209.
5. Berger, Stefan. 'The British and German labour movements before the second world war: the *Sonderweg* revisited', *Twentieth-Century British History* 3 (1992), 219–48.
6. Campbell, Alan. 'Communism in the Scottish coalfields, 1920–1936: a comparative analysis of implantation and rejection', *Tijdschift voor Sociale Geschiedenis* 18 (1992), 168–90.
7. Caunce, Stephen. 'Twentieth-century farm servants: the horselads of the East Riding of Yorkshire', *Agricultural History R.* 39 (1991), 143–66.
8. Channon, Cyril E.; Channon, Margaret L. 'Evacuation to Teignmouth, 1939–1945', *Devonshire Association Report & T.* 124 (1992), 135–51.
9. Davies, Andrew. *Leisure, gender and poverty: working-class culture in Salford and Manchester, 1900–1939.* Milton Keynes; Open University; 1992. Pp xi, 210.
10. Dickie, Marie. 'Town patriotism in Northampton, 1918–1939: an invented tradition?', *Midland History* 17 (1992), 109–17.
11. Duncan, Robert. 'Independent working class education and the formation of the labour college movement in Glasgow and the west of Scotland, 1915–1922', Bc175, 106–28.
12. Elliott, B.J. 'The social activities of the unemployed in an industrial city during the "great depression"', *T. of the Hunter Arch. Soc.* 16 (1991), 52–60.
13. Englander, David. 'The Fire Brigades union and its members', Bc7, 101–38.
14. Evans, Neil. 'Writing the social history of modern Wales: approaches, achievements and problems', *Social History* 17 (1992), 479–92.

15. Faulkner, Evelyn. '"Powerless to prevent him": attitudes of married working-class women in the 1920s and the rise of sexual power', *Local Population Studies* 49 (1992), 51–61.

16. Foster, John. 'A proletarian nation? Occupation and class since 1914', Bc138, 201–40.

17. Green, Joseph. *Social history of the Jewish East End in London, 1914–1939: a study of life, labour and liturgy.* Lampeter; Mellen; 1992. Pp 540.

18. Grundy, Emily M.D. 'Socio-demographic variation in notes of movement into institutions among elderly people in England and Wales: an analysis of linked census and mortality data, 1971–1985', *Population Studies* 46 (1992), 65–84.

19. Harris, José. 'War and social history: Britain and the home front during the second world war', *Contemporary European History* 1 (1992), 17–35.

20. Haste, Cate. *Rules of desire: sex in Britain since world war I.* London; Chatto & Windus; 1992. Pp xi, 356.

21. Imhof, Arthur E. 'The implications of increased life expectancy for family and social life', Bc23, 347–76.

22. Jones, R. Merfyn. 'Labour implantation: is there a Welsh dimension? *Tijdschift voor Sociale Geschiedenis* 18 (1992), 231–47.

23. Jones, Teuan Guynedd. 'The city and its villages' [Swansea], Bc212, 79–92.

24. Kay, Diana; Miles, Robert. *Refugees or migrant workers?: European volunteer workers in Britain, 1946–1951.* London; Routledge; 1992. Pp 229.

25. Knox, W. 'Class, work and trade unionism in Scotland', Bc138, 108–37.

26. Lewis, Jane. *Women in Britain since 1945: women, family, work and the state in the postwar years.* Oxford; Blackwell; 1992. Pp 160.

27. Lunn, Kenneth. 'Labour culture in dockyard towns: a study of Portsmouth, Plymouth and Chatham, 1900–1950', *Tijdschift voor Sociale Geschiedenis* 18 (1992), 275–93.

28. McCrone, D. 'Towards a principled elite: Scottish elites in the twentieth-century', Bc138, 174–200.

29. McGuckin, Ann. 'Moving stories: working-class women', Bc135, 197–220.

30. McIvor, Arthur J. 'Women and work in twentieth-century Scotland', Bc138, 138–73.

31. McShane, Harry. 'Glasgow's housing crisis', Bc175, 25–51.

32. Murphy, M. 'Economic models of fertility in postwar Britain—a conceptual and statistical re-interpretation', *Population Studies* 46 (1992), 235–58.

33. Pugh, Martin. *Women and the women's movement in Britain, 1914–1959.* London; Macmillan; 1992. Pp 368.

34. Roper, Michael. 'Yesterday's model: product fetishism and the British company man, 1945–1985', Bc223, 190–211.

35. Sander, William. 'Catholicism and the economics of fertility', *Population Studies* 46 (1992), 477–90.

36. Segars, Terry. 'War, women and the FBU', Bc7, 139–57.
37. Sheridan, D. (ed.). *Wartime women: a mass-observation*. London; Heinemann; new edn. 1990. Pp 256.
38. Taylor, Eric. *Forces sweethearts: service romances in world war II*. London; Hale; 1990. Pp 208.
39. Tebbutt, Melanie. '"You couldn't help but know": public and private space in the lives of working class women, 1918–1939', *Manchester Region History R.* 6 (1992), 72–79.
40. Thatcher, A.R. 'Trends in numbers and mortality at high ages in England and Wales', *Population Studies* 46 (1992), 411–26.
41. Thompson, F.M.L. 'English landed society in the twentieth century: III, self-help and outdoor relief', *T. of the Royal Historical Soc.* 6th ser. 2 (1992), 1–23.
42. Vincent, David. *Poor citizens: the state and the poor in twentieth-century Britain*. London; Longman; 1991. Pp viii, 258.
43. Webster, Charles. 'Psychiatry and the early National Health Service: the role of the Mental Health Standing Advisory Committee', Bc145, 103–16.

(h) *Social Policy*

1. Berridge, Virginia. 'The early years of AIDS in the United Kingdom, 1981–1986: historical perspectives', Bc25, 303–28.
2. Booth, Christopher C. 'The *British Medical Journal* and the twentieth-century consultant', Bc28, 248–62.
3. Brown, Frank E. 'Analysing small building plans: a morphological approach', Bc38, 259–76.
4. Bryder, Linda. '"Wonderlands of buttercup, clover and daisies": tuberculosis and the open-air school movement in Britain, 1907–1939', Bc57, 72–95.
5. Cantor, David. 'The aches of industry: philanthropy and rheumatism in inter-war Britain', Bc26, 225–45.
6. Crawshaw, Anthony. 'Aerial archaeology in Yorkshire: a "starfish" site', *Yorkshire Arch. J.* 64 (1992), 209–11.
7. Crick, Bernard. 'Citizenship and education', Bc200, 273–86.
8. Davies, T.G. 'Health and the hospitals' [in Swansea], Bc212, 165–78.
9. Davis, O.L. 'The invisible evacuees: England's urban teachers during the first autumn of war, 1939', *History of Education Soc. B.* 49 (1992), 53–60.
10. Finnimore, Brian. *Houses from the factory: system building and the welfare state, 1942–1974*. London; Rivers Oram; 1989. Pp ix, 278.
11. Foster, John. 'Red Clyde, red Scotland', Bc6, 106–24.
12. Goldin, Claudia. 'Marriage bars: discrimination against married women workers from the 1920s to the 1950s', Bc32, 511–38.
13. Grimes, Sharon Schildein. *The British National Health Service: state intervention in the medical marketplace, 1911–1948*. London; Garland; 1991. Pp viii, 239.

14. Harris, Bernard. 'Government and charity in the distressed mining areas of England and Wales, 1928–1930', Bc26, 207–24.
15. Hart, Julian Tudor. 'The *British Medical Journal*, general practitioners and the state, 1840–1990', Bc28, 228–47.
16. Hasegawa, Junichi. *Replanning the blitzed city centre: a comparative study of Bristol, Coventry and Southampton, 1941–1950*. Milton Keynes; Open University; 1992. Pp xii, 179.
17. Heater, Derek. 'Britain, Europe and citizenship', Bc200, 287–304.
18. Howarth, Alan. 'Political education: a government view', Bc200, 318–26.
19. Jones, David J.V. 'Crime, order and the police' [in Swansea], Bc212, 194–205.
20. Jones, Harriet. 'Beveridge's Trojan horse', History Today 42/10 (1992), 44–49.
21. Leneman, Leah. 'Ballencrieff: a tenant's opposition to land settlement', T. of the East Lothian Antiquarian Soc. 20 (1980 for 1988), 59–63.
22. Lewis, Jane. 'Models of equality for women: the case of state support for children in twentieth-century Britain', Bc44, 73–92.
23. Lewis, Jane. 'Providers, "consumers", the state and delivery of healthcare services in twentieth-century Britain', Bc23, 317–45.
24. Lewis, Jane. 'The medical journals and the politics of public health 1918–1990', Bc28, 207–27.
25. Liedtke, Rainer. 'Self-help in Manchester Jewry: the Provincial Independent Society', Manchester Region History R. 6 (1992), 62–71.
26. Macnicol, John. 'Welfare, wages and the family: child endowment in comparative perspective, 1900–1950', Bc57, 244–75.
27. Meyer, W.R. 'Schools vs parents in Leeds, 1902–1944', J. of Educational Administration & History 22/2 (1990), 16–26.
28. Peretz, Liz. 'Regional variation in maternal and child welfare between the wars: Merthyr Tydfil, Oxfordshire and Tottenham', Bc205, 133–49.
29. Powell, Martin. 'How adequate was hospital provision before the NHS?: an examination of the 1945 South Wales hospital survey', Local Population Studies 48 (1992), 22–32.
30. Roderick, Gordon W. 'Education in an industrial society' [in Swansea], Bc212, 179–93.
31. Slater, John. 'New curricula, new directions', Bc200, 305–17.
32. Smith, Harold L. 'The politics of Conservative reform: the equal pay for equal work issue, 1945–1955', Historical J. 35 (1992), 401–15.
33. Thane, Pat. 'Visions of gender in the making of the British welfare state: the case of women in the British Labour party and social policy, 1906–1945', Bc44, 93–118.
34. Thom, Deborah. 'Wishes, anxieties, play, and gestures: child guidance in inter-war England', Bc57, 200–19.
35. Urwin, Cathy; Sharland, Elaine. 'From bodies to minds in childcare literature: advice to parents in inter-war Britain', Bc57, 174–99.
36. Wrigley, Chris. 'Trade unionists, employers and the cause of industrial unity and peace, 1916–1921', Bc117, 155–84.
37. Yelling, J.A. *Slums and redevelopment: policy and practice in England,*

1918–1945, with particular reference to London. London; UCLP; 1992. Pp 232.

(i) *Naval and Military*

1. Andriessen, J.H.J. 'Facetten van de Britse blokkade van de Noordzee tijdens de Eerste Wereldoorlog: de terechtstelling van kapitein Fryatt' [Aspects of the British blockade of the North Sea during the first world war: the execution of Captain Fryatt], *Marineblad* 101 (1991), 177–82.
2. Annan, Noel. 'The work of the Berlin control commission: a personal view', Bc140, 1–16.
3. Bishop, Edward. *The debt we owe: the Royal Air Force benevolent fund, 1919–1989.* Shrewsbury; Airlife; 1989. Pp 200.
4. Bond, Brian. 'Alanbrooke and Britain's Mediterranean strategy, 1942–1944', Bc150, 175–93.
5. Boog, Horst. 'Introduction' [air war], Bc20, 1–6.
6. Boog, Horst. 'The Luftwaffe and indiscriminate bombing up to 1942', Bc20, 373–404.
7. Brooks, Stephen (ed.). *Montgomery and the Eighth army: a selection from the diaries, correspondence and other papers, 1942/1943.* London; Bodley Head (Army Records Soc. 7); 1991. Pp xii, 418p.
8. Bushaway, Bob. 'Name upon name: the great war and remembrance', Bc190, 136–67.
9. Carver, Michael. *Tightrope walking: British defence policy since 1945.* London; Hutchinson; 1992. Pp viii, 191.
10. Chorley, W.R. *RAF bomber command losses, vol. 1: 1939–1940.* Earl Shilton; Midland Counties; 1992. Pp 160.
11. Claxton, Eric Charles. *Struggle for peace: the story of the Casualties' union in the years following the second world war.* Lewes; Book Guild; 1992. Pp 350.
12. Cox, Sebastian. 'The sources and organisation of RAF intelligence and its influence on operations', Bc20, 553–79.
13. Croft, John. 'Palestine, September 1918', *J. of the Soc. for Army Historical Research* 70 (1992), 33–45.
14. Cross, J.P. *First in, last out: an unconventional British officer in Indo-China (1945–1946) and (1972–1976).* London; Brassey's; 1992. Pp 242.
15. Danchev, Alex. 'Being friends: the Combined Chiefs of Staff and the making of allied strategy in the second world war', Bc150, 195–210.
16. English, John A. *The Canadian army and the Normandy campaign: a study of failure in high command.* London; Praeger; 1991. Pp 295.
17. Ferris, John R. (ed.). *The British army and signals intelligence during the first world war.* Stroud; Sutton; 1992. Pp 368.
18. Firebrace, John. *British Empire campaigns and occupations in the Near East, 1914–1923: a postal history.* London; Christie's Robson Lowe; 1991. Pp xii, 460.
19. Fraser, David (ed.). *In good company: the first world war letters and diaries of the the Hon. William Fraser, Gordon Highlanders.* Salisbury; Russell; 1990. Pp x, 348.

20. French, David. 'Who knew what and when? The French army mutinies and the British decision to launch the third battle of Ypres', Bc150, 133–53.
21. Gooch, John. '"Hidden in the Rock": American military perceptions of Great Britain, 1919–1940', Bc150, 155–73.
22. Groehler, Olaf. 'The strategic air war and its impact on the German civilian population', Bc20, 279–97.
23. Gurney, David. *The Red Cross civilian postal message scheme with the Channel Islands during the Occupation, 1940–1945.* Ilford; C.I.S.S.; 1992. Pp 400.
24. Harris, Paul. 'Egypt: defence plans', Bc179, 61–78.
25. Hartfield, Alan. 'Lieutenant-General Sir John Sharman Fowler, KCB, KCMG, DSC, first colonel commandant, Royal Corps of Signals', *J. of the Soc. for Army Historical Research* 70 (1992), 67–70.
26. Hays Parks, W. 'Air war and the laws of war', Bc20, 20, 310–72.
27. Hinsley, Francis Harry; Simkins, C. Anthony G. *British intelligence in the second world war, vol. 4: security and counter-intelligence.* London; HMSO; 1990. Pp xii, 420.
28. Howse, Derek. *Radar at sea: the Royal Navy in world war 2.* London; Macmillan; 1992. [Pp 360].
29. Jackson, William G.F.; Bramall, Dwin. *The chiefs: the story of the United Kingdom Chiefs of Staff.* London; Brassey's; 1992. Pp xxii, 482.
30. Jones, R.V. 'Scientific intelligence of the Royal Air Force in the second world war', Bc20, 580–95.
31. Kemp, Anthony. *The Special Air Service regiment at war, 1941–1945.* London; Murray; 1991. Pp 268.
32. Kingsford, Paul. *After Alamein: prisoner of war diaries, 1942–1945.* Lewes; Book Guild; 1992. Pp 140.
33. Konvitz, Josef W. 'Bombs, cities, and submarines: allied bombing of the French ports, 1942–1943', *International History R.* 14 (1992), 23–44.
34. Laidlaw, Richard B. 'The OSS and the Burma Road 1942–1945', Bc18, 102–22.
35. Layman, R.D. '*Engadine* at Jutland', *Warship* (1990), 93–101.
36. Leandri, C.V. 'De quelques flous historiques qui entourent la sortie du Bismarck et la perte du Hood en mai 1941', *Neptunia* 183 (1991), 22–30.
37. Lowden, John L. *Silent wings at war: combat gliders in world war II.* Washington (DC); Smithsonian; 1992. Pp 288.
38. MacKenzie, S.P. *Politics and military morale: current affairs and citizenship education in the British army, 1914–1950.* Oxford; Oxford UP; 1992. Pp xiii, 245.
39. Merskey, Harold. 'Shell-shock', Bc145, 245–67.
40. Messerschmidt, Manfred. 'Strategic air war and international law', Bc20, 298–309.
41. Moshe, Tuvia Ben. *Churchill: strategy and history.* Hemel Hempstead; Harvester Wheatsheaf; 1992. Pp 397.
42. Murray, Williamson. 'The influence of pre-war Anglo-American doctrine on the air campaigns of the second world war', Bc20, 235–53.
43. Newell, J.Q.C. 'Learning the hard way: Allenby in Egypt and Palestine, 1917–1919', *J. of Strategic Studies* 14 (1991), 363–87.

44. Noakes, Jeremy. 'Introduction' [civilian in war], Bc188, 1–13.
45. O'Neill, Robert. 'Problems of command in limited warfare: thoughts from Korea and Vietnam', Bc150, 263–78.
46. Overy, Richard J. 'Air power in the second world war: historical themes and theories', Bc20, 7–28.
47. Prete, Roy A. 'Joffre and the question of Allied supreme command, 1914–1916', *P. of the Annual Meeting of the Western Soc. for French History* 16 (1989), 329–38.
48. Prior, Robin; Wilson, Trevor. *Command on the western front: the military career of Sir Henry Rawlinson, 1914–1918*. Oxford; Blackwell; 1991. Pp 352.
49. Probert, Henry A. 'The determination of RAF policy in the second world war', Bc20, 683–701.
50. Raugh, Harold E. *Wavell in the middle east, 1939–1941: a study in generalship*. London; Brassey's; 1992. Pp 270.
51. Richardson, F.C. 'Radio-navigation in the UK in world war II', *J. of Navigation* 45 (1992), 60–69.
52. Rossberg, Horst. 'A prisoner of war remembers Stobs camp', *Hawick Arch. Soc. T.* (1991), 10–16.
53. Rubin, Gerry R. *Durban, 1942: a British troopship revolt*. London; Hambledon; 1992. Pp 184.
54. Shores, C.F.; Franks, L.R.; Guest, R. (comp.). *Above the trenches: a complete record of the fighter aces and units of the British empire air forces, 1915–1920*. Stoney Creek (Ont); Fortress; 1990. Pp 368.
55. Smith, Claude. *The history of the Glider Pilot Regiment*. London; Cooper; 1992. Pp 184.
56. Smith, Malcolm. 'The air threat and British foreign and domestic policy: the background to the strategic air offensive', Bc20, 609–26.
57. Smith, Robert P. '"The unsung heroes of the air": logistics and the air war over Europe, 1941–1945', Bc20, 254–69.
58. Starkey, Pat. *I will not fight: conscientious objectors and pacifists in the north-west during the second world war*. Liverpool; Liverpool UP (Historical essays 7); 1992. Pp 40.
59. Sumida, Jon Tetsuro. '"The best laid plans": the development of Britsh battle-fleet tactics, 1919–1942', *International History R.* 14 (1992), 681–700.
60. Tavener, Nick. 'Capo, Kincardineshire: investigation of a second world war bomb crater', *Soc. of Antiquaries of Scotland P.* 121 (1991), 437–42.
61. Taylor, Philip M. *War and the media: propaganda and persuasion in the Gulf war*. Manchester; Manchester UP; 1992 Pp 288.
62. Terraine, John. 'Theory and practice of the air war: the Royal Air Force', Bc20, 467–95.
63. Travers, Tim. 'Could the tanks of 1918 have been war-winners for the British Expeditionary Force?', *J. of Contemporary History* 27 (1992), 389–406.
64. Vance, Jonathan F. 'The politics of camp life: the bargaining process in two German prison camps', *War & Soc.* 10 (1992), 109–26.

65. Wark, Wesley K. 'The air defence gap: British air doctrine and intelligence warnings in the 1930s', Bc20, 511–26.
66. Williams, Rhodri. 'Lord Kitchener and the battle of Loos: French politics and British strategy in the summer of 1915', Bc150, 117–32.
67. Woodward, David R. (ed.). *The military correspondence of Field-marshal Sir William Robertson, Chief of the Imperial General Staff, December 1915-February 1918.* London; Bodley Head; 1990. Pp x, 359.
68. Woodward, Sandy; Robinson, Patrick. *One hundred days: the memoirs of the Falklands battle group commander.* London; HarperCollins; 1992. Pp xix, 360.

(j) *Intellectual and Cultural*

1. Aldgate, Tony. 'Mr Capra goes to war: Frank Capra, the British Army Film Unit, and Anglo-American travails in the production of *Tunisian victory*', *Historical J. of Film, Radio & Television* 11 (1991), 21–39.
2. Allison, Keith J. 'W.H. St Quintin of Scampston, naturalist', Bc103, 23–39.
3. Allsobrook, David Ian. *Music for Wales: Walford Davies and the National Council of Music, 1918–1941.* Cardiff; Wales UP; 1992. Pp xii, 200.
4. Almond, J.K. 'Chemistry and the production of metals', Bc36, 161–80.
5. Amphlett, C.B. 'Chemistry and the development of the nuclear fuel cycle', Bc36, 223–34.
6. Baker, W.O. 'Materials for electronics', Bc36, 265–87.
7. Barkan, Elazar. *The retreat of scientific racism: changing concepts of race in Britain and the United States between the world wars.* London; Longman; 1992. Pp 440.
8. Benison, S.; Barger, A.C.; Wolfe, E.L. 'Walter B. Cannon and the mystery of shock: a study of Anglo-American co-operation in world war I', *Medical History* 35 (1991), 217–49.
9. Bennett, Douglas. 'The drive towards the community', Bc145, 321–32.
10. Berrios, German E. 'British psychopathology since the early 20th century', Bc145, 232–44.
11. Berry, Paul; Bostridge, Mark. *Vera Brittain: a life.* London; Chatto & Windus; 1992. Pp 416.
12. Beveridge, Allan. 'Thomas Clouston and the Edinburgh School of Psychiatry', Bc145, 359–88.
13. Bianchini, Franco; Schwengel, Hermann. 'Re-imagining the city', Bc149, 212–34; 257–63.
14. Birkett, Jennifer; Harvey, Elizabeth D. 'Introduction' [determined women], Bc40, 1–37.
15. Boyd, Kelly. 'Knowing your place; the tensions of manliness in boys' story papers, 1918–1939', Bc223, 145–67.
16. Bradley, Dick. *Understanding rock'n'roll: popular music in Britain, 1955–1964.* Milton Keynes; Open University; 1992. Pp 192.
17. Burnham, John. 'The *British Medical Journal* in America', Bc28, 165–87.
18. Campbell, W.C. 'Edward Leicester Atkinson: physician, parasitologist,

and adventurer', *J. of the History of Medicine & Allied Sciences* 46 (1991), 219–40.

19. Carruthers, Annette. *Edward Barnsley and his workshop: arts and crafts in the twentieth century.* Wendlebury; White Cockade; 1992. Pp 208.

20. Clark, Chris. 'An oral history of jazz in Britain', *Oral History* 18/1 (1991), 66–72.

21. Clayton, Robert J.; Algar, Joan (ed.). *A scientist's war: the war diary of. Sir Clifford Paterson, 1939–1945.* London; Peregrinus for the Institution of Electrical Engineers; 1991. Pp 680.

22. Collier, Simon. '"Hullo tango!" The English tango craze and its after-echoes', Bc136, 213–25.

23. Collins, John; Barker, Nicolas. *The two forgers: a biography of Harry Buxton Forman and Thomas James Wise.* Aldershot; Scolar; 1992. Pp 280.

24. Corner, John; Harvey, Sylvia. 'Introduction: Great Britain Ltd', Bc149, 1–20; 235–36.

25. Corner, John; Harvey, Sylvia. 'Mediating tradition and modernity: the heritage/enterprise couplet', Bc149, 45–75; 241–44.

26. Crammer, J.L. 'Extraordinary deaths of asylum inpatients during the 1914–1918 war', *Medical History* 36 (1992), 430–41.

27. Crawley, Eduardo. 'The unnoticed era', Bc136, 247–53.

28. Crowther, Jan; Noble, M. 'Adult education and the development of regional and local history: East Yorkshire and north Lincolnshire, *c.*1929–1985', Bc205, 150–72.

29. Davies, James A. 'Writing and drama' [in Swansea], Bc212, 206–17.

30. Devons, S. 'Rutherford and the science of his day', *Notes & Records of the Royal Soc. of London* 45 (1991), 221–42.

31. Duckworth, Jackie. 'Sylvia Pankhurst as an artist', Bc50, 36–57.

32. Dunthorne, Kirstine Brander. 'Art and artists' [in Swansea], Bc212, 229–43.

33. Dykes, David. *The university college of Swansea: an illustrated history.* Stroud; Sutton; 1992. Pp x, 244.

34. English, Barbara. 'The rebuilding of Sledmere house, 1911–1917', Bc103, 40–55.

35. Evans, D.A.; Lawson, K.R. 'A century of crop protection chemicals', Bc36, 68–87.

36. Fish, H. 'The contribution of chemistry to environmental health', Bc36, 17–26.

37. Fishburn, Evelyn. 'Borges and England', Bc136, 201–11.

38. Fore, H. 'Contributions of chemistry to food consumption', Bc36, 106–20.

39. Garnsey, Elizabeth. 'An early academic enterprise: a study of technology transfer', *Business History* 34/4 (1992), 79–98.

40. Gascoigne, R.M. 'Julian Huxley and biological process', *J. of Historical Biology* 24 (1991), 433–56.

41. Gelder, Michael. 'Adolf Meyer and his influence on British psychiatry', Bc145, 419–35.

42. Gray, G.W. 'Development of liquid crystal materials for information technology', Bc36, 288–303.

43. Harper, Sue. 'The representation of women in British feature film, 1945–1950', *Historical J. of Film, Radio & Television* 12 (1992), 217–30.

44. Harries, Phillip. 'Arthur Waley (1899–1966): poet and translator', Bc17, 214–22; 308–09.

45. Harvie, Christopher. 'Second thoughts of a Scotsman on the make: politics, nationalism and myth in John Buchan', *Scottish Historical R.* 70 (1991), 31–54.

46. Hewison, Robert. 'Commerce and culture', Bc149, 162–77; 253–54.

47. Hounshell, D. 'Chemistry and the development of natural and synthetic fibres', Bc36, 123–43.

48. Howells, John G. 'The establishment of the Royal College of Psychiatrists', Bc145, 117–34.

49. Jenner, F.A. 'Erwin Stengel: a personal memoir', Bc145, 436–44.

50. Kinnell, Margaret. 'Discretion, sobriety and wisdom: the teacher in children's books', Bc190, 168–91.

51. Kushner, Tony. 'James Parkes, the Jews, and conversionism: a model for multi-cultural Britain?', Bc189, 451–61.

52. Lawrence, Christopher. 'Experiment and experience in anaesthesia: Alfred Goodman Levy and chloroform death, 1910–1960', Bc24, 263–94.

53. Leonard, Jim. 'A people's plan: a study of the "Max Lock" survey and plan for Middlesbrough, 1943 and 1946', Bc205, 117–32.

54. Leontief, Wassily. 'The present state of economic science', Bc30, 141–45.

55. Lewis, Peter M. 'Mummy, matron and the maids: feminine presence and absence in male institutions, 1934–1963', Bc223, 168–89.

56. Light, Alison. *Forever England: femininity, literature and conservatism between the wars*. London; Routledge; 1991. Pp xvi, 281.

57. Lovell, Richard. 'Choosing people: an aspect of the life of lord Moran (1882–1977)', *Medical History* 36 (1992), 442–54.

58. McAleer, Joseph. *Popular reading and publishing in Britain, 1914–1950*. Oxford; Oxford UP; 1992. Pp 284.

59. McDowell, William H. *History of BBC broadcasting in Scotland, 1923–1983*. Edinburgh; Edinburgh UP; 1992. Pp 376.

60. McNeil, Maureen. 'The old and new worlds of information technology in Britain', Bc149, 116–36; 248–52.

61. McPherson, A. 'Schooling', Bc138, 80–107.

62. Mellor, Adrian. 'Enterprise and heritage in the dock', Bc149, 93–115; 246–48.

63. Moore, Jerrold Northrop (ed.). *Edward Elgar: letters of a lifetime*. Oxford; Oxford UP; 1990 Pp xix, 524.

64. Moriarty, Catherine. 'L'iconographie chrétienne des monuments aux morts de la première guerre mondiale dans le Royaume-Uni', *Guerres Mondiales et Conflits Contemporains* 167 (1992), 71–86.

65. Murray, Peter. 'Paternoster—post Holford', *London J.* 16 (1991), 129–39.

66. Nelson, Richard. 'The roles of firms in technical advance: a perspective from evolutionary theory', Bc134, 164–84.

67. Phillips, Catherine. *Robert Bridges: a biography*. Oxford; Oxford UP; 1992. Pp xiii, 364.
68. Pines, Malcolm. 'The development of the psychodynamic movement', Bc145, 206–31.
69. Pronay, Nicholas. 'The film industry', Bc131, 212–36.
70. Rapp, Dean. 'The early discovery of Freud by the British general educated public, 1912–1919', *Social History of Medicine* 3 (1990), 217–43.
71. Reeves, Nicholas; Taylor, John H. *Howard Carter and the quest for Tutankhamun*. London; British Museum; 1992. Pp 208.
72. Richards, Jeffrey. 'Popular imperialism and the image of the army in juvenile literature', Bc10, 80–108.
73. Robins, Kevin. 'Tradition and translation: national culture in its global context', Bc149, 21–44; 236–41.
74. Rosenberg, Eugene; Cork, Richard. *Architect's choice: art in architecture in Great Britain since 1945*. London; Thames & Hudson; 1992. Pp 176.
75. Rosenberg, Nathan. 'Science and technology in the twentieth century', Bc134, 63–96.
76. Rosie, George. 'Museumry and the heritage industry', Bc6, 157–70.
77. Samuel, Raphael. 'Mrs Thatcher's return to Victorian values', Bc196, 9–30.
78. Sayce, Roger B. *History of the Royal Agricultural College, Cheltenham*. Stroud; Sutton; 1992. Pp xii, 370.
79. Seymour, Miranda. *Ottoline Morrell: life on a grand scale*. London; Hodder & Stoughton; 1992. Pp 432.
80. Shepherd, Michael. 'Psychiatric journals and the evolution of psychological medicine', Bc28, 188–206.
81. Sillars, Stuart. *British romantic art and the second world war*. London; Macmillan; 1992. Pp 240.
82. Smith, John Sharwood. 'The Hall school, Weybridge: a pioneering contribution to international understanding', *History of Education Soc. B*. 43 (1989), 48–54.
83. Smout, T.C. 'Patterns of culture' [Scotland], Bc138, 261–81.
84. Spalding, Frances. 'The modern face, 1918–1960', Bc93, 387–421.
85. Storey, G.O. *History of physical medicine: the story of the British Association of Rheumatology and Rehabilitation*. London; Royal Soc. of Medicine; 1992. Pp 108.
86. Szasz, Ferenc M. *British scientists and the Manhattan project: Los Alamos years*. London; Macmillan; 1992. Pp xx, 167.
87. Tantam, Digby. 'The anti-psychiatry movement', Bc145, 333–47.
88. Thomas, John Hugh. 'Music' [in Swansea], Bc212, 218–28.
89. Thorne, Robert. 'The setting of St Paul's cathedral in the twentieth century', *London J*. 16 (1991), 117–28.
90. Toplis, Ian. 'Sir Albert Richardson in Wendover: a microcosm of an architectural career', *Records of Buckinghamshire* 32 (1991 for 1990), 137–46.
91. Walmsley, Tom. 'Psychiatry in Scotland', Bc145, 294–305.
92. Wheeler, Richard. 'The park and garden survey at Stowe: the replanting and restoration of the historic landscape', *Huntington Library Q*. 55 (1992), 527–32.

93. Wollen, Tana. 'Over our shoulders; nostalgic screen fictions for the 1980s', Bc149, 178–93; 255.
94. Worpole, Ken. 'The age of leisure', Bc149, 137–50; 252–53.

J. MEDIEVAL WALES

See also Aa22,79; Bc96

(a) *General*

1. Carr, A.D. 'A debatable land: Arwystli in the middle ages', *The Montgomeryshire Collections* 80 (1992), 39–54.
2. Courtney, Paul 'A native-welsh mediaeval settlement: excavations at Beili Bedw, St Harmon, Powys', *B. of the Board of Celtic Studies* 38 (1991), 233–55.
3. Fasham, P.J. 'Investigations in 1985 by R.B. White at Castle Meadows, Beaumaris', *T. of the Anglesey Antiquarian Soc. & Field Club* (1992), 123–30.
4. Longley, David. 'Excavations at Plas Berw, Anglesey, 1983–1984', *Archaeologia Cambrensis* 140 (1991), 102–19.
5. McCann, W.J. 'The Welsh view of the Normans in the 11th and 12th centuries', *Honourable Soc. of Cymmrodorion T.* (1991), 39–67.
6. Rees, Sian. *A guide to ancient and historic Wales: Dyfed.* London; HMSO for Cadw; 1992. Pp 241.
7. Siddons, Michael Powell. *The development of Welsh heraldry*, vol. 1. Aberystwyth; National Library of Wales; 1991. Pp xxvi, 433.
8. Walker, David; Walker, Margaret. 'An Anglo-Welsh town' [Swansea], Bc212, 1–16.
9. Whittle, Elizabeth. *A guide to ancient and historic Wales: Glamorgan and Gwent.* London; HMSO for Cadw; 1992. Pp 217.

(b) *Politics*

1. Crouch, David. 'The last adventure of Richard Siward', *Morgannwg* 35 (1991), 7–30.
2. Higham, Nicholas J. 'Medieval "overkingship" in Wales: the earliest evidence', *Welsh History R.* 16 (1992–3), 145–59.
3. Thornton, David Ewan. 'A neglected genealogy of Llywelyn ap Gruffudd', *Cambridge Medieval Celtic Studies* 23 (1992), 9–23.

(c) *Constitution, Administration and Law*

1. Walters, D.B. 'Roman and Romano-canonical law and procedure in Wales', *Recueil de mémoires et travaux publié par la Société d'histoire de droit et des institutions des anciens pays de droit écrit* 15 (Montpellier, 1991), 67–102.

(d) *External Affairs*

None

(e) *Religion*

1. Brook, Diane. 'The early Christian church east and west of Offa's dyke', Bc96, 77–89.
2. Davies, Wendy. 'The myth of the Celtic church', Bc96, 12–21.
3. Edwards, Nancy; Lane, Alan. 'The archaeology of the early church in Wales: an introduction', Bc96, 1–11.
4. Evans, J. Wyn. 'Aspects of the early church in Carmarthenshire', Bc101, 239–53.
5. Evans, J. Wyn. 'The survival of the *clas* as an institution in medieval Wales: some observations on Llanbadarn Fawr', Bc96, 33–40.
6. Gwynne-Jones, D.M. 'Brecon cathedral *c*.1093–1537, the church of the Holy Rood', *Brycheiniog* 24 (1990–2), 23–37.
7. Hockey, S.F. 'Llangua, alien priory of Lyre', *J. of the Historical Soc. of the Church in Wales* 27 (1990), 8–13.
8. Howlett, David. '*Orationes Moucani*: early Cambro-Latin prayers', *Cambridge Medieval Celtic Studies* 24 (1992), 55–74.
9. Huws, Daniel. 'The earliest Bangor missal', *National Library of Wales J.* 27 (1991–2), 113–30.
10. James, Heather. 'Early medieval cemeteries in Wales', Bc96, 90–103.
11. James, Terrence A. 'Air photography of ecclesiastical sites in south Wales', Bc96, 62–76.
12. Jones, G.R. 'Excavations at Capel Teilo, near Kidwelly', Bc101, 255–59.
13. Parry-Jones, E.; Owen, Sybil. 'Masons' marks in St Asaph cathedral', *Archaeologia Cambrensis* 140 (1991), 155–60.
14. Pryce, Huw. 'Ecclesiastical wealth in early medieval Wales', Bc96, 22–32.
15. Pryce, Huw. 'Pastoral care in early medieval Wales', Bc100, 41–62.
16. Roberts, Tomos. 'Welsh ecclesiastical place-names and archaeology', Bc96, 41–44.
17. Thomas, Charles. 'The early church in Wales and the West: concluding remarks', Bc96, 145–49.
18. Turner, Rick. *Lamphey bishop's palace, Llawhaden castle, Carswell medieval house, Carew cross*. Cardiff; Cadw; 1991. Pp 52.
19. Williams, Glanmor. 'Kidwelly priory', Bc101, 189–204.
20. Williams, Glanmor. 'Poets and pilgrims in fifteenth- and sixteenth-century Wales', *Honourable Soc. of Cymmrodorion T.* (1991), 69–98.

(f) *Economic Affairs*

1. Booth, P.H.W.; Carr, A.D. (ed.). *Account of master John de Burnham the younger, chamberlain of Chester, of the revenues of the counties of Chester and Flint, 1361–1362*. Manchester; Record Soc. of Lancashire and Cheshire vol. 125; 1991. Pp lxxix, 247.

2. Kissock, Jonathan. 'Farms, fields and hedges: aspects of the rural economy of north-east Gower, c.1300 to c.1650', *Archaeologia Cambrensis* 140 (1991), 130–47.
3. Redknap, Mark. 'Sea and lake use in Wales, 400–1100', Bc207, 107–18.
4. Silvester, R.J.; Jones, N.W. 'Excavations at Heol-y-dŵr, Hay-on-Wye, Breconshire', *Brycheiniog* 24 (1990–2), 11–19.

(g) *Social Structure and Population*

1. Courtney, Paul. 'Feudal hierarchies and urban origins in south-east Wales', Bc214, 215–20.
2. Graham-Campbell, James A. 'Dinas Powys metalwork and the dating of enamelled zoomorphic penannular brooches', *B. of the Board of Celtic Studies* 38 (1991), 220–32.
3. Griffiths, Ralph A. 'A tale of two towns: Llandeilo Fawr and Dinefwr in the middle ages', Bc101, 205–26.
4. Kissock, Jonathan. 'The origins of medieval nucleated rural settlements in Glamorgan: a conjectural model', *Morgannwg* 35 (1991), 31–49.
5. Miles, Trevor J. 'Chepstow port wall excavations, 1971', *The Monmouthshire Antiquary* 7 (1991), 5–15.
6. Morgan, Prys. 'Locative surnames in Wales: a preliminary list', *Nomina* 14 (1992 for 1990–1), 7–23.
7. Rees, David. 'Neuadd Wen: changing patterns of tenure', Bc101, 43–51.
8. Smith, Llinos Beverley. 'Fosterage, adoption and God-parenthood: ritual and fictive kinship in medieval Wales', *Welsh History R.* 16 (1992–3), 1–35.
9. Wilkinson, P. 'Hen Gastell, Briton Ferry, west Glamorgan (SS 7315 9403)', *Morgannwg* 35 (1991), 87–88.

(h) *Naval and Military*

1. Arnold, C.J. 'Excavation of Offa's dyke, Chirk castle', *T. of the Denbighshire Historical Soc.* 40 (1991), 93–97.
2. Avent, R. 'The castles built by King Edward I in Wales between 1277 and 1300', *Internationales Burgen Institut B.* 47 (1990–1), 49–58.
3. Butler, Lawrence A.S. 'Castles of the Welsh princes in north Wales', *Internationales Burgen Institut B.* 47 (1990–1), 41–48.
4. Gerrard, S. 'Carew castle, Pembrokeshire coast national park', *Archaeology in national parks* ed. R.F. White and R. Iles (Leyburn; 1991), 47–54.
5. Grant, Allan; Grant, Chris. *The castle companion: a guide to historical castles of south Wales.* Pontypool; Village; 1991. Pp 104.
6. Knight, Jeremy K. *Chepstow castle and port wall, Runton church, Chepstow bulwarks camp.* Cardiff; Cadw; rev. edn. 1991. Pp 52.
7. Knight, Jeremy K. *The three castles: Grosmont castle; Skenfrith castle, White castle, Hen Gwrt moated site.* Cardiff; Cadw; 1991. Pp 48.
8. Knight, Jeremy K. 'Newport castle', *The Monmouthshire Antiquary* 7 (1991), 17–42.

9. Manley, John. 'The outer enclosure on Caergwrle hill, Clwyd', *J. of the Flintshire Historical Soc.* 33 (1992), 13–20.
10. Salter, Mike. *The castles of Gwent, Glamorgan and Gower.* Malvern; Folly; 1992. Pp 76.
11. Taylor, A.J. 'The Hope castle account of 1282', *J. of the Flintshire Historical Soc.* 33 (1992), 21–53.
12. Wrathmell, Stuart. 'Penhow castle, Gwent: survey and excavation, 1976–1979: part one', *The Monmouthshire Antiquary* 6 (1990), 17–45.

(i) *Intellectual and Cultural*

1. Bachellery, Édouard. 'La poèsie de demande dans la littérature galloise', *Etudes Celtiques* 27 (1990), 285–300.
2. Bromwich, Rachel; Evans, D. Simon (ed.). *Culhwch and Olwen: an edition and study of the oldest Arthurian tale.* Cardiff; Wales UP; 1992. Pp lxxxiii, 226.
3. Burdett-Jones, M.T. 'A fragment of text in Llyfr Gwyn Rhydderch', *Cambridge Medieval Celtic Studies* 23 (1992), 7–8.
4. Cherry, John; Redknap, Mark. 'Medieval and Tudor finger rings found in Wales', *Archaeologia Cambrensis* 140 (1991), 120–29.
5. Crawford, T.D. 'The *toddaid* and *gwawdodyn byr* in the poetry of Dafydd ap Gwilym, with an appendix concerning the *traethodlau* attributed to him', *Etudes Celtiques* 27 (1990), 301–36.
6. Dark, Ken R. 'Epigraphic, art-historical, and historical approaches to the chronology of class I inscribed stones', Bc96, 51–61.
7. Duggan, E.J.M. 'Notes concerning the "Lily crucifixion" in the Llanbeblig hours', *National Library of Wales J.* 27 (1991–2), 39–48.
8. Gruffydd, R. Geraint. '*Englynion y Cusan* by Dafydd ap Gwilym', *Cambridge Medieval Celtic Studies* 23 (1992), 1–6.
9. Huws, Daniel. 'The Tintern abbey bible', *The Monmouthshire Antiquary* 6 (1990), 47–54.
10. Jones, Thomas (ed.). *Ystoryaeu seint greal, rhan I: y keis* [Stories of the holy grail, part I: the search]. Cardiff; Wales UP; 1992. Pp xxvi, 311.
11. Knight, Jeremy K. 'The pottery from Montgomery castle', *Mediaeval and Later Pottery in Wales* 12 (1990–1991), 1–100.

K. SCOTLAND BEFORE THE UNION

See also Aa14; Bc76,168; Lh3,6,13

(a) *General*

1. Barrow, G.W.S. 'The charters of David I', Bc27, 25–37.
2. Cox, Richard A.V. '*Allt Loch Dhaile Beaga*: place-name study in the west of Scotland', *Nomina* 14 (1992 for 1990–1), 83–97.
3. Torrie, Elizabeth P.D. 'Medieval Dundee: a town and its people', *Abertay Historical Soc. Publications* 30 (1990), 120.

(b) *Politics*

1. Adams, Simon. 'Two "missing" Lauderdale letters: Queen Mary to Robert Dudley, earl of Leicester, 5 June 1567 and Thomas Randolph and Francis Russell, earl of Bedford to Leicester, 23 November 1564', *Scottish Historical R.* 70 (1991), 55–57.
2. Brown, Keith M. *Kingdom or province?: Scotland and the regal union, 1603–1715.* London; Macmillan; 1992. Pp xi, 226.
3. Fry, Michael. 'The Whig interpretation of Scottish history', Bc6, 72–89.
4. Lenman, Bruce P. 'The poverty of political theory in the Scottish revolution of 1688–1690', Bc48, 244–59.
5. Macdonald, Alasdair A. 'Mary Stewart's entry to Edinburgh: an ambiguous triumph', *Innes R.* 42 (1991), 101–10.
6. Scott, P.H. *Andrew Fletcher and the treaty of Union.* Edinburgh; Donald; 1992. Pp ix, 274.
7. Whatley, Christopher A. 'An uninflammable people?', Bc6, 51–71.

(c) *Constitution, Administration and Law*

1. Black, Robert. 'Exporting Scottish decrees', Bc128, 1–12.
2. Chorus, Jeroen M.J. 'The judge's role in the conduct of civil proceedings: some continental and Scottish ideas before 1880', Bc128, 32–46.
3. Clive, E.M. '*Jus quaesitum tertio* and carriage of goods by sea', Bc128, 47–56.
4. Hunter, D.M. (ed.). *The court book of the barony and regality of Falkirk and Callendar, I: 1638–1656.* Edinburgh; Stair Soc. vol. 38; 1991. Pp xv, 340.
5. Leslie, R.D. 'The applicability of domestic law in cases with a foreign element', Bc128, 57–65.
6. McBryde, William W. 'Bourhill *v*. Young: the case of the pregnant fishwife', Bc128, 66–77.
7. Meyers, David W. 'Letting doctor and patient decide: the wisdom of Scots law', Bc128, 91–103.
8. Miller, David Carey. 'Systems of property: Grotius and Stair', Bc128, 13–31.
9. Munro, Jean. *The inventory of Chisholm writs, 1456–1810.* Edinburgh; Scottish Record Soc.; 1992. Pp 205.
10. Nêve, Paul. 'Disputations of Scots students attending universities in the northern Netherlands', Bc195, 95–108.
11. Schanze, Erich. 'Interpretation of wills: an essay critical and comparative', Bc128, 104–16.
12. Sellar, W.D.H. 'Marriage by cohabitation with habit and repute: review and requiem?', Bc128, 117–36.
13. Smith, Alexander McCall. 'Criminal law and the moral tradition', Bc128, 78–90.
14. Summerson, Henry. 'The early development of the laws of the Anglo-Scottish marches, 1249–1448', Bc195, 29–42.

15. Thomson, J.M. 'An island legacy—the delict of conspiracy', Bc128, 137–51.
16. Tompson, Richard S. 'James Greenshields and the House of Lords: a reappraisal', Bc195, 109–24.
17. Walters, D.B. 'Legal aid, access to justice and the rule of law', Bc128, 152–61.
18. Wilson, John B. 'Royal burgh of Lochmaben court and council book, 1612–1721', *Dumfriesshire & Galloway Natural History & Antiquarian Soc. T.* 65 (1990), 84–92.
19. Wilson, W.A. 'The importance of analysis', Bc128, 162–71.
20. Young, Margaret D. (ed.). *The parliament of Scotland*, vol. 1. Edinburgh; Scottish Academic; 1992. Pp lx, 403.

(d) *External Affairs*

1. Griffiths, N.E.S.; Reid, John G. 'New evidence on New Scotland, 1629', *William & Mary Q.* 3rd ser. 49 (1992), 492–508.

(e) *Religion*

1. Adam, R.J. (ed.). *The calendar of Fearn: text and additions, 1471–1667.* Edinburgh; Scottish History Soc.; 1991. Pp 277.
2. Barrell, A.D.M. 'The papacy and the regular clergy in Scotland in the fourteenth century', *Records of the Scottish Church History Soc.* 24 (1991), 103–21.
3. Black, Jeremy. 'The archives of the Scots college, Paris, on the eve of their destruction', *Innes R.* 43 (1992), 53–59.
4. Broun, Dauvit. 'The literary record of St Nynia: fact and fiction?', *Innes R.* 42 (1991), 143–50.
5. Caillet, Maurice. 'Scotland in the antiquarian collection of the library of the Irish college in Paris', *Innes R.* 43 (1992), 18–52.
6. Dawson, Jane E.A. 'The two John Knoxes: England, Scotland and the 1558 tracts', *J. of Ecclesiastical History* 42 (1991), 555–76.
7. Donald, Peter H. 'Archibald Johnston of Wariston and the politics of religion', *Records of the Scottish Church History Soc.* 24 (1991), 123–40.
8. Goldstein, R.James. 'The Scottish mission to Boniface VIII in 1301: a reconsideration of the context of the *Instructiones* and *Processus*', *Scottish Historical R.* 70 (1991), 1–15.
9. Healey, Robert M. 'John Knox's "history": a "compleat" sermon on Christian duty', *Church History* 61 (1992), 319–33.
10. Hewitt, George. 'Reformation to revolution', Bc6, 16–30.
11. Hill, Peter H.; Pollock, David C. 'The Northumbrian church at Whithorn', Bc206, 189–94.
12. Kirk, James. 'The Scottish Reformation', *Encyclopedia of the Reformed Faith* ed. Donald K. McKim (Louisville (Ke); Knox/Westminster; 1992), 345–47.

13. Kyle, R.G. 'The Christian commonwealth: John Knox's vision for Scotland', *J. of Religious History* 16 (1991), 247–59.
14. Lamb, R.G. 'Church and kingship in Pictish Orkney: a mirror for Carolingian continental Europe', Bc206, 101–06.
15. Macfarlane, Leslie J. 'The elevation of the diocese of Glasgow into an archbishopric in 1492', *Innes R.* 43 (1992), 99–118.
16. Macquarrie, Alan. 'Early Christian religious houses in Scotland: foundation and function', Bc100, 110–33.
17. Macqueen, John. 'The dear green place: St Mungo and Glasgow, 600–1966', *Innes R.* 43 (1992), 87–98.
18. Mitchell, A.M. 'Writings relating to the ruins of the old church of St Andrew in Gullane', *T. of the East Lothian Antiquarian Soc.* 20 (1989 for 1988), 1–10.
19. Moran, Peter A. 'The library of the Scots college, Douai', *Innes R.* 43 (1992), 65–69.
20. Murray, Douglas M. 'Martyrs or madmen? The Covenanters, Sir Walter Scott and Dr Thomas McCrie', *Innes R.* 43 (1992), 166–75.
21. Oram, Richard D. 'In obedience and reverence: Whithorn and York, c.1128–1250', *Innes R.* 42 (1991), 83–100.
22. Roberts, Alasdair F.B. 'The role of women in Scottish Catholic survival', *Scottish Historical R.* 70 (1991), 129–50.
23. Roberts, Alasdair F.B. 'Thomas Abernethy, Jesuit and covenanter', *Records of the Scottish Church History Soc.* 24 (1991), 141–60.
24. Sefton, Henry R. *John Knox.* St Andrews; St Andrews UP; 1992. Pp 160.
25. Sharpe, Richard. 'Roderick MacLean's *Life* of St Columba in Latin verse (1549)', *Innes R.* 42 (1991), 111–32.
26. Symms, Peter. 'A disputed altar: parish pump politics in a sixteenth century burgh', *Innes R.* 42 (1991), 133–36.
27. Watt, D.E.R.; *et al.* (ed.). *Series episcoporum ecclesiae catholicae occidentalis ab initio usque ad annum 1198, series 6, tomus 1: ecclesia Scotiana.* Stuttgart; Hiersemann; 1991. Pp 91.

(f) *Economic Affairs*

1. Anderson, Per Sveaas. 'When was regular, annual taxation introduced in the Norse islands of Britain?: a comparative study of assessment systems in north-western Europe', *Scandinavian J. of History* 16 (1991), 73–83.
2. Bateson, J.D.; Stott, P. 'A late 14th-century coin hoard from Tranent, East Lothian', *Soc. of Antiquaries of Scotland P.* 120 (1990), 161–68.
3. Donnachie, Ian. '"The enterprising Scot"', Bc6, 90–105.
4. Hansen, Steffen Stummann. 'Cultural contacts in the Faroe islands in the viking age', Bc209, 13–18.
5. Hill, Peter H. 'A thousand years of contact: the economy of Whithorn from 450–1450', Bc209, 19–24.
6. Hill, Peter H.; Kucharski, Karina. 'Early medieval ploughing at Whithorn and chronology of plough pebbles', *Dumfriesshire & Galloway Natural History & Antiquarian Soc. T.* 65 (1990), 73–83.

7. Marshall, Gordon. *Presbyteries and profits: Calvinism and the development of capitalism in Scotland, 1560–1707.* Edinburgh; Edinburgh UP; new edn. 1992. Pp 416.

8. Weber, Birthe. 'Norwegian exports in Orkney and Shetland during the viking and middle ages', Bc209, 159–68.

(g) *Social Structure and Population*

1. Caldwell, David H. 'Tantallon castle, East Lothian: a catalogue of the finds', *Soc. of Antiquaries of Scotland P.* 121 (1991), 335–57.

2. Cardy, Amanda. 'In the shadow of St Ninian: life and death in medieval Whithorn', Bc185, 93–98.

3. Corser, Peter. 'Medieval settlement and land-use in the uplands of southern Scotland', Bc210, 101–06.

4. Ewan, Elizabeth. 'Town and hinterland in medieval Scotland', Bc214, 113–20.

5. McKean, Charles. 'The house of Pitsligo', *Soc. of Antiquaries of Scotland P.* 121 (1991), 369–90.

6. Quine, Gillian. 'Medieval shielings on the Isle of Man: fact or fiction?', Bc210, 107–12.

7. Smith, Ian M. 'The nature and extent of early medieval estate structure in the eastern Scottish borders', Bc210, 85–94.

(h) *Naval and Military*

1. Alcock, Leslie; Alcock, Elizabeth A. 'Reconnaissance excavations on early historic fortifications and other royal sites in Scotland, 1974–1984: 4, excavations at Alt Clut, Clyde Rock, Strathclyde, 1974–1975', *Soc. of Antiquaries of Scotland P.* 120 (1990), 95–149 + fiche.

2. Bonner, Elizabeth A. 'Continuing the "auld alliance" in the sixteenth century: Scots in France and French in Scotland', Bc21, 31–46.

3. Contamine, Philippe. 'Scottish soldiers in France in the second half of the fifteenth century: mercenaries, immigrants or Frenchmen in the making?', Bc21, 16–30.

4. Dukes, Paul. 'The first Scottish soldiers in Russia', Bc21, 47–54.

5. Fedosov, Dmitry G. 'The first Russian Bruces', Bc21, 55–66.

6. Harris, Stuart. 'The fortifications and siege of Leith: a further study of the map of the siege in 1560', *Soc. of Antiquaries of Scotland P.* 121 (1991), 359–68.

7. Morrison, Ian A. 'Survival skills: an enterprising Highlander in the Low Countries with Marlborough', Bc21, 81–96.

8. Ruckley, N. 'Geological and geomorphological factors influencing the form and development of Edinburgh castle', *Edinburgh Geologist* 26 (1991), 18–25.

9. Simpson, Grant G. 'Introduction' [the Scottish soldier abroad], Bc21, vii-xii.

(i) *Intellectual and Cultural*

1. Alcock, Elizabeth A. 'Burials and cemeteries in Scotland', Bc96, 125–29.
2. Bawcutt, Priscilla. 'The earliest texts of Dunbar', Bc58, 183–98.
3. Bergeron, David M. 'Charles I's Edinburgh pageant (1633)', *Renaissance Studies* 6 (1992), 173–84.
4. Everist, Mark. 'From Paris to St Andrews: the origins of W1', *J. of the American Musicological Soc.* 43 (1990), 1–42.
5. Fradenburg, Louise Olga. *City, marriage, tournament: arts of rule in late medieval Scotland.* Madison (Wi); Wisconsin UP; 1991. Pp xv, 390.
6. Higgins, Paula. 'Parisian nobles, a Scottish princess, and the woman's voice in late medieval song', *Early Music History* 10 (1991), 145–200.
7. Maxwell-Irving, Alastair M.T. 'Lochwood castle', *Dumfriesshire & Galloway Natural History & Antiquarian Soc. T.* 65 (1990), 93–99.
8. Purser, John. *Scotland's music: a history of the traditional and classical music of Scotland from early times to the present day.* Edinburgh; Mainstream; 1992. Pp 311.
9. Ryder, M.L.; Gabra-Sanders, T. 'Textiles from Fast castle, Berwickshire, Scotland', *Textile History* 23 (1992), 5–22.
10. Samson, Ross. 'The reinterpretation of the Pictish symbols', *J. of the British Arch. Association* 145 (1992), 29–65.
11. Samson, Ross. 'The rise and fall of tower houses in post-Reformation Scotland', Bc38, 197–243.
1. Davey, Peter. 'Dutch clay tobacco pipes from Scotland', Bc191, 279–89.
2. Haggarty, George; Murray, Charles. 'A highly decorated equestrian roof-finial from Edinburgh', Bc191, 194–97.
3. Hamp, Eric P. 'A few St Kilda toponyms and forms', *Nomina* 14 (1992 for 1990–1), 73–76.

L. IRELAND TO c.1640

See also Me6,h3,i12

(a) *General*

1. Bardon, Jonathan. *A history of Ulster.* Belfast; Blackstaff; 1992. Pp 934.
2. Buckley, V.M.; Sweetman, P.D. *Archaeological survey of county Louth.* Dublin; Stationery Office; 1991.
3. Day, Angelique. 'The computer as a source for Irish history', Bc224, 101–08.
4. Flanagan, Laurence. *A dictionary of Irish archaeology.* Dublin; Gill & Macmillan; 1992. Pp 222.
5. Foster, Roy. 'History, locality and identity: a lecture to mark the centenary of the society', *J. of the Cork Historical & Arch. Soc.* 97 (1992), 1–10.
6. Frame, Robin. 'King Henry III and Ireland: the shaping of a peripheral lordship', Bc112, 179–202.

7. Galloway, Peter; Simms, Cormac. *The cathedrals of Ireland.* Belfast; Inst. of Irish Studies; 1992. Pp 231.
8. Grose, D. *The antiquities of Ireland: a supplement to Francis Grose* ed. R. Stalley. Dublin; 1991.
9. Power, Jan. 'Archives report: Galway diocesan archives', *Archivium Hibernicum* 46 (1992 for 1991–2), 135–38.
10. Smith, Brendan. 'British history: an Irish perspective', *Studies in Medieval & Renaissance History* ns 13 (1992), 161–74.

(b) *Politics*

1. Palmer, William. 'Gender, violence and rebellion in Tudor and early Stuart Ireland', *Sixteenth Century J.* 23 (1992), 699–712.

(c) *Constitution, Administration and Law*

1. Ellis, Steven G. 'Ionadaiocht I bParlaimint na hEireann ac deireadh na mean-aoise' [Attendance at late medieval Irish parliaments], *P. of the Royal Irish Academy* 91C (1991), 297–302.
2. Gerriets, Marilyn. 'Theft, penitentials and the compilation of the early Irish laws', *Celtica* 22 (1991), 18–32.
3. Lyle, Emily. 'A line of queens as the pivot of a cosmology', Bc91, 276–89.
4. Walton, Julian. C. *The royal charters of Waterford.* Waterford; Waterford Corporation; 1992. Pp 56.

(d) *External Affairs*

1. Gillingham, John. 'The beginnings of English imperialism', *J. of Historical Sociology* 5 (1992), 392–409.
2. Henry, Gráinne. *Irish military community in Spanish Flanders, 1586–1621.* Dublin; Irish Academic; 1992. Pp 240.
3. O'Doherty, Br)íd. 'Alsace and Ireland: medieval links', *Seanchas Ard Mhaca* 14 (1991), 161–78.
4. Power, Rosemary. 'Irish travellers in the Norse world', Bc224, 127–34.

(e) *Religion*

1. Bradley, Thomas; Walsh, John R. *A history of the Irish Church, 400–700.* Dublin; Columba; 1991. Pp 160.
2. Candon, Anthony. 'Barefaced effrontery: secular and ecclesiastical politics in early twelfth-century Ireland', *Seanchas Ard Mhaca* 14 (1991), 1–25.
3. Candon, Anthony. 'Belach Conglais and the diocese of Cork, AD 1111', *Peritia* 5 (1989 for 1986), 416–18.
4. Charles-Edwards, Thomas. 'The pastoral role of the church in the early Irish laws', Bc100, 63–80.

5. Etchingham, Colman. 'The early Irish church: some observations on pastoral care and dues', *Eriú* 42 (1991), 99–118.
6. Gillespie, Raymond; Cunningham, Bernadette. 'Holy Cross abbey and the counter-Reformation in Tipperary', *Tipperary Historical J.* 1992, 170–80.
7. Hamlin, Ann. 'The early Irish church: problems of identification', Bc96, 138–44.
8. Harbison, Peter. 'Early Christian antiquities at Clonmore, county Carlow', *P. of the Royal Irish Academy* 91C (1991), 177–200.
9. Herity, Michael. 'The hermitage on Ardoilean, county Galway', *Royal Soc. of Antiquaries of Ireland J.* 120 (1990), 65–101.
10. Lynch, Anthony. 'The administration of John Bole, archbishop of Armagh, 1447–1471', *Seanchas Ard Mhaca* 14 (1991), 39–108.
11. Mac Aodha, Breandan S. 'The priest and the Mass in Irish place-names', *Nomina* 14 (1992 for 1990–1), 77–82.
12. Martin, F.X. 'Murder in a Dublin monastery, 1379', *Keimalia: studies in medieval archaeology and history*, ed. G. MacNiocaill and Patrick F. Wallace (Galway; Galway UP; 1989), 468–99.
13. McKee, J. 'Pierre Drelincourt: l'exil en Angleterre et en Irlande', *B. de la Société de l'Histoire du Protestantisme Français* 137 (1991), 177–95.
14. Murray, James. 'The sources of clerical income in the Tudor diocese of Dublin, *c.*1530–1600', *Archivium Hibernicum* 46 (1992 for 1991–2), 139–60.
15. O'Neill, Assumpta. 'Waterford diocese, 1096–1363: ii, episcopal succession, 1135–1222', *Decies* 44 (1990), 5–16.
16. O'Neill, Assumpta. 'Waterford diocese, 1096–1363: iii, episcopal succession, 1222–1363', *Decies* 45 (1991), 34–45.
17. O'Neill, Assumpta. 'Waterford diocese, 1096–1363: iv, the dioscesan chapter and extents of the diocese', *Decies* 46 (1992), 21–39.
18. Osborough, W.N. 'Ecclesiastical law and the Reformation in Ireland', Bc125, 223–52.
19. Powell, Timothy E. 'Christianity or solar monotheism: the early religious beliefs of St Patrick', *J. of Ecclesiastical History* 43 (1992), 531–40.
20. Purcell, Mary (ed.). 'Dublin diocesan archives: Hamilton papers (3)', *Archivium Hibernicum* 46 (1992 for 1991–2), 22–134.
21. Sharpe, Richard. 'Churches and communities in early medieval Ireland: towards a pastoral model', Bc100, 81–109.
22. Smith, Brendan. 'The Armagh-Clogher dispute and the "Mellifont conspiracy": diocesan politics and monastic reform in early thirteenth-century Ireland', *Seanchas Ard Mhaca* 14 (1991), 26–38.
23. Walsh, Katherine. 'Richard FitzRalph of Armagh (d. 1360): professor-prelate-saint', *County Louth Arch. & Historical J.* 22 (1992 for 1990), 111–24.
24. Watt, J.A. 'The medieval chapter of Armagh Cathedral', Bc70, 219–45.

(f) *Economic Affairs*

1. Colfer, Billy. 'Medieval Wexford', *J. of the Wexford Historical Soc.* 13 (1990–1), 5–29.

2. Empey, Adrian. 'Medieval Thurles: origins and development', *Thurles, the cathedral town*, ed. William Crobett William and William Nolan (Dublin; Geography Publications 1989), 31–40.
3. Henderson, Julian; Ivens, Richard. 'Dunmisk and glass-making in early Christian Ireland', *Antiquity* 66 (1992), 52–64.
4. Holm, P. 'The slave trade of Dublin, IXth to XIIIth century', *Peritia* 5 (1989 for 1986), 317–45.
5. McGrail, Sean. 'Boats and ships in Dublin, 900–1300', Bc207, 119–24.
6. Meenan, Rosanne. 'A survey of late medieval and early post-medieval Iberian pottery from Ireland', Bc191, 186–93.
7. O'Brien, A.F. 'Medieval Youghal: the development of an Irish seaport trading town, *c.*1200- *c.*1500', *Peritia* 5 (1989 for 1986), 346–78.
8. Rynne, Colin. 'The early Irish watermill and its continental affinities', Bc208, 21–26.
9. Sayers, William. 'Anglo-Norman verse on New Ross and its founders', *Irish Historical Studies* 28 (1992), 113–23.
10. Simms, Katharine. 'Nomadry in medieval Ireland: the origins of the creaght or *caoraigheacht*', *Peritia* 5 (1989 for 1986), 379–91.
11. Wallace, Patrick F. *The viking age buildings of Dublin: medieval Dublin excavations, 1962–1981*. Dublin; Royal Irish Academy; 1992. 2 vols. Pp 205–18.
12. Walsh, Paul. 'An account of the town of Galway', *J. of the Galway Arch. & Historical Soc.* 44 (1992), 47–118.

(g) *Social Structure and Population*

1. Barry, Terry. 'Dispersed rural settlement in Ireland in the later middle ages', Bc210, 175–80.
2. Bradley, John. 'The topographical development of Scandinavian Dublin', Bc118, 43–56.
3. Lapoint, Elwyn C. 'Irish immunity to witch-hunting, 1534–1711', *Eire-Ireland* 22 (1992), 76–93.
4. McMahon, Mary. 'Archaeological excavations at Bridge Streeet Lower, Dublin', *P. of the Royal Irish Academy* 91C (1991), 41–71.
5. Ó Concheanainn, Tomás. 'Aidèd Náth I and Ui Fhiachrach genealogies', *Eigse* 25 (1991), 1–27.
6. O'Brien, Elizabeth. 'Pagan and Christian burial in Ireland during the first millennium AD: continuity and change', Bc96, 130–37.
7. O'Dowd, Mary. *Power, politics and land: Sligo, 1562–1688*. Belfast; Inst. of Irish Studies; 1991. Pp 196.
8. Simms, Anngret. 'Dublin: vom wikingischen Seehandelsplatz zur anglo-normannischen Rechtsstadt' [Dublin: from Viking trading port to Anglo-Norman borough], Bc126, 246–57.
9. Simms, Anngret; Fagan, Patricia. 'Villages in county Dublin: their origins and inheritance', Bc118, 79–119.
10. Smith, William J. 'Exploring the social and cultural topographies of sixteenth- and seventeenth-century county Dublin', Bc118, 121–79.

(h) *Naval and Military*

1. Cairns, Conrad. 'The Irish tower house—a military view', *Fortress* 11 (1991), 3–13.
2. Cumming, W. 'Grange castle', *J. of the County Kildare Arch. Soc.* 17 (1991), 222–25.
3. Duffy, Seán. 'The "continuation" of Nicholas Trevet: a new source for the Bruce invasion', *P. of the Royal Irish Academy* 91C (1991), 303–15.
4. Jordan, A.J. 'The date, chronology and evolution of the county Wexford tower house', *J. of the Wexford Historical Soc.* 13 (1990–1), 30–82.
5. Klingelhöfer, E. 'The Renaissance fortifications at Dunboy castle, 1602: a report of the 1989 excavations', *J. of the Cork Historical & Arch. Soc.* 97 (1992), 85–96.
6. Lydon, James. 'The Scottish soldier in medieval Ireland: the Bruce invasion and the galloglass', Bc21, 1–15.
7. McNeill, Tom E. 'The origin of tower houses', *Arch. Ireland* 6/1 (1992), 13–14.
8. O'Conor, Kieran. 'Irish earthwork castles', *Fortress* 12 (1992), 3–12.
9. O'Conor, Kieran. 'The later construction and use of motte and bailey castles in Ireland: new evidence from Leinster', *J. of the County Kildare Arch. Soc.* 17 (1991), 13–29.
10. O'Keeffe, Tadhg. 'Medieval frontiers and fortifications: the Pale and its evolution', Bc118, 57–77.
11. O'Sullivan, Harold. 'Military operations in county Louth in the run-up to Cromwell's storming of Drogheda', *County Louth Arch. & Historical J.* 22 (1992 for 1990), 187–208.
12. Ryan, Michael. 'The Sutton Hoo ship burial and Ireland: some Celtic perspectives', Bc97, 83–116.
13. Smith, Brendan. 'The Bruce invasion and county Louth, 1315–1318', *County Louth Arch. & Historical J.* 22 (1991 for 1989), 7–15.
14. Stout, Matthew. 'Ringforts in the south-west midlands of Ireland', *P. of the Royal Irish Academy* 91C (1991), 201–43.
15. Thomas, Avril. *The walled towns of Ireland.* Dublin; Irish Academic; 1992. 2 vols. Pp 214; 257.

(i) *Intellectual and Cultural*

1. Breeze, Andrew. 'Two bardic themes: *The Trinity in the Blessed Virgin's Womb* and *The rain of folly*', *Celtica* 22 (1991), 1–15.
2. Caerwyn Williams, J.E.; Ford, Patrick K. *The Irish literary tradition.* Cardiff; Wales UP; new translation, 1992. Pp xii, 355.
3. Carey, John. 'The Irish "Otherworld": Hiberno-Latin perspectives', *Éigse* 25 (1991), 154–59.
4. Herbert, Máire. 'Goddess and king: the sacred marriage in early Ireland', Bc91, 264–75.
5. Hogan, Arlene. *Kilmallock Dominican priory: an architectural perspective, 1291–1991.* Kilmallock; Kilmallock Historical Soc.; 1991. Pp72.
6. King, Heather A. 'The medieval and seventeenth century carved stone

collection in Kildare', *J. of the County Kildare Arch. Soc.* 17 (1987–91), 59–95.

7. Kottje, Raymond. 'Beiträge der frühmittelalterlichen Iren zum gemeinsamen europäischen Haus' [The contribution of the early medieval Irish to the common European tradition], *Historisches Jahrbuch* 112 (1992), 3–22.

8. Ó Cuív, Brian. 'Gregory and St Dunstan in a middle-Irish poem on the origins of liturgical chant', Bc3, 273–97.

9. Ó Meadhra, Uaínínn. 'A medieval Dublin talismanic portrait? An incised profile cut-out head from Christ Church Place, Dublin', *Cambridge Medieval Celtic Studies* 22 (1991), 39–54.

10. O'Keeffe, Tadhg. 'La façade romane en Irlande', *Cahiers de Civilisation Médiévale* 34 (1991), 357–65.

11. Richardson, Hilary; Scarey, John. *An introduction to Irish high crosses.* Cork; Mercier; 1990. Pp 152.

12. Schneiders, M. 'The Irish calendar in the Karlsruhe Bede', *Archiv fÅr Liturgiewissenschaft* 31 (1989), 33–78.

13. Simms-Williams, Patrick. 'The additional letters of the Ogam alphabet', *Cambridge Medieval Celtic Studies* 23 (1992), 29–76.

M. IRELAND SINCE c.1640

See also Aa111,141,158; La1

(a) *General*

1. Boyce, D. George. *Ireland, 1828–1923: from ascendancy to democracy.* Oxford; Blackwell; 1992. Pp ix, 123.

2. Craig, M. *Dublin, 1660–1860.* London; Penguin; new edn. 1992. Pp 464.

3. Larkin, Emmet (ed.). *Alexis de Tocqueville's journey in Ireland: July-August 1835.* Washington (DC); Catholic University of America Press; 1990. Pp xiv, 157.

4. Whelan, Kevin. 'Beyond a paper landscape: J.H. Andrews and Irish historical geography', Bc118, 379–424.

(b) *Politics*

1. Aughey, Arthur. 'Northern Ireland: a putting together of parts', Bc200, 256–69.

2. Bartlett, Thomas. *The fall and rise of the Irish nation: the Catholic question in Ireland, 1690–1830.* Dublin; Gill & Macmillan; 1992. Pp xi, 430.

3. Bottigheimer, Karl S. 'The Glorious revolution and Ireland', Bc48, 234–43.

4. Brennan, Niamh. 'The Ballagh barracks rioters', Bc187, 57–84.

5. Bruce, Steve. *The Red Hand: Protestant paramilitaries in Northern Ireland*. Oxford; Oxford UP; 1992. Pp xiv, 311.

6. Callanan, Frank. *The Parnell split, 1890–1891*. Cork; Cork UP; 1992. Pp xxiv, 327.

7. Callanan, Frank. 'After Parnell: the political consequences of Timothy Michael Healy', *Studies* 80 (1991), 371–76.

8. Callanan, Frank. '"Clerical dictation": reflections on the Catholic Church and the Parnell split', *Archivium Hibernicum* 45 (1990), 64–75.

9. Connolly, Sean J. *Religion, law and power: the making of Protestant Ireland, 1660–1760*. Oxford; Oxford UP; 1992. Pp xi, 346.

10. d'Alton, Ian. 'A perspective upon historical process: the case of southern Irish Unionism', Bc143, 70–91.

11. Davis, Richard. 'Patrick O'Donohoe: outcast of the exiles', Bc187, 246–83.

12. Dunleavy, John. 'Parnell: a most difficult pupil', *Studies* 80 (1991), 366–70.

13. Foster, Roy. 'Interpretations of Parnell', *Studies* 80 (1991), 349–57.

14. Gillespie, Raymond. 'The Irish Protestants and James II, 1688–1690', *Irish Historical Studies* 28 (1992), 124–33.

15. Hart, Peter. 'Michael Collins and the assassination of Sir Henry Wilson', *Irish Historical Studies* 28 (1992), 150–70.

16. Jackson, Alvin. 'Unionist myths, 1912–1985', *Past & Present* 136 (1992), 164–85.

17. Kelly, James. *Prelude to Union: Anglo-Irish politics in the 1780s*. Cork; Cork UP; 1992. Pp xi, 276.

18. Keogh, Dermot. 'Mannix, de Valera and Irish nationalism', Bc182, 196–225.

19. Larkin, Emmet. 'The fall of Parnell: personal tragedy, national triumph', *Studies* 80 (1991), 358–65.

20. MacDonagh, Oliver. *O'Connell: the life of Daniel O'Connell, 1775–1847*. London; Weidenfeld & Nicolson; 1991. Pp 712.

21. Munck, Ronnie. 'The making of the troubles in Northern Ireland', *J. of Contemporary History* 27 (1992), 211–29.

22. O'Connor, Emmet. *A Labour history of Ireland, 1824–1960*. Dublin; Gill & Macmillan; 1992. Pp xiii, 270.

23. O'Donnell, Ruan. 'General Joseph Holt', Bc187, 27–56.

24. Ohlmeyer, Jane H. 'The "Antrim plot" of 1641: a myth?', *Historical J.* 35 (1992), 905–19.

25. Powell, Frederick W. *The politics of Irish social policy, 1600–1990*. Lampeter; Mellen; 1992. Pp xiv, 374.

26. Smyth, James J. *The men of no property: Irish radicals and popular politics in the late eighteenth century*. London; Macmillan; 1992. Pp xi, 251.

27. TeBrake, Janet K. 'Irish peasant women in revolt: the Land League years', *Irish Historical Studies* 28 (1992), 63–80.

28. Travers, Pauric. 'The financial relations question, 1800–1914', Bc143, 41–69.

29. Tweedy, Hilda. *Link in the chain: the story of the Irish Housewives Association, 1942–1992.* Dublin; Attic; 1992. Pp 144.
30. Valiulis, Maryann Gialanella. *Portrait of a revolutionary General: Richard Mulcahy and the founding of the Irish Free State.* Dublin; Irish Academic; 1992. Pp xii, 289.
31. Vincent, Joan. 'A political orchestration of the Irish famine: county Fermanagh, May 1847', Bc124, 75–98.
32. Walker, Brian M. (ed.). *Parliamentary election results in Ireland, 1918–1992.* Dublin; Royal Irish Academy; 1992. Pp viii, 358.
33. Walker, Graham. '"The Irish Dr Goebbels": Frank Gallagher and Irish republican propaganda', *J. of Contemporary History* 27 (1992), 149–65.

(c) *Constitution, Administration and Law*

1. Kendle, John. *Ireland and the federal solution: the debate over the United Kingdom constitution, 1870–1920.* Kingston (Ont); McGill-Queen's UP; 1989. Pp viii, 295.
2. Keogh, Dermot. 'The Jesuits and the 1937 constitution', *Studies* 78 (1989), 82–95.
3. Townshend, Charles. 'Policing insurgency in Ireland, 1914–1923', Bc11, 22–41.

(d) *External Affairs*

1. Martin, Allan. 'An Australian Prime Minister in Ireland: R.G. Menzies, 1941', Bc143, 180–200.
2. Pilhorget, Rene. 'Louis XIV et l'Irlande', *Revue d'Histoire Diplomatique* 106 (1992), 7–26.

(e) *Religion*

1. Fenning, H. 'Dominicans in Ireland, 1841–1845', *Collectanea Hibernica* 33 (1991), 178–208.
2. Ford, Alan (ed.). 'Correspondence between archbishops Ussher and Laud', *Archivium Hibernicum* 46 (1992 for 1991–2), 5–21.
3. Giblin, C. 'Papers of Richard Joachim Hayes, O.F.M., in the Franciscan Library, Killiney: part 9, 1817–1820', *Collectanea Hibernica* 33 (1991), 97–177.
4. Harvey, A.D. 'Who were the Auxiliaries?', *Historical J.* 35 (1992), 665–69.
5. Hempton, David; Hill, Myrtle. *Evangelical Protestantism in Ulster society, 1740–1890.* London; Routledge; 1992. Pp xiv, 272.
6. Hösler, M. 'Irishmen ordained at Prague, 1629–1786', *Collectanea Hibernica* 33 (1991), 7–53.
7. Kerrigan, Colm. *Father Mathew and the Irish temperance movement, 1838–1849.* Cork; Cork UP; 1992. Pp vii, 249.

8. Lynch, Anthony. 'A meeting of priests in Co. Derry, 1669', *Collectanea Hibernica* 33 (1991), 93–96.
9. MacDermot, Brian (ed.). *Irish catholic petition of 1805: the diary of Denys Scully*. Dublin; Irish Academic; 1992. Pp xxii, 208.
10. Millet, B. 'Calendar of volume 15 of the *Fondo di Vienna* in Propaganda archives', *Collectanea Hibernica* 33 (1991), 54–92.
11. Quinn, John F. 'The "vagabond friar": Father Mathew's difficulties with the Irish bishops, 1840–1856', *Catholic Historical R.* 78 (1992), 542–56.
12. Taylor, Lawrence J. 'The languages of belief: nineteenth-century religious discourse in southwest Donegal', Bc124, 142–75.
13. Tierney, M. 'A short-title calendar of the papers of Archbishop Michael Slattery in Archbishop's House, Thurles: part 3, 1846', *Collectanea Hibernica* 33 (1991), 209–19.
14. Wigham, Maurice J. *Irish Quakers: a short history of the Religious Society of Friends in Ireland*. Dublin; Religious Soc. of Friends in Ireland, Historical Committee; 1992. Pp 176.

(f) *Economic Affairs*

1. Barnard, T.C. 'Reforming Irish manners: the religious societies in Dublin during the 1690s', *Historical J.* 35 (1992), 805–38.
2. Bell, Jonathan. *People and the land: farming in nineteenth century Ireland*. Belfast; Friar's Bush; 1992. Pp vi, 102.
3. Brown, Kenneth. 'Firemen's trade unionism in Northern Ireland', Bc7, 176–95.
4. Daly, Mary E. *Industrial development and Irish national identity, 1922–1939*. Dublin; Gill & Macmillan; 1992. Pp xv, 201.
5. Graham, Brian J.; Proudfoot, Lindsay J. 'Landlords, planning and urban growth in eighteenth and early nineteenth-century Ireland', *J. of Urban History*. 18 (1992), 308–29.
6. Guinnane, Timothy W. 'Intergenerational transfers, emigration, and the rural Irish household system', *Explorations in Economic History* 29 (1992), 456–76.
7. Harris, Richard I.D. *Regional economic policy in Northern Ireland, 1945–1988*. Aldershot; Avebury; 1991. Pp xiii, 207.
8. Kelly, James. 'Scarcity and poor relief in eighteenth-century Ireland: the subsistence crisis of 1782–1784', *Irish Historical Studies* 28 (1992), 38–62.
9. Killen, James. 'Transport in Dublin: past, present and future', Bc118, 305–25.
10. Lyne, Gerard J. 'Gross survey transcripts for county Galway', *Analecta Hibernica* 35 (1992), 159–86.
11. MacCarthy, Robert. *The Trinity college estates, 1800–1923*. Dundalk; Dundalgen; 1992. Pp ix, 273.
12. McGregor, Pat. 'The labor market and the distribution of landholdings in pre-famine Ireland', *Explorations in Economic History* 29 (1992), 477–93.

13. Ó Gráda, Cormac. 'Dublin's demography in the early nineteenth century: evidence from the Rotunda', *Population Studies* 45 (1991), 43–54.
14. Ó Gráda, Cormac. 'Literary sources and Irish economic history', *Studies* 80 (1991), 290–99.
15. Silverman, Marilyn. 'From fisher to poacher: public right and private property in the salmon fisheries of the river Nore in the nineteenth century', Bc124, 99–141.

(g) Social Structure and Population

1. Aalen, F.H.A. 'Health and housing in Dublin, c.1850–1921', Bc118, 279–304.
2. Birdwell-Pheasant, Donna. 'The early twentieth-century Irish stem family: a case study from county Kerry', Bc124, 205–35.
3. Budd, J.W.; Guinnane, Timothy W. 'Intentional age-misreporting, age-heaping, and the 1908 Old Age Pensions Act in Ireland', *Population Studies* 45 (1991), 497–518.
4. Canny, Nicholas. 'The marginal kingdom: Ireland as a problem in the first British empire', Bc2, 35–66.
5. Cohen, Marilyn. 'Peasant differentiation and proto-industrialization in the Ulster countryside: Tullylish, 1690–1825', *J. of Peasant Studies* 17 (1990), 413–32.
6. Cohen, Marilyn. 'Survival strategies in female-headed households: linen workers in Tullylish, County Down, 1901', *J. of Family History* 17 (1992), 303–18.
7. Collingwood, Judy. 'Irish work-house children in Australia', Bc182, 46–61.
8. Geoghegan, Vincent. 'Ralahine: an Irish Owenite community (1831–1833)', *International R. of Social History* 36 (1991), 377–411.
9. Grose, Kelvin. '"A strange compound of good and ill"', Bc187, 85–111.
10. Guinnane, Timothy W. 'Age at leaving home in rural Ireland, 1901–1911', *J. of Economic History* 52 (1992), 651–74.
11. Johnson, Keith; Flynn, Michael. 'The convicts of the *Queen*', Bc187, 10–26.
12. Kelly, Patrick. 'The improvement of Ireland', *Analecta Hibernica* 35 (1992), 45–84.
13. Keogh, Dermot. 'Argentina and the Falklands (Malvinas): the Irish connection', Bc136, 123–41.
14. MacGinley, M.E.R. 'The Irish in Queensland: an overview', Bc182, 103–19.
15. Nolan, William. 'Society and settlement in the valley of Glenasmole, c.1750- c.1900', Bc118, 181–228.
16. Parkhill, Trevor. 'Convicts, orphans, settlers: patterns of emigration from Ulster, 1790–1860', Bc182, 6–28.
17. Reece, Bob. 'Ned Kelly's father', Bc187, 218–46.
18. Reece, Bob. 'The ballad of Maitland jail', Bc187, 112–34.
19. Reece, Bob. 'The Connerys', Bc187, 184–17.

20. Reece, Bob. 'The *True history* of Bernard Reilly', Bc187, 135–50.
21. Richards, Eric. 'The importance of being Irish in colonial South Australia', Bc182, 62–102.
22. Rickard, John. 'H.B. Higgins: "one of those wild and irreconcilable Irishmen"', Bc182, 181–95.
23. Robins, Joseph. 'The emigration of Irish work-house children to Australia in the nineteenth century', Bc182, 29–45.
24. Smyth, William J. 'Making the documents of conquest speak: the transformation of property, society, and settlement in seventeenth-century counties Tipperary and Kilkenny', Bc124, 236–92.

(h) *Naval and Military*

1. Denman, Terence. *Ireland's unknown soldiers: 16th (Irish) Division in the Great War.* Dublin; Irish Academic; 1992. Pp 209.
2. Johnstone, Tom. *Orange, green and khaki: the story of the Irish regiments in the great war, 1914–1918.* Dublin; Gill & Macmillan; 1992. Pp 498.
3. Stewart, Richard W. 'The "Irish road": military supply and arms for Elizabeth's army during the O'Neill rebellion in Ireland, 1598–1601', Bc69, 16–37.
4. Wheeler, James Scott. 'Logistics and supply in Cromwell's conquest of Ireland', Bc69, 38–56.

(i) *Intellectual and Cultural*

1. Bolton, Geoffrey. 'The strange career of William de la Poer Beresford', Bc187, 284–303.
2. Boylan, Thomas A.; Foley, Timothy P. *Political economy and colonial Ireland: the propagation and ideological functions of economic discourse in the nineteenth century.* London; Routledge; 1992. Pp xiv, 208.
3. Caldicott, C.E.J. 'Patrick Darcy, an argument', Bc178, 191–320.
4. Cullen, Mary. 'Women's history in Ireland', Bc8, 429–41.
5. Dunne, Tom. 'Representations of rebellion: 1798 in literature', Bc143, 14–40.
6. Fitzpatrick, David. 'Review article: women, gender and the writing of Irish history', *Irish Historical Studies* 27 (1991), 267–73.
7. Grogan, Geraldine. 'Daniel O'Connell and European Catholic thought', *Studies* 80 (1991), 56–64.
8. Hand, Geoffrey. 'Aubrey Gwynn: the person', *Studies* 81 (1992), 375–84.
9. Jones, Greta. 'Eugenics in Ireland: the Belfast Eugenics Society, 1911–1915', *Irish Historical Studies* 28 (1992), 81–95.
10. Keating, Paul. 'Cultural values and entrepreneurial action: the case of the Irish republic', Bc217, 92–108.
11. Kingston, F.T. *Metaphysics of George Berkeley (1685–1753), Irish philosopher.* Lampeter; Mellen; 1992. Pp 220.

12. McGurk, John. 'Trinity college, Dublin, 1592–1992', *History Today* 42/3 (1992), 41–47.
13. O'Brien, Conor Cruise. *The great melody: a thematic biography of Edmund Burke.* London; Sinclair-Stevenson; 1992. Pp lxxv, 692.
14. Reece, Bob. 'Frank the poet', Bc187, 151–83.
15. Reece, Bob. 'Writing about the Irish in Australia', Bc182, 226–42.
16. Smith, F.B. 'Stalwarts of the garrison: some Irish academics in Australia', Bc182, 120–47.
17. Smyth, Alfred P. *Faith, famine and fatherland in the nineteenth-century Irish midlands: perceptions of a priest and historian, Anthony Cogan, 1826–1872.* Dublin; Four Courts; 1992. Pp 222.
18. Stalley, Roger (ed.). *Daniel Grose (c.1766–1838): the antiquities of Ireland, a supplement to Francis Grose.* Dublin; Irish Architectural Archive; 1991. Pp xxiv, 214.
19. Troy, Jakelin. '"Der mary this is fine cuntry is there is in the wourld": Irish-English and Irish in late eighteenth- and nineteenth-century Australia', Bc182, 148–80.
20. Walsh, Katherine. 'Aubrey Gwynn: the scholar', *Studies* 81 (1992), 385–92.
21. Warren, John. '*The Dublin Review*, its reviewers and a "philosophy of knowledge"', *Recusant History* 21 (1992), 86–98.
22. Woods, C.J. 'Review article: Irish travel writings as source material', *Irish Historical Studies* 28 (1992), 171–83.

(j) *Local History*

1. Cullen, Louis M. 'The growth of Dublin, 1600–1900: character and heritage', Bc118, 252–77.
2. Gillmor, Desmond. 'Dublin city's countryside', Bc118, 359–78.
3. Gulliver, P.H. 'Shopkeepers and farmers in south Kilkenny, 1840–1981', Bc124, 176–204.
4. Horner, Arnold. 'From city to city-region: Dublin from the 1930s to the 1990s', Bc118, 327–58.
5. Jordan, Alison. *Who cared? Charity in Victorian and Edwardian Belfast.* Belfast; Institute of Irish Studies; 1992. Pp xii, 262.
6. Ó Conchubhair, Fearghail. 'The evolution of Catholic parishes in Dublin city from the sixteenth to the nineteenth centuries', Bc118, 229–50.
7. Ó Dálaigh, B. (ed.). *Corporation book of Ennis.* Dublin; Irish Academic; 1990. Pp 455.

(k) *Science and Medicine*

1. Finnane, Mark. 'Irish psychiatry, part 1: the formation of a profession', Bc145, 306–13.
2. Healy, David. 'Irish psychiatry, part 2: use of the Medico-Psychological Association by its Irish members— *plus ça change!*', Bc145, 314–20.

3. Jones, Greta. 'Marie Stopes in Ireland: the mothers' clinic in Belfast, 1936–1947', *Social History of Medicine* 5 (1992), 255–77.

N. EMPIRE AND COMMONWEALTH POST 1783

See also Aa110; Bc11,130,182,187; Mg11,14,16–23,i15–16,19

(a) *General*

1. Boucher, Maurice; Penn, Nigel (ed.). *Britain at the Cape, 1795 to 1803.* Houghton (S. Africa); Brenthurst; 1992. Pp 264.
2. Brown, Judith M. *Winds of change.* Oxford; Oxford UP; 1991. Pp 22.
3. Cell, John W. *Hailey: a study in British imperialism, 1872–1969.* Cambridge; Cambridge UP; 1992. Pp 320.
4. Craton, Michael; Saunders, Gail. *Islanders in the stream. A history of the Bahamian people, vol. 1: from aboriginal times to the end of slavery.* Athens (Ga); Georgia UP; 1992. Pp 640.
5. Dean, D.W. 'Final exit? Britain, the Commonwealth and the repeal of the External Relations Act, 1945–1949', *J. of Imperial & Commonwealth History* 20 (1992), 391–418.
6. Dickason, Olive Patricia. *Canada's first nations: a history of the founding peoples.* Norman (Ok); Oklahoma UP; 1992. Pp 624.
7. Gough, Barry M. *The north-west coast: British navigation, trade, and discoveries to 1812.* Vancouver; UBC; 1992. Pp 288.
8. Hargreaves, John D. 'Habits of mind and forces of history: France, Britain and the de-colonization of Africa', Bc130, 207–19.
9. Johnson, Howard. *The Bahamas in slavery and freedom.* London; Currey; 1991. Pp viii, 184.
10. Kent, John. *Internationalization of colonialism: Britain, France and black Africa, 1939–1956.* Oxford; Oxford UP; 1992. Pp xi, 365.
11. Kirk-Greene, Anthony Hamilton Millard. 'De-colonization in British Africa', *History Today* 42/1 (1992), 44–50.
12. Lee, David. 'Australia, the British Commonwealth and the United States, 1950–1953', *J. of Imperial & Commonwealth History* 20 (1992), 445–69.
13. MacKenzie, John M. 'Postscript' [popular imperialism], Bc10, 221–22.
14. Martin, Ged W. 'The Canadian question and the late modern century', *British J. of Canadian Studies* 7 (1992), 215–47.
15. McAteer, William. *Rivals in Eden: a history of the French settlement and British conquest of the Seychelles Islands, 1742–1818.* Lewes; Book Guild; 1991. Pp 355.
16. Molyneux, Geoffrey. *British Columbia: an illustrated history.* Vancouver; Polestar; 1992. Pp 135.
17. Mommsen, W.J. 'Das britische Empire: zum Wandel der Ausübung imperialistischer Herrschaft im 19. und 20. Jahrhundert' [The British empire: changes in the exercise of imperial power in the 19th and 20th centuries], *Historische Mitteilungen* 3 (1990), 45–56.

18. Mullins, Patrick. *Retreat from Africa*. Edinburgh; Pentland; 1992. Pp vii, 108.
19. Newbury, Colin. 'The semantics of international influence: informal empires reconsidered', Bc130, 23–66.
20. Panteli, Stavros. *The making of modern Cyprus: from obscurity to statehood*. New Barnet; Interworld; 1990. Pp 287.
21. Robson, Lloyd. *A history of Tasmania, vol. 2: colony and state from 1856 to the 1980s*. Melbourne/Oxford; Oxford UP; 1992. Pp x, 663.
22. Sutherland, William. *Beyond the politics of race: an alternative history of Fiji to 1992*. Canberra; Department of Political and Social Change, Research School of Pacific Studies, Australian National University; 1992. Pp 251.
23. Twaddle, Michael. 'Imperialism and the state in the third world', Bc130, 1–22.
24. Webby, Elizabeth (ed.). *Colonial voices, letters, diaries, journalism and other accounts of nineteenth-century Australia*. Brisbane; Queensland UP; 1989. Pp xxiii, 484.
25. Whyte, Beverly. *Yesterday, today and tomorrow: a hundred year history of Zimbabwe*. Harare (Zimbabwe); Burke; 1990. Pp 245.

(b) *Politics*

1. Chakrabarty, Dipesh. 'Marxism and modern India', *History Today* 42/3 (1992), 48–51.
2. Chan, Stephen. *Twelve years of Commonwealth diplomatic history: Commonwealth summit meetings, 1979–1991*. Lampeter/Lewiston/ Queenston; Mellen; 1992. Pp 146.
3. Chippington, George. *Singapore: the inexcusable betrayal*. London; Self Publishing Association; 1992. Pp 256.
4. Cumming, Cliff. 'Scots radicals in Port Phillip, 1838–1851', *Australian J. of Politics & History* 37 (1991), 434–45.
5. de Brou, David. 'The rose, the shamrock and the cabbage: the battle for Irish voters in Upper-Town Quebec, 1827–1836', *Histoire Sociale/Social History* 24 (1991), 305–34.
6. Edwards, Peter; Pemberton, Gregory. *Crises and commitments: the politics and diplomacy of Australia's involvement in southeast Asian conflicts, 1948–1965*. London; Allen & Unwin; 1992. Pp 528.
7. Eldredge, Elizabeth A. 'Sources of conflict in South Africa, *c.*1800–1830: the *mfecane* reconsidered', *J. of African History* 33 (1992), 1–35.
8. Emudong, C.P. 'The political economy of the evolution of a new British colonial policy in Africa, 1938–1947', *Scandinavian J. of Development Alternatives* 9 (1990), 131–50.
9. Gangal, S.C. 'Gandhi and South Africa', *International Studies* 29 (1992), 187–97.
10. Groom, A. John R. 'Gibraltar: no end to "empire"', *Round Table* 323 (1992), 273–91.
11. Hamilton, Carolyn Anne. '"The character and objects of Chaka": a reconsideration of the making of Shaka as *mfecane* motor', *J. of African History* 33 (1992), 37–63.

12. Hudson, W.J. *Blind loyalty, Australia and the Suez crisis, 1956.* Melbourne; Melbourne UP; 1989. Pp 172.
13. Joshi, Shashi. *Struggle for hegemony in India, 1920–1947.* London; Sage; 1992. Pp 400.
14. Kennedy, Dane. 'Constructing the colonial myth of Mau Mau', *International J. of African Historical Studies* 25 (1992), 241–60.
15. Knox, Bruce. 'Democracy, aristocracy and empire: the provision of colonial honours, 1818–1870', *Australian Historical Studies* 25 (1992), 244–64.
16. Lamb, Alastair. *Kashmir: a disputed legacy, 1846–1990.* Hertingfordbury; Roxford; 1991. Pp xvi, 368.
17. Legault, Albert; Fortmann, Michel. *A diplomacy of hope: Canada and disarmament, 1945–1988.* Montreal; McGill-Queen's UP; 1992. Pp 632.
18. McLean, David. 'ANZUS origins: a reassessment', *Australian Historical Studies* 24 (1990), 64–82.
19. Meaney, N. 'Australia, the great powers and the coming of the cold war', *Australian J. of Politics & History* 38 (1992), 316–33.
20. Miners, J. 'The ending of British rule in Hong Kong', Bc130, 274–86.
21. Musambachime, M.C. 'Dauti Yamba's contribution to the rise and growth of nationalism in Zambia, 1941–1964', *African Affairs* 90 (1991), 259–81.
22. Ohadike, Don C. *The Ekumeku movement: western Igbo resistance to the British conquest of Nigeria, 1883–1914.* Athens (Oh); Ohio UP; 1991. Pp xi, 203.
23. Pandey, Gyanendra. *The construction of communalism in colonial north India.* Delhi; Oxford UP; 1991. Pp xvi, 297.
24. Richelson, Jeffery T.; Ball, Desmond. *The ties that bind: intelligence co-operation between UKUSA countries—the United Kingdom, the United States of America, Canada, Australia and New Zealand.* Sydney; Allen & Unwin; 1990. Pp xviii, 415.
25. Rizvi, Gowher. 'Gandhi and Nehru: an enduring legacy', *Round Table* 323 (1992), 363–68.
26. Sen, Sunanda. *Colonies and the empire: India, 1890–1914.* London; Sangam; 1992. Pp x, 227.
27. Singh, Iqbal. *Between two fires: towards an understanding of Jawaharlal Nehru's foreign policy,* vol. 1. New Delhi; Orient Longman; 1992. Pp x, 376.
28. Ticktin, Hillel. *The politics of race: discrimination in South Africa.* London; Pluto; 1991. Pp vii, 115.
29. Tremewan, Peter. 'The French alternative to the treaty of Waitangi', *New Zealand J. of History* 26 (1992), 99–104.
30. Trotter, Ann. 'Coming to terms with Japan: New Zealand's experience, 1943–1963', Bc19, 125–42.
31. Wood, Donald. 'Berbice and the unification of British Guiana', Bc130, 67–79.
32. Wright, John. 'Political mythology and the making of Natal's *mfecane*', *Canadian J. of African Studies* 23 (1989), 272–91.

33. Wright, Ray. *A people's counsel: a history of the parliament of Victoria, 1856–1990*. Melbourne; Oxford UP; 1992. Pp xv, 304.
34. Zweig, Ronald W. 'The Palestine problem in the context of colonial policy on the eve of the second world war', Bc179, 206–16.

(c) *Constitution, Administration and Law*

1. Anderson, David M. 'Policing and communal conflict: the Cyprus emergency, 1954–1960', Bc11, 187–217.
2. Arnold, David. 'Police power and the demise of British rule in India, 1930–1947', Bc11, 42–61.
3. Atkinson, Alan. 'Beating the bounds with lord Sydney, Evan Nepean and others', *Australian Historical Studies* 25 (1992), 217–19.
4. Bernier, Gérald; Salée, Daniel. *The shaping of Quebec politics and society: colonialism, power and the transition to capitalism in the 19th century*. Basingstoke; Taylor & Francis; 1992. Pp 250.
5. Byrne, Paula. *Criminal law and colonial subject: New South Wales, 1810–1830*. Cambridge; Cambridge UP; 1992. Pp 256.
6. Cadigan, Sean T. 'Paternalism and politics: Sir Francis Bond Head, the Orange order and the election of 1836', *Canadian Historical R.* 72 (1991), 319–47.
7. Cashmore, T.H.R. 'A random factor in British imperialism: district administration in colonial Kenya', Bc130, 124–35.
8. Cassidy, Cheryl M. 'Pro rege, lege and grege: newspaper accounts of the Jamaica insurrection and its aftermath', *Victorian Periodicals R.* 25 (1992), 3–6.
9. Chakrabarty, Rishi Ranjan. *Duleep Singh: the maharaja of Punjab and the raj*. Oldbury; Samara; 1988. Pp xiii, 201.
10. Chandavarkar, Rajnarayan. 'Plague panic and epidemic politics in India, 1896–1914', Bc25, 203–40.
11. Constantine, Stephen (ed.). *Dominions diary: the letters of E.J. Harding, 1913–1916*. Halifax; Ryburn; 1992. Pp 336.
12. Crais, Clifton C. *The making of the colonial order: white supremacy and black resistance in the Eastern Cape, 1770–1865*. Johannesburg; Witwatersrand; 1992. Pp 304.
13. Dare, Robert. 'Pauper's rights, governor Grey and the poor law in South Australia', *Australian Historical Studies* 25 (1992), 220–43.
14. Dick, Lyle. 'The Seven Oaks incident and the construction of a historical tradition, 1816 to 1970', *J. of the Canadian Historical Association* new ser. 2 (1991), 91–113.
15. Evans, Raymond; Thorpe, William. 'Power, punishment and penal labour: *Convict workers* and Moreton Bay', *Australian Historical Studies* 25 (1992), 90–111.
16. Fisher, Michael H. *Indirect rule in India: residents and the residency system, 1764–1857*. Delhi; Oxford UP; 1992. Pp xv, 516.
17. Forster, Stig. *Die mächtigen Diener der East India Company: Ursachen und Hintergründe der britischen Expansionspolitik in Südasien,*

1793–1819. [The powerful servants of the East India Company: the causes and background of British expansionist policy in South Asia]. Stuttgart; Steiner; 1992. Pp 416.

18. Frost, Alan. 'Historians, handling documents, transgressions and transportable offences', *Australian Historical Studies* 25 (1992), 192–213.

19. Furnivall, John S. *The fashioning of Leviathan: the beginnings of British rule in Burma.* Canberra; Australian National University, Dept. of Anthropology, Research School of Pacific Studies; 1991. Pp ii, 178.

20. Hardiman, David (ed.). *Peasants and resistance in India, 1858–1914.* Delhi; Oxford UP; 1992. Pp xi, 304.

21. Humphery, Kim. 'Objects of compassion: young male convicts in Van Diemen's Land, 1834–1850', *Australian Historical Studies* 25 (1992), 13–33.

22. Hyam, Ronald (ed.). *British documents on the end of empire, 1945–1951, series A, vol. 2: the Labour government and the end of empire, part 1: high policy and administration.* London; HMSO; 1992. Pp lxxxiv, 372.

23. Hyam, Ronald (ed.). *British documents on the end of empire, 1945–1951, series A, vol. 2: the Labour government and the end of empire, part 2: economics and international relations.* London; HMSO; 1992. Pp xxi, 498.

24. Hyam, Ronald (ed.). *British documents on the end of empire, 1945–1951, series A, vol. 2: the Labour government and the end of empire, part 3: strategy, politics and constitutional change.* London; HMSO; 1992. Pp xxi, 419.

25. Hyam, Ronald (ed.). *British documents on the end of empire, 1945–1951, series A, vol. 2: the Labour government and the end of empire, part 4: race relations and the Commonwealth.* London; HMSO; 1992. Pp xvii, 389.

26. Kent, John. 'The Ewe question, 1945–1956: French and British reactions to "nationalism" in West Africa', Bc130, 183–206.

27. Killingray, David; Anderson, David M. 'An orderly retreat? Policing the end of empire', Bc11, 1–21.

28. Kyle, Avril. 'Little depraved felons', *Australian Historical Studies* 25 (1992), 319–24.

29. Lamphear, John. *The scattering time: Turkana responses to colonial rule.* Oxford; Oxford UP; 1992. Pp xxiii, 308.

30. Lau, Albert. *The Malayan union controversy, 1942–1948.* Oxford; Oxford UP; 1991. Pp xvi, 308.

31. Lowe, Kate; McLaughlin, Eugene. 'Sir John Pope Hennessy and the "native race craze": colonial government in Hong Kong, 1877–1882', *J. of Imperial & Commonwealth History* 20 (1992), 223–47.

32. Mackay, David. 'Banished to Botany Bay: the fate of the relentless historian', *Australian Historical Studies* 25 (1992), 214–16.

33. Mann, Kristin; Roberts, Richard. *Law in colonial Africa.* London; Currey; 1991. Pp xv, 264.

34. Marks, Shula; Trapido, Stanley. 'Lord Milner and the South Africa state reconsidered', Bc130, 80–94.

35. McCracken, John, 'Authority and legitimacy in Malawi: policing and politics in a colonial state', Bc11, 158–86.

36. McCulloch, Michael. 'The death of whiggery: lower-Canadian British constitutionalism and the *tentation de l'histoire parallèle*', *J. of the Canadian Historical Association* new ser. 2 (1991), 195–213.

37. McHugh, Paul G. *The Maori magna carta: New Zealand law and the treaty of Waitangi*. Auckland/Oxford; Oxford UP; 1991. Pp xv, 392.

38. Misra, B.B. *The unification and division of India*. New York; Oxford UP; 1990. Pp xxxiv, 422.

39. Mungazi, Dickson A. *Colonial policy and conflict in Zimbabwe: a study of cultures in collision, 1890–1979*. London; Russack; 1991. Pp 208.

40. O'Brien, John. 'The British Civil Service and Australia between the wars', Bc143, 146–58.

41. Pottinger, George. *Mayo: Disraeli's viceroy*. Salisbury; Russell; 1990. Pp 224.

42. Rathbone, Richard J.A.R. 'Political intelligence and policing in Ghana in the late 1940s and 1950s', Bc11, 84–104.

43. Rathbone, Richard J.A.R. (ed.). *British documents on the end of empire, vol. 1: Ghana, parts 1–2: 1941–1957*. London; HMSO; 1992. 2 vols. Pp lxxxvii, 421; xxix, 443.

44. Rees, Gareth. 'Coal-mining communities: the Welsh and Australian experience compared', Bc121, 109–26.

45. Reesor, Bayard William. *The Canadian constitution in historical perspective*. Scarborough (Ont); Prentice-Hall Canada; 1991. Pp 350.

46. Reeves, Peter Dennis. *Landlords and governments in Uttar Pradesh: a study of their relations until zamindari abolition*. Oxford; Oxford UP; 1991. Pp xiii, 359.

47. Ross, Eric. *Full of hope and promise: the Canadas in 1841*. Buffalo (NY); McGill-Queen's UP; 1991. Pp xvii, 169.

48. Shaw, Alan George Lewers. 'James Stephen and colonial policy: the Australian experience', *J. of Imperial & Commonwealth History* 20 (1992), 11–34.

49. Sinclair, Keith. *Kinds of peace: Maori people after the wars, 1870–1885*, Auckland; Auckland UP; 1991. Pp 161.

50. Smith, K.J.M. 'Macaulay's "utilitarian" Indian penal code: an illustration of the accidental function of time, place and personalities in law making', Bc195, 145–64.

51. Spate, O.H.K. 'Thirty years ago: a view of the Fijian political scene: confidential report to the British Colonial Office, September 1959', *J. of Pacific History* 25 (1990), 103–24.

52. Stockwell, A.J. 'Policing during the Malayan emergency, 1948–1960: communism, communalism and de-colonization', Bc11, 105–26.

53. Throup, David. 'Crime, politics and the police in colonial Kenya, 1939–1963', Bc11, 127–57.

54. Travis Hanes III, W. 'Sir Hubert Huddleston and the independence of the Sudan', *J. of Imperial & Commonwealth History* 20 (1992), 248–73.

55. Zines, Leslie. *Constitutional change in the Commonwealth*. Cambridge; Cambridge UP; 1991. Pp 118.

(d) *Religion*

1. Bayly, Susan. *Saints, goddesses and kings: Muslims and Christians in southern Indian society, 1700–1900.* Cambridge; Cambridge UP; 1989. Pp xv, 502.
2. Cathcart, Michael; Griffiths, Tom; Watts, Lee; Anceschi, Vivian; Houghton, Greg; Goodman, David. *Mission to the south seas: the voyage of the Duff, 1796–1799.* Parkville; History Department, University of Melbourne; 1990. Pp ix, 172.
3. Chimhurdu, Herbert. 'Early missionaries and the ethnolinguistic factor during the "invention of tribalism" in Zimbabwe', *J. of African History* 33 (1992), 87–109.
4. Comaroff, Jean; Comaroff, John L. *Of revelation and revolution: Christianity, colonialism and consciousness.* Chicago/London; Chicago UP; 1991. Pp xx, 414.
5. Doll, Peter M. 'American high churchmanship and the establishment of the first colonial episcopate in the Church of England: Nova Scotia, 1787', *J. of Ecclesiastical History* 43 (1992), 35–59.
6. Embree, Ainslie T. 'Christianity and the state in Victorian India: confrontation and collaboration', Bc13, 151–65.
7. Hansen, Holger Bernt. 'Church and state in a colonial context', Bc130, 95–123.
8. Hough, Brenda. 'Prelates and pioneers: the Anglican church in Rupert's Land and English mission policy around 1840', *J. of the Canadian Church Historical Soc.* 33/1 (1991), 51–63.
9. Larsson, Birgitta. *Conversion to greater freedom? Women, church and social change in north-western Tanzania under colonial rule.* Uppsala; Almquist & Wiksell; 1991. Pp 230.
10. Mosothoane, Ephraim. 'John William Colenso: pioneer in the quest for an authentic African Christianity', *Scottish J. of Theology* 44 (1992), 215–36.
11. Omenka, Nicholas Ibeawuchi. *The school in the service of evangelization: the Catholic educational impact in eastern Nigeria, 1886–1950.* Leiden; Brill; 1989. Pp xv, 317.
12. Phillips, Walter W. 'Religious response to Darwin in Australia in the nineteenth century', *J. of Australian Studies* 26 (1990), 37–51.
13. Porter, Andrew N. 'Religion and empire: British expansion in the long nineteenth century, 1780–1914', *J. of Imperial & Commonwealth History* 20 (1992), 370–90.
14. Saunders, Graham. *Bishops and Brookes: the Anglican mission and the Brooke raj in Sarawak, 1848–1941.* Oxford; Oxford UP; 1992. Pp xvii, 290.
15. Scott, David. 'Conversion and demonism: colonial Christian discourse and religion in Sri Lanka', *Comparative Studies in Soc. & History* 34 (1992), 331–65.
16. Sindima, Harvey J. *The legacy of Scottish missionaries in Malawi.* Lampeter; Mellen; 1992. Pp 164.
17. Thomas, Nicholas. 'Colonial conversions: difference, hierarchy, and

history in early twentieth-century evangelical propaganda', *Comparative Studies in Soc. & History* 34 (1992), 366–89.

(e) *Economic Affairs*

1. Bhacker, M. Reda. *Trade and empire in Muscat and Zanzibar: the roots of British domination*. London; Routledge; 1992. Pp 224.
2. Cadigan, Sean T. 'Merchant capital, the state, and labour in a British colony: servant-master relations and capital accumulation in Newfoundland's northeast-coast fishery, 1775–1799', *J. of the Canadian Historical Association* new ser. 2 (1991), 17–42.
3. Chatterji, Basudev. *Trade, tariffs and empire: Lancashire and British policy in India, 1919–1939*. Delhi; Oxford UP; 1992. Pp xiv, 521.
4. Cleary, M.C. 'Plantation agriculture and the formulation of native land rights in British North Borneo, c.1880–1930', *Geographical J.* 158 (1992), 170–81.
5. Drabble, John H. *Malayan rubber: the inter-war years*. Basingstoke; Macmillan; 1991. Pp 384.
6. Duin, P. Van. 'White building workers and coloured competition in the South African labour market, c.1890–1940', *International R. of Social History* 37 (1992), 59–90.
7. Dyster, Barrie; Meredith, David. *Australia in the international economy in the twentieth century*. Melbourne; Cambridge UP; 1990. Pp 362.
8. Ferns, H.S. 'Argentina: part of an informal empire?', Bc136, 49–61.
9. Fieldhouse, David. 'War and the origins of the Gold Coast Cocoa Marketing Board, 1939–1940', Bc130, 153–82.
10. Guha, Sumit. *Growth, stagnation or decline? Agricultural productivity in British India*. Delhi; Oxford UP; 1992. Pp 288.
11. Hetherington, Penelope. 'Child labour in Swan River colony, 1829–1850', *Australian Historical Studies* 25 (1992), 34–52.
12. Johnson, David. 'Settler farmers and coerced African labour in Southern Rhodesia, 1936–1946', *J. of African History* 33 (1992), 111–28.
13. Jones, Geoffrey. 'International financial centres in Asia, the middle east and Australia: a historical perspective', Bc95, 405–28.
14. Jones, Stuart; Muller, Andre. *The South African economy, 1910–1990*. Basingstoke; Macmillan; 1992. Pp ix, 380.
15. Lee, David. 'Protecting the sterling area: the Chifley government's response to multilateralism, 1945–1949', *Australian J. of Political Science* 25 (1990), 178–95.
16. McKay, A.C. 'The establishment of the British trade agencies in Tibet: a survey', *J. of the Royal Asiatic Soc.* 2/3 (1992), 399–422.
17. Mendelsohn, Richard. *Sammy Marks, 'the uncrowned king of the Transvaal'*. Cape Town/Athens (Oh); Philip/Ohio UP; 1991. Pp x, 300.
18. Morgan, W.T.W. 'Britain's loss, Queensland's gain: a reassessment of the Queensland-British Food Corporation, 1948–1952', *Australian Studies* 6 (1992), 10–28.
19. Muirhead, B.W. 'Britain, Canada, and the collective approach to freer

trade and payments, 1952–1957', *J. of Imperial & Commonwealth History* 20 (1992), 108–26.

20. Napier, Priscilla. *Raven castle: Charles Napier in India, 1844–1851.* Salisbury; Michael Russell; 1991. Pp xix, 305.

21. Piva, Michael J. *The borrowing process: public finance in the province of Canada, 1840–1867.* Ottawa; Ottawa UP; 1992. Pp 233.

22. Prakash, Gyan (ed.). *The world of the rural labourer in colonial India.* Delhi; Oxford UP; 1992. Pp viii, 304.

23. Ray, Rajat Kanta (ed.). *Entrepreneurship and industry in India, 1800–1947.* Delhi; Oxford UP; 1992. Pp xi, 263.

24. Smith, Woodruff D. 'Complications of the commonplace: tea, sugar and imperialism', *J. of Interdisciplinary History* 23 (1992), 259–78.

25. Stein, Burton (ed.). *The making of agrarian policy in British India, 1770–1900.* Delhi; Oxford UP; 1992. Pp 249.

26. Thomas, Kenneth. 'The adventures of H. & G. Simonds Ltd in Malta and East Africa', *Business Archives* 62 (1991), 40–54.

27. Tsokhas, Kosmas. 'Protection, imperial preference, and Australian conservative politics, 1923–1929', *J. of Imperial & Commonwealth History* 20 (1992), 65–87.

(f) *Social Structure and Population*

1. Ambler, Charles H. *Kenyan communities in the age of imperialism: the central region in the late nineteenth century.* New Haven (Ct); Yale UP; 1988. Pp xii, 181.

2. Barrie, W.D. *Immigration to New Zealand, 1854–1938.* Canberra; Research School of Social Sciences, Australian National University; 1991. Pp 198.

3. Berman, Bruce; Lonsdale, John. *Unhappy valley: conflict in Kenya and Africa.* London; Currey; 1992. Pp xvi, 504.

4. Bessant, Leonard Leslie. 'Coercive development, land shortage, forced labour and colonial development in the Chiweshe reserve in colonial Zimbabwe, 1938–1946', *International J. of African Historical Studies* 25 (1992), 39–65.

5. Bourrie, Doris. 'British emigrants from New York to York, Upper Canada, 1817', *Families* 29 (1990), 231–41.

6. Bowen, Lynne. *Muddling through: the remarkable story of the Barr colonists.* Vancouver; Douglas & McIntyre; 1992. Pp 272.

7. Brunger, Alan G. 'The distribution of Scots and Irish in Upper Canada, 1851–1871', *Canadian Geographer* 34 (1990), 250–58.

8. Bryan, Patrick. *The Jamaican people, 1880–1902: race, class and social control.* Basingstoke; Macmillan; 1991. Pp xiv, 300.

9. Chan, W.K. *The making of Hong Kong society: three studies of class formation in early Hong Kong.* Oxford; Oxford UP; 1991. Pp xii, 251.

10. Charles, Persis. 'The name of the father: women, paternity and British rule in nineteenth-century Jamaica', *International Labor & Working Class History* 41 (1992), 4–22.

11. Cox, Jeffrey. 'Independent English women in Delhi and Lahore, 1860–1947', Bc13, 166–84.
12. Daniel, T.K. 'The scholars and the saboteurs: the wrecking of a South African Irish scheme, Paris, 1922', *Southern African-Irish Studies* 1 (1991), 162–75.
13. Dean, D.W. 'Conservative governments and the restrictions of Commonwealth immigration in the 1950s: the problems of constraint', *Historical J.* 35 (1992), 171–94.
14. Dhanagare, D.N. (ed.). *Peasant movements in India, 1920–1950.* Delhi; Oxford UP; 1992. Pp xiii, 254.
15. Duder, C.J.D. 'Beadoc: the British East Africa disabled officers' colony and the white frontier in Kenya', *Agricultural History R.* 40 (1992), 142–50.
16. Ernst, Waltraud. 'Colonial psychiatry: the European insane in British India, 1800–1858', Bc145, 152–62.
17. Fedorowich, Kent. 'The migration of British ex-servicemen to Canada and the role of the Naval and Military Emigration League', *Histoire Sociale/Social History* 25 (1992), 75–99.
18. Fender, Stephen. *Sea changes: British emigration and American literature.* Cambridge; Cambridge UP; 1992. Pp 300.
19. Haines, Robin F. 'Therapeutic emigration: some South Australian and Victorian experiences', *J. of Australian Studies* 33 (1992), 76–90.
20. Haines, Robin F.; Shlomowitz, Ralph. 'Immigration from the United Kingdom to colonial Australia: a statistical analysis', *J. of Australian Studies* 34 (1992), 43–52.
21. Hall, Catherine. 'In the name of which father?', *International Labor & Working Class History* 41 (1992), 23–28.
22. Hamalengwa, Munyonzwe. *Class struggle in Zambia, 1889–1989, and the fall of Kenneth Kaunda, 1990–1991.* Lanham; UP of America; 1992. Pp 194.
23. Hornsby, Stephen J. 'Patterns of Scottish emigration to Canada, 1750–1870', *J. of Historical Geography* 18 (1992), 397–416.
24. Hornsby, Stephen J. 'Scottish emigration and settlement in early nineteenth century Cape Breton', *People, places, patterns, processes: geographical perspectives on the Canadian past* ed. Graeme Wynn (Toronto; Copp, Clark, Pitman; 1990), 110–38.
25. Houston, Cecil J.; Smyth, William J. *Irish emigration and Canadian settlement: patterns, links and letters.* Toronto; Toronto UP and Ulster Historical Foundation; 1990. Pp viii, 370.
26. King, D.J. Francis. 'From Staplehurst to Wellington: six pauper families from Staplehurst who emigrated to Wellington, New Zealand, in 1839', *Archaeologia Cantiana* 109 (1992 for 1991), 185–89.
27. Lloyd, Lewis. 'Some cases of working-group migration to Australia', Bc121, 69–92.
28. Lockwood, Glen J. 'Irish immigrants and the critical years in eastern Ontario: the case of Montague township, 1821–1851', Bc127, 203–35.
29. Lyons, Maryinez. *The colonial disease: a social history of sleeping sickness in northern Zaire, 1900–1940.* Cambridge; Cambridge UP; 1992. Pp xvi, 335.

30. Macdonald, Charlotte. *A woman of good character: single women as immigrant settlers in nineteenth-century New Zealand*. Wellington; Bridget Williams; 1990. Pp 283.

31. Mackay, Donald. *Flight from famine: the coming of the Irish to Canada*. Toronto; McClelland & Stuart; 1990. Pp 336.

32. McCracken, Donal P. 'Irish settlement and identity in South Africa before 1910', *Irish Historical Studies* 28 (1992), 134–49.

33. McLean, Marianne. *The people of Glencarry: highlanders in transition, 1745–1820*. Montreal/Kingston; McGill-Queen's UP; 1991. Pp xxiii, 285.

34. McLean, Marianne. 'Peopling Glengarry county: the Scottish origins of a Canadian community', Bc127, 151–73.

35. Milliss, Roger. *Waterloo Creek: the Australia day massacre of 1838, George Gipps and the British conquest of New South Wales*. Melbourne; McPhee Grimble; 1992. Pp xxiii, 965.

36. Moore, Christopher. 'The disposition to settle: the Royal Highland emigrants and loyalist settlement in upper Canada, 1784', Bc127, 53–79.

37. Morgan, Sharon. *Land settlement in early Tasmania: creating an antipodean England*. Cambridge; Cambridge UP; 1992. Pp 270.

38. Mukonoweshuro, Eliphas G. *Colonialism, class formation and under-development in Sierra Leone*. Lanham; UP of America; 1991. Pp 268.

39. Presley, Cora Ann. *Kikuyu women, the Mau Mau rebellion, and social change in Kenya*. Boulder (Co); Westview; 1991. Pp 325.

40. Pybus, Cassandra. *Community of thieves*. Melbourne; Willian Heinemann; 1991. Pp xiv, 198.

41. Rowland, Joan G. *Jews in British India: identity in a colonial era*. Hanover (New Hampshire)/London; UP of New England for Brandeis UP; 1989. Pp xiii, 355.

42. Schreuder, Deryck. 'Empire by "plantation": bringing the Ulster option to Highveld South Africa, 1900–1905', Bc143, 92–114.

43. Shaw, Alan George Lewers. 'British policy towards the Australian aborigines, 1830–1850', *Australian Historical Studies* 25 (1992), 265–85.

44. Swaisland, Cecillie. *Servants and gentlewomen to the Golden Land: the emigration of single women from Britain to southern Africa, 1820–1939*. Oxford; Berg; 1992. Pp 200.

45. Tomich, Dale. 'Gender: production of social relations', *International Labor & Working Class History* 41 (1992), 37–41.

46. Trigger, David S. *Whitefella comin: aboriginal responses to colonialism in northern Australia*. Cambridge; Cambridge UP; 1992. Pp 268.

47. Vaughan, Megan. *Curing their ills: colonial power and African illness*. Oxford; Polity; 1991. Pp xii, 224.

48. Walvin, James. *Black ivory: a history of British slavery*. London; HarperCollins; 1992. Pp x, 365.

49. Walvin, James. *Slaves and slavery: the British colonial experience*. Manchester; Manchester UP; 1992. Pp 192.

50. Willis, John. 'Le Québec, l'Irlande et les migrations de la grande famine: origine, contexte et dénouement', *La grande mouvance* ed. Marcel Bellavance (Sillery, Québec; Septentrion; 1990), 115–45.

(g) *Naval and Military*

1. Barthorp, Michael. 'Army reform and the Egyptian war, 1882', *Soldiers of the Queen* 71 (1992), 8–12.

2. Baynes, Sir John. 'Scottish soldiers in the last years of empire, 1901–1967', Bc21, 147–59.

3. Brereton, John Maurice. 'Red shirts at Peshawar', *J. of the Soc. for Army Historical Research* 70 (1992), 87–100.

4. Carter, William S. *Anglo-Canadian wartime relations, 1936–1945: RAF Bomber Command and No. 6 (Canadian) group*. New York/London; Garland; 1991. Pp xv, 204.

5. Davies, Harry (ed.). *Allanson of the 6th: an account of the life of Colonel Cecil John Lyons Allanson, 6th Ghurka Rifles*. Worcester; Square One; 1990. Pp 256.

6. Day, David. *Reluctant nation: Australia and the Allied defeat of Japan, 1942–1945*. Oxford; Oxford UP; 1992. Pp x, 326.

7. Droogleever, R.W.F. *The road to Isandhlwana: Colonel Anthony Durnford in Natal and Zululand, 1873–1879*. London; Greenhill; 1992. Pp 272.

8. Eames, Aled. 'Wales and Australia: the maritime connection', Bc121, 44–68.

9. English, John. *The Canadian army and the Normandy campaign: a study in the failure of the high command*. Westport (Ct); Praeger; 1991. Pp 295.

10. Featherstone, Donald. *Victorian colonial warfare: Africa, 1842–1902*. London; Cassell; 1992. Pp 160.

11. Featherstone, Donald. *Victorian colonial warfare: from the conquest of Sind to the Indian Mutiny*. London; Blandford; 1992. Pp 160.

12. German, Tony. *The sea is at our gates: the history of the Canadian navy*. Markham (Ont); McCelland & Stewart; 1991. Pp 360.

13. Granatstein, Jack L. *Canada's war: the politics of the Mackenzie King government, 1939–1945*. Toronto; Toronto UP; 1990. Pp xvi, 436.

14. Grey, Jeffrey. *Australian brass: the career of lieutenant-general Sir Horace Robertson*. Cambridge; Cambridge UP; 1992. Pp 320.

15. Henshaw, Peter James. 'The transfer of Simonstown: Afrikaner nationalism, South African strategic dependence, and British global power', *J. of Imperial & Commonwealth History* 20 (1992), 419–44.

16. Hyslop, Robert. *Aye, aye minister: Australian naval administration, 1939–1959*. Canberra; Australian Government Publishing Service; 1990. Pp 229.

17. Jackson, Frances. 'The *Duke of Portland*, the *Traveller*, and the *Prince Regent*: three little-known ships from Australia', *Hawaiian J. of History* 26 (1992), 45–54.

18. James, Lawrence. '"The white man's burden"?: imperial wars in the 1890s', *History Today* 42/8 (1992), 45–51.

19. Kiernan, Victor. 'Scottish soldiers and the conquest of India', Bc21, 97–110.

20. Knight, Ian J. *British forces in Zululand, 1879*. London; Osprey; 1991. Pp 64.

21. Knight, Ian J. *The Zulu war: twilight of a warrior nation*. London; Osprey; 1992. Pp 96.

22. Knight, Ian J. *Zulu: Isandlwana and Rorke's Drift, 22–23 January 1879*. London; Windrow & Greene; 1992. Pp 144.

23. Knight, Ian J. (ed.). *By the orders of the great white queen: campaigning in Zululand through the eyes of the British soldier, 1879*. London; Greenhill; 1992. Pp 256.

24. Krebs, Paula M. '"The last of the gentlemen's wars": women in the Boer war concentration camp controversy', *History Workshop J*. 33 (1992), 38–56.

25. Laband, John. *Kingdom in crisis: the Zulu response to the British invasion of 1879*. Manchester; Manchester UP; 1992. Pp 288.

26. Laband, John; Mathews, J. *Isandlwana*. Pietermaritzburg; Natal Monuments Council; 1992. Pp 98.

27. Laband, John; Thompson, Paul. *Kingdom and colony at war: sixteen studies on the Anglo-Zulu war, 1879*. Pietermaritzburg/Cape Town; Natal UP; 1990. Pp xvii, 358.

28. Lunt, James. *Scarlet Lancers: from sabres to scimitars—history of 16th/5th Queen's Royal Lancers, 1689–1992*. London; Cooper; 1992. Pp 256.

29. MacKenzie, John M. 'Introduction: popular imperialism and the military', Bc10, 1–24.

30. Marshall, Oliver. *The Caribbean at war: British West Indies in world war II*. London; North Kensington Archive, Nottingdale Urban Studies Centre; 1992. Pp 30.

31. McGibbon, Ian. *Path to Gallipoli: defending New Zealand, 1840–1915*. Petone (NZ); Brooks; 1991. Pp xiv, 274.

32. Moreman, T.R. 'The British and Indian armies and north-west frontier warfare, 1849–1914', *J. of Imperial & Commonwealth History* 20 (1992), 35–64.

33. Morton, Desmond. *A military history of Canada; from Champlain to the Gulf war*. Markham (Ont); McCelland & Stewart; 1992. Pp 384.

34. Nasson, Bill. *Abraham Esau's war: a black South African war in the Cape, 1899–1902*. Cambridge; Cambridge UP; 1991. Pp xxvi, 237.

35. Pfeiffer, Rolf. 'Exercises in loyalty and trouble-making: Anglo-New Zealand friction at the time of the great war, 1914–1919', *Australian J. of Politics & History* 25 (1992), 178–92.

36. Pritchard, Allan. 'Letters of a Victorian naval officer: Edmund Verney in British Columbia, 1862–1865', *BC Studies* 86 (1990), 28–57.

37. Robson, Brian (ed.). 'The Kandahar letters of the Reverend Alfred Cane' [cont.], *J. of the Soc. for Army Historical Research* 69 (1991), 206–20.

38. Waller, John H. *Beyond the Khyber pass: the road to British disaster in the first Afghan war*. New York; Random House; 1990. Pp xxxiii, 329.

39. Young, John. *Like stones they fell: battles and casualties of the Zulu war, 1879*. London; Greenhill; 1991. Pp 224.

(h) *Intellectual and Cultural*

1. Basu, Aparna. 'Women's history in India: an historiographical survey', Bc8, 181–209.

2. Bell, Leonard. *Colonial constructs: European images of the Maori, 1840–1914*. Auckland; Auckland UP; 1991. Pp 316.

3. Bending, R. Bagulo. *A history of education in northern Ghana, 1907–1976*. Accra; Ghana UP; 1991. Pp 284.

4. Cannon, Garland; Grout, Andrew. 'British orientalists' co-operation: a new letter of Sir William Jones', *B. of the School of Oriental & African Studies* 55 (1992), 316–18.

5. Cassels, Nancy G. *Orientalism, evangelicalism and the military cantonment in early nineteenth-century India: a historiographical overview*. Lampeter/Lewiston/Queenston; Mellen; 1991. Pp 169.

6. Chandra, Sudhir (ed.). *The oppressive present: literature and social conscience in colonial India*. Delhi; Oxford UP; 1992. Pp xi, 192.

7. Clarke, Sonia. *'Vanity Fair' in southern Africa, 1869–1914*. Houghton (S. Africa); Brenthurst; 1991. Pp 224.

8. Comaroff, John L. 'Images of empire, contests of conscience: models of colonial dominion in South Africa', *American Ethnologist* 16 (1989), 661–85.

9. Cooper, L.; Stoler, Ann L. 'Tensions of empire: colonial control and visions of rule', *American Ethnologist* 16 (1989), 609–21.

10. Cottrell, Michael. 'St Patrick's Day parades in nineteenth-century Toronto: a study of immigrant adjustment and elite control', *Histoire Sociale/Social History* 25 (1992), 57–73.

11. Davies, Vincent. *The British cemeteries of Patna and Dinapore*. London; BACSA; 1989. Pp 24.

12. Dawson, Graham. 'The blond Bedouin: Lawrence of Arabia, imperial adventure and the imagining of English-British masculinity', Bc223, 113–44.

13. Delgoda, Sinharaja Tammita. '"Nabob, historian and orientalist." Robert Orme: the life and career of an East India Company servant (1728–1801)', *J. of the Royal Asiatic Soc.* 4 (1992), 363–76.

14. Duder, C.J.D. 'Love and the lions: the image of white settlement in Kenya in popular fiction, 1919–1939', *African Affairs* 90 (1991), 427–38.

15. Dutton, Geoffrey; Elder, David. *Colonel William Light: founder of a city*. Melbourne; Melbourne UP; 1991. Pp xvi, 313.

16. Edwards, Norman. *The Singapore house and residential life, 1819–1939*. Oxford; Oxford UP; 1990. Pp xviii, 281.

17. Greenwall, Ryno. *Artists and illustrators of the Anglo-Boer war*. Vlaeberg (South Africa); Fernwood; 1992. Pp 264.

18. Grimshaw, Patricia. 'Writing the history of Australian women', Bc8, 151–69.

19. Hamilton, M.H. *Turn the hour: a tale of life in colonial Kenya*. Lewes; Book Guild; 1991. Pp 285.

20. Haynes, Douglas E. *Rhetoric and ritual in colonial India: the shaping of a public culture in Surat city, 1852–1928*. Berkeley (Ca); California UP; 1991. Pp xi, 363.

21. Holt, Thomas. 'Gender in the service of bourgeois idealogy', *International Labor & Working Class History* 41 (1992), 29–36.

22. Hyams, Bernard. 'The Colonial office and educational policy in British North America and Australia before 1850', *Historical Studies in Education* 2 (1990), 323–38.

23. Inden, Ronald. *Imagining India.* Cambridge (Ma); Blackwell; 1990. Pp vii, 299.

24. Keath, Michael. *Herbert Baker: architecture and idealism, 1892–1913: the South African years.* Cape Town; Ashanti; 1992. Pp 248.

25. MacDonagh, Oliver. 'A history of the Australian bi-centennial history project', Bc182, 243–66.

26. Majeed, Javed. *Ungoverned imaginings: James Mill's 'The history of British India' and orientalism.* Oxford: Oxford UP; 1992. Pp 225.

27. McDonald, Ian. *The Boer war in postcards.* Stroud; Sutton; 1990. Pp 192.

28. Moore, Katherine. 'A divergence of interests: Canada's role in the politics and sport of the British empire in the 1920s', *Canadian J. of the History of Sport* 1 (1990), 21–29.

29. Pedersen, Susan. 'National bodies, unspeakable acts: the sexual politics of colonial policy-making', *J. of Modern History* 63 (1991), 647–80.

30. Pieterse, Jan P. Nederveen. *White on black: images of Africa and blacks in western popular culture.* New Haven/London; Yale UP; 1992. Pp 260.

31. Pratt, Mary Louise. *Imperial eyes: travel writing and transculturation.* London; Routledge; 1992. Pp xii, 257.

32. Ramusack, Barbara. 'Cultural missionaries, maternal imperialists, feminist allies: British women activists in India, 1865–1945', *Women's Studies International Forum* 13 (1990), 309–21.

33. Reece, Bob. 'The Welsh in Australian historical writing', Bc121, 12–43.

34. Rich, Paul B. *Hope and despair: English speaking intellectuals and South African politics, 1896–1976.* London; British Academic; 1992. Pp 224.

35. Robson, John M.; Moir, Martin; Moir, Zawahir (ed.). *Writings on India by John Stuart Mill.* London; Routledge; 1990. Pp lvii, 336.

36. Schoeman, Karel. 'Uit de versamelings van die Suid-Afrikaanse Biblioteek: *De Dagstar* van Philippolis—'n vergete koerant (1859)' [From the collection of the South African Library: the *Dagstar* of Philippolis, a forgotten newspaper], *Q. B. of the South African Library* 46 (1991), 100–05.

37. Smyth, Rosaleen. 'The post-war career of the Colonial Film Unit in Africa: 1946–1955', *Historical J. of Film, Radio & Television* 12 (1992), 163–77.

38. Suleri, Sara. *The rhetoric of English India.* Chicago; Chicago UP; 1992. Pp 232.

39. Tillotson, G.H.R. 'The Indian travels of William Hodges', *J. of the Royal Asiatic Soc.* 2 (1992), 377–98.

40. Whitehead, C. 'Sir Christopher Cox: an imperial patron of a different kind', *J. of Educational Administration & History* 21/1 (1989), 28–42.

INDEX OF AUTHORS

Aalen, F.H.A., Bc118; Mg1
Abbattista, Guido, Ac1
Abdy, Charles, Bb2
Aberg, Alan, Bc210
Aberth, John, Ec1
Abraham, Gerald, Bc77
Abulafia, Anna Sapir, Ee1
Abulafia, David, Bc70
Acheson, Eric, Eg1
Achinstein, Sharon, Fj1
Adam, James S., If1
Adam, R.J., Ke1
Adams, J.N., Aa1; Cb1
Adams, Jad, Ib1
Adams, James Eli, Hl1
Adams, R.J.Q., Id1
Adams, Simon, Aa2; Fb1; Kb1
Adamson, J.L., Bb3
Addison, Paul, Bb4
Addy, John, Fh1
Addyman, Peter, Bc214
Adman, Peter, Gb1
Aers, David, Ac2–3; Bc94
Agiakloglou, C., Ba136
Ailes, Adrian, Eh1
Aird, W.M., Ee2
Aitchison, J., Bb5
Alban, J.R., Ia1,g1
Alberigo, G., Ee37
Alblas, Jacques B.H., Fe1
Alcock, Elizabeth A., Kh1,i1
Alcock, Joan P., Cb2
Alcock, Leslie, Da1; Kh1
Aldcroft, Derek H., Hf1; If2
Alderman, Geoffrey, Ba1; Ib2
Aldgate, Tony, Ij1
Aldred, David, Ek1
Aldrich, Zoë, Hg1
Alexander, James, If3
Alexander, Philip S., Eg2
Algar, Joan, Ij21
Ali, Yasmin, Ib3
Allais, Maurice, Gf1
Allan, Elizabeth, Hh38
Allan, John, Ej1–2; Fg1
Allderidge, Patricia, Hk1
Allen, Brian, Gi1
Allen, D.F., Fd1

Allen, J.R.L., Ca1,21
Allen, Louis, Id2
Allen, Matthew, Hj1
Allen, Robert C., Ba2,b6
Allison, A.F., Gi2
Allison, Keith J., Ij2
Allmand, C.T., Ea1
Allnutt, Richard, Ef5
Allott, Stephen, Gi3
Allsobrook, David Ian, Ij3
Almond, J.K., Ij4
Alston, Lee J., Hf82
Alston, Robin C., Ac4
Altenburg, Detle, Ei13
Altenburg, Hans H. Steinhoff, Ei13
Alter, Peter, Hl2
Altoff, Gerard T., Gh1
Alvey, Norman, Hh1
Ambler, Charles H., Nf1
Ambrose, Peter, Ig2
Ambrosius, Gerold, Ha1
Amphlett, C.B., Ij5
Anceschi, Vivian, Nd2
Andersen, Henning Soby, Gd1
Anderson, David M., Bc11; Nc1,27
Anderson, J. Stuart, Ba3
Anderson, James D., Cb3
Anderson, James E., De1
Anderson, M., Ig3
Anderson, Olive, Hb1
Anderson, Patricia J., Hl3
Anderson, Per Sveaas, Kf1
Anderson, Perry, Ac5
Anderson, R.D., Ba4
Anderson, R.G.W., Hi1
Anderson, S.J., Ac94
Anderson, T., Ei1
Andrew, Donna T., Ge1,j1
Andrews, J.A., Ca2
Andrews, J.S., Ca2
Andrews, Kenneth R., Fi1
Andriessen, J.H.J., Ii1
Angerer, Thomas, Hd1
Anglo, Sydney, Fb2
Annan, Noel, Ii2
Anngoulvent, Anne-Laure, Fj2
Anon, Aa3–26,b1–3; Bb7–8,c5,73,83–
    84,178,d1–17; Fg2,k1; Hf2; If4–5

Index of Authors

Anstey, Caroline, Id41
Archer, Ian, Ba5
Archer, Rowena E., Ef1
Archibald, M.M., Dd1
Arcidiacono, Bruno, Id3
Arena, Richard, Hf3
Arestis, Philip, Ab4
Armit, Ian, Bc168
Armitage, David, Fj3
Armstrong, John, Hf4; If6
Arneson, Richard J., Bc16
Arnold, A.J., If7
Arnold, C.J., Jh1
Arnold, David, Nc2
Arnoult, Sharon L., Fe2
Arnstein, Walter L., He1
Arthurson, Ian, Eh2
Artis, M., If8
Ashbee, Andrew, Fk2–4
Ashcraft, Richard, Bc60; Fe3,j4
Ashcroft, Loraine, Gg1
Ashford, Nigel, Ib4
Ashley, Maurice, Fb3
Ashley, Raymond E., Ba6
Ashton, Owen R., Hb2
Ashworth, Lawrence, Hh35
Ashworth, William, If9
Aspinwall, Bernard, He2
Aston, Elaine, Hg2
Aston, Margaret, Fe4
Atkin, Malcolm, Fi2
Atkin, Malcolm A., Ef2
Atkins, P.J., Hk2
Atkinson, Alan, Nc3
Atkinson, C., Fg3
Atkinson, Diane, Hb3
Atkinson, J.B., Fg3
Attreed, Lorraine, Ec2
Aughey, Arthur, Mb1
Austern, Linda Phyllis, Fk5
Avent, R., Jh2
Axton, Marie, Aa27
Ayer, A.J., Bc15
Ayers, Brian S., Ef3,k2
Ayton, Andrew, Eh3,4

Babbidge, Adrian, Bb9
Bachellery, Edouard, Ji1
Backhouse, Janet, Fk6
Baggott, Rob, Ib5
Bagguley, Paul, If10
Bailey, Douglass W., Ba7
Bailey, Keith, Ec3

Bailey, Mark, Ef4–5
Bailey, P., Bb10
Bailey, Richard N., De2
Bailey, Richard W., Ba8
Bailey, Victor, Bc7; Ia2–3,f11–12
Bailyn, Bernard, Bc2; Fa1
Baines, Arnold H.J., Db1
Baines, Paul, Gc1
Baker, Jennifer, Hc11
Baker, Nigel, Bb11; Ek3
Baker, P.A., Aa28
Baker, W.O., Ij6
Baker, William, Hl4
Balachandran, G., Id4
Balard, Michel, Ea18
Balfour, Michael, If13
Ball, Desmond, Nb24
Ball, Stuart, Ib6
Ballaster, Rosalind, Fk7–8
Ballhatchet, Helen, Hl5
Ballhatchet, Joan, Hc1
Baltrusch-Schneider, Dagmar Beate, Dc1
Balzaretti, Ross, Dd2
Bamford, K.G., He3
Bammesberger, Alfred, De3–4
Band, Thomas A., He4
Barash, Carol, Fk9
Barber, A.J., Ca3
Barber, James, Fg1
Barber, Jill, Gf2
Barber, Sarah, Bc224; Fb4
Barbezat, D., If14
Barclay, Jean, Hh2
Bardon, Jonathan, La1
Barger, A.C., Ij8
Bargielowska, Ina-Maria Zweiniger, Ac6
Barkan, Elazar, Ij7
Barke, Michael, Hf5
Barker, Katherine, Bc47; Fg4
Barker, Nicolas, Ij23
Barker, Philip, Ek29
Barker, T.C., Ac7
Barlow, Frank, Da2
Barnard, John E., Gj2
Barnard, Leslie W., Fe5; Ge2
Barnard, T.C., Mf1
Barnes, John, Hj2
Barnes, Thomas Garden, Fb5
Barnett, R.D., Bd18
Barr, William, Hd2
Barrell, A.D.M., Ee3; Ke2
Barrell, John, Bc89; Gc2,i4–5
Barrell, Rex A., Fb6,e6

239

Index of Authors

Barrett, Geoffrey N., Bb12
Barrett, Gillian, Cb4
Barrie, W.D., Nf2
Barrie-Curien, Viviane, Ge3
Barrow, G.W.S., Bc76; Ka1
Barrow, Julia, Ee4,k4
Barry, Jonathan, Ba9–10,b13,c26,217
Barry, Terry, Lg1
Barthorp, Michael, Ng1
Bartle, G.F., Hi2
Bartlett, C.J., Id5
Bartlett, Kenneth, Fe7
Bartlett, Thomas, Mb2
Bartrip, Peter, Hc2,k3
Baskerville, E.J., Fe8
Baskerville, Stephen W., Gb1
Bassett, Steven, Ba11,c216; Dc2–3,f1
Basu, Aparna, Nh1
Bately, Janet, Da3,e5
Bates, David, Ea2,e5
Bates, Don G., Fl1
Bateson, J.D., Kf2
Bateson, Mark, Ec4
Baugh, Daniel A., Gh2
Baugh, G.C., Ic1
Baumber, Michael, Ge4
Bausell, R. Barker, Ab5
Bawcutt, Priscilla, Ki2
Baxter, Colin F., Aa29
Baxter, Ronald, Ej42
Baxter, Stephen B., Fj5
Bayley, Justine, Df2–4
Bayley, Susan N., Hi3
Baylis, John, Id6
Bayliss, Mary, Bd19
Bayly, Susan, Nd1
Baynes, Sir John, Ng2
Baynton-Williams, Ashley, Ab6
Beadle, Richard, Aa30; Ei2
Beal, Peter, Bc220
Beale, Georgia R., Fl2
Beales, Derek, Hb4
Bearman, Robert, Fk10
Beattie, Alastair G., Bd20
Beattie, John M., Fc1,2
Beattie, Margaret H., Bd20
Beaver, Dan, Fg5
Bechler, Zev, Fl3
Beck, Peter J., Id7
Beckerman, John S., Ee6
Beckett, Ian F.W., Hj3
Beckett, John V., Ba12,b14; Ga1
Beckford, J.A., Ba13

Beckley, Susan, Gf3
Beckwith, Sarah, Ec5
Beddoe, Deirdre, Ig4
Beedell, Ann, Ba14
Beedham, Katherine F., Gb1
Beer, Frances F., Ee7
Begg, Paul, Ba15
Behre, Goran, Gh3
Belchem, John, Bc65; Hb5–6,g3
Bell, David N., Ab7; Ei3
Bell, Jonathan, Mf2
Bell, Leonard, Nh2
Bell, Robert, Fg54
Bellamy, Joyce M., Ab8
Bellavance, Marcel, Nf50
Bence-Jones, Mark, Ba16
Bendall, A. Sarah, Aa31
Bending, R. Bagulo, Nh3
Bending, Stephen, Gi6
Benfield, Stephen, Ca4
Benison, S., Ij8
Benn, Caroline, Hb7
Bennett, Douglas, Ij9
Bennett, G.H., Ib7,d89
Bennett, John, Aa32
Bennett, Judith M., Ba17; Eg3
Bennett, Martyn, Fb7
Bennett, Ronan, Fb8
Bennett, Shelley M., Aa33
Benskin, Michael, Ei4
Benson, John, Bc35
Benson, Robert, Gi7
Bentley, Michael, Hb8
Bentley-Cranch, Dana, Ha2,l6
Beresford, Maurice, Ab9,c8; Bb15
Berg, Maxine, Ac9; Gf4–6
Berger, Stefan, Ig5
Bergeron, David M., Fb9,j6; Ki3
Berghoff, Hartmut, Hf6
Berman, Bruce, Nf3
Bernard, G.W., Bc49; Fb10–12
Bernhard, Michael, De35
Bernier, Gérald, Nc4
Berridge, Virginia, Aa34,c10; Ih1
Berrios, German E., Bc145; Hk4,5; Ij10
Berry, Diana, Hk6
Berry, Paul, Ij11
Bertie, David M., Hl7
Bertram, Marshall, Hd3
Bessant, Leonard Leslie, Nf4
Besson, Gisèle, Ea5
Bettess, F., Df5
Bettey, J.H., Bd21; Fe9,g6

240

Betts, I.M., Ca12
Betts, Robin, Hi4
Bevan, Bryan, Ea3
Beveridge, Allan, Ij12
Bevir, Mark, Hb9,g4
Beynon, Huw, Ac11
Bhacker, M. Reda, Ne1
Biagini, Eugenio F., Hb10
Biancalana, J., Ec6
Bianchini, Franco, Ij13
Biddle, Martin, Db2
Bidwell, Paul T., Bc215
Bidwell, William B., Fa2
Biffen, John, Hb11
Biggs, Barry J., Ge5
Bignamini, Ilaria, Bb16
Biller, P.P.A., Eg4
Binfield, Clyde, He5–6,l8
Bingham, Caroline, Bb17
Binney, M., Ek5
Binski, Paul, Ej3
Birch, Debra J., Ee8
Bird, Stephen, Aa35
Birdwell-Pheasant, Donna, Mg2
Birke, Adolf M., Bc140
Birkett, Jennifer, Bc40; Ij14
Birrell, Jean, Ef6,k6
Birrell, T.A., Hl9
Bishop, Edward, Ii3
Bishop, M.C., Eh5
Bissett, Allan, Hf7
Bizzarro, Tina Waldeier, Ej4
Black, Alistair, Hl10
Black, E.W., Ca5
Black, Jeremy, Aa36,c12–14; Ba18; Fd2;
    Ga2–4,b2–3,d2–9,e6,h4–7; Ke3
Black, Maggie, Ei5; Ia4
Black, Michael, Ba19
Black, Robert, Kc1
Blackburn, Mark, Dd3
Blackburn, Richard, Ac15
Blacker, Carmen, Hl11; Id8
Blackwell, Kenneth, Aa37
Blades, Nigel, Ef26
Blagg, T.F.C., Cb5
Blaikie, J.A.D., Hg5
Blair, Claude, Eh6
Blair, W. John, Bc100; Dc4–6
Blake, Norman F., Bc228; Ei6–7
Blakemore, Steven, Bc132; Gb4
Blamires, David, Ei8
Bland, Roger, Cb6
Blaug, M., If15

Blaylock, Stuart, Ej2
Bliese, John R.E., Eh7
Blood, N. Keith, Ek7–8
Blunt, C.E., Dd4
Bock, Gisela, Ac16; Bc44
Boddington, Andy, Df6
Boffey, Julia, Ei9
Bohstedt, John, Hb12
Bolton, Brenda M., Ed1
Bolton, Geoffrey, Mi1
Bolton, J.L., Ef7
Bonavia, M.R., If16
Bond, Brian, Ii4
Bond, C. James, Bb18
Bonfield, Lloyd, Ba20
Bonner, Elizabeth A., Kh2
Bonney, Margaret, Ef8
Bono, Barbara J., Fk11
Bonthrone, Mark A., Bd22
Boog, Horst, Bc20; Ii5–6
Boorman, David, Ia5
Boorman, W.H., Hk7
Boos, Florence S., Bc225; Hl12
Booth, Alan, Ia14
Booth, Christopher C., Ih2
Booth, P., Hg6
Booth, P.H.W., Jf1
Boothman, Lyn, Fh2
Borsay, A., Gi8
Bostridge, Mark, Ij11
Botrel, J.F., Ac17
Bottigheimer, Karl S., Mb3
Boucher, David, Ac18
Boucher, Maurice, Na1
Bourdillon, Jennifer, Df7
Bourrie, Doris, Nf5
Bovill, Donald G., Hi5–6
Bowden, Mark, Ek7
Bowden, Sue M., If17–18
Bowen, John Atherton, Ek41
Bowen, Lynne, Nf6
Bower, Jacqueline, Fg7
Bowers, Brian, Gd10
Bowers, Roger, Ei10; Fk12
Bowler, Catherine, Aa38
Bowler, D.P., Ff1
Bowler, Peter J., Hk8
Bowring, Richard, Hl13
Bowring, Walter, Id9
Boyce, D. George, Ma1
Boyce, Gordon, Hf8–9
Boyd, Kelly, Ij15
Boylan, Thomas A., Mi2

Boyle, S.D., Ca6
Boyns, Trevor, Hf10
Boynton, Lindsay, Ge7
Bozeman, Theodore Dwight, Fe10
Braches, Ernst, Aa39
Brachmann, Hans-Jurgen, Bc126
Brackenridge, J. Bruce, Fl4
Bradbury, J., Eh8
Bradley, Dick, Ij16
Bradley, Ian, Hl14
Bradley, James E., Ge8
Bradley, John, Lg2
Bradley, Margaret, Hk9
Bradley, Richard, Ca7
Bradley, Thomas, Le1
Bragshay, Mark, Hh3
Brailey, Nigel, Hd4
Brailsford, Dennis, Ba21
Brake, Laurel, Hl15
Bramall, Dwin, Ii29
Brambles, G.W., Bb53
Bramwell, Bill, Hh4
Brand, Paul, Ba22,c199; Ef9
Brandmuller, W., Eb7
Branigan, Keith, Ca46,b7
Brant, Clare, Bc39; Gi9
Bratton, J.S., Hg7
Brautigam, Dwight D., Bc74; Fa3,k47
Brearley, Margaret F., Ie1
Breeze, Andrew, Li1
Breeze, David J., Cb8–9
Breight, Curt, Fk13
Breitenbach, Esther, Bb19,c135
Brennan, Niamh, Mb4
Brenner, Y.S., Bc34
Brent, Richard, Hb13
Brereton, John Maurice, Ng3
Breslaw, E.G., Fd3
Brett, Martin, Ei11–12
Breuilly, John J., Ac19; Bc176,219;
    Ha3,b14–17,g8–9
Brewer, Anthony, Hf11
Brewer, D., Ei13
Brewster, T.C.M., Ef10
Bridbury, A.R., Bc114; Df8; Ea4,f11–12
Bridges, E.M., Ia6
Briggs, Asa, Hl16
Briggs, C. Stephen, Bb20–25,42,123,129,
    c129
Brighton, J. Trevor, Bb26; Gi10
Brimblecombe, Peter, Aa38
Brink, Daphne H., Ee9
Britnell, R.H., Aa40

Broadberry, S.N., Ia7,f19–20
Brock, Michael, Ac20
Brock, W.H., Hk10
Bromley, Allan, Aa41
Bromley, Rosemary D.F., Ia8
Bromwich, Rachel, Ji2
Brook, Diane, Je1
Brooke, Christopher N.L., Ac21;
    Ee10,i14
Brooke, Daphne, Df9
Brooke, Robert, Ba23
Brooke, Stephen, Ib8
Brooks, C.M., Ej38
Brooks, Howard, Ek9
Brooks, Nicholas P., Db3,c7
Brooks, Stephen, Ii7
Brosnahan, Leger, Ej5
Brossard-Dandre, Michele, Ea5
Brotherstone, Terry, Ib9
Broun, Dauvit, Ke4
Brown, Anthony E., Bb1,c64; Ek10
Brown, C.E., Aa42
Brown, Callum G., Bb27; Ie2
Brown, Dana Goodburn, Ef26
Brown, David, Bc193
Brown, David K., Hk11
Brown, Desmond H., Ba24
Brown, Duncan H., Ej58
Brown, Frank E., Ih3
Brown, Gordon, Gi11
Brown, Iain Gordon, Gi12–13
Brown, Judith M., Na2
Brown, Keith M., Fe11; Ga5; Kb2
Brown, Kenneth, Mf3
Brown, Kevin L., Gi14
Brown, Peter, Ba25
Brown, S.E., Bb28
Brown, S.F., Ei15
Brown, Stewart J., He7; Ie3
Brown, Terence W., Bb29
Brown, Vivien, Ef13
Brownsword, R., Ej6
Bruce, Steve, Ie4; Mb5
Bruce, Warren John, Bb30
Brumhead, Derek, Gf7
Brundage, James A., Ee11
Brunelle, Gayle K., Ff2
Brunger, Alan G., Nf7
Brunton, Deborah, Gg2
Bryan, Patrick, Nf8
Bryan, Tim, Hf12
Bryant, Julius, Gi16
Bryant, R.V., If21

Index of Authors

Bryce, J.C., Gi17
Bryder, Linda, Ih4
Bryson, Anne, Hg10
Buchanan, James M., Gf8
Buchanan, Tom, Ib10
Buchet, C., Ba26
Buchet, Luc, Df10
Buchwald, Jed C., Hk12
Buckle, Stephen, Gi18
Buckley, Roger, Id10
Buckley, V.M., La2
Buckley, W. Kemmis, He8
Budd, J.W., Mg3
Budgen, Chris, Hf13
Budny, Mildred, De6
Bulley, Anne, Gd11
Bullion, John L., Gb5,26,d12
Bullock, Ian, Bc50; Ib11
Bullough, D.A., Db4
Bulmer, Simon, Ic2
Burdett-Jones, M.T., Ji3
Burge, Alun, Ib12
Burgess, Clive, Ee12–13
Burgess, Glenn, Fj7–8
Burham, Peter, Id11
Burk, Kathleen, Hf14; If22
Burke, Peter, Ac22; Fj9
Burl, Aubrey, Fl5
Burness, Catriona, Ib13
Burnham, John, Ij17
Burnley, David, Ei16
Burnley, Raymond, Hf16–17
Burns, David, Bb31
Burns, J.H., Ec7
Bursey, Peter, Ea6
Burt, Roger, Hf15–17
Burtt, Shelly G., Fj10
Bush, Helen, Ba27,c211
Bush, M.L., Ac23; Bc43; Fb13
Bushaway, R.W. [Bob], Gi19; Ii8
Bushnell, O.A., Gd13
Bushrod, Emily, Hh5
Butler, David, Ib14
Butler, John, Fj11
Butler, Lawrence A.S., Bb32; Jh3
Butlin, Robin, Gf9
Butterworth, Christine A., Ca8–9
Buurman, Gary B., Hf18
Bynum, William F., Ba28,c28; Hk13–15
Byrne, Maurice, Gi20
Byrne, Paula, Nc5

Cadigan, Sean T., Nc6,e2

Caerwyn Williams, J.E., Li2
Cahn, Walter, Ej7
Caillet, Maurice, Ke5
Cain, T.G.S., Bc59; Fk14–15
Caine, Barbara, Hb18
Cairncross, Frances, Bc52
Cairncross, Sir Alec, Bc52; If22–25
Cairns, Conrad, Lh1
Cairns, John W., Gc3
Calder, Angus, Ga6
Calder, William M., III, Hl116
Caldicott, C.E.J., Mi3
Caldwell, David H., Kg1
Caldwell, John, Ba29
Callanan, Frank, Mb6–8
Cameron, James K., Fj12
Cameron, Kenneth, Ab10–11
Cameron, M.L., Dc8
Camp, Anthony J., Aa43
Campbell, Alan, Ig6
Campbell, Bruce M.S., Ef14,53
Campbell, D., Gf10
Campbell, James, Bc193; Da4
Campbell, P.F., Fd4
Campbell, Sheila, Bc90
Campbell, T., Ek54
Campbell, W.A., Ba30,c36
Campbell, W.C., Ij18
Campion, Peter, Hf19
Candon, Anthony, Le2–3
Canley, James P., Fk16
Cannadine, David, Ac24; Hl17
Cannata, Antonella, Id12
Cannon, Garland, Nh4
Canny, Nicholas, Mg4
Canovan, Margaret, Ac25
Cantor, David, Ac26; Ih5
Cantor, Geoffrey, Ge9
Capelle, Torsten, Df11
Capie, Forrest H., Bc144
Caple, C., Ba31
Cardwell, Donald S.L., Hk16
Cardy, Amanda, Kg2
Carey, John, Ba32; Li3
Carleton, Kenneth W.T., Fe12
Carling, Alan, Ac27,111; Ei17
Carlson, Eric Josef, Fe13,14
Carlsson, Stig, Ab12
Carlton, Charles, Fi3–4
Carpenter, Christine, Ea7
Carpenter, D.A., Eb1–2
Carr, A.D., Ja1,f1
Carroll, Peter N., Ac28

Carruthers, Annette, Ij19
Carson, G.M., Bd23
Carter, Alan, Ac29
Carter, Harold, Bb5; Hg11
Carter, Ian, Hg12
Carter, John Marshall, Ei18
Carter, Patrick, Fc3
Carter, William S., Ng4
Carter-Edwards, Dennis, Gh8
Caruana, I.D., Ca10
Carver, Martin O.H., Bc113,206–207; De7,f12–13
Carver, Michael, Ia9,i9
Cashman, Bernard, Hh6
Cashmore, T.H.R., Nc7
Cassell, Michael, Bb33
Cassels, Nancy G., Nh5
Cassidy, Cheryl M., Aa44; Nc8
Cassis, Youssef, Bc95; If26
Casteras, Susan P., Hl18–19
Castiglione, Dario, Ba33; Gi18
Catch, J.R., Hl20
Cathcart, Michael, Nd2
Catterall, Peter, Aa45
Caunce, Stephen, Ig7
Cavallo, Sandra, Ba34
Cawthon, Elisabeth A., Hl21
Cell, John W., Na3
Cerasano, S.P., Bc68; Fk17
Chadwick. Roger, He9
Chainey, Graham, Bb34–36
Chakrabarty, Dipesh, Nb1
Chakrabarty, Rishi Ranjan, Nc9
Challinor, Ray, Ac92
Challis, Christopher E., Ba35–36,c226; Ff3; If27
Chalmers, Hero, Fk18
Chamberlaine-Brothers, R.J., Aa46
Chambers, Douglas, Fk19
Champ, Judith F., Bb37
Champion, J.A.I., Fe15
Chan, Stephen, Nb2
Chan, W.K., Nf9
Chandavarkar, Rajnarayan, Nc10
Chandler, Alfred D., Jr, Hf20
Chandra, Sudhir, Nh6
Channon, Cyril E., Ig8
Channon, Margaret L., Ig8
Chapman, John, Bb38; Gf11; Hf21; If28
Chapman, Stanley D., Ba37
Charles, B.G., Ab13
Charles, Persis, Nf10
Charles-Edwards, Thomas, Le4

Charlton, Christopher, Hl22
Charlton-Jones, Richard, Bc93; Fk20
Charsley, Simon R., Ba38
Chase, Malcolm, Hg13
Chatterji, Basudev, Ne3
Chatterton, Matthew, Aa32
Cheape, H., Bb39
Checkland, Olive, Hf22
Chernin, Eli, Hk17–18
Cherry, Bridget, Ej8
Cherry, John, Ee14,j9; Ji4
Cherry, R., Hg14
Cherry, Steven, Bb40
Chibnall, Marjorie, Ac30; Bc27
Chick, Martin, If29
Chihiro, Hosoya, Id13
Chilcote, Paul Wesley, Ge10
Childs, Michael J., Hg15
Childs, Wendy R., Ea8,f15
Chimhurdu, Herbert, Nd3
Chinn, Carl, Hh7
Chippington, George, Nb3
Chitwood, P., Ek11
Chorley, W.R., Ii10
Chorus, Jeroen M.J., Kc2
Christensen, Arne-Emil, Ef16
Christie, Peter, Hl23
Christodoulou, Joan, Ge11
Church, Roy A., If30
Church, S.D., Ea9
Cipolla, Carlo M., Ac31
Clamp, J.D., Gi21
Clamp, Peter G., Hi7–8
Clanchy, M.T., Ei19
Clapinson, Mary, Aa47
Clapson, Mark, Ba39
Clark, Anna, Hg16
Clark, B.E., Gf12
Clark, Cecily, De8; Ei20–21
Clark, Chris, Ij20
Clark, Gregory, Ba40
Clark, J.F. McDiarmid, Hk19
Clark, J.O.M., Ba138
Clark, John, Hl24
Clark, Jonathan C.D., Ga7
Clark, M.A., Gf12
Clark, Samuel, Ac32
Clarke, D.V., Hl25
Clarke, George, Gi22
Clarke, Helen, Bc209
Clarke, John Stock, Hl26
Clarke, Norma, Hl27
Clarke, Peter F., If31–32

Index of Authors

Clarke, Sonia, Nh7
Clavin, Patricia, Id14,f33
Claxton, Eric Charles, Ii11
Clayton, Robert J., Ij21
Cleary, John, Ba142
Cleary, M.C., Ne4
Cleary, Simon Esmonde, Cb10
Clegg, Cyndia Susan, Fk21
Cleggett, David A.H., Bb41
Clemoes, Peter, De9
Clifton, Gloria C., Hc3
Clive, E.M., Kc3
Clough, Charlotte, Bb42
Coakley, J.F., He10
Coates, Richard, Ab14; Df14
Coats, Jerry, He11
Cobban, Alan B., Ei22–23
Cobet, Justus, Hl116
Cobham, D., If8
Cocke, Thomas, Ej10
Cockerill, Timothy, Bd33
Codignola, Luca, Aa48
Cody, Lisa, Gj3
Coggins, D., Df15
Cogswell, Thomas, Fb14
Cohen, Edward H., Aa49
Cohen, I. Bernard, Aa50
Cohen, Marilyn, Mg5–6
Cohen, Michael J., Bc179; Id15
Cohen, Sheldon S., Ge12
Colbourne, Amanda, Eg5
Coldham, Peter Wilson, Fh3
Cole, Ann, Ab15
Coleman, Bruce, He12
Coleman, D.C., Ba41,c151
Coleman, Delphine, Bb43
Coleman, Ron, Gh28
Colfer, Billy, Lf1
Colgrave, Bertram, Da5
Collard, Jane, Bb44
Colley, Linda, Ba42; Ga8
Collier, Simon, Ij22
Collin, Richard H., Hj4
Collingwood, Judy, Mg7
Collingwood, R.G., Ca11
Collini, Stefan, Hl28
Collins, John, Ij23
Collins, Michael, Hf23; If17
Colman, Pamela, Ga9
Colpi, Terri, Ba43
Colson, Jean, Aa51
Colwell, Stella, Ab16
Comaroff, Jean, Nd4

Comaroff, John L., Nd4,h8
Connolly, Sean J., Mb9
Constantine, Stephen, Nc11
Contamine, Philippe, Kh3
Cook, A., Fl6
Cook, G.C., Bb45
Cook, Hugh, Hj5–6
Cook, Kay, Hb19
Cooper, L., Nh9
Cooper, N.H., Bb46
Cooper, Sheila M., Fg8
Cooter, Roger, Bc57; Hk20
Copeland, Rita, Ei24
Coppens, Chris, Aa52
Coquillette, Daniel R., Fj13
Corbett, Julian S., Gd14–15
Corbett, Margery, Gd16
Corfield, Penny, Ba44
Cork, Richard, Ij74
Corner, John, Bc149; Ij24–25
Cornish, Rory T., Aa53
Corpe, Stella, Bb47
Corrêa, Alicia, Dc9
Corrigan, Philip, Ba45
Corser, Peter, Kg3
Cortazzi, Sir Hugh, Bc17; Hl29–30; Id16
Cosgel, Metin M., Ef17
Cosgrave, Patrick, Ib15
Coss, Edward J., Gh9
Coss, Peter R., Bc112
Costen, Michael D., Bb48; Da6,f16
Cottrell, Michael, Nh10
Cottrell, P.L., Hf24
Cottret, Bernard, Fe16
Coudert, Allison P., Fe17
Coull, J.R., Hf25
Coulson, Charles, Ek12
Coulton, Barbara, Eg6
Counihan, J., Ek13
Courtenay, Adrian, Hb20
Courtney, Paul, Bb49; Ja2,g1
Cowan, Ian B., Ac33
Coward, Barry, Fh4
Cowie, Leonard W., Aa54
Cowling, Mark, Ac34
Cox, Gary W., Hb21
Cox, Jeffrey, He13; Nf11
Cox, Jeffrey N., Gi23
Cox, R.W., Aa55
Cox, Richard A.V., Ka2
Cox, Sebastian, Ii12
Coxall, Bill, Ib16
Crabtree, Pam, Dd5

Crafts, N.F.R., Gf13; If19–20
Craig, B.L., Aa56
Craig, Bill, If34
Craig, J.S., Fh5
Craig, M., Ma2
Crais, Clifton C., Nc12
Crammer, J.L., Ij26
Cramp, Rosemary, De10
Crampton, R.J., Id17
Craton, Michael, Fd5; Na4
Craven, Kenneth, Gi24
Crawford, Anne, Ef18
Crawford, Barbara, Dc10
Crawford, Michael J., Ge13
Crawford, Patricia, Fg9
Crawford, T.D., Ji5
Crawley, Eduardo, Ij27
Crawshaw, Anthony, Ih6
Creaton, Heather J., Aa57
Credland, Arthur S., Bb50
Creed, Robert Payson, De11
Cressy, David, Ba46
Cretella, Louis Anthony, Id18
Crew, Peter, Bb51
Crewe, D.G., Gh10
Crick, Bernard, Ih7
Crick, Julia, Ei25
Croad, Stephen, Ac35
Crobett, William, Lf2
Crocker, Glenys, Bb52
Crocker, Richard, Bc78; Ei26–27
Croft, John, Ii13
Croft, Pauline, Fb15,c4
Croft, Stuart, Id19
Croham, Lord, If35
Croll, Andy, Hl31
Cromartie, Alan, Fc5
Cronin, James E., Ic3
Crook, David, Ea10
Croom, Jane N., Ek14
Crosby, Alan, Bd24
Crosby, Christina, Hl32
Cross, Claire, Fe18
Cross, J.P., Ii14
Cross, Paul J., Ab17
Crouch, David, Ea11–12,c8; Jb1
Crouzet, François M., Ba47; Ff4
Crowfoot, Elisabeth, Bc53
Crowley, N., Ca12
Crowther, Jan, Ha4; Ij28
Crowther, M. Anne, Ba48; Hh8,k24
Cruft, Catherine H., Ab18
Crummy, Philip, Bc161

Cubitt, Catherine, Dc11
Cull, Robert, Hf26
Cullen, Louis M., Mj1
Cullen, Mary, Mi4
Cullum, P.H., Eg7–8
Cumming, Cliff, Nb4
Cumming, W., Lh2
Cunliffe-Charlesworth, Hilary, Hb22
Cunningham, Bernadette, Le6
Cunningham, Colin, Hl33
Cunningham, Valentine, Hl34
Curran, John, If36
Currie, Christopher K., Ba49; Ca13; Ef19–20
Curry, Anne, Eh9
Cust, Richard, Fb16–18

Dahrendorf, Ralf, If37
Dailey, Barbara Ritter, Fd6
Dales, D.J., Dc12
Dalgleish, George R., Gi25–26
Dalton, John P., Ee15
Dalton, Paul, Ea13–14
Dalton, Roger, Gf14
d'Alton, Ian, Mb10
Dalwood, Hal, Bb11; Df17
Daly, Mary E., Mf4
Dalziel, Raewyn, Hg17
Danchev, Alex, Ii15
d'Andrea, Antonio, Fj14
d'Angelo, Michela, Gf15
Daniel, T.K., Nf12
Daniels, Gordon, Bc17; Hl30; Id20
Daniels, R., Dc13; Ef21
Daniels, Stephen, Hf27
Dare, Robert, Nc13
Dark, Ken R., Ca14; Ji6
Darragh, James, He14
Das, Sudipta, Gd17
Daunton, M.J., Hf28–29
Davenport, Peter, Bc171
Davenport, Trevor George, Bb159
Davenport-Hines, R.P.T., Ba50; If38
Davey, Peter, Kj1
Davey, Roger, Gf16
David, Andrew, Ab19
David, R.G., Bb53
Davidoff, Leonore, Ac36
Davidson, Hilda Ellis, Df18–19
Davie, Donald, Fe19
Davies, Andrew, Ba51; Ig9
Davies, Brian, Bb54
Davies, C.S.L., Ea15

Index of Authors

Davies, David, Ba52
Davies, G.J., Gg3
Davies, Gareth Alban, Fk22
Davies, Glenys, Cb11
Davies, H. Rhodri, Hg18
Davies, Harry, Ng5
Davies, James A., Ij29
Davies, John Whittaker, Ab20
Davies, Julian, Fe20
Davies, K.M., Aa86
Davies, Katherine, Fg10
Davies, M.J., Aa1
Davies, Mansel, Hk21
Davies, Russell, Hg19
Davies, Sam, Ib17–18
Davies, T.G., Ih8
Davies, Vincent, Nh11
Davies, Wendy, Je2
Davis, G., Hg20
Davis, J. Colin, Fe21
Davis, Jill, Hg21–22
Davis, John A., Bc22; Hf60
Davis, Lance E., Hf26
Davis, O.L., Ih9
Davis, R.H.C., Ac37
Davis, R.W., Bc13; Hb23,e15
Davis, Richard, Mb11
Davis, Richard W., Hg23
Davis, Tracy C., Hf30–31,g22
Davison, Alan, Bb55
Davison, Lee, Bc82; Fb49; Gc4
Davison, Leigh M., Hi9
Dawson, Charles, Hf32
Dawson, Graham, Nh12
Dawson, Jane E.A., Ke6
Dawson, Michael, Ib19
Day, Angelique, La3
Day, David, Ng6
Day, Helen, Hg24
Day, Kenneth, Hk22
de Brou, David, Nb5
de Bruyn, Frans, Gi15
de Divitiis, G. Pagano, Fd7
de Hamel, Christopher, Ej11
de Krey, Gary S., Fb19
de la Bédoyère, Guy, Cb12
de Medeiros, Marie-Therèse, Ei28
de Micheli, C., Ca15
de Ridder-Symoens, H., Bc72
de Schepper, Marcus, Aa52
de Vegvar, Carol Neuman, Df20
Dean, D.W., Na5,f13
Dean, David, Fb20–21

Dean, Dennis, Gj4; Ib20
Dean, Martin, If70
Dearne, M.J., Cb7
Deighton, Anne, Id21
Dekar, Paul R., He16
Delgoda, Sinharaja Tammita, Nh13
Dell, Edmund, If39
Dellheim, Charles, Hl35
Dening, Greg, Gh11
Denman, Terence, Mh1
Denton, Jeffrey, Ec4,e16
Denver, David, Ib21
Denyer, Susan, Bb56
Derdak, Thomas, Ab21
Derolez, Rene, De12
Derry, John W., Hb24
Desmond, Adrian J., Hk23
Dever, Mark E., Fe22
Devlin-Thorp, Sheila, Gi27
Devons, S., Ij30
Dew, Barbara, Hi10
Dewar, Peter Beauclerk, Bd25
Dewhurst, M., Hf33
DeWindt, Anne Reiber, Ec9
Dhanagare, D.N., Nf14
Di Sciullo, Franco M., Ga10
Dick, Lyle, Nc14
Dickason, Olive Patricia, Na6
Dickie, Marie, Ig10
Dickinson, H.T., Fb22
Dickinson, J.R., Fa4
Dickinson, Tania M., Bc185; Df21–22
Dickson, A.D.R., Bc138; Ia10
Diederiks, Herman, Ac38; Bc42
Digby, Anne, Hg25
Dillistone, F.W., Bb57
Dinn, Robert, Ek15
Dintenfass, Michael, Ba53
Dinwiddy, J.R., Bc153
Ditchfield, G.M., Ge14–17
Dix, Brian, Bb58
Dixey, Richard, Hk40
Dixon, Philip, Ek16
Dixon, Piers, Id22
Dobash, Russell P., Hc4
Dobbs, Betty Jo Teeter, Fl7
Dobson, Alan, Id23
Dobson, Barrie, Ee17,g9,h10
Dobson, Brian, Cb22
Dockray, Keith, Eh11
Dodd, A., Ek17
Dodds, Klaus-John, Hb25
Dodds, M.G., If88

Index of Authors

Dodgshon, Robert A., Gf17
Dodgson, J. McN, Ab22–23
Doering-Manteuffel, Anselm, Hd5
Doggett, Alison, Bb93
Doggett, Maeve E., Hc5
Dohar, William J., Ee18
Dolan, J.E., If40
Doll, Peter M., Nd5
Donahue, Charles, Jnr, Ec10
Donald, Archie, Bb59
Donald, Peter H., Ke7
Donnachie, Ian, Aa58; Bb60,c6; Hc6; Kf3
Donnelly, J.F., Ba54
Dooley, Allan C., Hl36
Dorman, Marianne, Fe23
Dosi, Giovanni, Ba55,c134
Douglas, W.A.B., Gh12
Dow, Derek A., Bb189
Dow, Helen Jeanette, Fk23
Dowden, M.J., Gc5
Downes, Richard, If41
Downie, R.S., Gi28
Doyle, A.I., Dc14
Doyle, William, Ac39
Drabble, John H., Ne5
Draisey, J., Fg18–20
Drake, Frederick C., Gh13–14
Draper, Peter, Ej12
Dray, William H., Fa5
Droogleever, R.W.F., Ng7
Duckworth, Jackie, Ij31
Duder, C.J.D., Nf15,h14
Duffy, Eamon, Ee19
Duffy, Michael, Ba56,c173; Gd18,h15
Duffy, Seán, Lh3
Duggan, E.J.M., Ji7
Duin, P. Van, Ne6
Dukes, Paul, Kh4
Dumville, David N., Bc154; Da7,e13
Dunbar, John G., Ac40; Fg10
Duncan, A.A.M., Ac41
Duncan, Colin A.M., Hf34
Duncan, J.A., Ac94
Duncan, J.G., Gf18
Duncan, Robert, Bb61,c175; Ig11
Dunford, J.E., Ba73
Dunleavy, John, Mb12
Dunlop, David, Fb23
Dunn, C.J., Ek67
Dunn, John, Bc15
Dunne, Tom, Ac42; Mi5
Dunning, Robert W., Bc119; Ee20
Dunthorne, Kirstine Brander, Ij32

Dupree, Marguerite W., Hk24; If42
Durham, Martin, Ib22
Durston, Christopher G., Fg11,i5
Dutton, David, Hb26; Ib23–24
Dutton, Geoffrey, Nh15
d'Uzer, Vincenette, Fe24
Dwyer, John, Gi29
Dybikowksi, James, Ge18
Dyck, Ian, Hg26
Dyer, Alan, Ef22; Fg12
Dyer, Christopher C., Ba11,b62; Ef23,g10,k1
Dyer, G.P., Ba57
Dykes, David, Ij33
Dymkowski, Christine, Hg27
Dyson, Brian, Aa59
Dyson, Tony, Bc165
Dyster, Barrie, Ne7

Eaglen, R.J., Ef24
Eales, Jacqueline, Fe25
Eames, Aled, Ng8
Earle, Peter, Fi6
Easby, Rebecca J., Hl37
Eastwood, David, Gb6; Hb27
Ebersold, Bernd, Ib25
Edgerton, David, If43–44
Edmonds, Jill, Hg28
Edmunds, R. David, Gh16
Edwards, B.J.N., Df23
Edwards, Francis, Aa60
Edwards, Gavin, Bc121
Edwards, Geoffrey, Ic4
Edwards, John, Ei29
Edwards, Nancy, Bc96; Je3
Edwards, Norman, Nh16
Edwards, Peter, Nb6
Egan, Geoffrey, Ef25–26,j13–14
Ehrlich, Cyril, Hl38
Eichengreen, Barry, If45
Eisenach, Eldon, Fe26
Elder, David, Nh15
Eldred, Nelson B., Bd26
Eldredge, Elizabeth A., Nb7
Eldrid, Trevor, Bd26
Eley, Geoff, Ba58
Elkin, P.W., If46
Elliot, Walter, Ca16
Elliott, B.J., Ig12
Elliott, John R. Jr, Fk24
Ellis, E.L., Ia11
Ellis, Joyce, Aa61
Ellis, Steven G., Fb24; Lc1

248

Index of Authors

Ellmers, Chris, Bb63
Elm, Kaspar, Dc1; He54
Elrington, Christopher, Aa62
Elton, Sir Geoffrey R., Ba59; Fa6,i7
Elvin, Raymond, Bb64
Embree, Ainslie T., Nd6
Emerick, K., Ek18
Emmerson, Robin, Ba60
Emmison, F.G., Fg13
Emms, Margaret, Bb65
Empey, Adrian, Lf2
Emsley, Clive, Ba61
Emudong, C.P., Nb8
Englander, David, Id53,g13
English, Barbara, Aa63,c43; Bb66,c103; Fi8; Ij34
English, John, Ng9
English, John A., Ii16
English, John C., Ge19
English, Judie, Bb67
Enoch, D.G., Hh9
Erlich, Haggai, Id24
Ernst, Waltraud, Nf16
Erskine, Audrey, Aa64; Bc218
Etchingham, Colman, Le5
Evans, A.K.B., Aa65; Ee21
Evans, Chris, Gf19–20
Evans, D.A., Ij35
Evans, D.H., Bc106
Evans, D.R., Bc186
Evans, D. Simon, Ji2
Evans, J. Daryll, Bb68
Evans, J. Wyn, Je4–5
Evans, Jeremy, Cb13; Da8
Evans, Mary, Ia29
Evans, Michael R., Ed2
Evans, Muriel Bowen, Gg4
Evans, Neil, Bb69; Hb19; Ig14
Evans, Nesta, Fg14
Evans, R.H., Fe27
Evans, Raymond, Nc15
Evans, Richard J., Hk25
Evans, Robert Rees, Fb25
Evans, W. Gareth, Hi11
Everist, Mark, Ki4
Everson, Paul L., Ba62; Df24
Ewan, Elizabeth, Kg4
Ezell, Margaret J., Fk25

Faber, Sir Richard, Fg15
Fagan, Patricia, Lg9
Fairclough, K.R., Ff5
Fallon, John P., Ab24

Falvey, Heather, Ek19
Faraday, Michael, Bb70
Farley, Michael, Ek20
Farmer, David L., Ef27
Farrant, John H., Bb71
Farrell, Robert T., Bc97; Da9
Farrington, Susan Maria, Bd27
Fasham, P.J., Ja3
Faulkner, Evelyn, Ig15
Featherstone, Donald, Ng10–11
Feaver, George, Hl39
Fedorak, Charles John, Gb7
Fedorowich, Kent, Nf17
Fedosov, Dmitry G., Kh5
Fee, Elizabeth, Hk26
Feenstra, R., Bb72
Feigenbaum, Lenore, Gj5
Feldman, David, He17
Feliciano Ramos, H.R., Gf21
Fell, Christine, De14
Fellows-Jensen, Gillian, Aa66
Fells, I., If47
Feltes, N.N., Hf35
Fender, Stephen, Nf18
Fenning, H., Me1
Fenoaltea, Stefano, Ef28
Ferguson, Moira, Ba63
Fermer, H., If48
Fernie, Eric C., Bb73; Df25
Ferns, H.S., Hf36; Ne8
Ferris, I.M., Ca17
Ferris, Ina, Hl40
Ferris, John, Fb26
Ferris, John R., Id25,i17
Ferris, Lesley, Hg29
Feuchtwanger, Edgar J., Hb28–29
Fforde, John S., If49
Field, David, Bb67
Field, F.N., Ca18
Fieldhouse, David, Ne9
Fielding, Steven, Ba64; Ib26–28
Figala, Karin, Aa67
Filipiuk, Marion, Hl41
Filmer-Sankey, William, Bc193; Df26
Fingard, Judith, Hg30
Finlay, Richard J., Ib29
Finn, Dallas, Hl42
Finn, Margot, Hb30
Finnane, Mark, Mk1
Finnimore, Brian, Ih10
Firebrace, John, Ii18
Firth, Anthony E., Hi12
Firth, Gary, Bb74

249

Fischer, Olga, Ei30
Fish, H., Ij36
Fishburn, Evelyn, Ij37
Fisher, J.R., Hf37
Fisher, Michael H., Nc16
Fisher, Trevor, Hl43
Fisk, Malcolm J., Hh32
Fissel, Mark Charles, Bc69; Fb27
Fissell, Mary E., Ga11,j6
Fitzpatrick, David, Mi6
Fitzpatrick, Martin, Gb8,e18,20
Fitzsimmons, Linda, Hg31
Flanagan, Laurence, La4
Flannery, S., Ee22
Flavell, Julie M., Gb9–10
Fletcher, Alan J., Ee23
Fletcher, David H., Gf22
Fletcher, Ian, Gh17–18
Fletcher, J.M., Bc72; Fg16,k26
Flint, James, Hb31
Flockhart, Jim, If50
Flynn, Michael, Mg11
Fogelman, Aaron, Gg5
Foley, Timothy P., Mi2
Foot, Mirjam Michaela, Ab47
Foot, Sarah, Dc15–16
Ford, Alan, Me2
Ford, Deborah, Bb75
Ford, Judy Ann, Ej15
Fore, H., Ij38
Foreman, M., Ef61
Foreman-Peck, James, Ia12
Foréville, Raymonde, Eg11
Forman, Paul, Ac44
Forster, G.C.F., Aa68
Forster, Geoffrey, Aa69
Forster, Stig, Nc17
Forsyth, Richard K., Ec19
Fortmann, Michel, Nb17
Foster, Charles F., Gg6
Foster, David, Bc205; Hc7
Foster, John, Ig16,h11
Foster, Meryl R., Ee24
Foster, Roy, La5; Mb13
Foster, S.M., Da10
Foster, Stewart, Aa70; He18–19
Fowler, Alastair, Fk27
Fowler, D.H., Fl8
Fowler, Peter J., Ac45
Fowler, Simon, Aa71
Fox, Frank, Fi9
Foy, J.D., Ab25
Fradenburg, Louise Olga, Ba65,c91;

Fb28; Ki5
Frame, Robin, La6
Frank, Robert, Ac46
Frank, Roberta, De15–17
Frankis, John, Ee25
Franklin, Michael, Bc70; Ee26
Franklin, R. William, He20–21
Franks, L.R., Ii54
Frantzen, Allen J., Da11
Frantzen, J., If51
Fraser, Antonia, Fb29
Fraser, Constance Mary, Bb76
Fraser, David, Ii19
Fraser, Kevin J., Gj7
Fraser, T.G., Bc19
Freedman, Lawrence, Bc150; Ia13
Freeman, Eric F., De18
Freeman, Hugh, Bc145; Hk5,27
French, Christopher J., Gf23
French, David, Ii20
French, Katherine I., Ei31
French, Roger, Gj8
French, T.W., Ej16
Frend, W.H.C., Cb14
Frere, Sheppard S., Ca11,19
Freyer, Tony Allan, Ba66
Friar, Stephen, Bd28
Friedlander, Dov, Hg32
Friedman, Jerome, Fg17
Friedman, John B., Ei32
Frost, Alan, Nc18
Fry, Michael, Bb77,c30; Kb3
Frye, Susan, Fb30
Fukuda, Haruko, Id26
Fulford, Michael G., Bc192;
Ca1,20–21,b15
Fuller, Sarah, Ei33
Fullinwider, S.P., Hk28
Fulton, Richard, Hl44
Furniss, Tom, Gi30
Furnivall, John S., Nc19
Futrell, Robert Frank, Id27
Fyfe, Janet, Ba67

Gabbey, Alan, Fb31,l9
Gabra-Sanders, T., Ki9
Gaffney, V., Ca22
Gaimster, David, Bc191
Galloway, James A., Ef14
Galloway, Peter, La7
Gamby, Erik, Db5
Gameson, Richard, De19–21
Gangal, S.C., Nb9

Index of Authors

Gardiner, Juliet, Id28
Gardiner, Mark, Ek21
Gardner, Viv, Bc63; Hg33
Garner, Les, Hb32
Garnett, E., Hg34
Garnsey, Elizabeth, Ij39
Garrigan, Kristine Ottesen, Hg35
Garrod, Simon, Ca4
Garside, Patricia L., Hl45
Gascoigne, R.M., Ij40
Gaspar, P.P., Ba57
Gatch, Milton McC, Dc17,e22
Gaudefoy, Ghislain, Ea16
Gauthiez, Bernard, Ek22
Gay, Peter, He22
Geake, H., Df27
Gearson, John P.S., Id29
Gelder, Michael, Ij41
Gelderblom, Arie-Jan, Fj15
Gelling, Margaret, Aa72,c47; Ea17
Gem, Richard D.H., De23,f28
Gênet, Jean-Philippe, Ea18
Gentles, Ian, Fi10
Geoghegan, Vincent, Mg8
George, C.H., Fe28
George, James H., Id30
George, Stephen, Bc201; Ic5,d31–32
Gerchow, Jan, Dc18
German, Tony, Ng12
Geroski, P.A., If52
Gerrard, Bill, If53
Gerrard, S., Bb25; Jh4
Gerriets, Marilyn, Lc2
Geselowitz, Michael N., De27
Gethyn-Jones, Eric, Ca23; Gj9
Getz, Faye Marie, Eg12,i34–36
Giannetti, Renato, Ba55,c134
Gibbs, Graham, Aa73
Giblin, C., Me3
Gibson, A., Gc6
Gibson, J.S.W., Hf38
Gibson, Jeremy, Aa74
Gibson, William, Ba68
Gibson, William T., Aa75; He23–24
Giddings, Robert, Ba69
Gidman, Jean M., Eg13
Gidney, L., Ek65
Gifford, E.W.H., Df29
Gilbert, Bentley Brinkerhoff, Hb33
Gilbert, David, Hf39
Gilbert, Mark, Id33
Gilchrist, Roberta, Ee27,g14,k23
Gill, Sean, He25

Gillespie, Raymond, Le6; Mb14
Gilliam, Elizabeth, Fe29
Gillingham, John, Ea19,i37; Ld1
Gillmor, Desmond, Mj2
Gilmour, Ian, Ga12
Ginter, Donald E., Ba70
Girouard, Mark, Hg36
Gittings, Clare, Ba71; Ff6
Gittos, Brian, Df30
Gittos, Moira, Df30
Glaister, R.T.D., Gi31
Glanville, Philippa, Fk28
Glasscock, Robin E., Ba72,c181; Ek24–25
Glees, Anthony, Id34
Glendinning, Victoria, Hl46
Glennie, Paul, Ff7
Glover, Elizabeth, Bb78
Glynn, Sean, Ia14
Gneuss, Helmut, De24
Gobetti, Daniela, Fj16–17
Godden, Malcolm R., Bc180; De25–26
Godwin, Jeremy, Bb79
Goering, Joseph, Ee28,i38
Goff, Moira, Gi32
Gohler, Gerhard, Gi40
Goldberg, P.J.P., Bc92; Ef29–30,g15
Goldie, Mark, Fj18
Goldin, Claudia, Ih12
Golding, Arthur, Bb184
Goldstein, R.James, Ke8
Golinski, Jan, Gj10
Gooch, John, Ii21
Good, G.H., Bc41
Goodall, Francis, Aa76
Goodall, John, Bd29
Goodburn, Damian M., Ca24; Ef31–32
Goodchild, John, Ea20
Gooding, David, Hk68
Goodings, R.F., Ba73
Goodman, A.T., Hc8
Goodman, Anthony E., Bc4; Ea21–22
Goodman, David, Nd2
Goodman, Jordan, Ba88
Goose, Nigel, Aa77
Gordon, Eleanor, Bb19;c135
Gordon, Eleanora C., Eg16
Gordon, W.M., Bc195
Gorham, Deborah, Ia15
Goring, Jeremy, Bd30
Goring, Rosemary, Ab26
Gorst, Anthony, Id35
Gough, Barry M., Gd19; Na7

Index of Authors

Gough, Harold, Df31
Gourvish, Terry, Ab27
Gow, Ian, Bb80; Gi33
Gowers, Ian, Fe30
Gowing, Ronald, Gj11
Grace, Frank, Hg37
Grafsky, V.G., Ac48
Graham, Andrew, If54
Graham, Brian J., Mf5
Graham, Helena, Ef33
Graham, Jenny, Gb11
Graham, T.W., Aa78
Graham-Campbell, James A., Bc108; Ca25; Dd6–8; Jg2
Granatstein, Jack L., Ng13
Gransden, Antonia, Bc152
Granshaw, Lindsay, Hk29–30
Grant, Alexander, Eh12
Grant, Alison, Ff8
Grant, Allan, Jh5
Grant, Brian, Ba74
Grant, Chris, Jh5
Grant, Don, Ac49
Grant, Wyn, Bc37; If55
Grauer, Anne, Eg17
Gray, A. Stuart, Bb140
Gray, Charles M., Fc6
Gray, G.W., Ij42
Gray, Madeleine, Bb49; Fb32,c7,f9
Gray, Todd, Bc218; Fd8,f10,g18–20
Greasley, David, Hf40
Greatrex, Joan, Ee29–30
Greaves, Richard L., Bc198; Fe31–32
Green, E.H.H., Hf41
Green, Joseph, Ig17
Green, Judith A., Ea23–24
Green, Miranda J., Ab28
Green, Muriel M., Hg38
Green, Paul G., Ge21
Green, S.J.D., He26,g39
Greenall, R.L., Hg40
Greenaway, John R., Ib30,c6
Greene, J. Patrick, Ee31
Greene, Jack P., Fd9
Greene, Robert A., Fc8
Greene, Thomas R., Ie5
Greenwall, Ryno, Nh17
Greenwood, Sean, Id36
Greeves, Tom, Ff11
Gregory, Alexis, Hl47
Gregory, Desmond, Gd20
Gregory, Roger, Hc9
Gregory, Tony, Ca26

Greis, Gloria Polizzotte, De27
Grenter, Stephen, Bb81
Grenville, Jane, Bc206
Gress, F., Hb29
Gretsch, Mechthild, Dc19
Grew, Francis, Ca27
Grey, Jeffrey, Ng14
Griffin, Colin, Hf42
Griffin, Dustin, Gi34
Griffin, Nicholas, Hl48
Griffiths, Antony, Gi35
Griffiths, David, Dd9–10
Griffiths, Dennis, Ga4
Griffiths, Gwym, Bb82
Griffiths, Jeremy, Bc220
Griffiths, John C., Hf43
Griffiths, N.E.S., Kd1
Griffiths, Nick, Ca27
Griffiths, Ralph A., Bc212; Jg3
Griffiths, Rhidian, Aa79
Griffiths, Tom, Nd2
Griffiths, Trevor, Gf24
Grijp, P. van der, Gd21
Grimes, Sharon Schildein, Ih13
Grimshaw, Patricia, Nh18
Groarke, Leo, Gi36
Groehler, Olaf, Ii22
Grogan, Geraldine, Mi7
Groom, A. John R., Nb10
Gros, Daniel, Id37
Grose, D., La8
Grose, Kelvin, Mg9
Gross, Anthony, Ed3
Grout, Andrew, Nh4
Grubb, Farley, Fg21
Gruenfelder, John K., Fg22
Gruffudd, Ceris, Ac50
Gruffydd, R. Geraint, Ji8
Grundy, Emily M.D., Ig18
Grundy, Isobel, Bc62; Fg23
Gruner, Wolf D., Gd22
Guelzo, Allen C., He27
Guerrero, Ana Clara, Gd23
Guest, R., Ii54
Guha, Sumit, Ne10
Guinnane, Timothy W., Mf6,g3,10
Gulliver, P.H., Ac97; Bc124; Mj3
Gunn, Steven J., Bc1; Fb33–35,d10
Gurney, David, Ii23
Gustafsson, Bo, Bc142; If56
Gutwein, Daniel, Hb34
Guy, John, Fb36,37
Guy, Susanna, Ab29

Gwyn, Julian, Hf44
Gwynne-Jones, D.M., Je6

Hackett, Helen, Fk29
Hagerty, R.F., Ee32
Haggarty, George, Kj2
Haigh, R.H., Id81
Haines, Robin F., Nf19–20
Hainsworth, D.R., Fg24
Hakfoort, C., Ac51
Hald, Anders, Fl10
Haldenby, David, Df32
Hale, Matthew, Aa80
Haley, E., Cb16
Haley, K.H.D., Fd11
Halfpenny, Peter, Ac52
Halkon, Peter, Ca28
Hall, A. Rupert, Fl11–12
Hall, Bert, Bc90
Hall, Catherine, Bc197; Nf21
Hall, Christopher D., Gh19
Hall, David, Bc170
Hall, Mary, Hh10
Hall, Mary B., Bb83
Hall, Richard A., Bb84,c209; Ek26
Hallaway, H.R., Ga13
Halsall, Guy, Da12
Halsey, A.H., If57
Halstead, John, Hb35
Hamalengwa, Munyonzwe, Nf22
Hamburger, Joseph, Hl49
Hamerow, Helena, Df33
Hamersley, Lydia, Fk30
Hamilton, Carolyn Anne, Nb11
Hamilton, Ian W.F., He28
Hamilton, James T., Bb85
Hamilton, K.A., Id89,109
Hamilton, M.H., Nh19
Hamilton, Valerie, Id38
Hamlin, A., Ek30
Hamlin, Ann, Le7
Hamlin, Christopher, Hk31
Hammer, Paul E.J., Fb38
Hamowy, Ronald, Fj19
Hamp, Eric P., Kj3
Hampsher-Monk, Iain, Ba75
Hanawalt, Barbara A., Eg18
Hand, Geoffrey, Mi8
Hands, Gordon, Ib21
Hannah, Gavin, Ge22
Hansen, Holger Bernt, Nd7
Hansen, Steffen Stummann, Kf4
Hansford-Miller, Frank, Ba76–77; Ee33;

Fe33–37
Hanson, W.S., Cb17
Harbison, Peter, Le8
Harborne, Leslie R., Bb86
Hardiman, David, Nc20
Hardin, Richard F., Fa7
Harding, Richard, Gb12
Harding, Vanessa, Ek27
Hardwick, Kit, Ea25
Hardy, Sheila, Hg41
Hare, J.N., Eg19
Haren, Michael J., Ee34
Hareven, T.K., Ba78
Hargreaves, John D., Na8
Härke, Heinrich G., Df21,34
Harley, Basil, Gh20; Hl50
Harley, C.K., Gf13
Harley, Jessie, Hl50
Harling, Philip, Hh11
Harman, P.M., Bc87; Hk32
Harper, Sue, Ij43
Harper-Bill, Christopher, Bc51,98; Ee35
Harpham, Edward J., Bc155; Fk31
Harries, Jill, Cb18
Harries, Phillip, Ij44
Harrington, Brian, Hi13
Harrington, C.R., Hd3
Harrington, P., Fi11
Harris, Bernard, Ih14
Harris, Frances, Gb16
Harris, Geraldine, Hl51
Harris, John R., Bc86; Gf25
Harris, José, Hb36,f45,h12; Ig19
Harris, Kate, Aa81
Harris, Michael, Hk33
Harris, P.R., Ab30; Bc61
Harris, Paul, Ii24
Harris, Richard I.D., Mf7
Harris, Robert, Gb13
Harris, S.D., Bb87
Harris, Stuart, Kh6
Harris, Tim, Ac53
Harrison, A.C., Ca29
Harrison, D.F., Ba79
Harrison, Jennifer, Fg25
Harrison, John, Aa67
Hart, Cyril R., Bc99; Da13,b6–21,f35;
  Ea26,c11–15
Hart, Elizabeth, Aa82; Ge23
Hart, Jenifer, Ba80
Hart, Julian Tudor, Ih15
Hart, M.W., Ia16
Hart, Marjolein 't, Fd12

Hart, Peter, Mb15
Hart, Vaughan, Gi37
Hartfield, Alan, Ii25
Hartley, T.E., Fb39
Hartman, Joan E., Fj20
Harvey, A.D., Ba81; Me4
Harvey, Barbara K., Fg26
Harvey, Charles, Aa83; If58
Harvey, Elizabeth D., Bc40; Fk32; Ij14
Harvey, John H., Ed4
Harvey, Margaret M., Ed5–6,e36,i39
Harvey, P.D.A., Aa84; Ei40
Harvey, Ruth, Bc98
Harvey, Sylvia, Bc149; Ij24–25
Harvie, Christopher, Ib31,j45
Hasan, F., Fd13
Hasegawa, Junichi, Ih16
Haskett, Timothy S., Ec16
Haslam, Graham, Fc9,10; Gf26
Haslam, Jeremy, Df36
Hassall, M.W.C., Ca30
Hast, Adele, Ab31–33
Haste, Cate, Ig20
Hatcher, Stephen, He29
Hatchwell, Richard, Bb88
Haughton, Joseph P., Ac54
Hawkes, Sonia Chadwick, Bc193; Df37
Hawkey, D., Bb89
Hawkings, David T., Aa85
Hawkins, Richard, Aa80
Haworth, Richard, Aa86
Hay, Tempest, Bb90
Haycock, Lorna, Gi38
Hayes, P.P., Bc169
Hayes, Paul, Bc150; Hj7
Hayfield, Colin, Ef10,j17; Gf27
Haynes, Alan, Fc11
Haynes, Barry, Hg42
Haynes, Douglas E., Nh20
Hays Parks, W., Ii26
Hayward, Valerie, Bd31
Hazlehurst, Cameron, Ib32
Head-König, Anne-Lise, Ba82
Heal, Veryan, Bc207
Healey, Edna, Ba83
Healey, Robert M., Ke9
Healy, David, Mk2
Heater, Derek, Ih17
Hebden, J.R., Fg27
Heffer, Eric S., Ac92
Heikkila, H., Id39
Heim, Carol E., Hf46
Hellinga-Querido, Lotte, Fk33

Helmholz, Richard H., Ba84,c125
Helmstadter, R.J., Bc13; He15,30
Hemingway, Andrew, Hl52
Hempton, David, Me5
Henderson, Alan C., Ab34
Henderson, C.G., Ff12
Henderson, Diana E., Fk34
Henderson, Diana M., Bb91
Henderson, George, De28
Henderson, Julian, Lf3
Henderson, Robert, Hl53
Hendrick, Harry, Hk34,l54
Hendry, John, If59
Hendry, Joy, Ac55
Henig, M., Ca3
Hennessy, Alistair, Ba85–86,c136; Id40
Hennessy, Elizabeth, If60
Hennessy, Peter, Ia17,b33,d41
Henry, Avril, Ej18
Henry, Gráinne, Ld2
Henrywood, R.K., Bb92
Henshaw, Peter James, Ng15
Hepple, L.W., Bb93
Herbert, D., Hl55
Herbert, Méire, Li4
Herbert, Sandra, He31
Herity, Michael, Le9
Herlihy, David, Eg20
Herman, Gerald, Ia18
Herren, Michael W., Aa87
Herrmann, Joachim, Bc126
Hersh, George K., Aa113
Heslop, T.A., Ee37,j19,k28
Hetherington, Penelope, Ne11
Hewison, Robert, Ij46
Hewitt, David, Hj8
Hewitt, George, Ke10
Hexter, J.H., Bc80; Fa8,b40
Hey, D., Fc12
Heywood, Brenda, Ca31
Hibbard, Caroline M., Fi12
Hicks, Carola, Bc75; Ej20
Hicks, Michael, Ee38
Higgins, David, Fg1
Higgins, Paula, Ki6
Higginson, J.H., Hi14
Higgs, Edward, Aa88
Higham, Nicholas J., Cb19–21; Db22–23; Jb2
Higham, Robert, Ek29–30
Higonnet, Patrice, Bc32
Hiley, David, Bc78
Hill, Bridget, Gi39

Index of Authors

Hill, Christopher, Fe38
Hill, F.J., Ab35
Hill, James Michael, Bb94
Hill, John, Bb95–96
Hill, Joyce, Dc20
Hill, Myrtle, Bc224; Me5
Hill, P.R., Cb22
Hill, Peter H., Ke11,f5–6
Hill, Rosalind, Ac30
Hillewaert, Bieke, Ej21
Hillsman, Walter, Hl56–57
Hilton, Rodney H., Ef34,g21
Hinchcliffe, John, Bc163
Hinchcliffe, Tanis, Bb97
Hinchliff, Peter, He32
Hinchliffe, G., Gg7
Hinde, Thomas, Bb98
Hinde, Wendy, Hb37
Hinds, Hilary, Fj21
Hines, J., Df38
Hingley, Richard, Cb23–24
Hinshelwood, R.D., Hk35
Hinsley, Francis Harry, Ii27
Hinton, James, Ib34
Hirst, Derek M., Fb41
Hislop, M.J.B., Ek31
Hitchcock, Tim, Bc82; Ga14
Hobby, Elaine, Fg28
Hobelt, Lothar, Fd14
Hobsbawm, E.J., Hl58
Hocart, R., Bb99
Hockey, S.F., Ee39; Je7
Hodges, Richard M., Bc209; Dd11–12;
    Ef35
Hodgson, N., Cb37
Hodson, Sara S., Aa89
Hoey, Laurence, Ej22
Hofstetter, Walter, De29
Hoftijzer, P.G., Gf28
Hogan, Arlene, Li5
Hogan, John, Hb38
Hogg, Malcolm, Aa61
Hogg, Richard M., Bc227; De30–31
Hohenberg, Paul M., Ac38; Bc42
Holbrook, Neil, Bc215
Holdsworth, Christopher J., Ee40–42
Hollis, Stephanie, Aa90; Dc21
Holm, P., Lf4
Holman, Nigel, Hl59
Holmes, Clive, Fb42
Holmes, Colin, Id42
Holmes, Ken, Ba87
Hölscher, Wolfgang, Id43

Holstun, James, Bc203; Fj22–23
Holt, Geoffrey, S. J., Bb100
Holt, J.C., Ea27
Holt, Richard, Ba11,b11; Ef36
Holt, Thomas, Nh21
Homer, R.F., Ej6
Honeyman, Katrina, Ba88
Hood, Christine, Bd32
Hood, Gervase, Fk35
Hook, David, Ei41
Hooke, Della, Db24; Ek32
Hope, Annette, Bb101
Hope, John, Ib35
Hopkins, John T., Fk36
Horden, Peregrine, Df39
Horn, Joyce M., Bb102
Horn, Pamela, Hg43
Horner, Arnold, Mj4
Horner, John, Ia19
Hornix, Willem J., Hk36
Hornsby, Stephen J., Nf23–24
Horowitz, Elliott, Fe39
Horrell, Sara, Gf29
Horridge, Glenn K., He33
Horrox, Rosemary, Eb3
Horsey, I.P., Df40; Ef37
Horwitz, Henry, Ba89
Hosgood, Chris, Hg44
Hösler, M., Me6
Hostettler, John, Hb39
Hough, Brenda, Aa91; Nd8
Houghton, Greg, Nd2
Hounshell, D., Ij47
Houston, Alan Craig, Fj24
Houston, Cecil J., Nf25
Houston, R.A., Bb103; Fh6
Howard-Davis, C., Ek65
Howard-Hill, T.H., Aa92
Howarth, Alan, Ih18
Howarth, David, Fk37
Howe, A.C., Hf47
Howe, Elizabeth, Fk38
Howell, David W., Hf48,h13
Howell, Margaret, Ea28
Howells, Gwyn, Id44
Howells, John G., Ij48
Howes, Lesley, Bb104
Howkins, Alun, Ba90; Hg45
Howlett, David, Je8
Howsam, Leslie, He34
Howse, Derek, Ii28
Hoyle, Richard W., Bc122;
    Fb43–44,c13–17,e40,g29–30

255

Index of Authors

Hubbard, William H., Ha1
Huber, Gerhard, Gi40
Huberman, Michael, Hf49
Hudleston, C. Roy, Bd33
Hudson, Anne, Ei42
Hudson, Benjamin T., Db25
Hudson, Elizabeth K., Fe41
Hudson, Hazel, Eg22
Hudson, John G.H., Ec17,i43
Hudson, L.W., Bb3
Hudson, Pat, Gf6,30; Hg46
Hudson, Robert P., Ei95
Hudson, W.J., Nb12
Huelin, Gordon, Bb105
Huffman, Joan B., He35
Huggins, P.J., Df41
Hughes, Ann, Fe42,h7
Hughes, Colin, Bb106
Hughes, Geoffrey, Ba91
Hughes, Jill, Ee43
Hughes, Jonathan, Ee44,g23
Hughes, Pat, Fg31
Hughes, Peter, Aa93
Hughes, R. Elwyn, Fl13
Hughes, Stephen, Bb107–108
Hughes, Thomas P., Ba92
Hulbert, Anna, Ej23
Hull, F., Fg32
Hulse, Lynn, Fk39
Hume, R., Gi54
Humphery, Kim, Nc21
Humphries, Enoch, Ia20
Humphries, Jane, Gf29
Hundert, Edward, Gi41
Hunt, David, Bb109
Hunt, John Dixon, Fl14
Hunt, Margaret, Gg8
Hunt, Philip A., Gf24
Hunt, Tamara L., Hb40
Hunt, Tony, Eh15,i44
Hunt, W.M., Hf50
Hunt, William, Fd15
Hunter, D.M., Kc4
Hunter, David, Gi42
Hunter, James, If61
Hunter, Janet, Id45
Hunter, Michael, Fe43,k40
Hussey, John, Ac56; Fi13
Hüttenberger, Peter, Id46
Hutton, Sarah, Fe44,k41
Huws, Daniel, Je9,i9
Huws, Gwilym, Aa94
Hyam, Ronald, Nc22–25

Hyams, Bernard, Nh22
Hyslop, Robert, Ng16

Iles, R., Bb18; Jh4
Imhof, Arthur E., Ig21
Immerwahr, John, Gi43
Imray, Jean, Bb110
Ince, Laurence, Bd34
Inden, Ronald, Nh23
Ingle, H. Larry, Fe45–46
Ingle, Stephen, Ib36
Inglin, Oswald, Id47
Inglis, K.S., Ia21
Ingram, Edward, Ba93,c147
Innes, Joanna, Gb14
Innes of Edingight, Sir Malcolm, Ga15
Inskip, Peter, Gi44
Ion, A.H., Hl60
Ireson, Tony, Ia22
Irwin, D.A., Ff13
Isaac, Michael J., Ia23
Israel, Jonathan I., Fb45
Isserlin, Raphael, Ee45
Ittmann, Karl, Hg47
Itzkowitz, David C., He36
Ivens, Richard, Lf3
Ives, E.W., Fb46–48

Jackson, Alvin, Mb16
Jackson, Frances, Ng17
Jackson, Gordon, Bb111
Jackson, Peter, Dc22
Jackson, R.V., Gf31
Jackson, William G.F., Ii29
Jacob, James R., Ge24
Jacob, Margaret C., Ge24
Jacob, W.M., He37
Jacobs, Derek, Gg9
Jacobs, John, Hh14
Jacobsen, M., Id48
Jacobson, David L., Ge25
Jacyna, Stephen, Gj12
Jaffe, James A., Hg48
Jaggard, E.K.G., Hb41
Jaki, S.L., He38
James, David, Aa95
James, Edward, Bc185,206
James, Frank A.J.L., Gj13
James, Harold, Hf51; If62
James, Heather, Bc101; Cb25; Je10
James, Jude, Gf32
James, Lawrence, Gh21; Ng18
James, Susan E., Ej24

256

Index of Authors

James, Terrence A., Je11
Jamieson, Alan G., Hf52
Jancar, Joze, Hk22
Jansen, Henrik M., Dd13
Jansen, Virginia, Ej25
Jansson, Maija, Fa2
Jarnut, Jorg, Ei13
Jarvis, A.E., Aa96
Jarvis, Adrian, Hf53
Jarvis, P., Ca31–32
Jefford, Michael, Ia24
Jeffreys-Jones, Rhodri, Bc18
Jenkins, A.P., Ge26
Jenkins, E.W., Hi15
Jenkins, Geraint H., Bb112; Fe47
Jenkins, H.J.K., Ef38
Jenkins, Ian D., Hl61
Jenkins, K., Ac57
Jenkins, T.A., Hb42
Jenks, Stuart, Ed7
Jenner, F.A., Ij49
Jennings, Sarah, Bc208; Ef39,j26
Jermy, K.E., Cb26
Jervis, Martin R., Hf54
Jessula, Georges, Id49
Jewell, Helen M., Ec18
Jewitt, A. Crispin, Aa97
Jobey, George, Bb113
John, Angela V., Bc66; Ha5,h15
John, Eric, Df42
Johnman, Lewis, If63–64
Johns, Catherine, Cb39
Johnson, Christine, Bb114
Johnson, Christopher, Bb115
Johnson, David, Ne12
Johnson, Graham, If65
Johnson, Howard, Na9
Johnson, J.K., Bc127
Johnson, Keith, Mg11
Johnson, Mary Orr, Hk37
Johnson, Matthew, Hl62
Johnson-South, Ted, Db26
Johnston, Dorothy B., Gi45
Johnston, Henry Butler McKenzie, Hd6
Johnston, J.A., Gg10
Johnstone, Tom, Mh2
Jolly, Margaret, He39
Jones, Alex I., De32
Jones, Andrew, Bb116
Jones, B.L., Bb117
Jones, Bill, Bc200; Ib37
Jones, Charles A., Id50
Jones, Clyve, Fg33; Gb15–16

Jones, Colin, Ba9,c26
Jones, D.R.L., He40
Jones, David J.V., Hc10; Ih19
Jones, Dilwyn, Ca33
Jones, Donald, Bb118
Jones, Dot, Hg49
Jones, E.D., Eg24–25
Jones, Emyr Wyn, Hl63
Jones, G.D.B., Ca34
Jones, G.R., Je12
Jones, Geoffrey, If58; Ne13
Jones, Glanville, Df43
Jones, Greta, Mi9,k3
Jones, Gwen, Ek21
Jones, Gwenith, Ab36
Jones, H.S., Hl64
Jones, Harriet, Ih20
Jones, Ieuan Gwynedd, Hg50
Jones, J.Graham, Hb43; Ib38
Jones, Kathleen, Hh16–17
Jones, Malcolm, Ej27
Jones, Maldwyn A., Fd16
Jones, Martin H., Ei45
Jones, Martin J., Cb27–28
Jones, Michael, Ed8
Jones, Michael John, Gf33
Jones, Michael K., Ea29
Jones, N.W., Jf4
Jones, Peter, Aa98; Bc29; Gf34,i46
Jones, R.H., Bc41; Ef40
Jones, R. Merfyn, Bb119; Ig22
Jones, R.V., Ii30
Jones, Richard F.J., Bc184; Ca17,b29–30
Jones, Rosemary A.N., Hh18
Jones, Stuart, Ne14
Jones, Teuan Guynedd, Ig23
Jones, Thomas, Ji10
Jonsson, Kenneth, Bc139; Dd14–15
Jordan, A.J., Lh4
Jordan, Alison, Mj5
Jordon, T.E., Hg51
Jorgensen, Lise Bender, Bc174
Joshi, Shashi, Nb13
Joyce, Patrick, Hg52
Judge, R., Hd7

Kadish, Sharman, Ib39
Kaelble, Hartmut, Ba94,c34
Kahan, Alan, Hb44
Kahn, Deborah, Bc105; Ej28–29
Kain, Roger J.P., Aa99; Bc47; Hc11
Kaiser, Thomas E., Gf35
Kaiser-Lahme, Angela, Aa100; Id51

Index of Authors

Kalinowski, Lech, Ej30
Kamaryc, R.M., Ab37
Kamminga, Harmke, Bb207
Kane, Paula M., He41
Kanth, Rajani K., Hf55
Karsor, Eugene L., Aa101
Kastovsky, Dieter, De33
Katz, David S., Fe48
Kavanagh, Dennis, Ac58; Ib40–41
Kavanagh, Gaynor, Aa102
Kay, Anne, Id52
Kay, Diana, Ig24
Kean, Hilda, Hb45
Kearney, Anthony, Hi16–17
Kearney, Hugh, Ba95
Kearns, Gerry, Bc55; Hg53,h19
Keath, Michael, Nh24
Keating, Paul, Mi10
Keats-Rohan, K.S.B., Ea30
Keeble, N.H., Fa9
Keeble, S.P., If66
Keen, Richard, Bb120–121
Keen, Rosemary, Aa103
Keevill, Graham D., Ca35; Df44
Keirn, Tim, Bc82; Fb49,c18
Keith-Lucas, Bryan, Ge17
Keller, Josepher, Id53
Kelly, Francis, Bc104
Kelly, James, Mb17,f8
Kelly, John Thomas, Fl15
Kelly, Patrick, Mg12
Kelly, Susan, Dd16
Kemp, Anthony, Ac59; Ii31
Kendall, Calvin B., Bc183; Da14
Kendle, John, Mc1
Kennedy, Dane, Nb14
Kennedy, Paul, Ib42
Kenney, Alison, Aa104
Kenny, Anthony, Hl65
Kent, Christopher A., Ac60; Hl66
Kent, John, Na10,c26
Kent, Timothy, Bb122
Kenward, H.K., Bb84
Kenyon, John, Ac61
Keogh, Dermot, Mb18,c2,g13
Kepos, Paula, Ab38
Ker, N.R., Aa105
Kerkham, Caroline, Bb123
Kerr, Margaret H., Ec19
Kerridge, Eric, Ba96
Kerrigan, Colm, Me7
Kershaw, Stephen E., Fb50
Kettenacker, Lothar, Id54–55

Kettle, Michael, Id56
Key, Michael, Bb124
Keynes, Simon, Db27–28
Kicklighter, Joseph, Ed9
Kiernan, Victor, Ng19
Kiesewetter, Hubert, Hd8
Killen, James, Mf9
Killey, Kate, Fk42
Killingray, David, Bc11; Nc27
Kindleberger, Charles, If67
King, Alec Hyatt, Ab39–40
King, Anthony, Ca36,b31
King, D.J. Francis, Nf26
King, Edmund, Ec20
King, Elspeth, Hb46
King, Heather A., Li6
King, John, Ba97,c136
King, Peter, Gc7
King, Steve, Gg11
King, T.J., Fk43
King, Walter, Fg34
Kingsford, Paul, Ii32
Kingsley, Nicholas, Bb125; Gf36
Kingston, F.T., Mi11
Kinnear, Kate, Bb111
Kinnell, Margaret, Ij50
Kinzer, Bruce L., Hb47–48
Kirby, M.W., Ia25
Kirk, James, Ke12
Kirkbridge, C., Bb25
Kirk-Greene, Anthony Hamilton Millard,
  Na11
Kirk-Smith, Harold, Fd17
Kissock, Jonathan, Jf2,g4
Kitch, Malcolm, Fh8
Kitson, Peter J., Gi47
Kiyoshi, Ikeda, Id57
Kjølbye-Biddle, Birthe, Bb126;Db2
Klausner, David, Bc90
Klein, Benjamin, Fk44
Klein, Lawrence R., Gf37,g12
Kleinman, Paul, If68
Klempere, Bill, Df45
Klingelhöfer, E., Lh5
Klottrup, Alan Coates, He42
Kluge, Bernd, Dd17
Knapp, James, If69
Knapp, Jeremy, Fd18
Knight, Alan, Ac62
Knight, David, Hk38
Knight, Frances, He43
Knight, Ian J., Ng20–23
Knight, Jeremy K., Bb127; Ca37; Jh6–8,i11

258

Knotter, A, Ib43
Knowlden, Patricia E., Fb51
Knowles, James, Fk45
Knowles, Richard, Eh13
Knox, Bruce, Nb15
Knox, R., Bb128
Knox, Thomas, Gb17
Knox, W., Ig25
Knüsel, Christopher, Df46
Kochavi, Arieh J., Id58
Kolinsky, Martin, Bc179; Ib44
Kollar, Rene, He44–45
Kollerstrom, Nick, Gj14
Konvitz, Josef W., Ii33
Korhammer, Michael, Bc222; De34
Kornicki, P.F., Hd9,l67
Kossock, Manfred, Hb49
Kostiner, Joseph, Id59
Kottje, Raymond, Li7
Kramer, Alan, Id60
Kramer, Sigfrid, De35
Krebs, Paula M., Ng24
Kroll, Richard, Bc60; Fe49
Krontiris, Tina, Fk46
Kross, Editha, Hb49
Krueger, Christine L., Hl68
Kruppa, Patricia S., He46
Kruse, S.E., Dd18
Kucharski, Karina, Kf6
Kuklick, Henrika, Ba98
Kunz, Diane, Id61
Kunze, Bonnelyn Young, Bc74; Fe50,k47
Kushner, Tony, Aa106; Ij51
Kusunoki, Akiko, Fe51
Kyle, Avril, Nc28
Kyle, R.G., Ke13

Laband, John, Ng25–27
Laborderie, Olivier de, Ei46
Lacy, Brian, Ba99
Lahme, Rainer, Hd10
Laidlar, John F., Hl69
Laidlaw, Richard B., Ii34
Laine, Michael, Bc123
Laing, Margaret, Ei47
Laing, Stuart, Ba100
Laithwaite, Michael, Hg54
Lake, Peter, Fe52
Lamb, Alastair, Nb16
Lamb, H.H., Ba101
Lamb, R.G., Ke14
Lambert, Frank, Ge27
Lambrick, George, Ca38

Lamoine, Georges, Ba102
Lamphear, John, Nc29
Land, Andrew, Aa107
Landau, Norma, Gg13
Landes, David S., Ba103,c32
Landon, Theodore Luke Giffard, Bd35
Lane, Alan, Bc96; Je3
Lane, Ann J., Id62,109
Lane, Joan, Gj15
Lane, T.W., Bc169
Lang, S.J., Ei48
Langley, Harold D., Gh22
Lanning, George, Gh23
Lapidge, Michael, Bc180; Dc23–25,e35–36
Lapierre, Laurier, Gh24
Lapoint, Elwyn C., Lg3
Large, Peter, Ff14
Larkham, Peter J., Hl70
Larkin, Emmet, Ma3,b19
Larson, John, Ej31
Larsson, Birgitta, Nd9
Lasdun, Susan, Ba104
Lasko, Peter, Ej32
Laslett, Peter, Bd36
Lass, Roger, Ei49
Latham, John, Hk39
Lau, Albert, Nc30
Laughlin, Wayne, Fi2
Law, Jules, Hl71
Law, Robin, Ff15
Lawrence, Anne, Fi14
Lawrence, Christopher, Ac63; Bc24; Hk40; Ij52
Lawrence, Ghislaine, Ba105
Lawrence, Jon, Hb50; If70
Lawson, Jo, Ek20
Lawson, K.R., Ij35
Lawson, M.K., Db29
Lawson, Philip, Gb27,d24
Lawson, Z., Hg55
Lawton, Richard, Ba106
Laybourn, Keith, Ba107
Layman, R.D., Ii35
Lazonick, William, Ba108; If71
Le Fevre, Peter, Fi15
le Patourel, Jean, Ac64
Leach, Douglas Edward, Ba109
Leach, John, Fa16
Leandri, C.V., Ii36
Leather, John, Hl72
Lebas, Elizabeth, Hh20
Ledbury, Chris, Bb129

Index of Authors

Lee, C.H., Ba110
Lee, Colin, Fe53
Leedham-Green, E.S., Ff16
Lees, Michael, Id63
Lefferts, Peter M., Ei50
Legault, Albert, Nb17
Legault, Roch, Gh25
Leicester, Colin R., If21
Leimus, Ivar, Dd19
Leinster-Mackay, Donald, Hi18
Leitch, Roger, Bb130
Lemay, J.A. Leo, Fd19
Lemire, Beverly, Ba111; Gf38
Lendinara, Patrizia, De37
Leneman, Leah, Hc12; Ih21
Lenman, Bruce P., Ga16; Ib45; Kb4
Lentin, Antony, Id64
Leonard, Jim, Ij53
Leontief, Wassily, Ij54
Lepine, David N., Ee46,g26
Leppard, M.J., Ek33
Leslie, Michael, Bc204; Fk48–49
Leslie, R.D., Kc5
Letwin, Shirley Robin, Ib46
Levene, Mark, Id65
Leventhal, F.M., Ib47
Levine, Andrew, Aa177
Levine, David, Bb131
Levine, Joseph M., Fe54,k50–51
Levy, Kenneth, Dc26
Lewis, C.P., Ea31; Gf39
Lewis, C. Roy, Hg11
Lewis, Carenza, Eg27
Lewis, G.B., Ia26
Lewis, James A., Gh26
Lewis, Jane, Ig26,h22–24
Lewis, Neil, Ei51
Lewis, Peter M., Ij55
Lewit, Tamara, Cb32
Liedtke, Rainer, Ih25
Lieven, Dominic, Hg56
Light, Alison, Ij56
Ligou, Daniel, Fb52
Lilley, Jane, Df47
Lilley, Kate, Fk52
Lindert, Peter H., Ba112
Lindley, Keith, Fh9
Lindley, Phillip G., Bc1; Fb35,k53–54
Lindsay, A., Aa161
Lineham, P., Ee68
Ling, Roger, Ca39
Lingham, B.F., Hg57
Linkman, A.E., Hl73

Lippincott, Louise, Gi48
Little, Alan, Hh21
Little, Eddie, Ib48
Little, Walter, Id66
Littlejohn, J.H., Hl74
Livingstone, D.N., Ba113
Llewellyn, Howard, Aa108
Llewellyn-Edwards, Tam, Fe55
Lloyd, Andrew, Aa109
Lloyd, John, If72
Lloyd, Lewis, Nf27
Lloyd, Simon D., Bc112; Eh14–15
Lloyd, Thomas, Hl75
Lloyd, Trevor, Hl76
Lloyd-Morgan, Ceridwen, Hh22
Loades, D.M., Fb53,e56,i16
Lobb, Susan J., Ca9
Locherbie-Cameron, M.A.L., De38
Lock, Stephen, Ba28,c28
Lockett, N.J., Ca51
Lockwood, Glen J., Nf28
Lodge, R.A., Ei52
Loengard, Janet Senderowitz, Ec21
Loft, L., Ge28
Lomas, Michael J., Gi49–50
Longfield-Jones, G.M., Fi17
Longley, David, Ja4
Longuet-Higgins, H. Christopher, Gj16
Lonsdale, John, Nf3
Loock, Hans D., He54
Lopez, Miguel Martinez, Ed4
Lord, Evelyn, Hh23
Lord, Mary Louise, Ei53
Lord, P., Hl77
Loschky, David, Ba114
Loudon, Irvine, Ba115–116; Gj17; Hk41
Loudon, Jean, Hk41
Loughlin, James, Hb51
Louis, Cameron, Fe57,k55
Lovatt, Roger, Ei54
Lovell, John C., Hf56,h24; If73
Lovell, Richard, Ij57
Low, Anthony, Fk56
Lowden, John L., Ii37
Lowe, Jeremy, Hh25
Lowe, Kate, Nc31
Lowe, Lisa, Ba117
Lowe, Michael C., Bb132; Gf40
Lowe, N.F., Gj18
Lowe, Peter, Bc19; Id67
Lowe, Rodney, Aa107
Lowe, William C., Gb18
Lowenthal, D., Ba118

Index of Authors

Lowerson, John, Ba119
Lownie, Andrew, Bc18
Loyn, Henry R., Db30; Ec22
Lucas, Colin, Gd25
Lucas, Ronald Norman, Ca40
Lucíč, Josip, Ed10
Luckin, Bill, Aa61
Luddy, Maria, Ac65
Ludlow, Neil, Bb133
Lüdtke, Hartwig, Ej33
Luff, P.A., Gh27
Lukitz, Liora, Id68
Lund, William R., Fj25
Lunn, Kenneth, Ig27
Lunt, James, Ng28
Lydon, James, Lh6
Lyle, Emily, Lc3
Lynch, Anthony, Le10; Me8
Lyne, Gerard J., Mf10
Lynn, Martin, Aa110; Hf57
Lyon, C.S.C., Dd4
Lyons, Maryinez, Nf29

Mac Aodha, Breandan S., Le11
MacCaffrey, Wallace Trevithie, Fb54
MacCarthy, Robert, Mf11
Maccioni, P.A., Ee47
MacCurtain, Margaret, Ac65
MacDermot, Brian, Me9
MacDonagh, Oliver, Ac66; Mb20; Nh25
Macdonald, Alasdair A., Kb5
Macdonald, Callum A., Id69
Macdonald, Charlotte, Nf30
Macdonald, Graham, Ac67
MacDonald, Michael, Fe58
MacDougall, Ian, Id70
Macfarlane, Leslie J., Ke15
MacGinley, M.E.R., Mg14
Machin, G.I.T., Ie6
Macinnes, L., Cb17
Mackay, David, Nc32
Mackay, Donald, Nf31
MacKay, John, Gh28
Mackenzie, Campbell, Hk6
Mackenzie, David, If74
MacKenzie, John M., Ba120,c10; Na13, g29
MacKenzie, S.P., Ii38
Mackie, Peter, Hi19
Mackreth, D.F., Ca41
Maclear, J.F., Ge29
MacLeod, Christine, Hf58
Macnicol, John, Ih26

MacNiocaill, G., Le12
Macpherson-Grant, Nigel, Ej34
Macquarrie, Alan, Ke16
Macqueen, John, Ke17
Maddern, Philippa C., Eg28
Maddicott, J.R., Dd20
Magdalino, Paul, Bc45
Mager, Olaf, Id71
Magnusson, Lars, Gf41
Magri, Susanna, Hh20
Maguire, Laurie E., Fk57
Maguire, Nancy Klein, Fk58
Mahood, Linda, Hh26
Mainman, Ailsa J., Dd21,f48
Maissonneuve, B., Gh29
Maissonneuve, M., Gh29
Majeed, Javed, Nh26
Malcolm, Joyce Lee, Fb55
Malcolmson, A.P.W., Aa111
Malim, Tim, Df49
Mallinson, Allan, Hj9
Mallmann, Wilhelm E., Hd11
Manchester, A.H., Ab41
Manchester, Keith, Ba121
Mandler, Peter, Ba122; Ia27
Mangan, J.A., He47
Manley, John, Jh9
Manley, Keith A., Ab42
Manley, T.R., If75
Mann, John C., Ca42,b33–34
Mann, Kristin, Nc33
Manners, Jon, Ac34
Manning, Brian, Fb56
Manning, Roger B., Fc19
Mantello, F.A.C., Ei38
Marcombe, David, Ee48
Marglin, Stephen A., Ba123
Marks, Lara, Hg58
Marks, Robert, Ac30
Marland, Hilary, Hk42–43
Marley, David F., Gh30
Marlow, A., Bb89
Marmoy, C.F.A., Fl16
Marriott, J. Stuart, Hi20
Marsden, Ben, Hk44
Marsden, Peter, Bb134; Cb35
Marsh, Joss Lutz, He48
Marshall, Anne, Df50
Marshall, Gary, Df50
Marshall, Gordon, Kf7
Marshall, J.D., Bb135
Marshall, John, Fe59

Marshall, Oliver, Ba124; Ng30
Marshall, P.J., Gh31
Martel, Gordon, Ba125
Martin, Allan, Md1
Martin, David, Ek21
Martin, F.X., Le12
Martin, G.H., Bc109–111
Martin, Ged W., Na14
Martindale, Andrew, Ej35
Martindale, Jane, Ea32,e49
Martinich, A.P., Fj26
Martyn, Isolde, Ee50
Marx, Steven, Fk59
Marzo, Alessandro, Gh32
Mason, Emma, Ec23
Mason, Joan, Hk45
Mason, Tony, Ib49
Masschaele, James, Ef41
Massie, Robert K., Hd12
Mate, Mavis, Eg29
Mather, F.C., Ge30
Mathews, J., Ng26
Mathias, Peter, Bc22; Hf59–60
Mavor, Elizabeth, Gd26
Maxted, Ian, Aa112
Maxwell-Irving, Alastair M.T., Ki7
Mayer, David, Bb136
Mayfield, David, Ac68
Mayhew, Nicholas J., Ef42–43
Maynard, W.B., He49
Mayor, S.H., Fe60
Mayr, Ernst, Hk46
Mayring, Eva A., Bc140; Id72
McAleer, J. Philip, Ej36–37
McAleer, Joseph, Ij58
McAllister, William, Bb137
McAteer, William, Na15
McBride, W.M., Gj19
McBryde, William W., Kc6
McCalman, I.D., He50
McCann, Timothy J., He51
McCann, W.J., Ja5
McCarthy, M.R., Ej38,k65
McClelland, Keith, Hg59
McClelland, V. Alan, He52
McCloskey, Donald N., Aa113
McConkey, Kenneth, Hl78
McCoog, Thomas M., Fe61
McCorry, Helen C., Gh33–35
McCracken, Donal P., Nf32
McCracken, John,, Nc35
McCray Beier, Lucinda, Fl17
McCrea, Adriana, Fk60

McCrone, D., Ig28
McCulloch, Michael, Nc36
McDonald, Ian, Nh27
McDonald, John, Hf61
McDonnell, John, Ef44
McDonough, Frank, Ib50
McDowell, William H., Ij59
McElligott, Ignatius, He53
McEntee, Ann Marie, Fj27
McGann, Jerome, Hl79
McGibbon, Ian, Ng31
McGrade, Arthur Stephen, Ei55
McGrail, Sean, Lf5
McGregor, Pat, Mf12
McGuckin, Ann, Ig29
McGurk, John, Mi12
McGurk, Patrick, Aa73
McHardy, Alison K., Eb4,e51,g30
McHugh, Paul G., Nc37
McIvor, Arthur J., Bb61,c175; Ib51,g30
McKay, A.C., Ne16
McKay, Peter K., Gi51
McKean, Charles, Kg5
McKee, J., Le13
McKercher, B.J.C., Id73–74
McKim, Donald K., Ke12
McKinlay, Alan, Aa114
McKinley, Richard, Ei56
McKitterick, David, Fl18
McLaren, A., Ba126
McLaren, Mary-Rose, Ei57
McLaughlin, Eugene, Nc31
McLaughlin, Pat, Hc4
McLean, David, Nb18
McLean, Ian, Hb52
McLean, Marianne, Nf33–34
McLelland, D., Aa115
McLeod, C., Ba127
McLeod, Hugh, He54–55
McMahon, Marie P., Gb19
McMahon, Mary, Lg4
McMurry, Sally, Hf62
McNair, Philip, Fe62
McNamee, Colm, Eh16
McNeil, Maureen, Ij60
McNeill, Tom E., Ek34; Lh7
McNiven, Peter, Aa116
McNulty, J. Bard, Ej39
McPherson, A., Ij61
McRae, Andrew, Ff17
McRee, Ben R., Eg18,31
McShane, Harry, Ig31
McVeigh, C., If76

Index of Authors

Mead, Helen, Ba128
Meale, Carol M., Ei9
Meaney, Audrey L., Dc27,f51
Meaney, N., Nb19
Means, Laurel, Ei58
Medlycott, Mervyn, Aa74
Medoff, Jeslyn, Fk61
Mee, Jon, Gi52
Meenan, Rosanne, Lf6
Meers, Sharon I., Id75
Meeson, Robert, Bc158
Meikle, Maureen M., Fi18–19
Mein, Margaret, Ia28
Melissen, Jan, Id76
Mellar, Harvey, Ac69
Melling, Joseph, Ba10,c217; Hg60,h27
Mellor, Adrian, Ij62
Melman, Billie, Ba129
Menard, Russell R., Ba130
Mendelsohn, J. Andrew, Fl19
Mendelsohn, Richard, Ne17
Mendle, Michael, Fc20
Mennell, Stephen, Ba131
Mentgen, G., Ei59
Mercer, Helen, Bc131; Ib52,f77–78
Meredith, David, Ne7
Meredith, Peter, Ei60
Merrell, James H., Fd20
Merskey, Harold, Ii39
Messerschmidt, Manfred, Ii40
Messinger, Gary S., Ib53
Metcalf, D.M., Dd22–23
Metcalf, Vivienne M., Bc186; Ek35
Metcalfe, John F., Hk47
Metzger-Court, Sarah, Id77
Mews, Stuart, Ie7
Meyer, Arline, Gi53
Meyer, W.R., Aa117; Ib54,h27
Meyers, David W., Kc7
Michie, Ranald, C., Bb138
Middleton, Roger, Aa51,c70
Middleton-Stewart, J., Ei61
Midgley, Clare, Ba132
Milburn, Geoffrey E., Ge31,g14
Miles, B.E., Hk48
Miles, Robert, Ig24
Miles, Trevor J., Jg5
Milford, Anna, Fi20
Milhous, J., Gi54
Milis, Ludo J.R., Ee52
Millar, Mary S., Hb53
Miller, C., Hf63
Miller, David Carey, Bc128; Kc8

Miller, George C., Bb139
Miller, Keith, Bb66
Miller, Leta E., Fl20
Miller, Margaret, Hf64
Miller, Mervyn, Bb140
Miller, Norman, Hf64
Millet, B., Me10
Milliss, Roger, Nf35
Mills, A.D., Ab43
Mills, Dennis, Gg22
Mills, Frederick V., Ge32
Milne, Gustav, Bc166–167,172; Ek36
Milner, N.E., If79
Milroy, James, Ei62
Milsom, John, Fk62
Minay, Priscilla, Gi55
Miners, J., Nb20
Mingay, Gordon, Gi56
Mini, Peter V., If80
Mirowski, Philip, Hf46
Misra, B.B., Nc38
Mitch, David F., Hj10
Mitchell, A.M., Ke18
Mitchell, Dennis J., Hb54
Mitchell, L.G., Gb20
Mitchell, L.M., Aa118
Moberg, Ove, Db31
Modigliani, Franco, Gf42
Moggridge, Donald Edward, If81
Mohr, Peter D., Hh28
Moir, Martin, Nh35
Moir, Zawahir, Nh35
Molleson, Theya, Cb36
Molteni de Villermont, Claude, Hf65
Molyneux, Geoffrey, Na16
Mommsen, W.J., Na17
Monfasani, J., Fg43
Moody, David, Ab44; Gg15
Mooers, C., Ba133
Moore, B., Ej40
Moore, Bob, Id78
Moore, Christopher, Nf36
Moore, D.T., Ek65
Moore, J., Hb55
Moore, Jerrold Northrop, Ij63
Moore, John S., Eg32
Moore, Katherine, Nh28
Moore, Kevin, Hb56
Moore, Lindy, Bb141
Moore, Pam, Bb142
Moore, Susan, Bb143
Moore-Colyer, Richard J., Bb144; Gf43; Hl80

Index of Authors

Moorehead, Caroline, Ba134
Moorhouse, Stephen, Ek37
Moorwood, R.D., Cb37
Moradiellos, Enrique, Id79
Moran, Madge, Ek38
Moran, Peter A., Bb145; Ke19
More, Charles, Bb146
Morehen, John, Fk63
Moreman, T.R., Ng32
Moreton, C.E., Eg33
Morewood, Steven, Id80
Morgan, Austen, Ib55
Morgan, Carol E., Hg61
Morgan, D. Densil, Ge33–34
Morgan, David, Ia29
Morgan, Derec Llwyd, Ge35
Morgan, Gerald, Gf44
Morgan, Huw, Ia6
Morgan, Kenneth, Ga17,f45
Morgan, Kenneth O., Bb147; Ib56
Morgan, Kevin, Aa119
Morgan, Nicholas, Aa120
Morgan, Paul, Aa121; Gi57
Morgan, Philip D., Bc2; Fa1,d21
Morgan, Prys, Jg6
Morgan, Raine, Aa122
Morgan, Roy, Bb148
Morgan, Sharon, Nf37
Morgan, W.T.W., Ne18
Moriarty, Catherine, Ij64
Morland, S.G., Bc120
Morrill, John S., Bc79; Fg35–36
Morris, Bernard, Ia30
Morris, Christopher J., Ea33; Ia31
Morris, David S., Id81
Morris, Frankie, Hb57
Morris, G.C.R., Fl21
Morris, Jeremy Noah, He56–57
Morris, Margaret, If82
Morris, P.J.T., Ba30,c36
Morris, Peter, Hj11
Morris, R.J., Ac71; Ha6,f66
Morris, Richard K., Ej41
Morris, V.J., Aa123,124
Morrison, Ian A., Kh7
Morrison, Kathryn, Ej42
Morriss, Roger, Hj12
Morrissey, Thomas E., Ee53
Mortimer, Cath, Dd24
Mortimer, Jean E., Fe63; Gi58
Morton, Alan D., Bc146; Df52–53
Morton, Desmond, Ng33
Morton, Jane, Hh29

Morus, Iwan Rhys, Hk49–50
Moshe, Tuvia Ben, Ii41
Mosothoane, Ephraim, Nd10
Moss, David J., Hf67,l81
Moss, P., Ek17
Mougel, François-Charles, If83
Moure, Kenneth, If84
Muckle, James, Bb64
Muirhead, B.W., Ia32; Ne19
Mukonoweshuro, Eliphas G., Nf38
Muller, Andre, Ne14
Muller, Rainer A., Hl2
Mullett, Michael, Aa125
Mullins, Patrick, Na18
Munck, Ronnie, Mb21
Mungazi, Dickson A., Nc39
Munn, Charles W., Ab45
Munro, Jane A., Hl82
Munro, Jean, Kc9
Munro, John H.A., Bc102
Munsell, F. Darrell, Hl83
Munster, M.B. zu, Ee54
Murdoch, Alexander J., Gb21,h36
Murdoch, Norman H., Hc13,e58
Murdoch, Tessa, Fe64
Murfet, G.J., Bb149
Murfin, A., If52
Murphy, A.E., Gf46
Murphy, Elinor, Dc28
Murphy, M., Ig32
Murphy, Margaret, Ef14
Murphy, Martin, Fe65
Murphy, Michael, Ab46; Bb150
Murphy, Peter, Ba135
Murray, Bruce K., Hb58
Murray, Charles, Kj2
Murray, Douglas M., Ke20
Murray, Hugh, Ac72; Ej43
Murray, James, Le14
Murray, Mary Charles, Ej44
Murray, Peter, Ij65
Murray, Williamson, Ii42
Musambachime, M.C., Nb21
Musset, Lucien, Ej45
Musto, R.G., Fg43
Myat-Price, E.M., Aa126
Myerly, Scott Hughes, Hl84
Mynors, R.A.B., Da5
Mytum, Harold, Bc210

Naish, John, Gh37
Nannestad, Eleanor, Hi21
Napier, C.J., If85

264

Napier, Priscilla, Ne20
Nardinelli, Clark, Hh30
Nash, David S., Hb59
Nash, Gerallt D., Bb151; Hi22
Nasson, Bill, Ng34
Neal, Larry, Gf47
Neale, Frances, Eg22
Neave, David, Bb152
Neaverson, Peter, Bb156
Nederman, Cary J., Ei63
Neillands, Robin, Ea34
Neilson, Keith, Hd13
Nekkers, Jan, Bc37
Nelson, Byron, Fj28
Nelson, Janet L., Bc213; Dd2; Fh10
Nelson, Richard, Ij66
Nenadic, Stana, Aa127
Nenk, Beverley, Ej46
Nenner, Howard, Fb57
Neuman de Vegvar, Carol L., Bc97
Nêve, Paul, Kc10
Neville, C.J., Eh17
Nevo, Joseph, Id82
Newall, Christopher, Hl85
Newbold, P., Ba136
Newbury, Colin, Na19
Newell, J.Q.C., Ii43
Newhouse, Neville H., Ge36
Newman, Caron, Ef45
Newman, J., Df54
Newman, John, Fg37
Newman, Keith A., Fe66
Newman, Stephen L., Fj29
Newsome, David, He59
Newsome, Eileen Betty, Aa128
Newton, Douglas, Id83
Newton, Judith, Hg62
Newton, Sam, Db32
Nicholas, Jacqueline M., Hg63
Nicholas, Stephen J., Hg63
Nicholls, David, Hd14
Nicholls, M.K., He60
Nicholls, Mark, Ff18
Nichols, Elizabeth, Bc192
Nicholson, Helen, Ei64
Nicholson, Margaria Hope, Fk41
Nicholson, Timothy Robin, Ba137
Nicol, Alexandra, Aa129
Nicolson, Malcolm, Gj20
Niedhard, Goltfried, Bc176
Niles, John D., Df55
Nilson, Bengt, Id84
Nish, Ian, Hd15

Nixon, Howard Millar, Ab47
Nixon, Malcolm I., Aa130
Noakes, Jeremy, Bc188; Ii44
Noble, M., Ij28
Noble, Peter, Ei65
Noel, E.B., Ba138
Nolan, John S., Fi21
Nolan, William, Lf2; Mg15
Noone, Timothy B., Ei66
Noonkester, Myron C., Ff19
Norris, Malcolm, Ej47
Norris, Pippa, Ib57
North, Douglass C., Ba139
Norton, Philip, Ib58
Norton, Rictor, Ba140
Norton, Wayne, Hh31
Nuding, Gertrude Prescott, Hl86
Nugent, Neill, Ib59
Nunokawa, Jeff, Hl87
Nuttall, Geoffrey F., Fe67
Nye, John Vincent, Id85

Ó Catháin, Séamus, Aa131
Ó Concheanainn, Tomás, Lg5
Ó Conchubhair, Fearghail, Mj6
Ó Cúiv, Brian, Li8
Ó Dalaigh, B., Mj7
Ó Gráda, Cormac, Mf13–14
Ó Meadhra, Uaínínn, Li9
Oakeshott, Ewart, Eh18
Oakley, Anne M., Bb47
Oates, John C.T., Bc88
Oberdorfer, L., Id86
O'Brien, A.F., Lf7
O'Brien, C., Ef46
O'Brien, Conor Cruise, Mi13
O'Brien, Elizabeth, Lg6
O'Brien, John, Bc182; Nc40
O'Brien, O., Bd37
O'Brien, Patrick K., Gf24,j21; Hf68
O'Brien, Susan, He61
O'Connell, Sheila, Fg38
O'Connor, Bernard, Hf69
O'Connor, Emmet, Mb22
O'Conor, Kieran, Lh8–9
O'Day, Alan, Aa132
O'Day, Rosemary, Fe68
Oddy, W.A., Dd31
O'Doherty, Bríd, Ld3
O'Donnell, E.E., Aa133
O'Donnell, Ruan, Mb23
O'Dowd, Mary, Ac65; Lg7
Offen, Karen, Ac73; Bc8

Index of Authors

Oggins, Robin S., Ef47
Oggins, Virginia Darrow, Ef47
O'Gorman, Christopher, He62
O'Gorman, Frank, Gb22–23
Ohadike, Don C., Nb22
O'Hara, Diana, Fg39
Ohlmeyer, Jane H., Mb24
Okasha, Elisabeth, De39
O'Keefee, Katherine O'Brien, De40
O'Keeffe, Tadhg, Lh10,i10
Oldfield, Geoffrey, Hb60
Oldfield, J.R., Gb24
O'Leary, Paul, Aa79
Oliver, R.C.B., Ha7
Oliver, Richard, Hc11,14
Olson, Alison G., Ga18–19
Olson, Sherri, Eg34
Omenka, Nicholas Ibeawuchi, Nd11
Omissi, David Enrico, Id87,f86
O'Neill, Assumpta, Le15–17
O'Neill, Robert, Ac74; Bc150; Ii45
Oram, Richard D., Ke21
O'Reilly, Jennifer, De41
Orme, Nicholas I., Ba141,c148;
    Dc29–30; Ee55–56,j48; Ie8
Orr, Bridget, Fg40
Ortenberg, Veronica, Dc31,f56
Orton, D.J., Aa123–124
Orton, Michael, Ba142
Osborough, W.N., Le18
Osler, Margaret J., Fl22
Osterhammel, Jürgen, Ac75
O'Sullivan, Deirdre, Da15
O'Sullivan, Harold, Lh11
Ottaway, Patrick J., Ba143,c157;
    Dd25
Ouen-John, Henry, Ac76
Ousby, Ian, Ga20
Outhwaite, R.B., Ba144; Gg16
Outram, Q., If30
Ovenden, K., Hl88
Overy, Richard J., Ii46
Owen, A.E.B., Eg35
Owen, Nicholas, Id88
Owen, Sybil, Je13
Owen, Tim, Ab48
Oxley, John, Bb153

Packer, Brian, Bb154
Paddock, J., Ca3
Padel, O.J., Ab51
Pagan, Hugh, Dd26
Page, R.I., Fk64

Page, Stephen J., Hg64
Paisey, David, Aa134
Palliser, David M., Df57; Fa10
Palmer, Bryan, Ac77
Palmer, J.J.N., Aa63,b22–23,c43
Palmer, Marilyn, Bb155–156
Palmer, William, Lb1
Palmer-Brown, C.P.H., Ca18
Pam, David, Hg65
Pandey, Gyanendra, Nb23
Pankhurst, Richard, Bc50; Ib60
Panteli, Stavros, Na20
Panton, F.H., Gc8
Pardoe, Jon, Hj13
Parfitt, Keith, Eg36
Paris, Michael, Hj14
Parisse, Michel, Dc1
Parish, Debra L., Fe69
Park, Katharine, Ba145
Parker, M., Bb157
Parker, M.S., Db33
Parkhill, Trevor, Mg16
Parkinson, Anthony, Hf70
Parkinson, John A., Hl89
Parnell, Geoffrey, Bb158
Parr, Anthony, Fk65
Parry, Ann, Hl90
Parry, Graham, Fj30,l23
Parry-Jones, E., Je13
Parsons, David, Ej49
Parsons, John Carmi, Ec24
Parsons, Maggy, Hg66
Partridge, Colin, Bb159
Partridge, Michael, Aa80
Pasqualucci, Paulo, Fj31
Paterson, Hugh, Ib51
Patterson, Annabel, Fc21
Patterson, Catherine F., Fb58
Patterson, Lee, Ei67
Patterson, Robert B., Ea35,i68
Pattison, Robert, He63
Paulmann, Johannes, Ib61
Paxton, Frederick, Dc32
Payling, S.J., Eg37
Pearce, Malcolm L., Ba146
Pearce, R.D., Hi23
Pears, Iain, Hb61
Pearsall, Derek, Ei69–70
Pearson, Jane, Bb160
Pearson, M.N., Ba147
Pearson, Robin O., Gf48; Hf71
Peaty, Ian P., Bb161
Peck, Linda Levy, Fg41

Index of Authors

Peddie, John, Gh38
Pedersen, Susan, Nh29
Pelaez, Manuel J., Ba141
Pelly, M.E., Id89
Peltonen, Markku, Fj32
Pemberton, Gregory, Nb6
Penn, Nigel, Na1
Penn, Simon A.C., Ef48
Pennell, C.R., Fd22
Penny, D. Andrew, Fe70
Pepper, Sarah, Hk51
Percy, Clayre, Ib69
Pereiro, James, He64
Peretz, Liz, Ih28
Perkins, D.R.J., Df58
Perks, Rob, Aa135–136
Perren, Richard, Hf72
Perrin, Fernand, Hk9
Perry, C.R., Hc15
Persson, Karl Gunnar, Ac78
Peteri, Gyorgy, If87
Petersen, Tore Tingvold, Id90
Peterson, J.W.M., Cb38
Petersson, H. Bertil A., Dd27
Pettegree, Andrew, Ac79
Petzold, Ulrich, Aa67
Pfaff, Richard W., Ee57–59
Pfeiffer, Rolf, Ng35
Phibbs, J.L., Ba148
Philip, Alan Butt, Ib62
Phillips, C.B., Fh11
Phillips, Catherine, Ij67
Phillips, Gordon, Ib63
Phillips, John A., Hb62–63
Phillips, Martin, Hf73
Phillips, Walter W., Nd12
Phillipson, Nicholas T., Gc9
Philp, Brian J., Ca43
Philp, Mark, Gb25,i59
Phythian-Adams, Charles, Ac80; Ba149
Pickering, Oliver, Aa137
Pickering, Paul A., Hb64
Pickford, Christopher J., Ab49
Pickstone, John V., Hk52
Pierrepont, J.R., Ee60
Pierson, Ruth Roach, Ac73; Bc8
Pieterse, Jan P. Nederveen, Nh30
Pigman, Geoffrey Allen, Hf68
Pilbeam, Elaine, Ab48
Pilhorget, Rene, Md2
Pimlott, Ben, Ib64
Pincus, Steven C.A., Fb59,d23
Pines, Malcolm, Ij68

Pinto, David, Fk66
Piper, Alan J., Aa105; Ef49
Piva, Michael J., Ne21
Plaisse, Andre, Eh19
Plank, Frans, Gf49
Pleuger, Gilbert, Gc10
Plyley, Michael J., Ec19
Pocock, J.G.A., Ac81; Fb60
Podmore, Colin J., Ge37
Pollard, A.J., Ea36
Pollard, Sidney, Ba150; Gf50; Hf74; Ib65
Pollins, Harold, Hf75
Pollock, David C., Ke11
Ponko, Vincent, Aa138
Ponsford, M.W., Bc41
Pooley, Colin G., Ba106; Hc16
Porta, Pier Luigi, Hl91
Porter, Andrew N., Nd13
Porter, Dorothy, Hk26
Porter, J.H., Hc17
Porter, Roy, Ac82; Ba28,151–154,c28, 190,221; Fa14; Ga21,j22
Porter, Thomas W., Hl92
Postle, Martin, Gi60
Postles, David A., Ef50–52,g38–40
Potter, Geoff, Ek39,40
Potter, Karen, Ib66
Potter, Lois, Fk67
Potter, Timothy W., Cb39
Pottinger, George, Nc41
Potts, Cassandra, Ea37
Pounds, N.J.G., Bb162
Poussou, J.-P, Fd24; Ga22
Powell, Brian, Hl93
Powell, Christopher, Gc14; Hh32
Powell, David, Ba155
Powell, Dilys, Bb163
Powell, Frederick W., Mb25
Powell, Geoffrey, Ba156
Powell, Joyce, Bb164
Powell, Martin, Ih29
Powell, Timothy E., Le19
Powell, W.R., Hb65
Power, Jan, La9
Power, John P., Ef53
Power, M.J., Hf76
Power, Rosemary, Ld4
Prakash, Gyan, Ne22
Prasch, R.E., Gf51
Pratt, Hubert T., Gj23
Pratt, Mary Louise, Nh31
Prescott, Elizabeth, Eg41

Index of Authors

Presley, Cora Ann, Nf39
Press, Jon, Aa83
Prest, Wilfred, Fb61
Preston, Jill, Ic7
Preston-Jones, Ann, Df59
Prestwich, J.O., Eh20
Prestwich, Michael, Ea38,h21
Prete, Roy A., Ii47
Price, B.J., If88
Price, Clifford, Ej50
Price, David Trevor William, Ie9
Price, Jacob M., Bd38; Ga23
Priddy, Deborah, Bb165
Priestley, Ursula, Bb166
Prior, Robin, Ii48
Pritchard, Allan, Ng36
Pritchard, Frances, Bc53; Ej14
Probert, Henry A., Ii49
Prochaska, F.K., Bb167
Procter, M.R., Ab50
Procter, Timothy, Aa169
Proctor, Ray, Bb168
Pronay, Nicholas, Ij69
Proud, Edward Baxby, Ia33
Proudfoot, Lindsay J., Mf5
Proudlock, Noel, Hf77
Pryce, Huw, Je14,15
Pugh, Martin, Ib67,g33
Pugh, Patricia, Aa139
Pugh, T.B., Fb62
Pugsley, Steven, Fh12
Purcell, Mary, Aa140; Le20
Purkiss, Diane, Bc39; Fe71,g42
Purser, John, Ki8
Purvis, Martin, Hf78
Pybus, Cassandra, Nf40

Quinault, Roland, Hb66,c18
Quine, Gillian, Kg6
Quiney, Anthony, He65
Quinn, David B., Ed11
Quinn, Dermot A., Aa141; He66
Quinn, John F., Me11

Rack, Henry D., Ba157; Ge38
Rady, Jonathan, Ek41
Rahtz, Philip, Bc158,164; Df60
Ralegh Radford, C.A., Eh22
Ralph, Elizabeth, Ff20
Ramsay, Nigel, Bc3; Dc33; Ej51
Ramsbottom, John D., Fe72
Ramusack, Barbara, Nh32
Rands, Susan, Da16

Ranft, Bryan, Hj15
Ranger, Terence, Bc25
Rankin, Susan, Ei71
Raphael, D.D., Gf52
Rapp, Dean, Ij70
Rappaport, S., Fg43
Rasor, Eugene L., Aa142
Rataboul, Louis J., Ge39
Rathbone, Richard J.A.R., Nc42–43
Raugh, Harold E., Ii50
Raven, James, Gf53
Ravenhill, William L.D., Ab51; Fk68; Ga24
Raw, Barbara C., Db34,e42–43
Rawding, Charles, Hg67
Rawes, Bernard, Ca44
Rawley, James A., Ga25
Rawlinson, George, Ib68
Ray, Anthony, Ej52
Ray, Michael, Hi24
Ray, Rajat Kanta, Ne23
Raylor, Timothy, Bc204; Fj33,k49
Raymond, Stuart, Aa143–147
Read, Donald, Hf79
Reddy, William M., Ac83
Redknap, Mark, Bc191; Jf3,i4
Reece, Bob, Bc187; Mg17–20,i14–15; Nh33
Reece, Richard, Cb40–41
Reed, James, Ei72
Reed, Michael, Hg68
Reed, Peter, Hf80
Reeder, David A., Ac84
Reedy, Gerard, Fe73
Rees, David, Jg7
Rees, Gareth, Nc44
Rees, Sian, Ja6
Reesor, Bayard William, Nc45
Reeves, A. Compton, Eg42
Reeves, Nicholas, Ij71
Reeves, Peter Dennis, Nc46
Reichl, Karl, De44
Reid, Alastair, Hg69–70
Reid, Christopher, Gi61
Reid, John G., Kd1
Reilly, John W., Bb169
Reinharz, Jehuda, Ia34,d91–92
Rendall, Jane, Ac73,85; Bc8
Rendel, Rosemary, Aa148
Renfrew, Jane M., Fk69
Renvoize, Edward, Hg71
Resnick, David, Fk70
Reusch, Ulrich, Id93

268

Index of Authors

Reuter, Timothy, Bc71
Rex, Richard, Fe74–75
Reynolds, Susan, Ac86; Db35
Rhodes, Dennis E., Aa149–150; Fd25
Rhodes, R.A.W., Ic8
Rice, Dorothy, Bd39
Rice, Geoffrey W., Gd27
Rich, J., Ha8
Rich, Paul B., Nh34
Richards, Eric, Ga26; Mg21
Richards, Jeffrey, Ij72
Richards, Julian D., Bc172; De45
Richards, M.P., Ei73
Richards, R., Gh39
Richards, Sandra, Ba158
Richardson, Christine, Gf54
Richardson, David, Gf55
Richardson, F.C., Ii51
Richardson, Hilary, Li11
Richardson, Malcolm, Ec25
Richardson, R.C., Bc107; Fh13
Richardson, Ruth, Hk53
Richardson, William A.R., Ab52
Richelson, Jeffery T., Nb24
Richer, A.F., Hc19
Richetti, J., Gi62
Richey, William, Gi63
Richmond, Colin, Ea39,b5–6,e61,g43
Rickard, John, Mg22
Riddy, Felicity, Bc58
Riden, Philip, Bb170
Ridgeway, Huw, Eg44
Ridgway, Maurice H., Fe76
Ridley, F.F., Ic9
Ridley, Jane, Hb67; Ib69
Riley, D., Fh14
Risse, Guenter B., Ab53; Gj24
Ritchie, Daniel E., Gi64
Ritchie, L.A., Ab54
Ritvo, Harriet, Gi65
Rizvi, Gowher, Nb25
Roach, Susan, Bc133
Robb, George, Ba159
Robbins, Christopher A., Fb63
Robbins, Keith, Ib70
Robbins, Michael, Ac87; Fk69
Robert, Jean-Louis, If70
Roberts, Alasdair F.B., Bb171; He67;
  Ke22–23
Roberts, Brian, If89
Roberts, D. Hywel E., Aa94
Roberts, Edward, Bb172
Roberts, H.L., Bc36

Roberts, J., De46
Roberts, Phyllis B., Aa151
Roberts, R. Frederick, Hf81
Roberts, R.J., Fj34
Roberts, Richard, Ba160; If90; Nc33
Roberts, Stephen, Fe77; Hb68
Roberts, Tomos, Je16
Robertson, Alex, Hi25
Robertson, Paul L., Hf82
Robertson, Una A., Fc22
Robins, Joseph, Mg23
Robins, Kevin, Ij73
Robins, Lynton J., Bc200; Id94
Robins, Roger, Ge40
Robinson, Andrew, Ge41
Robinson, Austin, If91
Robinson, Fred C., De47
Robinson, Ken, Bc59; Fg44,k15
Robinson, Mark, Ab55
Robinson, Patrick, Ii68
Robinson, Paul, Df61
Robinson, W.R.B., Fb64
Robinson, Wendy, Hi26
Robson, Ann P., Hl94
Robson, Brian, Ng37
Robson, John M., Nh35
Robson, Lloyd, Na21
Robson, M., Ee62
Rockley, Pete, If92
Roderick, Gordon W., Hi27–29; Ih30
Rodger, Nicholas A.M., Gh40–41
Rodger, Richard, Ac88; Hf83
Rodwell, K.A., Bb173
Roeber, A.G., Fd26
Roesdahl, Else, Bc159; Df62
Roffe, David, Ef54,k42
Rogal, Samuel J., Aa152
Rogers, David, Aa153
Rogers, G.A.J., Fj35
Rogers, Nicholas, Ac89; Ee63; Gc11–12
Rogers, T.D., Aa47
Rogow, Arnold A., Fj36
Rollason, David, Dc34; Ee64
Rollin, Henry R., Hk54
Rollings, Neil, Bc131; Ib52,71
Rollison, David, Bb174
Rooth, Tim, If93
Roots, Ivan, Fb65
Roper, Michael, Ba161,c223; Ig34
Rose, J.K.H., Bb175
Rose, Jonathan, Ba162
Rose, Louise, Aa154
Rose, Sonya O., Hg72

Index of Authors

Rosen, Frederick, Hb69
Rosen, Ulla, Ba163
Rosenberg, Eugene, Ij74
Rosenberg, Nathan, Ij75
Rösener, Werner, Eg45
Rosenthal, Jane, De48
Rosenthal, Joel T., Bc177; De49; Eg46
Rosie, George, Ij76
Roskams, Steve P., Bc214; Cb42
Rosner, Lisa, Gj25
Rosovsky, Henry, Bc32
Ross, Eric, Nc47
Ross, M.S., Hf84
Ross, Seamus, Ac90
Rossberg, Horst, Ii52
Rosser, Gervase, Dc35
Rothschild, Emma, Gf56
Rothwell, Victor, Ib72
Rothwell, W., Ei74
Rouse, Mary A., Aa155
Rouse, Richard H., Aa155
Rowe, J., Bb176
Rowe, Margaret, Bc218
Rowe, Margery, Aa154; Fk68
Rowe, Violet, Hb70
Rowell, Geoffrey, He68
Rowland, A.B., Bb177
Rowland, Joan G., Nf41
Rowse, A.L., Bd42
Roxburgh, Bob, If94
Roy, Ian, Fh15,i22
Royden, Michael W., Bb178
Royle, Edward, He69,g73
Ruane, Joseph, Ba164
Rubin, Gerry R., Ii53
Rubin, Miri, Bc70; Ee65–67,i75
Rubinstein, W.D., Hf85,g74–75
Ruckley, N., Kh8
Rudling, David, Bb179
Rudoe, Judy, Aa156
Ruggles, Richard I., Ba165
Ruggles, Stephen, Ba166
Rule, John C., Fd27; Ga27–28,f57; Hg76
Rumbold, Valerie, Fb66
Russell, Alice, Hb71
Russell, Dave, Hj16
Russell, Georgina, Ej53
Russell, Pamela B., Df63
Russell-Gebbett, Jean P., Hi30
Russett, V.E.J., Ef55
Ruston, Alan, Aa157; He70–71
Rutherford, Susan, Bc63; Hg77
Ryan, Alan, Hb72

Ryan, James G., Aa158
Ryan, Michael, Lh12
Ryder, M.L., Ki9
Ryder, Peter F., Eh23,j54,k43
Rynne, Colin, Lf8

Saaler, Mary, Ef56
Sacks, David Harris, Fb67,e78,f21
Sakata, Toshio, Ef57
Salée, Daniel, Nc4
Salgado, Gamini, Fg45
Salisbury, C.R., Ef58
Salisbury, Joyce E., Ab56
Salt, Denis, Ba167
Salter, John, Gi66
Salter, Mike, Jh10
Sambrook, James, Gi67
Sample, Paul, Bb180
Sampson, Anthony, Ia35
Sampson, K.J., Aa118
Samson, Ross, Ba168,c38; Cb43;
    Ki10–11
Samuel, Raphael, Ij77
Samuels, Michael L., Ei76
Samuelson, Paul A., Gf58
Sander, William, Ig35
Sanders, Richard, Ac91
Sanderson, Margaret H.B., Gi68
Sandred, K.I., Df64
Sanger, Chesley W., Ba169
Santoro, Verio, Da17
Sapiro, Virginia, Gi69
Sapsford, D., If114
Sargent, Mark L., Fb68
Sauer, Hans, De50
Sauerländer, Willibald, Ej55
Saul, Nigel, Bc46; Ei77–78
Saunders, Andrew, Ek44
Saunders, Ann, Bb110
Saunders, Gail, Na4
Saunders, Graham, Nd14
Saunders, Tom, Ek45
Savage, Hugh, Ac92
Savage, Stephen P., Ic10
Saville, John, Ab8; Ia36,b73
Sawers, Larry, Gf59
Sawyer, Malcolm, Ab4
Saxby, Richard, Gh42
Sayce, Roger B., Ij78
Sayers, Jane E., Ee68
Sayers, William, Lf9
Scaife, Robert G., Ba135
Scarey, John, Li11

Index of Authors

Scase, Wendy, Ei79
Scattergood, John, Fk71
Schanze, Erich, Kc11
Scharer, Anton, Dc36
Schiavone, Guiseppe, Fj37
Schmidt, Roger, Gi70
Schneiders, M., Li12
Schochet, Gordon J., Fe79,j38
Schoeman, Karel, Nh36
Schoenfeld, Edward, Dd28
Schofield, John, Bc165; Eg47
Schofield, M.M., Gf55
Schofield, Philip, Gd28
Schreuder, Deryck, Nf42
Schreuder, M. Wilhelmina H., Aa159
Schulenburg, Jane Tibbetts, De51
Schulman, Jane, Dd28
Schultz, Theodore W., Gf60
Schulze, Rainer, Id95
Schürer, Kevin, Ac93–94
Schwarz, Bill, Ia37,b74
Schwarz, L.D., Bb181
Schweizer, Karl W., Gb26–27
Schwengel, Hermann, Ij13
Schwieso, Joshua J., He72
Schwoerer, Lois G., Bc48; Fb69–70
Scola, Roger, Bb182
Scotland, N.A.D., He73
Scotland, Nigel, Bb183
Scott, Andrew Murray, Fb71
Scott, David, Fe80,h16; Nd15
Scott, Eleanor, Cb44
Scott, Geoffrey, Ge42
Scott, H.M., Gd29
Scott, Jonathan, Fj39
Scott, Miriam, Bb184
Scott, P.H., Kb6
Scourfield, E., Bb185
Scragg, Donald G., Dc37,e52–53
Screen, J.O.E., Gh43
Scull, Christopher J., Db36,f65
Searle, G.R., Hb73
Seaver, Paul S., Fg46
Seccombe, Wally, Hg78
Seebold, Elmar, Da18
Seed, John, Gg17
Seed, Patricia, Fd28
Seeliger, Sylvia, If28
Sefton, Henry R., Ge43; Ke24
Segars, Terry, Ig36
Seidman, Steven, Hl95
Seipp, David J., Fc23
Sekules, Veronica, Ei80,j56

Self, Robert, Ib75
Sell, Alan P.F., Fe81; Hl96
Sellar, W.D.H., Ba170; Kc12
Sen, Sunanda, Nb26
Seymour, Miranda, Ij79
Seymour-Jones, Carole, Hb74
Shannon, Richard, Hb75
Shapin, S., Fg47
Shapiro, Alan E., Ac95; Bc87
Shapiro, B., Fj40
Sharland, Elaine, Ih35
Sharman, Frank, Hc9
Sharp, Buchanan, Fh17
Sharp, Rita M., Hk55
Sharpe, J.A., Fh18
Sharpe, Kevin, Fb72
Sharpe, Pamela, Fg48
Sharpe, Richard, Bc100; Dc6,38; Ke25; Le21
Sharrock, Catherine, Fk72
Shaw, Alan George Lewers, Nc48,f43
Shaw, Gareth, Ac96; Bc35; Hf86–87
Shaw, John P., Bb186; If95
Shaw, Marion, Hl97
Shaw, Watkins, Fk73
Sheail, John, If96
Shell, Donald R., Ib76
Shephard, Amanda, Fj41
Shephard, Robert, Fb73
Shepherd, John, Bc117; Ib77
Shepherd, Michael, Ij80
Shepherd, Robert, Ba171
Sheppard, June A., Hg79
Sheridan, D., Ig37
Sherlock, David, Ek46
Sherlock, Stephen J., Bc160; Ek47
Sherrington, Emlyn, Hl98
Sherwood, Roy, Fb74
Shiman, Lilian Lewis, Hh33
Shires, Linda M., Bc56; Hl99–100
Shirland, Ann, Ei81–82
Shlomowitz, Ralph, Hf61; Nf20
Shoemaker, Robert B., Bb187,c82
Shoemaker, Susan Turnbull, Hh34
Shores, C.F., Ii54
Short, Brian, Ba172–173,c54
Short, Ian, Ei83
Shpayer-Makov, Haia, Ba174; Hk56
Shuttleworth, Sally, Hg80
Siddons, Michael Powell, Ja7
Siegenthaler, David, Fe82
Silagi, Michael, Hb76
Silke, John J., Aa160

Index of Authors

Sillars, Stuart, Ij81
Sillitoe, Paul J., Gf61
Silverman, Marilyn, Ac97; Bc124; Mf15
Silverstein, Arthur M., Ba175
Silvester, R.J., Jf4
Simkins, C. Anthony G., Ii27
Simm, Geoff, Bb188
Simmons, A. John, Fj42
Simms, Anngret, Lg8–9
Simms, Cormac, La7
Simms, Katharine, Lf10
Simms-Williams, Patrick, Li13
Simons, Janice, Bd40
Simonton, Dean Keith, Ac98
Simpson, A.D.C., Hk57
Simpson, Grant G., Bc21; Kh9
Simpson, J., Ba176
Sims-Williams, Patrick, De54
Sinclair, Eddie, Ej57
Sinclair, Keith, Nc49
Sinclair, Robert C., Hf88
Sindima, Harvey J., Nd16
Singer, Aubrey, Gd30
Singer, Peter, Bc14
Singh, Iqbal, Nb27
Singleton, John, He74
Skaggs, David Curtis, Bc85
Skerpan, Elizabeth, Fb75
Skidelsky, Robert, If97
Skinner, Andrew S., Bc29; Gf62,i71
Skinner, Keith, Ba15
Skinner, Quentin, Bc14
Skinner, Raymond J., Hc20
Skretkowicz, Victor, Fk74
Slack, Paul, Ba177,c25; Fg49
Slater, John, Ih31
Slater, S.C., Bb189
Slattery, J.F., Id96
Slinn, Judy, If38
Slocombe, Pamela M., Ek48
Smail, John, Ff22; Gg18
Smallbone, Linda, Aa32,b57
Smart, Nick, Ib78
Smart, Veronica, Ab58; Dd29; Ef59
Smith, Alexander McCall, Kc13
Smith, Anthony D., If98
Smith, Brendan, La10,e22,h13
Smith, Catherine, Gf63
Smith, Charles, Id97
Smith, Claude, Ii55
Smith, D.N., If30
Smith, David L., Fb76,c24
Smith, David M., Bc202

Smith, Dennis, Hf89,90; Id98
Smith, E.A., Hb77–78
Smith, F.B., Bc143; Hk58; Mi16
Smith, Harold L., Ih32
Smith, Ian M., Df66; Kg7
Smith, J.T., Ba178
Smith, Jeremy J., Ei84
Smith, John Sharwood, Ij82
Smith, Julia M.H., Dc39
Smith, K.E., Ac99
Smith, K.J.M., Nc50
Smith, Lance, Ek49
Smith, Lesley, Bc115; Ei85
Smith, Lindsay, Hl101
Smith, Llinos Beverley, Jg8
Smith, M.M., Aa161
Smith, Malcolm, Ii56
Smith, Nigel, Fa11
Smith, Richard M., Eg48
Smith, Robert P., Ii57
Smith, Roger, Hc21
Smith, Victor T.C., Ca45
Smith, William J., Lg10
Smith, Woodruff D., Ne24
Smithson, Peter, Ca46
Smout, T.C., Bb190,c196; Hl102; Ij83
Smyth, Alfred P., Mi17
Smyth, James J., Ib79; Mb26
Smyth, Rosaleen, Nh37
Smyth, William J., Mg24; Nf25
Snell, Keith D.M., Ba179; Gg19–20
Snider, A., Fl24
Snow, Vernon F., Fb77
Sober, Elliott, Aa177
Soden, Iain, Ee69
Soffe, Grahame, Ca36
Soffer, Reba N., He75,i31
Sokol, B.J., Fc25
Solomon, Graham, Gi36
Solow, Robert, If99
Sommerville, C. John, Fe83,g50
Sommerville, Johann P., Fb78,j43
Southall, Humphrey, Hb79; If100
Southgate, Beverley C., Fj44
Sowrey, F., If101
Spadoni, Carl, Aa37
Spahr, F., Hk59
Spalding, Frances, Ij84
Spargo, P.E., Aa162
Sparks, Carol, Fd29
Sparks, Margaret, Bc3
Spate, O.H.K., Nc51
Speake, G., Df22

Index of Authors

Spearman, Margaret, Bb184
Spearman, R.M., Ca47
Speck, William A., Fb79; Ga29,b28
Spence, Alistair, Ib80
Spence, Richard T., Ff23,i23
Spiegel, Gabriele M., Ac101
Spiers, Edward M., Hj17
Spinosa, Charles D., Ec26
Sprakes, Brian, Bb26
Stacey, Robert C., Ee70
Stacey, Stephen, Aa163
Stacey, William R., Fc26
Stafford, Pauline, Eg49
Stagg, D., Bb191
Stahl, Alan M., Dd30–31; Ef60
Stait, Bruce A., If102
Stalley, Roger, La8; Mi18
Staniland, Kay, Bc53
Stanley, Brian, Ba180
Stanlis, Peter J., Gi72
Staples, Peter, Ie10
Stapleton, Barry, Bc116; Fg51
Stapleton, Julia, Hl103
Starkey, David, Fa12
Starkey, Pat, Ii58
Stater, Victor, Fb80
Stavely, Keith W.F., Fj45
Steane, J.M., Ef61
Steane, Kate, Ek50
Stearn, Roger T., Hj18
Steedman, Carolyn, Hl104
Steedman, Ken, Bc165
Steele, Ian K., Gh44
Stein, Burton, Ne25
Stephens, John Russell, Hl105
Stephens, M.D., Hi29
Stephens, W.B., Ff24
Stephenson, Jayne D., Bb27
Steppler, Glenn A., Bb192
Stern, Virginia F., Fb81
Sternberg, Ilse, Ab59
Stetz, Margaret Diane, He76
Stevens, John, Ei86–87
Stevens, Lawrence, Ek51
Stevens, Patricia, Ek51
Stevens, Paul L., Gd31
Stevens, Terry, Ia38
Stevens, Wesley M., De55
Stevenson, David, Fa13
Stevenson, Jane, Dc40
Stevenson, Janet H., Ek52; Gb29
Stevenson, John, Aa132,c100; Ga30; Ib81

Stewart, Geoffrey, Ba146
Stewart, Ian, Ba181; Dd32
Stewart, Richard W., Fi24; Mh3
Stidder, Derek, Bb193
Stillinger, Jack, Hl106
Stinchcombe, Owen, He77,i32
Stobbart, Lorainne G., Fd30
Stocker, Margarita, Fe84
Stockwell, A.J., Nc52
Stoker, David A., Gc13,e44
Stoler, Ann L., Nh9
Stone, Lawrence, Ac101; Ba182
Stone, Richard, Gf64
Stoneman, Paul L., Hf91
Storey, G.O., Ij85
Storey, R.L., Eb7,e71–72
Storrs, Christopher, Fd31
Stott, P., Kf2
Stout, Geoffrey, Bb194
Stout, Harry S., Ge45
Stout, Matthew, Lh14
Stowell, Sheila, Hg81
Straub, K., Gi73
Street, Sarah, Aa164
Street, Sean, Hl107
Strickland, Matthew, Bc141; Eh24–25
Strong, Sir Roy, Ba183
Stroud, Patricia, Aa73
Stuart, Denis, Aa165
Stuart, Robert, Ca48
Stuart-Macadam, Patty, Cb45
Studd, Robin, Ek53
Sturdy, Steve, Hk60
Sturgess, Roy, Hf92
Sturm, P., Id99
Sturman, Christopher, Hg82
Suleri, Sara, Nh38
Sumida, Jon Tetsuro, Ii59
Summers, Anne, Hb80,k61
Summers, David W., Bb195; Hf93
Summerson, Henry, Ee73,h26; Kc14
Sumner, Graham, Bc121
Sundback, Esa, Id100
Surtees, John, Bb196
Sutcliffe, Ray, Bc207
Sutherland, William, Na22
Sutton, Anne F., Ei88–92
Swain, S.A.M., Ca51
Swaisland, Cecillie, Nf44
Swan, Philip, Bc205; Hk43
Swanson, Keith, Ec27
Swanson, Robert N., Ee74,f62
Sweet, P.R., Ac102

Index of Authors

Sweetman, P.D., La2
Swift, Roger, Hc22
Symms, Peter, Ke26
Symons, Lenore, Gd10
Symons, Malcolm, Bb197
Szarmach, Paul E., Aa166
Szasz, Ferenc M., Ij86
Szechi, Daniel, Fb82
Szymanski, J.E., Ek18,54

Tadmor, Naomi, Fg52
Talvio, Tuuka, Dd33
Tang, James Tuck-Hong, Id101
Tanner, Duncan, Ib82
Tanner, Heather J., Ed12
Tantam, Digby, Hk27; Ij87
Taplin, Eric, Hf94
Targetti, Ferdinando, If103
Tarn, John Nelson, Ie11
Tatton-Brown, Tim, Bb198,c3; Df67;
  Ek41,55–56
Tavener, Nick, Ii60
Taylor, A.J., Jh11
Taylor, Andrew J., Ib83
Taylor, Anne, Hb81
Taylor, Arnold J., Ek57
Taylor, Barbara, Gi74
Taylor, Barry, Aa167
Taylor, Brian, He78; Ie12
Taylor, C.M., Da19
Taylor, Christopher C., Ba184,b28; Ek8,10
Taylor, Clare, Hl108
Taylor, Daniel S., Ee28
Taylor, Eric, Ig38
Taylor, Ian, Ib84
Taylor, John, Ea8
Taylor, John H., Ij71
Taylor, Lawrence J., Me12
Taylor, Michael W., Hb82
Taylor, Miles, Ba185
Taylor, Pamela, Db37
Taylor, Peter, Hg83
Taylor, Philip M., Ii61
Taylor, Robert, Bb199
Taylor, Rogan, Ba186
Taylor, Stephen, Gb30,e46
Taylor, Tony, Bc176; Hb83
Taylor, W. Robert, Bb200
Teague, Frances, Fb83
Tebbutt, Melanie, Ig39
TeBrake, Janet K., Mb27
Teed, Peter, Ab60
Teich, Mikuláš, Bc221; Fa14

Temperley, Nicholas, Hl109–111
Terraine, John, Ii62
Teviotdale, E.C., Aa168; De56
Thacker, Alan T., Bc81; Dc41–42; Ea40
Thain, Colin, If104
Thane, Pat, Bc44; Ih33
Thatcher, A.R., Ig40
Thatcher, Ian D., Ib85
Thijssen, J.M.M.H., Gi75
Thirsk, Joan, Ff25–26,h19,l25
Thistlewaite, Nicholas, Hl112
Thom, Deborah, Ih34
Thomas, Avril, Lh15
Thomas, Charles, Je17
Thomas, D.L.B., Bb201
Thomas, David, Ff27–28
Thomas, David Oswald, Gb31
Thomas, James, Gi38
Thomas, John Hugh, Ij88
Thomas, Kenneth, Ne26
Thomas, Mark, Ba94,c34
Thomas, Nicholas, Nd17
Thomas, P., Gg21
Thomas, P.W., Fa15
Thomas, Peter D.G., Gb32,d32
Thomas, William, Bc14
Thompson, Andrew, Hd16
Thompson, Benjamin, Ee75
Thompson, Dorothy, Ac103
Thompson, F.M.L., Hg84; Ig41
Thompson, M.W., Ek58
Thompson, Paul, Ng27
Thompson, Pauline, Df68
Thompson, Peter, Ib49
Thompson, Ruby Reid, Fk75
Thompson, Willie, Ib86
Thomson, J.M., Kc15
Thomson, Robert G., Ej58
Thor, Jon, If105
Thorne, Robert, Ij89
Thorne, Susan, Ac68
Thornes, Robin, Fa16
Thornthwaite, S.E., Gf65
Thornton, Christopher, Ef63
Thornton, David Ewan, Jb3
Thorpe, Andrew, Ia39–40
Thorpe, William, Nc15
Throup, David, Nc53
Thrush, Andrew, Fi25
Thulesius, Olav, Fl26
Thurlby, Malcolm, Ej59
Thurley, Simon, Fk76
Thurmer, John, Ej60; He79

Index of Authors

Thwaites, W., Gf66
Thygesen, Niels, Id37
Ticktin, Hillel, Nb28
Tierney, M., Me13
Tighe, W.J., Fe29
Tiller, Kate, Ab61
Tillotson, G.H.R., Nh39
Tilney, Ruth,, Gg22
Tilson, Barbara, Bb202
Timbell, Martin, Hf15
Tinbergen, Jan, If106
Tingle, M., Ca22
Tiratsoo, Nick, Bc67; Ia41,b87,f107
Tite, Colin G., Fk77
Tittler, Robert, Bb203; Fh20
Tobin, James, If108
Tobriner, A., Fg53
Todd, John, Hh35
Todd, Margo, Fe85
Todisci, O., Ei93
Toft, L.A., Cb46
Tolley, R.J., Ek59
Tolliday, Steven W., Bc33
Tomaselli, Sylvana, Ac104; Gi76
Tomich, Dale, Nf45
Tomlin, R.S.O., Ca11,30,b47
Tomlinson, D.G., Bc106
Tomlinson, J.D., Bc131;
    Ib52,88,f109–110
Tomlinson, S., If111
Tomlinson, Sophie, Fk78
Tompson, Richard S., Kc16
Toninelli, Pier Angelo, Ba55,c134
Tonks, Eric Sidney, Bb204
Toon, Thomas E., De57
Topham, Jonathan, Hk62
Toplis, Ian, Ij90
Topulov, Christian, Hh20
Torrie, Elizabeth P.D., Ka3
Torvell, David, Ab62
Tosh, John, Ba161,c223; Hg85
Townley, Simon, Ee76
Townshend, Charles, Mc3
Tracy, Charles, Ej27,61–62
Tracy, James D., Ba187,c31
Trahern, Joseph B., Jr, De58
Trapido, Stanley, Nc34
Trapp, Joseph Burney, Fk79
Traugott, Elizabeth Closs, De59
Travers, Pauric, Bc182; Mb28
Travers, Tim, Ii63
Travis, Anthony S., Hk63–64; If112
Travis Hanes III, W.₈, Nc54

Treble, James H., Ia10
Treble, John H., Bc138
Trela, D.J., Ac105; Hl113
Tremewan, Peter, Nb29
Tressider, George A., Gi77
Trevett, Christine, Fe86
Trigger, David S., Nf46
Tritton, Paul, Hk65
Trotter, Ann, Nb30
Trowles, T., Aa57
Troy, Jakelin, Mi19
Trubowitz, Rachel, Fh21
Tsokhas, Kosmas, Id102,f113; Ne27
Tsoulouhas, Theofanis C., Ba188
Tsurushima, H., Ee77
Tsushima, Jean, Ba189
Tuck, Anthony, Bc4; Eg50
Tuck, Richard, Bc14
Tunbridge, Paul, Hk66
Turnbull, Deborah, Bb152
Turnbull, P., If114
Turnbull, Percival, Ek60
Turner, Barbara Carpenter, Bb205
Turner, Cheryl, Gi78
Turner, Garth, Ie13
Turner, Michael, Hf95
Turner, P., If18
Turner, R.C., Ef64
Turner, Rick, Je18
Turner, Trevor, Hk67
Turner, Wesley B., Gh14,45
Turville-Petre, Thorlac, Ee78
Twaddle, Michael, Bc130; Na23
Tweddle, Dominic, Bc156
Tweedale, Geoffrey, Aa169; Hf96; If115
Tweedy, Hilda, Mb29
Tweney, Ryan D., Hk68
Tylden-Wright, David, Fk80
Tyson, Blake, Gf67
Tyson, Colin, Eh27

Ulin, Donald, Hl114
Underdown, David, Fe87
Underwood, Malcolm G., Ea29
Unwin, Tim, Df69
Upton, C.A., Fg16,k26,81
Urmson, J.O., Bc15
Urwin, Cathy, Ih35
Usher, Brett, Fe88,k82
Usherwood, Paul, Hl115

Vaio, John, Hl116
Valiulis, Maryann Gialanella, Mb30

van Caenegem, Raoul C., Ec28
van der Kiste, John, Bd41
van der Meer, Gay, Dd15
van Heertum, Cis, Fk83
van Houts, Elisabeth M.C., Da20; Ea41,e79; Fk84
van Selm, Bert, Aa170
van Strien, C.D., Fd32
van Waarden, Frans, Bc37; If116–117
Vance, Jonathan F., Ii64
Vanhulst, Henri, Fe89
Vaughan, Megan, Nf47
Vaughan, Sir Edgar, Hi33
Vaughn, Karen Iversen, Fj46
Vecchioni, Domenico, Fi26
Venn, P., Bb206
Verduyn, Anthony, Ec29
Veseth, Michael, Ba190
Vickers, Brian, Fj47
Vickers, John A., Ge47–48
Vickery, Amanda, Ga31
Vidal, Daniel, Fe90
Vigier, Anil de Silva, Eb8
Vince, Alan, Bb115,c208; Fg54
Vincent, David, Ig42
Vincent, Joan, Mb31
Vincent, Nicholas C., Eb9–10,c30
Viner, David, Gc14
Virgoe, Roger, Ei94
Vis, G.N.M., Ee47
Visser-Fuchs, Livia, Ed13,i89–92
Voeltz, Richard A., Ia42
Voigts, Linda E., Ei95
vom Bruch, Rudiger, Hl2
Von Arx, Jeffrey P., Hl117
Vorlander, H., Hb29

Waas, B., If118
Waddams, S.M., He80
Waddell, D.A.G., Hd17
Waddington, G.T., Id103
Wagenaar, Michael, Bc42
Wagner, Hans-Peter, Gi79
Wagner, Sir Anthony, Bd42
Wahrman, Dror, Ac106; Gb33
Wainwright, Jonathan P., Fk85
Waite, Peter B., Ac107; Hf16–17
Waldron, Ronald, Ei96
Walker, Brian M., Mb32
Walker, David, Ek61; Ja8
Walker, G.T., Ca3
Walker, Graham, Ib89; Mb33
Walker, Greg, Fb84,e91

Walker, John, Id104
Walker, Margaret, Ja8
Walker, Pamela J., He81
Walker, R.F., Fg55
Walker, Ralph S., Gi80
Walker, Simon, Db38
Walkowitz, Judith, Hh36
Wallace, Patrick F., Le12,f11
Waller, John H., Ng38
Wallis, Helen, Fk86
Wallwork, Janet, Ac108
Wallwork, S.C., Ba191
Walmsley, Tom, Ij91
Walsh, Deborah, Ek60
Walsh, John R., Le1
Walsh, Katherine, Le23; Mi20
Walsh, Paul, Lf12
Walters, D.B., Jc1; Kc17
Walton, Julian. C., Lc4
Walton, Whitney, Hf97
Walvin, James, Nf48–49
Wanko, C., Fk87
Ward, Benedicta, Bc115; Dc43; Ee80
Ward, Ian, Ff29
Ward, Jennifer C., Ea42
Ward, Philip, Fb85
Ward, S.G.P., Gh46
Ward, William Reginald, Ge49
Wardle, Christopher J., Hh37
Wardley, Peter, Aa51,c70
Ward-Perkins, Bryan, Cb48
Wark, Wesley K., Aa171; Ii65
Warneford, Francis E., Bd43
Warner, Isabel, Id109
Warner, John Harley, Ab53
Warner, Malcolm, Hl118
Warner, Marina, Ba192
Warner, Sir Fred, Id105
Warnicke, Retha M., Ek62
Warren, John, Ej63; Mi21
Wasserstein, Bernard, Ba193
Waterhouse, Prudence, Hl33
Waterman, Anthony M.C., Ge50; Hf98
Waters, Chris, Hl119
Wathey, Andrew, Ei97–98
Watkin, Thomas Glyn, Db39
Watson, Alan, Hh38
Watson, B., Ek63
Watson, M.I., Hi34
Watt, D.E.R., Ke27
Watt, J.A., Ec31; Le24
Watts, D.G., Eg51
Watts, Lee, Nd2

Index of Authors

Waugh, Mary, Hf99
Wayment, Hilary, Fk88
Weale, C., Ek64
Wear, Andrew, Ba194,c23; Gj26
Weatherall, Mark, Bb207
Webb, Cliff, Aa172–174
Webb, R.K., Hl120
Webber, M.T.J., Ei99
Webber, Teresa, Ei100–101
Webber-Mortiboys, Christine, Hi35
Webby, Elizabeth, Na24
Weber, Birthe, Kf8
Weber, William, Gi81
Webster, Charles, Ig43
Webster, Leslie, Df70
Webster, Norman William, Ea43
Webster, P.V., Ca31–32
Weil, Rachel J., Fb86; Gg23
Weindling, Paul, Hk69–70
Welch, Edwin, Ge51
Welch, Martin G., Bc160; Da21
Wells, Peter S., Bc183; Da14
Wells, Roger, Hb84,g86,h39
Welsh, William Jeffrey, Bc85
Wenisch, Franz, Dc44
Wenley, Robert, Fk89
Wenzel, Siegfried, De60
Werhane, Patricia H., Ba195
Werner, Alex, Bb63
Werner, Joachim, Da22
Werner, Wolfram, Id106
Wernham, Richard B., Fi27
Wesselius, J.W., Fk90
West, Barbara, Cb35
West, John, Fi28
Westcott, Margaret, Fa17
Westrate, Bruce, Id107
Wetherell, Charles, Hb63
Wetherly, Paul, Ac109–111; Bc9
Whaples, R., Ac112
Whatley, Christopher A., Aa58;
    Bb60,208,c6; Kb7
Wheeler, Geoffrey, Bb209
Wheeler, James Scott, Fi29; Mh4
Wheeler, Michael, Hl121
Wheeler, Richard, Ij92
Whelan, Kevin, Bc118; Ma4
Whimster, Rowan, Ab63
Whipp, Richard, If119
Whitbread, Helena, Hg87
White, A., Gf68
White, Dan S., Ib90
White, E.M., Hg88

White, Norman, Hl122
White, Peter, Fe92
White, R.F., Jh4
White, R.H., Df71
White, Robin L.W., Bb86
Whitehead, C., Nh40
Whitehouse, David, Da23
Whiteley, Nigel, Hl121
Whiteley, Patrick, Ib91
Whiteside, D.T., Fl27
Whiteside, Noel, Aa107
Whiting, Richard, Ib92
Whittle, Elizabeth, Ja9
Whitwell, John Benjamin, Ca33,b49
Whitworth, A.M., Ca49
Whyte, Beverly, Na25
Wickenden, N.P., Bc162
Wicker, Nancy L. Hatch, Df72
Wickham, Chris, Df73
Wickham-Crowley, Kelley, De61
Wickham-Jones, M., If8
Wiebe, M.G., Hb53
Wieland, Gernot, Da24
Wigglesworth, Neil, Ba196
Wigham, Maurice J., Me14
Wijffels, Alain, Aa175
Wijnberg, Nachoem M., Ba197
Wilcox, Helen, Fk91
Wilcox, Jonathan, Ee81
Wild, F., Ek65
Wildgoose, Martin, Ef35
Wildy, Tom, Ib93
Wilkins, Mira, Hf100
Wilkinson, P., Jg9
Wilkinson, S., Ec32
William, Eurwyn, Bb210
Williams, A.H., Ge52
Williams, Ann, Bb81,c109–111;
    Db40,f74
Williams, David W., Bb211
Williams, E.H.D., Ba198,b212–213
Williams, F.J., Hg89
Williams, Frances, Bc163
Williams, Glanmor, Ac113; Bb214;
    Je19–20
Williams, Glyn, Bb215
Williams, Herbert, Hf101
Williams, I.L., Fh22
Williams, J.D., Gg24
Williams, John H., Bc163; Ek65
Williams, Merfyn, Bb51
Williams, Michael E., Fd33
Williams, Naomi, Hk71

277

Index of Authors

Williams, Rhodri, Ii66
Williams, Robin, Bb216
Williams, Sian Rhiannon, Hl123
Williamson, Jeffrey G., Ac114; Hf102,g90
Williamson, John, If120
Williamson, Philip, Ib94
Willis, John, Nf50
Willis, R.J., If121
Wilmot, Sarah, Aa99
Wilson, Arthur J., Ba199
Wilson, Barbara, Ek66
Wilson, C., Ba200
Wilson, D.R., Ba201
Wilson, David M., Ab64; Bc159; Da25,f75
Wilson, Eunice, Aa176
Wilson, G., Bc127; Gf69
Wilson, Janice C., Hk15
Wilson, Jason, Hk72
Wilson, Jean, Fg56
Wilson, John, Bb217; Gi82
Wilson, John B., Kc18
Wilson, Philip K., Gj27
Wilson, Ted, Hf103
Wilson, Trevor, Ii48
Wilson, W.A., Kc19
Wilson-North, W.R., Ek67
Wilthew, P., Ca47
Winch, Donald, Gf70
Winchester, Angus, Fe93
Winder, J.M., Df40
Winslow, Barbara, Ib95
Winstanley, Ian G., Bb188
Winstanley, Michael, Bb218
Winstead, Karen A., Ee82
Winter, J.M., Ie14
Winter, Michael, Ie15
Wise, P., Ca50
Wiseman, Susan, Bc62; Fg57,j48
Withers, Charles W.J., Bc55; Ga32; Hg53,91
Witkin, Robert, Ba202
Witney, K.P., Ef65
Wogan-Browne, Jocelyn, Ee83
Wolfe, E.L., Ij8
Wollen, Tana, Ij93
Womack, Peter, Fk92
Womersley, David, Gi83
Wood, Charles T., Ea44
Wood, Diana, Bc12,189,194; Ei102
Wood, Donald, Nb31
Wood, Florence, Bb219

Wood, George O., Bb220
Wood, I.N., Da26
Wood, Jason, Ee84
Wood, Kenneth, Bb219
Wood, L., Gi84
Wood, Marcus, Gb34
Wood, Neal, Fj49
Woodcock, Sally, Ej64
Woodfield, Paul, Ba203
Woodhouse, D.G., Ia43
Woodiwiss, Simon, Bc137
Woodland, Patrick, Gb35
Woods, C.J., Mi22
Woods, Robert, Ba200; Hg92–93
Woodward, David R., Ii67
Woodward, Sandy, Ii68
Woolf, Daniel R., Ba204; Fg58
Woolgar, C.M., Aa118; Ef66
Woolley, James, Gi85
Woolliscroft, D.J., Ca51
Wormald, Jenny, Fc27
Wormald, Patrick, De62; Ec33
Worpole, Ken, Ij94
Worsley, G., Ek68
Worswick, David, If122
Wrathmell, Stuart, Jh12
Wright, Deborah Kempf, Gi86
Wright, Elizabeth F., Hg94
Wright, Erik Olin, Aa177
Wright, H. Bunker, Gi86
Wright, John, Nb32
Wright, Michael, Aa90
Wright, Nicholas, Eh28
Wright, Patricia, Hl124
Wright, Ray, Nb33
Wright, Richard P., Ca11
Wright, Rosemary Muir, De63
Wright, Susan, Ba205
Wright, Sylvia, Ej65
Wrightson, Keith, Bb131
Wrigley, Chris, Bc117; Hb85; Ih36
Wu, Duncan, Aa178
Wyatt, Grace, Gg25
Wykes, David L., Fe94; Ge53
Wynn, Graeme, Nf24
Wynne-Davies, Marion, Bc68; Fk17,93

Xiang, Lanxin, Id108

Yallop, H.J., Bb221
Yardley, Bruce, Fe95
Yasamee, H.J., Id89,109
Yates, Nigel, Bb222

278

Index of Authors

Yelling, J.A., Ih37
Yeo, R., Ac115
Yorke, Barbara, Df76
Young, Anne Steel, Fb77
Young, Craig, Hf104
Young, David, Hk73,l125
Young, James D., Ib96
Young, John, Ng39
Young, Margaret D., Kc20

Zachs, William, Gi87
Zagorin, Perez, Bc60; Fj50
Zakai, Avihu, Fe96–97

Zaller, Robert, Fb87
Zebrowski, Martha K., Gb36
Zeepvat, R.J., Ca52
Ziegler, Dieter, Hf74,105
Ziegler, Georgianna, Fk94
Zier, Mark, Ei103
Zimmermann, Albert, Ei15
Zines, Leslie, Nc55
Zook, Melinda, Fb88
Zulueta, Julian de, Gh47
Zvelebil, Marek, Ba27,c211
Zweig, Ronald W., Nb34
Zweiniger-Bargielowskan, Ina, If123
Zwicker, Steven N., Fj51

INDEX OF PERSONAL NAMES

Abbott, Benjamin (d. 1870), Gj13

Abernethy, Thomas (covenanter), Ke23

Adam of Buckfield (fl. 1300?), Ei66

Adam, Robert (d. 1792), Gi12–13,16, 26,68

Adams, W.E. (d. 1906), Hb2

Addison, Lancelot (d. 1703), Fe39

Addison, Paul (historian), Ac6

Ælfgifu (fl. 960s), Db20

Ælfgyva (in Bayeux tapestry), De18

Ælfric (homilist), Dc20,e19,32

Æthelred II (king), Db29,d19

Æthelstan (king), Da7

Æthelweard (chronicler), Da20

Æthelwold, bishop of Winchester, St, Dc19,e35

Agricola, Gaius Julius, Cb8

Alanbrooke, viscount, see Brooke

Alcuin (scholar), Dc27

Alexander, Michael Solomon, bishop of Jerusalem, He78

Alfred (king), Da7,b26,c28,d2,20,e9

Allanson, Colonel Cecil John Lyons (d. 1943), Ng5

Allen, G.C. (d. 1982), Id77

Allenby, Field Marshal Edmund Henry Hynman, viscount Allenby of Megiddo, Ii43

Allestree, Richard (d. 1681), Fe1

Allinson, Thomas (d. 1918), Hk51

Andrewes, Lancelot, bishop of Winchester, Fe23

Andrews, Evan (d. 1869), Hl75

Andrews, J.H. (geographer), Aa86,c54; Bc118; Ma4

Anselm, archbishop of Canterbury, St, Ee1

Anstis, John (herald), Bd42

Antrim, earl of, see MacDonnell

Armagh, archbishop of, see Bole, FitzRalph, Usher

Arthur (king), Ji2,10

Arundel, Thomas, bishop of Ely, Ee11

Ashwell, Henry (manufacturer), Bb149

Askew family, Bd33

Asquith, Herbert Henry (d. 1928), Hb66; Ib32

Astell, Mary (d. 1731), Fk9,72

Aston, William George (d. 1911), Hl67

Athelstan "half king" (ealdorman), Db7

Atkinson, Edward Leicester (physician), Ij18

Attlee, Clement, 1st earl (d. 1967), Bc67

Aubrey, John (d. 1697), Fk80

Aughtie, Robert (d. 1901), Hl50

Augusta (princess), Gb5

Aurangzeb (Mughal emperor), Fd13

Austen, Jane (d. 1817), Hl55

Aylesford, earl of, see Finch

Ayrton, Heartha (d. 1923), Hk45

Ayscough, Francis (d. 1763), Gb13

Babbage, Charles (d. 1871), Aa41

Bacon, Francis, 1st viscount St Albans, Fj6,13,32,47,l24

Bacon, Roger (d. 1294), Ei36

Bagwell, Philip (historian), Bc117

Bain, Alexander (d. 1903), Hi13

Baker, Elizabeth (playwright), Hg31

Baker, Frank (historian), Aa82

Baker, Herbert (d. 1946), Nh24

Baker, Thomas (consul), Fd22

Bale, John, bishop of Ossory, Fe8

Balfour, Arthur James, 1st earl, Ib69,d65

Ballard, Colin (d. 1941), Hj5–6

Balogh, Thomas (d. 1985), If54

Bamburgh, house of, Ea33

Barberi, Dominic (d. 1876), He53

Barclay, Robert (d. 1690), Fe81

Barclay, Robert Herriot (fl. 1812), Gh12

Barker, Joseph (d. 1875), Hb68

Barnard, John (d. 1784), Gj2

Barnsley, Edward (artist), Ij19

Barraclough, Geoffrey (historian), Aa141, c37; Bc81

Barry, James (d. 1806), Gi5

Basan, Eadui (scribe), Ee57

Bass, W. Graham (glass maker), Hl25

Baxter, David (historian), Bb83

Bayly, Nicholas (d. 1782), Gb32

Beaconsfield, earl of, see Disraeli

Beale, John (d. 1683?), Fk19,48

Beattie, James (d. 1803), Gi45

Beaufort, lady Margaret, countess of Richmond, Ea29

Beaumont, Joan (d. 1469), Eg6
Becket, St Thomas, archbishop of Canterbury, Aa151; Ee47,80,j11
Bectun (abbot), Dc28
Bede, the Venerable, Da5,e21,f8; Li12
Bedford, earl of, see Russell
Beer, Samuel (political scientist), Ic3
Behn, Aphra (d. 1689), Fk8,61
Bennett, Charles, 2nd baron Ossulton, Fg33
Bentham, Jeremy (d. 1832), Hb69
Bentley, Thomas (fl. 1580s), Fg3
Beresford, baron, see de la Poer
Berg, Maxine (historian), Gf13
Berkeley family, Hb20
Berkeley, George, bishop of Cloyne, Bc15; Mi11
Bernoull, Johann I (d. 1742), Gj5
Bertano, Gurone, (papal diplomat), Fe7
Besant, Annie (d. 1933), Hb81,g73
Beveridge, William, 1st baron, Ih20
Bevin, Ernest (d. 1951), Ib77
Binns, Joseph (surgeon), Fl17
Birchensha, John (fl. 1664–72), Fl20
Bisset, Baldred (fl. 1300), Ke8
Black, Robert A. (historian), Hf46
Blacman, John (fl. 1450), Ei54
Blake, William (d. 1827), Gi52,63
Blanche of Lancaster (d. 1368), Ea43
Bligh, William (d. 1817), Gh11
Blomefield, Francis (d. 1752), Ge44
Blundell, Henry (d. 1810), Cb11
Bodley, Thomas (d. 1613), Fk26
Bohn, Henry (d. 1884), Ab42
Bole, John, archbishop of Armagh, Le10
Boleyn, Anne (queen), Fb12,47
Bolingbroke, viscount, see St John
Boniface, St, Df14
Boniface VIII (pope), Ke8
Bonville family, Ec2
Borges, Jorge Luis (author), Ij37
Boscawen, Admiral Edward (d. 1761), Gh40
Boulogne, count of, see Eustace II
Boulton, Matthew (d. 1809), Gf36
Bourchier, Henry, 2nd earl of Essex, Fb34
Bourke, Richard Southwell, 6th earl Mayo, Nc41
Boyer, Abel (Huguenot), Fe6
Boyle, Charles, 4th earl of Orrery, Fb22
Boyle, Robert (d. 1691), Fl22
Bradmore, John (fl. 1400), Ei48

Breage, St, Dc30
Breadalbane, earl of, Gc6
Brenner, Robert (historian), Ei17
Brewster, David (d. 1868), Hk57
Brewster, William (d. 1644), Fd17
Bridges, Robert (d. 1930), Ij67
Brierley, W.H. (architect), Ij34
Bright, Richard (d. 1858), Hk6
Brittain, Vera (d. 1970), Ia15,j11
Britton, John (d. 1857), Bb88
Brontë, Patrick (d. 1861), Ge4
Brook, baron, see Greville
Brooke, Alan Francis, viscount Alanbrooke, Ii4
Brooke, Christopher N.L. (historian), Aa16,c30; Bc70
Brooke, Sir Charles Anthony Johnson (d. 1917), Nd14
Brooke, Sir Charles Vyner (d. 1963), Nd14
Brooke, Sir James (d. 1868), Nd14
Brooks, John Gent (fl. 1850), Hh5
Browne, Edward (d. 1708), Fd32
Browne, Frank (d. 1960), Aa133
Bruce, Edward (d. 1318), Lh3,6,13
Bruce family, Ek47
Bruce, William (Russian), Kh5
Brunton, Richard Henry (d. 1901), Hf22
Buchan, John (d. 1940), Ij45
Buckingham, duke of, see Villiers
Buckley, James (d. 1839), He8
Buk family, Eg34
Bulkeley family, Gb32
Bull, George, bishop of St David's, Fe5
Bull, William (d. 1814), Gi20
Bunyan, John (d. 1688), Bc198
Burckhardt, Jacob (d. 1897), Hb44
Burgh, Hubert de, 1st earl of Kent, Eb10
Burgh, James (d. 1775), Gb36
Burgundy, duchess of, see Margaret; duke of, see Charles
Burke, Edmund (d. 1797), Bc132; Gb4,25,d28,i15,30,61,63–64,72; Mi13
Burley, Walter (d. 1345?), Ei63
Burne-Jones, Edward, 1st baronet, Hl82
Bute, earl of, see Stuart
Butler, Captain Richard (d. 1611), Fd33
Butler, Josephine (d. 1906), Hb18
Butterfield, Sir Herbert (historian), Fa6
Byron, George Gordon, 6th baron, Hb69

Cadwallon (king of Gwynedd), Jb2
Caesar, Mary (Jacobite), Fb66
Caffarelli, Antonio (lawyer), Ed5
Calvin, Jean (d. 1564), Fe70
Camden, William (d. 1623), Fk84
Cameron, Rondo (historian), Hf74
Campbell, John (d. 1775), Ac1
Campbell, John Henry, 4th earl of
  Loudoun, Gh44
Cane, Rev. Alfred Granger, Ng37
Canfield, Benedict (d. 1611), Fe90
Canning, George (d. 1827), Hd6
Cannon, Walter B. (psychiatrist), Ij8
Canterbury, archbishop of, see Anselm,
  Becket, Chichele, Cranmer, Dunstan,
  Islip, Lanfranc, Laud, Morton, Secker,
  Theodore, Whitgift
Cantilupe, Walter de, bishop of
  Worcester, Ee28
Capgrave, John (d. 1464), Ee82
Carew, Thomas (d. 1639?), Fk56
Carleton, Guy (d. 1808), Gd31
Carleton, Mary (d. 1673), Fk18
Carlisle, bishop of, see Everdon, Kirkby,
  Ross
Carlton, Robert (d. 1742), Gc13
Carlyle, abbot Aelred (d. 1955), He45
Carlyle, Thomas (d. 1881), Ac105;
  Hl6,12,27,113
Carnegy, James (merchant), Gf18
Caroline of Brunswick (queen), Hb40
Carpenter, John (d. 1442), Ei79
Carter, Howard (d. 1939), Ij71
Caryll, John (d. 1711), Fd1
Cashel, archbishop of, see Slattery
Castlemaine, earl of, see Palmer
Cathcart, Charles, 8th baron, Gb12
Cavendish, Margaret, duchess of
  Newcastle, Fg57,k78
Cavendish, Spencer Compton, 8th duke
  of Devonshire, Hb42
Caxton, William (d. 1492), Fk33
Cearl (king of Mercia), Db22
Cecil, Edgar, viscount Cecil of Chelwood,
  Aa10
Cecil, Robert, 1st earl of Salisury, Fk39
Cecil, Robert Arthur Talbot Gascoyne,
  3rd marquis of Salisbury, Hd4
Chaceporc, Peter (d. 1254), Ee32
Chalmers, Thomas (d. 1847), He7,f98
Chamberlain, Basil Hall (d. 1935), Hl13
Chamberlain, Joseph (d. 1914), Hb42,51

Chamberlain, Neville (d. 1940), Aa163
Chapman, John (d. 1778), Gf39
Charax, A. Claudius (commander), Cb9
Charles I (king), Fa13,b3,16,27,42,
  67,72,e20,66
Charles II (king), Fb55,g35
Charles the Bold, duke of Burgundy, Ed13
Charlesworth, May (imposter), Hc20
Charlesworth, Miriam (d. 1920), Hc20
Chaucer, Geoffrey (d. 1400), Ee10,g2,i69,
  74,j5,65
Cheriton family (of Nymet Rowland),
  Hl23
Chester, earl of, see Hugh, Ranulf II
Chichele, Henry, archbishop of
  Canterbury, Ed5
Chichester, bishop of, see Pecock
Chifley, Joseph Benedict (d. 1951), Ne15
Chisholm, clan, Kc9
Cholmondeley family, of Vale Royal
  (Chs), Fk36
Chudleigh, Mary (authoress), Fk9
Churchill, John, 1st duke of
  Marlborough, Fi13; Kh7
Churchill, Sir Winston, Aa9; Bb4;
  Ia28,b70,d3,48,56,i41
Cibber, Colley (d. 1758), Gi73
Clanvowe, Sir John (d. 1391), Ei67
Clarendon, earl of, see Hyde
Clark, William (d. 1763), Gf12
Clement, St, Dc10
Clifford family, earls of Cumberland,
  Fb44,f23
Clouston, Thomas (d. 1915), Ij12
Cloyne, bishop of, see Berkeley
Cnut (king), Db25,29,c18,d22
Cobbe, Frances Power (d. 1904), Hb18
Cobbett, William (d. 1835), Hg26
Cockburn, Admiral Sir George (d. 1853),
  Hj12
Cogan, Anthony (d. 1872), Mi17
Cohen, Gerald (political theorist), Ac34;
  Ei17
Coke, Sir Edward (d. 1634), Fc5,6
Colenso, John William (d. 1883), Nd10
Colet, John (d. 1519), Fk79
Collier, Arthur (d. 1732), Gi7
Collins, Michael (d. 1922), Mb15
Colop, John (bibliophile), Ei79
Columba, St, Ke25; Li7
Columbanus (d. 615), Li7
Columbus, Christopher, Ed11

Compton, Charles, 7th earl of
Northampton, Gi51
Comte, Auguste (d. 1857), Hl66
Conder, Josiah (d. 1920), Hl42
Congreve, Captain William (d. 1814),
Gh7
Connolly, James (d. 1916), Ib96
Conway, Anne, viscountess (d. 1679),
Fk41
Conway, Henry Seymour (d. 1795), Gd12
Cook, Captain James (d. 1779), Ab19;
Gd13,21
Cooke, William (fl. 1773), Gi38
Cooper, Anthony Ashley, 3rd earl of
Shaftesbury, Fb6,g38
Cornwallis, Charles, 1st marquis, Gh31
Coryat(e), Thomas (d. 1617), Fk65
Cotes, Roger (d. 1716), Fl8; Gj11
Cotton, Elizabeth (d. 1892), Hg41
Cotton, Sir Robert (d. 1631), Aa168;
Fc19,j30,k6,16,37,77,84
Courtenay family, Ec2,g26
Courtenay, Katherine, countess of Devon,
Fa17
Coutts family, Ba83
Coventry & Lichfield, bishop of, see
Langton
Cowan, Ian B. (d. 1990), Ac41
Cowper, William, 1st earl, Gb15
Cox, Sir Christopher William Machell (d.
1982), Nh40
Crafts, N.F.R. (historian), Gf13
Cragg family, Bd33
Craig, Edy (actress), Hg27
Cranmer, Thomas, archbishop of
Canterbury, Fe56
Craufurd, Robert, (d. 1812), Gh17
Crichton-Browne, James (d. 1938), Hh35
Cromwell, Oliver (d. 1657), Ac105;
Fb3–4,59,c24,d23,i3,28–29; Mh4
Cromwell, Thomas, 1st earl of Essex,
Fb85
Cross, Richard Assheton, (d. 1914), Hb54
Crosthwaite, Peter (d. 1808), Gj23
Cruickshanks, Eveline (historian), Gb15
Cubitt, James (d. 1912), Hl8
Cudworth, Ralph (d. 1688), Fb31,j50
Culhwch, (Arthurian hero), Ji2
Culpeper, Nicholas (d. 1654), Fl26
Cumberland, earl of, see Clifford
Cunningham, Allan (historian), Bc147
Curlew, Gilbert (d. 1908), Hh28
Curll, Edmund (d. 1747), Gc1

Custance, Admiral Sir Reginald Neville
(d. 1935), Hj1
Cuthbert, St, Dc36,e32; Ee2

Dacre, Thomas, 3rd lord (d. 1525), Fb24
Dafydd ap Gwilym (poet), Ji5,8
Dalgairns, John Dobree (d. 1849), He53
Dalrymple, James, 1st viscount Stair, Kc8
Dalrymple, John, 2nd earl of Stair, Gb12
Dalton, John (d. 1844), Gj23
Darcy, Patrick (d. 1668), Mi3
Dare, Joseph (d. 1883), Hg42
Darwin, Charles (d. 1882),
Hb55,e31,k28,46,72,l114; Nd12
David I (king of Scots), Ka1
Davies, David (d. 1890), Hf101
Davies, Emily (d. 1921), Hb18
Davies, Walford (composer), Ij3
Davy, Sir Humphrey (d. 1829), Gd10;
Hk38
Dawkins, Professor Sir William Boyd (d.
1929), Aa169
Dayman, Rev. Alfred Jeken (d. 1875),
He4
de Bray, Emile Frederic de (fl 1850s), Hd2
de Burnham, John (the younger), Jf1
de Gaulle, Charles (d. 1970), Id49
de la Poer, William, baron Beresford, Mi1
de Monmort, Pierre Rémond (d. 1719),
Gj5
Dee, John (d. 1608), Fj34
Deiniol, St, Je9
Delaval, Elizabeth (d. 1735), Fk25
Delaval, Sir John Hussey, baron (d.
1808), Gi84
Dening, Sir Esler (d. 1977), Id10
Derby, earl of, see Ferrers, Stanley
des Roches, Peter, bishop of Winchester,
Eb9
de Valera, Eamon (d. 1975), Mb18
Devereux, Robert, 2nd earl of Essex,
Fb54
Devlin, Lord Justice, If114
Devon, countess of, see Courtenay
Devonshire, duke of, see Cavendish
Dewar family, Bd25
Disraeli, Benjamin, 1st earl of
Beaconsfield, Hb53,75,e23
Dodd, Catherine I. (d. 1932), Hi25
Doddridge, Sir John (d. 1628), Fc4
Donnchadh Mór Ó Dálaigh (fl. c.1400),
Li1

Index of Personal Names

Dorchester, bishop of, *see* Eadnoth I
Dorset, earl of, *see* Sackville
Drake, Sir Francis (d. 1596), Fi27
Drelincourt, Pierre (exile), Le13
Drusius, Johannes (Hebraist), Fk26,90
Dudley, Ambrose, 1st earl of Warwick, Fb1
Dudley, Robert, 1st earl of Leicester, Aa2; Fb1,58; Kb1
Dunbar, William (fl. 1480), Ki2
Dundas family, Bb77
Dundee, viscount, *see* Graham
Dunhill, Alfred (industrialist), If13
Duns Scotus, John (d. 1308), Ei93
Dunstan, archbishop of Canterbury, Bc3; Da6,c7,12,33–34,41,e6,36,48,f67; Ee37; Li8
Dupin, Charles (d. 1873), Hk9
Durnford, Colonel Anthony W. (d. 1879), Ng7
Dwynwen, St, Je9

Eadnoth I, bishop of Dorchester, Db9
Eadred (king), Df31
Eastlake, George (fl. 1813), Gh46
Ebbut, Norman (d. 1968), Ib50
Eden, Sir Anthony (d. 1977), Ib72,d22
Edgar (king), Db6,d32
Edmonds, Sir James (d. 1956), Ac56
Edmund (king), Da7
Edmunds, Henry (inventor), Hk65
Edward I (king), Ea10,c4,e16,60,h17; Jh2
Edward III (king), Ea3,c1,h4,i63,97
Edward IV (king), Ea25,36,39,d13
Edward the Confessor (king), Da2,d33; Ec23
Edward V (king), Fb33
Edward VI (king), Fb53
Edward VII (king), Ha2
Edward VIII (king), Bd41
Edwards, Sir Owen Morgan (d. 1920), Hl98
Egerton, Rev. John Coker (priest), Hg86
Egerton, Sarah Fyge (authoress), Fk9
Elcho, Lady Mary (correspondent), Ib69
Eldred family, Bd26
Eleanor of Aquitaine (queen), Ea32
Eleanor of Castile (queen), Ea10
Eleanor of Provence (queen), Ea28
Elgar, Edward (d. 1934), Ij63
Eliot, George (d. 1880), Hl6,71
Eliot, Sir Charles (d. 1931), Id98

Elizabeth I (queen), Bc68,122; Fa10,b30, 39,54,83,c21,f27,k17,34
Elizabeth of York (queen), Ea44
Ellis, Henry Havelock (d. 1939), Hb80
Elster, Jon (political theorist), Ac110
Elton, Sir Geoffrey (historian), Fa6
Ely, bishop of, *see* Arundel, Lisle
Emma (queen), De18
Eochaidh Ó Eoghusa (d. 1612), Li1
Erasmus, Desiderius (d. 1536), Fk79
Erskine of Cambo family, Ga15
Esau, Abraham (d. 1901), Ng34
Essex, earl of, *see* Bourchier, Cromwell, Devereux
Eugenius IV (pope), Ed5
Eustace II, count of Boulogne, Ed12
Evans, Christmas (d. 1838), Ge33–34
Evans, M.C.S. (historian), Bc101
Evans, Theophilus (d. 1757), Ac113
Evelyn, John (d. 1706), Fk19,l23
Everdon, Silvester de, bishop of Carlisle, Ee73
Exeter, bishop of, *see* Grandisson, Moorman, Quinel
Eyre, Edward John (d. 1901), Nc8
Eyres family, of Warrington, Bd37

Fagel, Hendrik (d. 1838), Aa39
Fairfax, Thomas, 3rd lord, Fi10
Fantosme, Jordan (chronicler), Eh24
Faraday, Michael (d. 1867), Gd10,j13; Hk68
Farnworth, Richard (d. 1666), Fe55
Farrar, John (lawyer), Gf10
Fastolf, Sir John (d. 1459), Eg23
Fawcett, Millicent Garrett (d. 1929), Hb18
Fawcett, Sir William (d. 1804), Gh6
Felton, John (d. 1434), Ee23
Ferrers, John (d. 1324), Ec4
Ferrers, Robert de, 6th earl of Derby, Ec4
Fielding, Henry (d. 1754), Gi41,62
Filmer, Sir Robert (d. 1653), Fj7
Finch, Heneage, 6th earl of Aylesford, Gf22
Fisher, Jazeb Maud (d. 1779), Ga17
Fisher, John, bishop of Rochester, Fe74
Fisher, Marshall (d. 1899), Hl59
FitzGerald, Gerald, 8th earl of Kildare, Lc1
FitzNigel, Richard (d. 1198), Ec17
FitzRalph, Richard, archbishop of Armagh, Le23

Index of Personal Names

Fleming, Richard, bishop of Lincoln, Ee53

Fletcher, Andrew (d. 1716), Kb6

Fletcher, John (d. 1625), Fk60

Foote, George William (d. 1915), He48

Forbes, Sir William (d. 1806), Gi27

Forgan, Dr Robert (fascist), Ib2

Forman, Harry Buxton (d. 1917), Ij23

Forster, Sir John (d. 1602), Fi18

Forsyth, David (d. 1934), Hi15

Fortescue, Sir John (d. 1933), Ac56

Foulis, Sir John, of Ravelston (d. 1706), Fc22

Fowler, Lieutenant-General Sir John Sharman (d. 1939), Ii25

Fox, Charles James (d. 1806), Gb20

Fox, George (d. 1691), Fe27,32,45

Franklin, Lady Jane (d. 1879), Hg82

Franklin, Sir John (d. 1847), Hd2

Fraser, Alexander Campbell (d. 1914), Hl96

Fraser, Hon. William (soldier), Ii19

Freud, Sigmund (d. 1939), Ij70

Froissart, Jean (d. 1410), Ei28

Froude, William (d. 1879), Hk11

Fryatt, Captain (naval officer), Ii1

Gage, John (d. 1842), Hl9

Gallagher, Frank (d. 1962), Mb33

Gandhi, Mahondas Karamchand (d. 1948), Nb9,25

Gascoyne, Joel (cartographer), Ab51

Gaskell, Elizabeth (d. 1865), Hl68

Gaskell, Samuel (d. 1886), Hk27

Gaunt, John of, 2nd duke of Lancaster, Ea21,b8

Geoffrey of Monmouth (d. 1154), Ei25; Fa7

George, Dr William (d. 1967), Hb43

George, Henry (d. 1897), Hb76

George, Henry (d. 1916), Hf18

George III (king), Gb18,26

George IV (king), Hb40

George V (king), Bd41

George VI (king), Bd41

Germanus, abbot of Cholsey, Dc23

Gibbon, Edward (d. 1794), Gi83

Gibson, Margaret (historian), Bc115

Giffard family, Ek17

Gigur, John (fl. 1480), Ea39

Gilbert, Sir W.S. (d. 1911), Hl17

Gildas (monk), Cb20

Gillows, Robert (d. 1772), Ge7

Gilmore, Claire G. (historian), Hf46

Gipps, Sir George (d. 1847), Nf35

Gladstone, W.E., Hb10,l116

Gladstone, William Henry (d. 1891), Hb8

Gloucester, bishop of, see Warburton

Godfrey, Garrett (fl. 1530), Ff16

Godiva, lady (of Coventry), Ei31

Gonville, Edmund (d. 1351), Ee10

Gordon, Thomas (d. 1750), Gb19

Gore, George Ormsby, 3rd baron Harlech (d. 1938), Hb11

Gough, Brigadier General Sir John Edmond (d. 1915), Hj3

Gould, F.J. (d. 1938), Hb59

Gower, John (d. 1408), Ej65

Graham, John, 1st viscount Dundee, Fb71

Grandisson, John, bishop of Exeter, Ee14,55

Grant, A.J. (historian), Hi20

Grant, Sir Alexander (d. 1937), If1

Green, Thomas Hill (d. 1882), Hb1

Grenshields, Rev. James (clergyman), Kc16

Gregory I (pope), Li8

Grenville, George (d. 1770), Aa53; Gb27

Greville, Fulke, 1st baron Brook, Fk74

Grey, Charles, 2nd earl, Hb24

Grey, Sir George (d. 1898), Nc13

Griffin, Sir John, 4th lord Howard de Walden, Gg24

Grimshaw, William (d. 1763), Ge4

Grose, Daniel (d. 1838), Mi18

Grose, Francis (d. 1791), La8

Grosseteste, Robert, bishop of Lincoln, Ei38,51

Grotius, Hugo (jurist), Kc8

Grove, William Robert (d. 1896), Hk49

Gruffudd Fychan ap Gruffudd ab Ednyfed (poet), Ji3

Guibert of Nogent (chronicler), Ee1

Guilbert, Yvette (actress), Hl51

Guilford, earl of, see North

Guthlac, St, of Crowland, Dc8

Gwyn of Dyffryn family, Gf3

Gwynn, Aubrey (historian), Mi8,20

Hadden, Thomas (metalworker), Hg94

Hailey, William Malcolm, baron (d. 1969), Na3

Halley, Edmond (d. 1742), Fl6

Halloran, Laurence Hynes (d. 1831), Mg9

Hamilton, Cicely (dramatist), Hg81

Index of Personal Names

Hamilton, James Archibald (d. 1815), Aa140

Hamilton, Thomas (d. 1842), Hj8

Hampden, Renn Dickson (d. 1868), He63

Hanboys, John (fl. 1470), Ei50

Handlo, Robert (fl. 1526), Ei50

Hardie, James Keir (d. 1915), Hb7

Harding, Sir Edward John (d. 1954), Nc11

Harding, Thomas (d. 1572), Fe89

Harlech, baron, *see* Gore

Harley, C.K. (historian), Gf13

Harley, Sir Robert (d. 1656), Fe25

Harold II Godwineson (king), Dd26

Harris family, Bd39

Harris, Richard (slaver), Ga25

Harriss family, Gi20

Hartlib, Samuel (d. 1670?), Fj33

Hartnett, Sir Laurence (d. 1986), Ha8

Harvey, William (d. 1657), Fl1

Hastings, Henry, 3rd earl of Huntingdon, Fb58

Hastings, Selina, countess of Huntingdon, Ge51

Hatton family, Fk66

Hawes, Stephen (d. 1523?), Ei24

Hayes family, Bb124

Hayes, Richard Joachim, OFM, (d. 1824), Me3

Head, Sir Francis Bond (d. 1875), Nc6

Headlam, Sir Cuthbert (d. 1964), Ib6

Healy, Timothy Michael (politician), Mb7

Heath, Robert (d. 1800), Ge12

Heathcoat, John (d. 1861), Hh3

Hellins, Rev. John (d. 1827), Gi21

Henry I (king), Ea30,c22

Henry II (king), Ea19,35,c28,h24

Henry III (king), Ea28,b2,9–10,e30,73, g44; La6

Henry V (king), Ea1

Henry VI (king), Ed3

Henry VII (king), Ea15,36; Fa6,b62,j6,k23

Henry VIII (king), Fb29,46,i7

Henry, prince of Wales (d. 1612), Bc122; Fb15

Henson family, Eg34

Herbert, Arthur, 1st earl of Torrington, Fi15

Herbert, Edward, 1st lord Herbert of Chirbury, Fj11

Hereward the Wake (outlaw), Ea26

Hey, John (d. 1815), Ge50

Hickman, A.J., Hc9

Higgins, H.B. (d. 1930), Mg22

Hildebrand, Bror Emil (numismatist), Bc139

Hill, Christopher (historian), Fa5

Hobbes, Thomas (d. 1679), Bc14; Fj2,15,25–26,31,36,43,50

Hoccleve, Thomas (d. 1450), Ej65

Hodges, William (d. 1797), Nh39

Hodgkinson, Richard (d. 1847), Bb219

Hofmann, August Wilhelm (scientist), Hk64

Hogarth, William (d. 1764), Gi79,j18

Holt, Joseph (d. 1826), Mb23

Home, John (d. 1808), Ge43

Hood, Robin (outlaw), Eh3

Hooker, John (d. 1601), Fk68

Hoover, Herbert (d. 1964), Id30

Hopkins, Gerard Manley (poet), Hl107, 122

Hoppit, Julian (historian), Gf13

Hopton, Elizabeth (d. 1498), Eg6

Horner, John (b. 1911), Ia19

Horsley, Samuel, bishop of Rochester, Ge30,41

Hoskins, W.G. (historian), Ac8,80

Hotham, Sir John, the elder (d. 1645), Fi8

Houghton, baron, *see* Milnes

Howard, John, 1st duke of Norfolk, Ef18

Howard, John (d. 1790), Gc10

Howard, Sir Michael (historian), Aa13, c20,74; Bc150

Hubberthorne, Richard (pamphleteer), Fe46

Huddleston, Sir Hubert (d. 1950), Nc54

Hudson, Pat (historian), Gf13

Hudson, W.H. (d. 1922), Hk72

Hugh atte Fenne (bibliophile), Ei94

Hugh of Avranches, 1st earl of Chester, Ea31

Hughes, Thomas (d. 1896), He22

Hume, David (d. 1776), Ac75; Bc15; Gi13,18,36,43,75

Hungerford family, Ee38

Hunter, John (d. 1793), Gj12

Huntingdon, countess of, *see* Hastings; earl of, *see* Hastings, Hertford, Seymour

Hurst, John G. (archaeologist), Aa20,c64; Bc191

Hussey, Walter (musician), Ie13

Hutcheson, Francis (d. 1746), Fj16

Hutchinson, Thomas (regicide), Fb68

Hutton, James (d. 1797), Gj4

Huxley, Sir Julian (d. 1975), Ij40

Hyde, Edward, 1st earl of Clarendon, Fj20

Innocent III (pope), Ed1
Islip, Simon, archbishop of Canterbury, Eb7

Jackson, Margery (d. 1812), Ga13
Jackson, R.V. (historian), Gf13
James, David Bloomfield (d. 1900), He5
James I (king), Fa13,b9,j6,k35
James II (king), Fb86,e60,94,j18; Mb14
James IV (king of Scotland), Fb28
Jebb, John (d. 1786), Ge18
Jeffrey, George (d. 1685), Fk85
Jenkinson, Charles (d. 1808), Gb27
Jenner, Edward (d. 1823), Gj9
Jerusalem, bishop of, see Alexander
Jewel, John, bishop of Salisbury, Fe89
Joffre, General Joseph (d. 1931), Ii47
John (king), Ea9,27,c30,d1
John of Salisbury (d. 1180), Ei14
Johnson, Lyndon (d. 1973), Id61
Johnson, Samuel (d. 1784), Gd8
Johnson, William (Jesuit), Fe65
Johnston, Archibald, lord Wariston (d. 1663), Ke7
Jones, Captain Frederick (d. 1834), Ha7
Jones, Dr Thomas (d. 1955), Ia11
Jones, Sir William (d. 1794), Nh4
Jones, Theophilus (d. 1812), Ac113
Joscelyn, John (d. 1603), Da3
Joule, James Prescott (d. 1889), Hk16

Kaldor, Nicholas (economist), If103
Katsunoshin, Admiral Yamanashi (d. 1967), Id26
Kaunda, Kenneth (politician), Nf22
Kaye, Sir Arthur (d. 1726), Fb82
Kearney, Felix (d. 1830), Mg18
Keill, John (d. 1721), Gi75
Kelly, John (d. 1866), Mg17
Kelly, Ned (d. 1886), Mg17
Kelvin, baron, see Thomson
Kemp, Cardinal John, archbishop of York, Ed5
Kennedy, Captain William (d. 1876), Hf32
Kennedy, Malcolm Duncan (d. 1935), Hj13
Kennett-Barrington, Sir Vincent (d. 1903), Hj11
Kent, earl of, see Burgh
Keynes, John Maynard (d. 1946), Hf3; If31,53,80–81,97
Kildare, earl of, see FitzGerald

King, William Lyon Mackenzie (d. 1950), Ng13
Kipling, Rudyard (d. 1936), Hl6
Kirkby, John, bishop of Carlisle, Ee72
Kitchener, Horatio Herbert, 1st earl, Ii66
Kneller, Sir Godfrey (painter), Fk20
Knight family, Bd34
Knox, John (d. 1572), Ke6,9,13,24
Kuehnrich, Paul Richard (d. 1932), Hf96
Kyozo, Kikuchi (d. 1942), Id45

Lancaster, duke of, see Gaunt; earl of, see Thomas
Lancaster, Joseph (d. 1838), Hi33
Lancaster, Thomas, 2nd earl of, Ei29
Landon family, Bd35
Lanfranc, archbishop of Canterbury, Ee58,i12
Langhorn, Joseph (shoemaker), Hg40
Langton, Walter, bishop of Coventry & Lichfield, Ee43
Lansbury, George (d. 1940), Ib77
Larkin, James (revolutionary), Ib96
Latimer, Hugh, bishop of Worcester, Fe68
Latimer, William, Lord (d. 1381), Ed8
Latus family, Bd33
Laud, William, archbishop of Canterbury, Fe20,52,92; Me2
Law, John (d. 1729), Gf35,46
Lawrence family, He71
Lawrence, T.E. (d. 1935), Nh12
le Strange, Henry Styleman (d. 1862), He37
Lee, Nelson James (d. 1923), Aa157
Lee, Richard Nelson (d. 1872), Aa157
Leicester, earl of, see Dudley, Montfort
Lely, Sir Peter (painter), Fk20
Leofwine, Db1
Lessius, Leonardus (of Louvain), Gi2
Levy, Alfred Goodman (anaesthetist), Ij52
Lewes, G.H. (d. 1878), Hl6
Lewthwaite family, Bd33
Leyser, Karl (historian), Bc71
Light, Colonel William (d. 1839), Nh15
Lincoln, bishop of, see Fleming, Grosseteste, Remigius
Lindsey, Theophilus (d. 1808), Ge16
Lisle, Thomas de, bishop of Ely, Ec1
Lister, Anne (d. 1840), Hg38,87
Lister, Joseph, 1st baron, Hk40
Liston, Henrietta (d. 1828), Hl108
Liston, Sir Robert (d. 1836), Hl108

Index of Personal Names

Llandaff, bishop of, *see* Watson
Llewelyn, Mary (d. 1874), He40
Llewelyn, Richard Pendrill (d. 1891), He40
Lloyd George, David (d. 1945), Hb33,43, 58,85; Ia11
Llwyd, Morgan (d. 1659), Fe67
Llywelyn ap Gruffudd, prince of Wales, Jb3
Lock, Max (town planner), Ij53
Locke, John (d. 1704), Bc15,155; Fd32, e26,59,79,j4,16,18–19,29,35,42,46,k1, 31,70
Lockwood, Caroline (philanthropist), Bb64
London, bishop of, *see* Theodred
Longespe, William II (d. 1250), Eh14–15
Loudoun, earl of, *see* Campbell
Louis IX (king of France), Eh14
Lovelace, Richard (d. 1658?), Fk56
Lugard, Frederick Dealtry, baron Lugard of Abinger, Aa139
Lushington, Stepehen (d. 1873), He80
Lydgate, John (d. 1450), Ei24,70
Lyster, George Fosberry (d. 1899), Hf53

Macanty, Rose (d. 1958), Hl6
Macartney, Lord George (d. 1806), Gd30
Macaulay, Catharine (d. 1791), Gi39
Macaulay, T.B., 1st baron, Ac75; Nc50
MacDonagh, Oliver (historian), Aa21,c42, 66; Bc143
MacDonnell, Randal, 2nd earl of Antrim, Mb24
Machiavelli, Niccolò (d. 1527), Bc14
MacLean, John (d. 1923), Ib85,96
Macmillan, Harold, 1st earl, Id49
Maelgwn Gwynedd, Jb2
Maguire, Hugh (d. 1600), Li1
Mair, John (d. 1550), Fj12
Malleson, Elizabeth (d. 1916), He77
Malthus, Thomas (d. 1834), Hl91
Manhall, Alfred (d. 1924), Hf3
Manley, Mary Delarivière (d. 1724), Fk7–8
Manning, Cardinal Henry Edward (d. 1892), He51–52,59,62,64,66,68,l117
Mannix, Daniel (d. 1963), Mb18
Manson, Sir Patrick (d. 1922), Hk17
Marbeck, John (d. 1585), Fe12,k62
Marchant, George (engineer), Gf61
Margaret of York, duchess of Burgundy, Fk33
Margaret Tudor (queen of Scotland), Fb28
Marks, Sammy (d. 1920), Ne17

Marlborough, duchess of, *see* Sarah; duke of, *see* Churchill
Marlborough, Sarah, duchess of, Gb16
Marston, John (d. 1634), Fk45
Martin, Kingsley (d. 1969), Ib47
Martin V (pope), Ed6
Martineau, James (d. 1900), Hl96
Martyn, Sir William (d. 1504), Eg5
Marvell, Andrew (d. 1678), Fk60
Marx, Karl (d. 1883), Bc14; Ei17; Hg4
Mary I (queen), Fb53,e35
Mary II (queen), Fb70,79,88
Mary Magdalene, St, He46
Mary (queen of Scots), Kb1,5
Masham family, Bd36
Mathew, Fr. Theobald (d. 1856), Me7,11
Matilda, abbess of Essen, Da20
Maudsley, Henry (d. 1918), Hk54
Maxwell, James Clerk (d. 1879), Hk32
Mayo, earl of, *see* Bourke
McBane, Donald, Kh7
McCrie, Dr Thomas (d. 1835), Ke20
McIntosh, William (d. 1877), He67
McMillan, Margaret (d. 1931), Hl104
McNamara, Francis (poet), Mi14
McNeill, Malcolm (fl. 1890), Hh31
McShane, Harry (d. 1988), Ac92; Bc175
Menzies, Sir Robert (d. 1978), Md1
Meredith, George (d. 1909), He48
Meyer, Adolf (psychiatrist), Ij41
Middleton, Sir Gilbert de (d. 1318), Ea38
Mildred, St, Ee54
Miles, Alice Catherine (debutante), Hg66
Mill, James (d. 1836), Nh26
Mill, John Stuart (d. 1873), Bc14,123; Hb29,44,47–48,72,l1,28,41,49,64, 76,94,106; Nh35
Milner, Alfred, 1st viscount, Nc34,f42
Milnes, Richard Monckton, 1st baron Houghton, Hl6
Milthrith, St, Dc38
Minton, Herbert (d. 1858), Hg6
Mitchell, Thomas (d. 1860), He14
Molesworth, Dr J.E.N. (d. 1877), He3
Molette, Charles (historian), Aa60
Montagu, Basil (d. 1851), Aa178
Montcalm, marquis de, Gh44
Montfort, Simon de, 5th earl of Leicester, Eb2,10
Montgomery, Field Marshal Bernard Law, 1st viscount Montgomery of Alamein, Ii7
Moore-Stevens, J.C. (d. 1903), Hf7

Moorman, John, bishop of Exeter, Ee62
Moran, Charles, 1st baron, Aa28; Ij57
More, Henry (d. 1687), Fb31,e17,44,k41
More, Sir Thomas (d. 1535), Ek62; Fk79
Morein, Mary (actress), Fk87
Morgagni, Giovanni Battista (physician), Gj20
Morgan family (of Tredegar), Gc5
Morrell, Lady Ottoline (d. 1937), Ij79
Morris, William (d. 1896), Aa83; Hl12,79
Morris, William H. (historian), Bc101
Morstede, Thomas (d. 1450), Ei48
Morton, Cardinal John, archbishop of Canterbury, Ee50
Moucan (cleric), Je8
Moulin, Jacques de, FRS, Fl21
Mountbatten, Louis, of Burma, 1st earl, Aa118
Mulcahy, Richard (d. 1971), Mb30
Mungo, St, Ke17
Murdoch, William (d. 1839), Hf43
Murford, Nicholas (salter), Fg22
Murray, Albert Victor (Methodist), Gg14
Murray, Lindley (d. 1826), Gi3
Musgrave, Sir William (d. 1800), Gi35,53

Napier, General Sir Charles (d. 1853), Ne20
Napoleon (emperor), Hb61
Nehru, Jawaharlal (d. 1964), Nb25,27
Nepean, Evan (d. 1822), Nc3
Nesbitt, Arnold (d. 1779), Gb29
Netter, Thomas (d. 1430), Ei39
Neville, George, archbishop of York, Ek19
Neville, Hugh de, Ec30
Neville, Ralph 4th baron (d. 1388), Ek31
Newcastle, duchess of, see Cavendish; duke of, see Pelham-Holles
Newman, Cardinal John Henry (d. 1890), He1,11,18,38,63
Newton, James (fl. 1736–86), Ge22
Newton, Sir Isaac (d. 1727), Aa50,67,98, 162,c95; Bc87; 95; Bc87; Fl3–4,6–8, 11–12,27
Nicholson, Harold (d. 1965), Hl6
Ninian, St, Ke4
Nish, Ian (historian), Bc19
Norfolk, duke of, see Howard
Norman, Montagu Cullet, 1st baron, If84,87
Normandy, duke of, see Richard II
Norris, Sir John (d. 1597), Fi27

North, Frederick, lord, 2nd earl of Guilford, Gb10
North, Roger (d. 1734), Gi70
Northampton, earl of, see Compton
Northumberland, earl of, see Percy
Nynia, St, Ke4

Oakeshott, Michael (d. 1990), Ac18
Oates, John C.T. (d. 1990), Aa150
O'Brien, James Brankerne (d. 1864), Hg4
Ochino, Bernardino, (d. 1564), Fe62
Ockham, William of (d. 1349), Ei93
O'Connell, Daniel (d. 1847), Mb20,i7
O'Donohoe, Patrick (d. 1854), Mb11
Olwen, (Arthurian hero), Ji2
O'Neill, Hugh, 3rd earl of Tyrone, Mh3
Orford, earl of, see Walpole
Orme, Robert (d. 1801), Nh13
Ormerod, Eleanor (d. 1901), Hk19
Orrery, earl of, see Boyle
Ossory, bishop of, see Bale
Ossulton, baron, see Bennett
Osulf Thein (moneyer), Dd29

Paine, Thomas (d. 1809), Gb25
Paley, William (theologian), Ge50
Palmer, Roger, 1st earl of Castlemaine, Gd16
Panizzi, Sir Anthony (d. 1879), Ab42
Pankhurst, Sylvia (d. 1960), Aa159; Bc50; Hb22,32; Ib11,60,95,j31
Paris, Matthew (d. 1259), Ei40,64
Parkes, James (d. 1981), Ij51
Parnell, Charles Stuart (d. 1891), Mb6–8,12–13,19
Parr family, Ej24
Parry, Terry (trade unionist), Ia36
Passfield, baron, see Webb; baroness, see Webb
Paston, Robert, 1st earl of Yarmouth, Fk89
Paterson, Sir Clifford (d. 1948), Ij21
Patrick, St, Le19
Pattison, James (d. 1805), Gh32
Payne, Edward John (d. 1904), Hl20
Peckitt, William (d. 1795), Bb26; Gi10
Pecock, Reginald, bishop of Chichester, Ei79
Peel, Sir Robert, Hb27
Pelham-Holles, Thomas, 1st duke of Newcastle, Gb26
Percy, Algernon, 10th earl of Northumberland, Fi1
Percy family, Eg50

Percy, Henry, 6th earl of
  Northumberland, Fb43
Percy, Henry, 9th earl of
  Northumberland, Ff18
Perón family, of Argentina, Id44
Perry family, Bd38
Philip II Augustus (king of France), Ed1
Philip, Wilberforce Buxton (d. 1888),
  Nh36
Piggott, Major General FSG (d. 1966),
  Hl11
Piggott, Sir Francis Taylor (d. 1925),
  Hl11
Pitt, William, the younger (d. 1806),
  Gb7,d28
Plomer, William (d. 1974), Id2
Plumptre, James (d. 1832), Ga20
Pocahontas (princess), Fd19
Pond, Arthur (d. 1758), Gi48
Pope, Alexander (d. 1744), Gi73
Pope, John Adolphus (mariner), Gd11
Pope-Hennessy, Sir John (d. 1891), Nc31
Popham family, Fi22
Postan, Sir M.M. (historian), Ba40
Powle, Sir Stephen (civil servant), Fb81
Prestage, Edgar (d. 1951), Hl69
Prevost family, Bd19
Price, Joseph (d. 1807), Ge17
Price, Richard (d. 1791), Gb8,31
Price, William (seaman), Fd3
Priestley, Joseph (d. 1804), Gb11,e20
Prince, Henry James (preacher), He72
Prior, Matthew (d. 1721), Gi86
Provost family, Bd19
Prynne, William, Fe92
Ptolemy (geographer), Cb8,34
Pue, Gwilym (monk), Fl13
Puleston, John (d. 1524), Fb64
Pusey, Henry Bouverie (d. 1882), He1
Pym, John (d. 1643), Fb77

Quian Long (emperor of China), Gd30
Quinel, Peter, bishop of Exeter, Ee28

Rae, John (d. 1871), Hf11
Rae, Thomas Ian (d. 1989), Ac33
Rainborow, William (d. 1642), Fi1
Ralegh, Sir Walter (d. 1618), Fk51
Randolph, Thomas (d. 1590), Kb1
Ranulf II, 2nd earl of Chester, Ea13–14,23
Rawlinson, Henry, baron Rawlinson (d.
  1925), Ii48
Readwald (king), Db27

Reed, Rev. Andrew (d. 1862), He30
Reilly, Bernard (emigrant), Mg20
Remigius, bishop of Lincoln, Ee5
Rhodes, John M. (d. 1909), Hh2
Ricardo, David (d. 1823), Hf55,l91
Rich, John (d. 1761), Gi54
Richard I (king), Bc213; Ea5,19,c28,d10,
  h20,i37,41,45–46,59
Richard II, duke of Normaindy, Ek22
Richard II (king), Eb4
Richard III (king), Ea15,i89–91,92
Richard of Devizes (chronicler), Ei59
Richardson, Sir Albert (d. 1964), Ij90
Richmond, countess of, see Beaufort
Ridge family, Bd30
Rievaulx, baron, see Wilson
Robert I (king of Scots), Eh16
Robertson, Field Marshal Sir William (d.
  1923), Ii67
Robertson, General Sir Horace
  (Australian), Ng14
Robinson, Fr. Cutts (d. 1922), He52
Robinson, Kenneth (historian), Bc130
Robinson, Rear Admiral Tancred (d.
  1754), Gg7
Robson, J.M. (historian), Bc123
Rochester, bishop of, see Fisher, Horsley,
  Sprat
Rochford, earl of, see Zuylestein
Ross, John, bishop of Carlisle, Ee72
Ross, John (d. 1607), Fa7
Round, John Horace (d. 1928), Hb65
Rous, John (fl. 1812), Gh18
Rousseau, Jean Jacques (d. 1778),
  Gi64,72
Rowland, David (translator), Fk22
Rowlands, Daniel (d. 1790), Ge35
Rubinstein, W.D. (historian), Hf6,g84
Ruskin, John (d. 1900), Hl121
Russell, Bertrand, 3rd earl, Aa37; Ba134;
  Hl48
Russell, Francis, 2nd earl of Bedford, Kb1
Rustat, Tobias (d. 1693), Fk69
Rutherford, Ernest, 1st baron, Ij30

Sackville, Edward, 4th earl of Dorset,
  Fb76
Saddler, Michael (d. 1941), Hi14
Salisbury, bishop of, see Jewel; earl of, see
  Cecil; marquis of, see Cecil
Samuel, Herbert, viscount (d. 1963),
  Ba193
Sansom, Sir George (d. 1965), Id20

Satow, Sir Ernest (d. 1929), Hd4,9
Schliemann, Heinrich (d. 1890), Hl116
Schroder family, Ba160
Scott, Sir Walter (d. 1832), Hj8; Ke20
Scrope family, of Masham, Ej43
Scrope, Stephen, of Castlecombe (d. 1472), Eg23
Scully, Denys (d. 1830), Me9
Secker, Thomas, archbishop of Canterbury, Ge2,26
Seeley, John Robert (d. 1895), He75
Selden, John (d. 1654), Fj3
Seton-Watson, Robert William (d. 1951), Hd1
Seymour, Edward, 1st earl of Hertford, Fk13
Seymour, Sir Thomas, 1st baron Seymour of Sudley, Fb11
Shaftesbury, earl of, see Cooper
Shaka (king of the Zulus), Nb11
Shakespeare, William (d. 1616), Ei46; Fk11,43,59
Shaw, Benjamin (mechanic), Bd24
Shaw, George Bernard (d. 1950), Hb9
Shaw, Thomas George (d. 1927), Hk4
Shepherd, William (merchant), Gf18
Sheridan, Thomas (d. 1738), Gi85
Shirley, James (d. 1666), Fk24
Shoyo, Tsubouchi (d. 1935), Hl93
Shrewsbury, earl of, see Talbot
Sibbes, Richard (divine), Fe22
Siddons, Sarah (d. 1831), Gi61
Sidney, Algernon (d. 1683), Fj24,39
Sidney, Sir Philip (d. 1586), Fk11,74,81
Sigsworth, Eric M. (historian), Bc205
Simon, John Allsebrook, 1st viscount, Ib23
Simoni, Anna E.C. (historian), Aa149; Bc133
Simpson, Thomas (d. 1761), Gj14
Singh, Duleep (d. 1893), Nc9
Sir Galahad (Arthurian knight), Ji10
Siward, Richard (d. 1248), Jb1
Skelton, John (d. 1529), Fb84,k71
Skipwith, Sir William (d. 1610), Fk45
Skotkonung, Olaf (king of Sweden), Db5,31
Slattery, Michael, archbishop of Cashel, Me13
Sloper, George (d. 1810), Ga9
Smiles, Samuel (d. 1904), Aa96
Smith, Adam (d. 1790), Ba195,c29–30; Gc3,f1,8,10,34,42,49,51–52,56,58,

60,62,64,70,i14,17,28–29,40,46,66, 71,j16; Hf11
Smith, John (d. 1631), Fd19
Smollett, Tobias (d. 1771), Gi62
Soddy, Frederick (d. 1956), Hk21
Sondes family, Fg15
South, Robert (d. 1716), Fe73
Southcott, Joanna (d. 1814), Ge40
Southey, Robert (d. 1843), Gb6
Spence, Thomas (d. 1814), Gb34
Spencer, Herbert (d. 1903), Hb82
Spenser, Edmund (d. 1599), Fc21
Spira, Francis (d. 1548), Fe58
Sprat, Thomas, bishop of Rochester, Fe43
Spurgeon, Charles Haddon (d. 1892), He60
St Albans, viscount, see Bacon
St David's, bishop of, see Bull, Thirlwall
St John, Henry, 1st viscount Bolingbroke, Fb22
St Quintin, William Henry (d. 1933), Ij2
Stair, earl of, see Dalrymple; viscount, see Dalrymple
Stalin, Joseph, Id3
Standen, Anthony (spy), Fb38
Stanley, Charles, 8th earl of Derby, Fa4
Stanley, Edward Henry, 15th earl of Derby, Hd11
Stanley, Henry Morton (d. 1904), Hl63
Stanley, Sir William (d. 1495), Eg6,13
Stansfield family, Gg18
Stedman Jones, Gareth (historian), Ac68
Steed, Henry Wickham (d. 1956), Hd1
Stengel, Erwin (psychiatrist), Ij49
Stephen (king), Ea13–14,23–24,30,c20
Stephen, Sir James (d. 1859), Nc48
Stephen, Sir Leslie (d. 1904), Hi17
Stephenson, Sir Augustus (d. 1904), He9
Stevenson, John, Gi20
Steward, Dugald (d. 1828), Gf56
Stiles, Ezra (author), Fb68
Stillingfleet, Edward (divine), Fe44
Stiphel, William du, Ei32
Stokes, George Gabriel (d. 1903), Hk12
Stokes, William (d. 1881), Hd14
Stopes, Marie (d. 1958), Id8; Mk3
Strange, Albert (d. 1917), Hl72
Stuart, Gilbert (d. 1786), Gi87
Stuart, John, 3rd earl of Bute, Gb5
Stukeley, William (d. 1765), Gj7
Sullivan, Sir A.S. (d. 1900), Hl17
Summer, John Bird (d. 1862), He73
Sweetapple, William (b. 1688/9), Gg16

Swift, Jonathan (d. 1745), Gi24,34,37,85
Swire, Joseph (journalist), Id17
Sydney, viscount, *see* Townshend
Sykes, Sir Mark (d. 1919), Ij34
Symeon of Durham (chronicler), Ee64
Symonds, John (d. 1807), Gi56

Tadasu, Hayashi (d. 1913), Hd15
Talbot, George, 6th earl of Shrewsbury, Fb50
Taltarum, Thomas (litigant), Ec26
Tatham, Joseph (d. 1786), Gi58
Tathwell, Cornewall (medical student), Gi11
Tatsui, Baba (d. 1888), Hl5
Taylor, Brook (d. 1721), Gj5
Taylor, Clementia (d. 1908), He70
Taylor, Helen (d. 1907), Hl94
Thatcher, Margaret, baroness, Id19,j77
Theodore, archbishop of Canterbury, Dc27
Theodred, bishop of London, Db37
Thirlwall, Connop, bishop of St David's, Aa75
Thomas, 2nd earl of Lancaster, Ec4
Thomas of Westoe (bibliophile), Ee24
Thomas of Oxford (d. 1427), Bb26
Thompson, Alexander Hamilton (historian), Hi20
Thompson, F.M.L. (historian), Ba122; Hg74
Thomson, James (d. 1748), Gi67
Thomson, William, 1st baron Kelvin, Hk66
Thoresby, Ralph (d. 1725), Fe63
Thwaites family, Bd33
Tierney Clark, William (d. 1852), Hf90
Tito, Josip (d. 1980), Id62–63
Tocqueville, Alexis de (d. 1859), Hb44; Ma3
Toft, Mary (fl. 1726), Gj3
Toland, John (d. 1722), Fb25
Toole, Joseph (d. 1945), Ib48
Torrington, earl of, *see* Herbert
Townshend family, Eg33
Townshend, Thomas, 1st viscount Sydney, Nc3
Travers, John (apostate), Fe61
Trenchard, John (d. 1723), Gb19
Trevelyan, G.M. (d. 1962), Ac24
Trevet, Nicholas (d. 1334?), Ei53; Lh3
Trevisa, John (d. 1412), Ei96
Trollope, Anthony (d. 1882), Hl46

Tryggvasson, Olaf (king of Norway), Db5,31
Tuke, Daniel Hack (d. 1895), Hk14
Turner, William (divine), Fe14
Twining, Thomas (d. 1804), Gi80
Tyrone, earl of, *see* O'Neill

Udern, William, (vicar), Fe57
Unwin, George (d. 1926), Hl103
Urbicus, Q. Lollius (commander), Cb9
Usher, James, archbishop of Armagh, Me2

Valence, William de (d. 1296), Eg44
van der Boxe, Willem Christiaens (translator), Fk83
van Haestens, Henrick (d. c.1629), Aa52
Vanbrugh, Sir John (d. 1726), Gi37
Varigon, Pierre (d. 1722), Gj11
Verney, Sir Edmund Hope (d. 1910), Ng36
Vernon, Edward, admiral (d. 1757), Gh47
Verstegan, Richard (fl. 1590), Fk22
Vertue, George (d. 1756), Bb16
Victor Amadeus II (king of Savoy), Fd31
Victoria (queen), Hk65,l124
Villiers, George, 2nd duke of Buckingham, Fe95
Vinogradoff, Paul (historian), Ac48
Virgil (poet), Ei53
Voltaire, Françoise de, Fb52
von Kienbusch, C. Otto (collector), Bc5

Waddington, George (d. 1869), He42
Wainfleet, William, bishop of Winchester, Ea39
Waley, Arthur (d. 1966), Ij44
Walker, Rev. John (d. 1878), He18
Walpole, Robert, 1st earl of Orford, Gb12
War, Samuel (diarist), Fe85
Warbeck, Perkin (d. 1499), Fb23
Warburton, William, bishop of Gloucester, Ge46
Ward, Sir Henry George (d. 1860), Hd6
Warne, James (d. 1773), Gf32
Warwick, earl of, *see* Dudley
Waterhouse, Alfred (d. 1905), Hl33
Watkin, Sir Edward William (d. 1901), Hk63
Watson, Richard, bishop of Llandaff, Ge50

Index of Personal Names

Watt, James (d. 1848), Gf36
Watts, Isaac (d. 1748), Ge29
Wavell, Field Marshall Archibald (d. 1950), Ii50
Webb, Beatrice, baroness Passfield, Hb74,l39; Id42
Webb, General Daniel, Gh44
Webb, R.K. (historian), Bc13
Webb, Sidney, 1st baron Passfield, Hl39; Id42
Wedgwood Benn, Anthony, (politician), Ib1
Weizmann, Chaim (d. 1952), Ia34,d91–92
Wellesley, Arthur, 1st duke of Wellington, Gh21; Hb61,l83
Wellington, duke of, see Wellesley
Wesley, Charles (d. 1788), Ge31
Wesley, John (d. 1791), Ge10,19,37,39,47,52
Wesley, Susan (d. 1742), Ge23
Weston, Walter (d. 1940), Hl60
Wharton, Thomas, 1st marquis of, Fb63
Whewell, William (d. 1866), Hk47
Whichcote, Benjamin (d. 1683), Fc8
Whidbey, Joseph (d. 1833), Gh37
White, John (divine), Fe87
White, Richard (exile), Fd25
White, Thomas (d. 1676), Fj44
Whitefield, George (d. 1770), Ge27,45,47
Whiteside, Tom (historian), Aa23; Bc87
Whitgift, John, archbishop of Canterbury, Fe29
Wilberforce, William (d. 1833), Aa54
Wilks, Michael (historian), Aa73; Bc12
William I (king), Ea2,16,k22
William III (king), Fb57,69–70,79,d11,j5, 10
William of Drogheda (fl. 1230), Ee68
William of Falaise (b. c.1055), Ea6
William of Jumièges (historian), Ea41
William of Malmesbury (d. c.1143), Ld1
William of Moulins (d. 1100), Ea6
Williams, Charles Hanbury (d. 1759), Ga3
Williams, David (d. 1816), Ge18

Williams, Roger (d. 1595), Fj45
Williams, William (d. 1791), Ge35
Wilmot, Katherine (traveller), Gd26
Wilson, Field Marshall Sir Henry (d. 1922), Mb15
Wilson, Harold, baron Wilson of Rievaulx, Ib55,64,f54
Winchester, bishop of, see Æthelwold, Andrewes, des Roches, Wainfleet
Winfrey, Sir Richard (d. 1944), Ia22
Winstanley, Gerrard (d. 1660?), Fj23,37
Wirgman, Charles (d. 1891), Hl24
Wise, Thomas James (d. 1937), Ij23
Wiseman family, Bd23
Witney, Thomas (mason), Ej41
Wollstonecraft, Mary (d. 1797), Gi69,74,76
Wolsey, Cardinal Thomas, archbishop of York, Bc1; Fb35,37,48,84,d10,e11,75, 91,g37,k12,28,53,76,88
Wood, George (d. 1757), Gh36
Woodville, Elizabeth (queen), Ea44
Woodward, Admiral Sir John (b. 1932), Ii68
Woolf, Leonard (d. 1969), Ib47
Worcester, bishop of, see Cantilupe, Latimer
Wren, Sir Christopher (d. 1723), Fk40
Wroth, Lady Mary (fl. 1621), Fk29
Wulfstan, archbishop of York, Ee81
Wulfstan II, archbishop of York, Db29
Wycliffe, John (d. 1384), Ei42,55,102
Wyndham, John, of Felbrigg (d. 1475), Eg43
Wynflæd (litigant), Db1

Yamba, Dauti Lawton (politician), Nb21
Yarmouth, earl of, see Paston
York, archbishop of, see Kemp, Neville, Wolsey, Wulfstan I, Wulfstan II
Youings, Joyce (historian), Aa112

Zagorin, Perez (historian), Bc74
Zita, St (d. 1272), Ee25,78
Zuylestein, William N.H., 4th earl of Rochford, Gd27

COUNTY ABBREVIATIONS

ENGLAND

Bedfordshire	Bdf	Berkshire		Brk
Buckinghamshire	Bkm	Cambridgeshire		Cam
Cheshire	Chs	Cornwall		Con
Cumberland	Cul	Derbyshire		Dby
Devon	Dev	Dorset		Dor
Durham	Dur	Essex		Ess
Gloucestershire	Gls	Hampshire		Ham
Herefordshire	Hef	Hertfordshire		Hrt
Huntingdonshire	Hun	Isle of Wight		Iow
Isles of Scilly	Ios	Kent		Ken
Lancashire	Lan	Leicestershire		Lec
Lincolnshire	Lin	Middlesex		Mdx
Norfolk	Nfk	Northamptonshire		Nth
Northumberland	Nbl	Nottinghamshire		Ntt
Oxfordshire	Oxf	Rutland		Rut
Shropshire	Shr	Somerset		Som
Staffordshire	Sts	Suffolk		Sfk
Surrey	Sry	Sussex		Ssx
Warwickshire	War	Westmorland		Wes
Wiltshire	Wil	Worcestershire		Wor
Yorkshire	Yks			
Isle of Man	IoM			

SCOTLAND

Aberdeen	Abd	Angus		Ans
Argyll	Arl	Ayr		Ayr
Banff	Ban	Berwick		Bew
Bute	But	Caithness		Cai
Clackmannan	Clk	Dumfries		Dfs
Dunbarton	Dnb	East Lothian		Eln
Fife	Fif	Hebrides		Wis
Inverness	Inv	Kinross	Krs	
Kincardine	Kcd	Kirkcudbright		Kkd
Lanark	Lks	Midlothian		Mln
Moray	Mor	Nairn		Nai
Orkney	Ork	Peebles		Pee
Perth	Per	Renfrew		Rfw
Ross & Cromarty	Roc	Roxburgh		Rox
Selkirk	Sel	Shetland		Zet
Stirling	Sti	Sutherland		Sut
West Lothian	Wln	Wigtown		Wig

County Abbreviations

WALES

Anglesey	Agy	Brecknock		Bre
Caernarvon	Cae	Cardigan		Cgn
Carmarthen	Cmn	Denbigh		Den
Flint	Fln	Glamorgan		Gla
Merioneth	Mer	Monmouth		Mon
Montgomery	Mgy	Pembroke		Pem
Radnorshire	Rad			

IRELAND

Carlow	Car	Cavan		Cav
Clare	Cla	Cork		Cor
Donegal	Don	Dublin		Dub
Galway	Gal	Kerry		Ker
Kildare	Kid	Kilkenny		Kik
Laois	Las	Leitrim		Let
Limerick	Lim	Longford		Log
Louth	Lou	Mayo		May
Meath	Mea	Monaghan		Mog
Offaly	Off	Roscommon		Ros
Sligo	Sli	Tipperary		Tip
Waterford	Wat	Westmeath		Wem
Wexford	Wex	Wicklow		Wic

NORTHERN IRELAND

Antrim	Ant	Armagh		Arm
Derry	Dry	Down		Dow
Fermanagh	Fer	Tyrone		Tyr

INDEX OF PLACES

Aberdeen, Gi45

Aberdeenshire, Bb195; Hg12

Aberglaslyn (Car), Hk39

Aberystwyth, Gf44; university college of, Hk21

Abinger (Sry), Bb67

Abyssinia, Id15

Afghanistan, Ng38

Africa, Fd21,22; Na8,10–11,18,b8,c33, f47,g10, h30–31,37; east, Id9; Ne26; north, Ii32; South, Hj2; Na1,b9,28, c11–12,34,e6,14,17,f12,32,42,g15,24, 34,h8,17,24,27,34,36; southern, Nb7, 11,f44,h7; West, Aa110; Ff15; Hf57; Nc26

Aisgill (Cul), Hf48

Alderney (Channel Islands), Bb159

Algeria, Id49

Almondsbury, Lower (Gls), Bb173

Alsace, Ld3

Alt Clut (Dnb), Kh1

Ambleside (Wes), Ca2

America, Fd18–19,30,h3; Ga18; British North, Nh22; colonial, Bc125; Fd17; North, Aa48,c1,77; Ba20,109,127; Fd16,e31,96; Ga17,23,d12,31–32, e27,45,49,f59, g5,h44; South, Hd17; Ij22; Nh31; United States of, Ba66,190, c18,85; Gh1,8,12–13,16,22,45; Hd3, f14,20,26,62,100,h20,j4,k18,26; Id27,f41,i21,j7,17; Na12,b18,24

Amsterdam, Gf47

Anglesey, Bb22,117; Gb32,e34; Jb2

Antarctic, Id7; ocean, Ab19

Appleton (Chs), Gg6

Appletree (Cul), Ca49

Arabia, Id90

Ardfest (Ken), Li10

Ardiolean (Gal), Le9

Ardmore (Wat), Li10

Argentina, Ba85–86,97,124,b215,c136; Hf36; Id7,40,44,50,66,69,104,j22,27, 37; Mg13; Ne8

Argyll, Bb7

Arkholme (Lan), Hg34

Armagh, Le2,10,22–24

Arwystli (Mgy), Ja1

Ashby-de-la-Zouche (Lec), Ge51

Asia, Hl39; Id67; Ne13; central, Ba93; east, Bc19

Asthall Barrow (Oxf), Df22

Athelhampton (Dor), Eg5

Atlantic, south, Aa142

Auchtermuchty (Fif), Bd22

Audley End (Ess), Gg24

Australia, Aa21,c42,49–50,66; Ba14, c121,143,182,187; Hc6,f61; Id102, f113; Mb11,g7,9,17–18,20–23,i1, 14–16,19; Na12,24,b6,12,15,18–19, 24,c11,28, 40,44,48,d12,e7,13,15,27, f20,27,43, 46,g6,8,14,16–17,h18,22, 25,33; South, Nc13,f19; Western, Ne11

Austria, Ge49; Hd5

Avebury (Wil), Fl5

Avon, Bb18

Axe, river, Ef55

Bahamas, Gh26; Na4,9

Bakewell (Dby), church, Ee60

Balkans, Id3

Ballagh (Tip), Mb4

Baltic, Id100

Bamburgh (Nbl), castle, Ek68

Bampton (Dev), castle, Ek30

Banbury (Oxf), Hf38

Bangor (Cae), Bd1; Je9

Barbados, Fd4,6

Barrington (Cam), Df49

Barrington, Great (Gls), Gc14

Basle (Switzerland), Ee36

Bassingbourn (Cam), Cb2

Bath (Som), Bc171; Cb1; Gi8; Hg20

Bavant, North (Wil), Ca8

Beaulieu (Ham), abbey, Ee42

Beaumaris (Agy), Ja3

Beaurain-sur-Canche (France), castle, Ek57

Bedford, Bb89; Hh6

Bedfordshire, Aa173,b49; Ek10; Hc19

Bedlington (Nbl), Gf20

Belfast, Hf88; Mi9,j5,k3

Berbice (Guyana), Nb31

Berkeley (Gls), Gj9; castle, Ea35

Berkshire, Bb86

Berlin (Germany), Hk70; Id29,f70,i2

Berwick-upon-Tweed (Nbl), Eh5,23; Gg14

Index of Places

Betchworth (Sry), Bb211
Beverley (Yks), Bb50,c103; Ej40; Ha4
Bibury (Gls), Gc14
Binchester (Dur), Ca17
Binfield (Brk), Ca35
Birmingham (War), Bb37,169,200,202,
  c55; Fe94; Hf89,h4–5,7
Bisley (Gls), Gc14
Bitton (Gls), Gf19
Blackburn (Lan), Bb139
Blaensawdde (Bre and Cmn), Gf3
Blaina (Gwe), If89
Boddam (Abd), Hf93
Bodiam (Ssx), castle, Ek12
Bohadoon (Wat), Mg19
Bohemia, Ge49
Bolton (Lan), Hg83
Bonchurch (IoW), Df14
Borneo, north, Ne4
Borup (Sweden), Dd14
Bosham (Ssx), church, Ej42
Botany Bay (Australia), Nc3,18,32
Bowness-on-Solway (Cul), Ca34
Boxmoor (Hrt), Bd9
Brabourne (Ken), Ge17
Bradford (Yks), Bb74; Hg47
Brafield-on-the-Green (Nth), Db6
Bramfield (Hrt), Bb95
Bramley (Sry), Hf13
Brandon (Sfk), Dd5
Brazil, Ba47; If41
Brecon (Wales), Ac113; Ha7; Je6;
  Beacons Bb129
Brest (France), Gh42
Bridgnorth (Shr), Ek14
Bridgwater (Som), Ef55
Brighton (Ssx), Hh14,i24; If48
Brimpsfield (Gls), castle, Ek17
Brinkworth (Wil), Ca13
Bristol (Gls), Aa145; Bb44,92,118; Ed11,
  e12,f40,48; Fe78,f20–21; Ga11,f45,j6;
  If46,h16; castle, Ef48
British Columbia (Canada), Na16,g36
British Guiana, Nb31
Briton Ferry (Gla), Jg9
Brittany, Ed8
Brougham (Cul), castle, Ek65
Bruges (Belgium), He19
Brunanburh, Db38
Brunswick town, Hi24
Buckinghamshire, Bb86; Ec3; Hf69,l20
Budworth (Chs), Gg6
Bulgaria, Id17

Burghfield (Brk), Ca9
Burgundy, Bd29
Burma, Ii34; Nc19
Burnley (Lan), Hb71
Burwash (Ssx), Hg86
Bury St Edmunds (Sfk), Ek15; Fh5
Buxton (Dby), Ca46; Fa16

Cadbury (Som), Bc164
Cadiz (Portugal), Fi24
Caergwrle (Fln), hill, Jh9
Caerleon (Mon), Bc186
Caerphilly (Gla), Ge34
Caerwys (Fln), Bd2
Caesaromagus (Ess), Bc162
Caistor (Nfk), Ca26
Calais (France), Ef42; Fb85
Caldey Island (Pem), He45
Callendar (Sti), Kc4
Calverley (Yks), Gg11
Cambridge, Aa30,c21; Ba19,b35–36;
  Ei94; Ia21; Churchill archives centre,
  Aa9; Jesus college, Fk69; King's
  college, Bb34; Newnham college, Hi31;
  Pembroke college, Ei22; St Edward
  parish, Ee9; St Radegund's priory,
  Ee79; university of, Ac30; Bb207;
  Fl18; Ge50; He47,i31
Cambridgeshire, Aa31; Bb199,c170
Canada, Ba165,c85,127; Fd16; Gd24,h1,
  8,12–13,16,22,24–25,45; Hd2; Ia32,
  i16; Na6–7,14,b17,24,c11,14,45,e19,
  21,f6,17,23,25,31,33–34,g4,9,12–13,
  33, h22,28; lower, Nc36,47; upper,
  Nc6,47,f5,7,36
Canterbury (Ken): Bb47,90,98; Dc33,
  38,e20,f28,67; Fe8,53; Gc8;
  archbishop's palace, Ek41; archdiocese
  of, Aa151; Dc7; Christ Church
  cathedral, Ee57; diocese of, Fg39; St
  Augustine's abbey, Ek56; St John's
  hospital, Eg36
Cape Breton, Nf24
Cape Colony, Na1,c12,g34
Capel Teilo (Cmm), Je12
Capo (Ked), Ii60
Cardiff (Gla), Aa108; Ib12
Cardiganshire, Bb123; Gf2
Carew (Pem), castle, Jh4; cross, Je18
Caribbean, Gf21,h47; Hj4
Carlisle (Cul), Ca10,b47; Eh26; Ga13;
  Hf64; bishopric, Ee73; diocese of, Ee72
Carlton, Castle (Lin), Eg35

297

Index of Places

Carmarthen, Bb185; Hl77
Carmarthenshire, Bc101; Cb25; Hg19,88
Carsington (Dby), Ca39
Carswell (Pem), Je18
Cartagena, Gh47
Castell Dwyran (Cmn), Je4
Castle Bolton (Yks), Ek5
Castleton (Yks), Ek47
Cawood (Yks), Ek8
Channel Islands, Aa27,b14; Ii23
Charlton Kings (Gls), Ca44
Charlton, North (Nbl), Ek7
Chatsworth (Dby), Hl50
Chelmsford (Ess), Bc162
Cheltenham (Gls), Bb146; Hb20,i32; Ij78
Chepstow (Mon), Jg5; castle, Jh6
Cherbourg (France), Eh19
Chesham (Bkm), Fg14
Cheshire, Ab50; Ea31,40,c8; Gb1; Hf17; Jf1
Chester (Chs), Ef64; diocese of, Ge21; earldom of, Ac37; Bc81; Ea13,23,31, 40,c8,i43,99,j19
Chetsworth (Sfk), Db15
Chichester (Ssx), Bb148; Ca7; Ie13
Chiddingly (Ssx), Hg79
Chidham (Ssx), Ca7
Chignall St James (Ess), Ek9
Chilterns, Bb93
China, Gd30; Id108
Chirk (Den), castle, Jh1
Choppington (Nbl), Gg14
Cirencester (Gls), Ca3; Gc14; abbey, Ee21
Clare, county, Mg8
Clifton (Gls), Bb118
Clogher (Lou), Le22
Clonlisk (Off), Lh14
Clonmore (Car), Le8
Clwyd, river, Hf81
Clyde, river, Kh1
Clydeside, Ac92; Bb61,c175; Hf2; Ib9, 51,68,96,g11,31,h11
Coker, East (Som), Df30
Colchester (Ess), Bc161; Ca4; Gi80; Hb65
Connacht, Lg5
Constance (Germany), Ee53
Cork, Le3
Cornwall, Aa15,b51; Bb13,c148; Dc29–30,f59; Ee41,56; Fd8; Ga24,g21,h40; Hb41,e12,79,f99,g76; Ie15; duchy of, Bc122; Fc9–10; Gf26; earldom of, Ek44
Corsham (Wil), Ca41

Cottam (Yks), Df32
Court (Ken), Fg15
Coventry (War), Ba142,d14; Ee69,i31; Fh7; Ih16
Cowbridge (Gla), Jb1
Cowlam (Yks), Df32
Crediton (Dev), church, Eh6,22
Crowley (Chs), Gg6
Croydon (Sry), He56
Cuerdale (Lan), Bc108; Dd1,7–8
Cumberland, Eh16,26
Cumbria, Ej38
Cyprus, Na20,c1

Dartmouth (Dev), Ff10
Datchet (Bkm), Db1
Deeside, Bb171
Delhi (India), Nf11
Denbigh, Hf16
Denmark, Ba163; Gd1
Dent (Yks), Bd24
Derbyshire, Ec32,f35; Hf92; Ib7
Derry, city, Ba99; county, Me8
Devizes (Wil), Gi38
Devon, Aa15,112,154; Bb13,177,201, c148,218; Dc29; Ee41,56,f50; Fb20, e30,f8,11,24,g18–20,h12; Ga24,f40; Hc17,e12,79,f99; Ie8
Didcot (Gls), Hg57
Dinapore (India), Nh11
Dinefwr (Cmn), Jg3
Diss (Nfk), Gc13
Docklands, Ia37
Donegal, Me12
Dorchester (Dor), Fe87
Dorchester-on-Thames (Oxf), Db9
Dorset, Aa146; Bb184; Gf32,g3,h23,38
Douai (France), Ke19
Dover (Ken), Fe53
Drogheda (Lou), Lh11
Droitwich (Wor), Bc137
Dublin, Aa86,c54; Bc118; Le12,20,f4–5, 11,g2,4,8–10,h10,i9; Ma2,f9,13,g1, j1–2,4,6; diocese of, Aa140; Le14; Trinity college, Aa39; Mf11,i12
Dubrovnik (Croatia), Ed10
Dumbarton (Dnb), Kh1
Dumfries, Gb21
Dunboy (Cor), castle, Lh5
Dundee (Ans), Bb111,208; Ka3
Dunwich (Sfk), Df36; Ef5
Durban (South Africa), Ii53
Durham, Bb76,113; He42,49; bishopric, Ee2; cathedral priory, Ee24,64,f49, h10,i32

Index of Places

Dyfed, Ja6,e11
Dymock (Gls), Ca23

Eakring (Ntt), Bb206
East Anglia, Ab12; Bb40; Db16,32; kingdom of, Db16,27,36,e7
East Indies, Gd11
Eastbourne (Ssx), Bb196; Hi23
Edgbaston (War), Bd11
Edinburgh, Ga15,i11,25–27,33,55; Hi19; If1,j12; Kb5,i3,j2; castle, Kh8; university of, Gj25
Egginton (Dby), Gf14
Egypt, Eh14,15; Id24,i24,43,j71; Ng1
Eire, Na5
Ellington (Hun), Eg34
Elmet, kingdom of, Da19
Ely (Cam), Bb102; Hl59; bishopric, Ee11
Enfield (Mdx), Hg65
England, north-east, Bb69; Hi5–6; northern, Aa68; Cb21; southern, Bc116; Gi49; Hb84,h39; south-west, Bc47; Hc14,f15; west of, Hc11
Ennis (Cla), Mj7
Eskdale (Cul), Hh10
Eske (Yks), Bb66
Essen (Germany), Da20
Essex, Bb161; Db19; Ff5,g13; ealdordom of, Db14
Europe, Ba18; Df73; Ga2,d2,10; Hb14; Id37,f62; central, If87; eastern, Id89
Ewell (Sry), Bb2
Exeter (Dev), Bb46,132,c215; Ec2; Ff12, k68; Ga28,f69; cathedral, Aa64; Bc104; Ee14,46,g26,j1–2,8,18,23,25, 37,41, 48,53,56–57,59–61
Eye (Sfk), priory, Ef13
Eynsham (Oxf), abbey, Df44

Falkirk (Sti), Kc4
Falkland Islands, Aa142; Gd8; Id66; Mg13
Far Cotton (Nth), Hh24
Faroe Islands, Kf4
Fasham (Yks), Bb157
Fast (Bew), castle, Ki9
Fearn (Roc), abbey, Ke1
Fenlands, Bc169–170
Fermanagh, county, Mb31
Fiji, Nc51; islands, Na22
Flanders, Ed12; Hl9; Ii20; Ld2
Fledborough (Ntt), Gg16
Flint, Hf16; Jf1

Florence (Italy), Ba190; Ee36
Fonthill (Wil), Db28
Fordham (Ess), Gi80
Fort William Henry (NY), Gh44
Foss, river, Ek26
Fountains (Yks), abbey, Ek18
France, Ac39,62; Ba117,127,133,c86, d29; Ed9; Fd12,27; Gd22,26,28–29, f25,35,47; He20,f28,45,97,h20,j4,l41; Id64,85,e5,i20; Kh2–3; Na10
Fresfield (Nfk), Ge44
Furness (Lan), abbey, Ee84

Gainsborough (Lin), Fk54
Gallipoli (Turkey), Ng31
Galway, La9,f12; county, Mf10
Gascony, Ed9
Gaul, Ca37
Gedling (Ntt), Bb10
Genoa (Italy), Gd9
Germany, Aa100,134,c19,102; Ba133, c140; Da24; Ei45; Ge49; Ha3,b15,17, 28,d5,8,12,e21,f20,28,45,86,g8–9,56, j4,7,l107; Ia43,d33,43,46,51,54,59, 71–72,93,95,103,106,g5,i22,64
Ghana, Nc42–43,h3
Gibraltar, Id81,91; Nb10
Gillingham (Dor), Hf84
Glamorgan, Ja9,h10
Glasgow, Bb189,c175; Ge11; He2,f64, k37,44; If50,g11,31; Ke17; archbishopric, of, Ke15
Glastonbury (Som), Bc120; Da6; abbey, Aa81
Glenasmole, valley, Mg15
Gloucester, Fi2; Hf19
Gloucestershire, Aa145; Bb125,174,183; Ek61; Fg5; Gf36; Hf63
Godalming (Sry), Bb52
Gold Coast, Nc43,e9
Gothenburg (Sweden), Gf18,h3
Gower (Gla), Jf2,h10
Grafton Regis (Nth), Bb1
Greece, Hb69
Greenock (Rfw), Bb85
Grimsby (Lin), Bb217
Grinstead, East (Ssx), Ek33
Grisy (France), Bb145
Grosmont (Mon), castle, Jh7
Guadelupe, Hd6
Guatemala, Id75
Guernsey (Channel Islands), Bb99
Guildford (Sry), Gj3

299

Gullane (Eln), Ke18
Guyana, Fd3; Hd3
Gwaenysgor (Fln), Bd3
Gwent, Bb68,c186; Ca32; Ja9,h10
Gwynedd, Jb2

Haddon hall (Dby), Ej27
Hafod (Cgn), Bb123
Hale (Chs), Bb209
Halifax (Yks), Ff22; Gg18; Hg30,38,87
Halkyn (Fln), Bd4
Halstock (Dor), Ca40
Hamburg (Germany), Hb16; Id60
Hammersmith, Hf90
Hampstead (Mdx), Bb140
Hamwic (Ham), Bc146; Dd12,f53
Hanbury (Wor), Bb62
Handcross (Ssx), If101
Hardshaw (Lan), Ge36
Hartlepool (Dur), Ef21; monastery, Dc13
Harwich (Ess), Gj2
Hatfield (Yks), Db33
Havana (Cuba), Gh30
Hawaii, Gd13
Haworth (Yks), Ge4
Hayling Island (Ham), Ca36
Hay-on-Wye (Bre), Jf4
Helmsdale (Sut), Ca47
Hendon (Mdx), If86
Henley-in-Arden (War), Hi35
Hereford, diocese, Ee76
Herefordshire, Db24; Hk48
Herstmonceux (Ssx), Hg18
Hertford, Ek64; Hb70
Hertfordshire, Ba178
High Park (Ntt), Hf42
Hinton (Som), priory, Ee20
Hockham, Great (Nfk), Bb55
Holland, Bb72; Fd12; Hl9
Holme on Spalding Moor (Yks),
  Ca28
Holt (Wor), castle, Eg13
Holy Cross, (Tip) abbey, Le6
Holy Island (Nbl), Df43
Holy Land, Ei64
Home counties, Cb38
Hong Kong, Nb20,c31,f9
Honiton (Dev), Bb221
Hook Norton (Oxf), Ej44
Hope (Fln), Bd5; castle, Jh11
Hovingham (Yks), Df43
Hoxton (London), Hg22
Huddersfield (Yks), Hk42

Hull (Yks), Bb50; Fi8; He29; Ia43,b87;
  university of, Aa59
Hulton (Sts), Df45
Huntingdon, Db13
Huntingdonshire, Aa173

Iberia, Lf6
Ibstock (Lec), Bb33
Iceland, Ed11; If105
Ikerrin (Tip), Lh14
Ilfracombe (Dev), Hg54
India, Fd13; Gd17,h31; Id4,88; Nb1,13,
  23,25–27,c2,10,16–17,20,38,41,50,
  d1,6,e3,10,20,22–23,25,f14,16,41,
  g11,19,32,h1,4–6,13,23,26,32,35,
  38–39
Indian ocean, Gh39
Indo-China, Ii14; Nb6
Inverness, Bb137
Invernessshire, Bd20
Iona (Arl), Ke25
Ipswich (Skf), Gj2
Iran, Ba93
Iraq, Id68
Ireland, Aa21,111,131–132,158,160,c32,
  42,65–66,89,97,108; Ba14,95,164,
  167,b94,c124,143,187,224; Dc32,d9;
  Ed11; Ff3,h6; Gi85; Hb37; Ib96;
  Nf50; Northern, Mb21,f7
Irish sea, Bc108
Isle of Man, Fa4; Hi7–8; Kg6
Isle of Wight (Ham), Bb142; Eh3
Islington (Mdx), Hl8
Israel, Ia34
Italy, Bb106; Fd7; Hh20; Id18,59
Ivinghoe (Bkm), church, Ee32

Jamaica, Nc8,f8,10,21,45,h21
Japan, Bc17; Hd4,f22,100,l5,60; Id2;
  Nb30,g6
Jersey, Ea15
Jutland, Aa101; Ii35

Kandahar (Afghanistan), Ng37
Karlsrühe (Germany), Li12
Kashmir (India), Nb16
Kendal (Wes), Ca2; Gf12; church, Ej24
Kenilworth (War), Bb164
Kent, Ab34; Bb198,c109; Da18;
  Ef65,j15; Fg7,32; Gf22,g13
Kenya, Nb14,c7,29,53,f1,3,15,39,h14,
  19,29
Kerry, county, Mg2

Index of Places

Keston (Ken), Ca43
Kibworth Harcourt (Lec), Eg38
Kidwelly (Cmn), Bb133,143; Hl75; Je19
Kildare, Li6; county, Lh2
Kilkenny, county, Mg24,j3
Killiney (Dub), Me3
Kilmallock (Lim), Li5
Kings Lynn (Nfk), Bd40
Kingston-on-Thames (Sry), Ek39–40; Hh23
Kinwarton (War), Ca50; Ek32
Kirkbymoorside (Yks), Df43
Kirmington (Lin), Ca33
Korea, Id109,i45

Lahore (India), Nf11
Lake District, Bb56
Lambourn, Upper (Brk), Ca22
Lamphey (Pem), bishop's palace, Je18
Lancashire, Ab50; Bb219,d24; Df63; Ea23; Fg34,h14; Hi34; Ne3
Lancaster, Bb218; Gf68
Land's End (Con), Ab15,52,62
Largs (Ayr), Bb83
Laughton Place (Ssx), Bb71
Layston (Hrt), Bb96
Lea, river, Ff5
Leeds (Ken), Bb41
Leeds (Yks), Aa117,137; Bb15,c84; Fe63; Gi58; Hf77,i15,20; Ih27
Leek (Sfs), Bd19
Leicester, Hb59,g42,64
Leicestershire, Aa65,b36; Bb192; Eg1,40; Fe27
Leiden (Netherlands), Gf28
Leinster, Lh9
Leverstock Green (Hrt), church, Bd31
Lewes (Ssx), Bb179,d30
Leyhill (Bkm), Ek20
Lichfield (Sts), cathedral, Ee60,74
Lincoln, Bb115,c105; Cb28; Ec29,k11; Hf50,i21; bishopric of, Ee5; cathedral, Ej10,29,31; diocese, Eg30
Lincolnshire, Ab10,11; Bb14,204,c111, 169; Ca18,b49; Db18; Ea13; Gg10; Hg67,82; Ij28
Lindsey (Lin), Ec11
Linton (Cam), Bb28
Lisbon, Fd33
Littlecote (Wil), Fi22
Liverpool (Lan), Bb57,178,c65;Hb5–6,12, 56,c13,f76,88,94,g3,10,h21,34;Ib17, e11; docks, Hf53; university of, Ac30

Llanbadarn Fawr (Cgn), Je5
Llanbeblig (Cae), Ji7
Llanbleddian (Gla), Jb1
Llandeilo Fawr (Cmn), Je4,g3
Llanelidan (Den), Bd6
Llanelli (Cmn), Bb197
Llanfair D.C. (Mgy), Bd7
Llangua (Mon), Je7
Llangynwyd (Gla), He40
Llantrisant (Gla), If27
Llanuwchllyn (Mer), Hg18
Llawhaden (Pem), castle, Je18
Lochaber (Inv), Bd20
Lochmaben (Dfs), Kc18
Lochtayside (Per), Gc6
Lochwood (Dfs) castle of, Ki7
Locke (Dby), Ee48
Loddiswell (Dev), Ek67
London, Aa56–57,b3,24; Bb3,8,16,45, 63,78,110,124,138,167,181,187,c53, 166,d18–19,38; Ca12,24,b35,42; Df64; Ef25–26,32,g47,i9,11,57,76,79, 88, k27,36; Fc1,g25,h9,k87; Ga18–19, b9–10,c12,f23,47,g8,i48,j1; Hb3,c1,3, e35,f47,97,h1,36,i18,k2,23,56,70,l4, 45,83; Ia37,f68,70,h37; Billingsgate, Bc165; bishopric of, Db19; British library, Aa10,168,170,b42; British Museum, Aa156,b17,30,35,39–40,59; Bc61; Gi35; Hl53,61; Buckingham palace, Hl124; City of, Gb35; Hb34, f24,28,41,g58; county council, Bb3; Covent Garden theatre, Gi54; diocese of, Ge3; Ealing, abbey, He44; East End, Ig17; Goldsmith's college, Hi12; Inner Temple, Aa3; Leadenhall court, Bc167; Leicester house, Gb13; London library, Hl4,6; Ludgate hill, Ek63; Montagu house, Ab30; Fe64; Mount street, Aa60; Old Bailey, Gc1; Paternoster square, Ij65; Public Record Office, Aa7,107,129,b16; Science Museum Library, Aa41; Sotheby's, Aa162; Spitalfields, Bd35; St Augustine's church, Kilburn, He65; St Paul's cathedral, Ec4; Ij89; St Thomas's hospital, Gj27; Tower Hamlets, Bb150; Tower hill, Ba57; Tower of, Bb158; Ek55; university of, Ac30; Hi12; Westminster, Bb102,203; Westminster abbey, Ec23; Fk23
Londonderry, Ba99
Longleat House (Wil), Aa81

301

Longney (Gls), Ca1
Lothian, East, Bb186;Gg15
Louth, Lh11,13; county, La2
Louvain (Belgium), Aa52
Low Countries, Aa134,149,170; Bc72, 102,133
Ludgershall (Wil), castle, Ek52
Ludlow (Shr), Bb70
Lyre (Normandy), Je7

Magee Island (Ant), Lg3
Malawi, Nc35,d16
Malaya, Nb6,c30,52,e5
Maldon (Ess), Db11,f74
Malham moor (Yks), Ef2
Malmesbury (Wil), Ej30
Malta, Gf15; Ne26
Manchester (Lan), Aa109; Bb30,136, 182; Ge38; Hb16,d14,e33,f49,71,103, h2,28; Ib48,g9,39,h25; Gaiety theatre, Hg2; John Rylands university library, Aa116,169,c108; society of architects, Aa104; university of, Hi25
Manorbier (Pem), Fg55
Mansfield (Ntt), Gf63
Margam (Gla), abbey, Ei68
Marlborough (Wil), castle, Ek52
Maudsley (London), hospital, Hk1
Mauritius, Ba14
Mediterranean, Id18,87,i4
Melbourne, Nh15
Mellifont (Lou), Le22
Melton, High (Yks), Bb26
Mercia, Db23; kingdom of, Db22
Mersea (Ess), Db17
Mersey region, Ab50; river, Gf61
Merseyside, If94
Merthyr Tydfil (Gla), Bb160; Hl31; Ih28
Mexico, Ac62; gulf of, Gf21
Michelham (Ssx), priory, Ek51
Middle east, Aa138; Bc179; He10; Id15, 82,87,90,107; Ne13
Middlesbrough (Yks), Ij53
Midlands, Bb6; east, Bb14,156; west, Bb175; Ea17
Milburn tower (Edinburgh), Hl108
Mill Green (Ess), Ej46
Millom (Cul), Bd33
Minehead (Som), Ef55
Minorca, Gd20
Minster (Ken), Dc38; abbey, Ee54
Mold (Fln), Bd8
Montague (Ont), Nf28
Montgomery, Ab20; castle, Ji11

Montgomeryshire, Bb147; Hf17
Morant Bay (Jamaica), Nc8
Moreton Bay (Queensland), Nc15
Morpeth (Nbl), castle, Ek43
Motcombe (Dor), Hf84
Much Wenlock (Shr), Ek14
Murton (Cul), Gf67
Muscat (Oman), Ne1

Nant Melyn (Cmn), Gg4
Nantwich (Chs), Gg25
Nantyglo (Gwe), If89
Naples (Italy), Fg38
Natal (South Africa), Nb32,d10,g7,22,27
Near east, Ii18
Needham Market (Sfk), Df71
Nene, river, Ef38
Netherlands, Fd11,23–24,32,e1; Ib43, d78; Kc10,h7
Neuadd Wen (Cmn), Jg7
New England, Ge13
New Forest (Ham), Bb191
New Mills (Dby), Gf7
New South Wales, Mg11; Nc3,5,18,32, f35
New Zealand, Hg17; Nb24,29–30,c11, 37,49,f2,26,30,g31, 35,h2
Newark (Ntt), Gf63
Newbury (Brk), Fg11
Newby Hall (Yks), Gg7
Newcastle-under-Lyme (Sts), Ek53
Newcastle-upon-Tyne (Nbl), Ef46; Gb17, e31; Hf5
Newchurch (IoW), Fh22
Newfoundland, Nc11,e2
Newmarket (Cam), Gf39
Newport (Mon), Jh8
Newstead (Rox), Ca5,16
Niagara peninsula, Gh14
Nigeria, Nb22,d11
Nore, river, Mf15
Norfolk, Aa167,b12; Eg28,33,i2; Ge44
Normandy (France), Aa29; Ea37,41,d2, h9,j22; Ii16; Ng9
North sea, Ii1
Northampton, Hg40; Ie13,g10
Northamptonshire, Aa174; Bb199
Northop (Fln), Bd10
Northumberland, Bb113; Eg50; Fi19; Gb28,f65
Northumbria, Db23,c14; Ea33; kingdom of, Df9
Norton (Yks), Bc160

Norway, Gd25; Kf8
Norwich (Nfk), Bb12,73,102,166,d32;
    Ef3,g31,i81–82,k2; castle, Ej6
Nottingham, Ec12; Hb60,i30
Nottinghamshire, Ea10; Ga1; Hf37,39
Nova Scotia, Kd1; Nd5
Nuneham Courtenay (Oxf), Ge22
Nymet Rowland (Dev), Hl23

Ontario (Canada), Aa56
Orcop (Hef.), Bb43
Orford (Sfk), castle, Ek28
Orkney, Ke14,f1,8
Ormesby, North (Yks), Bb194
Oswestry (Shr), Hb11
Oundle (Nth), Db10
Ouse, river, Ef47,k26
Oxford, Bb97; Ee23,68; Fh15; Gf66,i57;
    Balliol college, Hi31; Bodleian library,
    Aa47,153; Christ Church, Ge19;
    diocese of, Ge26; Hart hall, Ei78;
    Magdalen college, Gf33; Merton
    college, Aa175; Fg16,k26; New
    college, Bb26; Rhodes House library,
    Aa139; university of, Aa121; Ee24,i15;
    He47,i31
Oxfordshire, Hi10; Ih28

Pacific ocean, Ab19; south, Nd2
Padiham (Lan), Hf54
Painshill Park (Sry), Bb104
Palestine, Hb34; Ib44,d53,65,82,91–92,
    97,i13,43; Nb34
Paris, Ei66; Hk70; If70; Nf12; Irish
    college, Ke5; Scots college, Ke3
Patagonia, Bb215
Patna (India), Nh11
Pembroke (Wales), Ab13
Pembrokeshire, Bb25
Penhow (Mon), Jh12
Pennard, West (Som), Da16
Penydarren (Gla), Ca31
Perth, Ff1
Peshawar (Pakistan), Bd27; Ng3
Peterborough (Nth), Bb124; abbey, Ef54;
    cathedral, Ef38
Peterhead (Abd), Hf25,l7
Philadelphia (USA), Bc5; Gj25; Hh34
Pitsligo (Abd), chateau, Kg5
Plas Berw (Agy), Ja4
Plymouth (Dev), Ge12,h37
Polynesia, Gd21
Poole (Dor), Df40; Ef37; Ia24

Port Phillip (Victoria), Nb4
Portugal, Ef15; Fi27; Gh18; Hl69
Potter Brompton (Yks), Ef10
Potterspury (Nth), Gi21
Poundbury (Dor), camp, Cb36,45
Prague (Czechoslovakia), Me6
Preston (Lan), Bb109; Hg55
Prussia, Ed7
Punjab, Nc9

Quarr (IoW), abbey, Ee39
Quebec, Gd24; Nb5,c4,f50
Queensland, Mg14; Ne18

Raby (Nbl), castle, Ek31
Radford Woodhouse (Ntt), Bb64
Radnorshire, Ha7
Ramsey (Hun), Ec9
Ratley (War), castle, Ek50
Ravenstone (Bkm), priory, Ee32
Ravenstonedale (Cul), Ek60
Reading (Brk), Hb25
Reculver (Ken), Df31
Repton (Dby), Db2
Rewley (Oxf), Ee42
Rhodesia, Nf4; Northern, Nf22;
    Southern, Ne12
Rhondda (Gla), valley, Hg49,h32
Rickerby (Cul), Bb79
Rickmansworth (Hrt), Ek19
Rimpton (Som), Ef63
Roberton (Lks), Bb220
Rochdale (Lan), He3
Rochester (Ken), Ca29; Db3; Ei1;
    cathedral priory, Ee77,i73
Roel (Gls), Ek1
Rome (Italy), Df56; Ee8; Gd16; Sacred
    Congregation, Aa48; Vatican, Me10
Romney (Ham), Ef45
Romney Marsh, Hc9
Roscrea (Tip), Li10
Ross, New (Wex), Lc4,f9
Rouen (France), Dd12; Ek22
Ruabon (Clwyd), Bd12
Rudgwick (Ssx), Hf13
Ruislip (Mdx), Gg9
Rupert's Land (Canada), Nd8
Russia, Gd7,26; Hg56; Kh4–5
Rutland, Bb192

Sacomb (Hrt), Bb95
Salehurst (Ssx), Ek21
Salford (Lan), Hh28; Ig9

Salisbury (Wil), Dd4; cathedral, Ei100–101
Saltash (Con), Hc8
Sandwich (Ken), Fe53
Sarawak, Nd14
Sark, Aa27
Sarre (Ken), Df58
Saudi Arabia, Id59
Scampston (Yks), Ij2
Scandinavia, Df4; Id86
Scotland, Aa14,58,78,115,b18,26,44, c12,33,40,55; Ba169,b17,19,27,39,60, 72,77,80,91,94,103,114,130,145,171, 190,222,c4,6,21,55,76,83,128,135, 138,168; Ca5,16,25,48,b9,17,34; Dc10,f9,66; Ea22,h12,17; Fa13,b9,27, c27; Ga5–6,16,26,b3,c3,e11,13,49,g2, h33–35,i31,68,87,j25; Ha6,b46,c4,6, 12,e7,14,19,28,f64,104,g5,91,h38, k24,l25,74; Ia10,b13,29,31,82,89, e3–4,f61,95,g6,25,28,30,h11,21,i52, j12,45,59,61,76,83,91; Lh3,6
Scottish Highlands, Cb8; Ga32,f17; He67,h31; Kc9,h7
Selkirk, Ke26
Serbia, Id63
Sevenoaks (Ken), Bb59
Severn, river, Ca1; Ef40
Seychelles, Na15
Shaftesbury (Dor), abbey, Dc28; Fe9
Sheffield (Yks), Bc106; Fc12; Hf96,k60, 71; Ig12
Sherbourne (Dor), Fg4
Shetland (Zet), Kf1,8
Shipston-on-Stour (Wor), Ef23
Shiptonthorpe (Yks), Ca28
Shoreham, New (Ssx), church, Ej64
Shrewsbury (Shr), Df1
Shropshire, Db24; Hf17
Sierra Leone, Nf38
Simonstown (South Africa), Ng15
Singapore, Id57; Nb3,h16
Sion (Mdx), Bb105
Skenfrith (Mon), castle, Jh7
Skipton (Yks), Fi23
Skye (Inv), Bd20
Sledmere (Yks), Ij34
Sligo, Lg7
Snape (Ess), Df26
Snowdonia (Wales), Bb51
Somerset, Aa147; Bb18,48,212–213, c119,d21; Da6; Ef55
Southampton (Ham), Bc146; Dd12,f7,

52–53; Ej58; Ih16; university of, Aa118
Sowerby (Yks), Gg11
Spain, Ei41,j52; Fi27; Gd23,h9,18,46; Id79,81
Spalding (Lin), priory, Eg24–25
Spaxton (Som), He72
Split (Croatia), Gi12
Springhead (Ken), Ca45
Sproatley (Yks), Bd13
Sprouston (Rox), Df66
Sri Lanka, Nd15
St Albans (Hrt), abbey, Ej3,62
St Andrews (Fif), Bb162
St Asaph (Fln), Je13
St Combs (Abd), Hf93
St David's (Pem), Aa75
St Harmon (Rad), Ja2
St Helens (Lan), Bb188
St Kilda (Wis), Kj3
Staffordshire, Bb75; Ek6; Hf17,92
Stainfield (Lin), Ca15
Stamford (Lin), Bb124
Stapehill (Dor), abbey, Bb100
Stapleford (Hrt), Bb95
Staplehurst (Ken), Nf26
Staxton (Yks), Ef10
St-Florent de Saumur (Anjou), priory, Ee49
Stokesay (Shr), castle, Ek59
Stone-in-Oxney (Ken), Bb87
Stow, West (Sfk), Dd5
Stowe (Bkm), Gi22,44; Ij92
Stratford-upon-Avon (War), Fk10; He4
Sudan, Nc54
Suez, Id80
Suffolk, Aa72,b12; Bb128,c113; Df54; Ef4,g28; Hg41
Surat (India), Nh20
Surrey, Aa126; Bb31,193; He57
Sussex, Ab34; Eg29; Fh10; Gf16; West, If28
Sutton Hoo (Sfk), Bc97,113,183; Da1,4, 14,22–23,25,d28,31,e2,11,15,27,61, f12–13,20; Lh12
Swan, river (Australia), Ne11
Swanage (Dor), Bb65
Swansea (Gla), Ac76; Bb214,c212; Hf10; Ia1,5–6,8,23,26,30,38,b56,g1,23,h8, 19,30,j29,32,88; Ja8; university of, Ij33
Sweden, Ba163; Gd25
Swinderby (Lin), Gg22

Index of Places

Switzerland, Id47
Sydenhams moat (War), Ek49

Tamworth (Sts), Bc158
Tantallon (Eln), castle, Kg1
Tanzania, Nd9
Tasmania, Mb11; Na21,c21,f37,40
Tattershall (Lin), church, Ea39
Teesside, Hg13
Teignmouth (Dev), Ig8
Tenterden (Ken), Bb154
Thames, river, Ca38; Ef25,k36,39
Thorner (Yks), Bb29
Thurles (Tip), Lf2; Me13
Tibet, Ne16
Tichborne (Dor), Bb172
Tickhill (Yks), Fe55
Tillingdown (Sry), Ef56
Timworth (Sfk), Gc7
Tintern (Mon), Ji9; abbey, Bb49
Tipperary, county, Mg24
Tiverton (Dev), Hh3
Toronto (Canada), Nh10
Torrington, Great (Dev), Hf7
Tottenham (Mdx), Ih28
Tranent (Eln), Gh36; Kf2
Tredegar (Mon), Gc5
Tremeirchion (Fln), Bd15
Trent, river, Ca18; Ef58
Tripoli, Fd22
Tullylish (Dow), Mg5–6
Turkey, Bc147
Tyburn (Mdx), Fc2
Tyne, river, Ef46

Ulster, Hb51; La1,e7; Me5
United Provinces, Fb59; Gf47
Utah (USA), Hc9
Uttar Pradesh (India), Nc46

Vale Royal (Chs), abbey, Ee16
Vanuatu, He39
Venezuela, Hd3,i33
Venice, Gh32
Victoria, Hd6; Nb33,f19
Vienna, university of, Hk28
Vietnam, Ii45; Nb6
Viste, Nf42

Wakefield (Yks), Ea20,h11,13
Wales, Aa19,22,79,94,172,c50; Ba13,
   b5,9,20–21,23–24,38,42,54,81–82,
   106–108,112,119–121,127,144,151,
155,163,168,210,215–216,222,c66,
   96,121,129; Cb17; Fb32,64,e47;
   Gf43; Ha5,b19,c10,f81,101,g50,h13,
   15,18,22,25,i2,11,22,29,l80,98,123;
   Ib38,e9,f6,89,123,g4,14,18,22,40,
   h14,j3; Nc44,f27,g8,h33; south, Bb69;
   Hf21,39–40,i27,28; Ih29; south-east,
   Hh9
Walkern (Hrt), Hk55
Walshcroft (Lin), Ab11
Waltham (Ess), abbey, Ee63
Warrington (Lan), Bc163
Warwickshire, Aa46; Ea7
Waterford, Lc4,e15–17
Wedmore (Som), church, Eg22
Wellington (NZ), Nf26
Welton (Yks), Bd16
Wenlock, Much (Shr), Ek38
Wensum, river, Ef3
West Indies, Fd5,i17; Gh10; Ng30
Westmorland, Bb53; Ff23; Gg1; Hc16
Wexford, Lf1,h4
Weybridge (Sry), Ij82
Wharram le Street (Yks), Hl70
Wharram Percy (Yks), Bc172
Whickham (Dur), Bb131
White (Mon), castle, Jh7
Whitehaven (Cul), Gf55
Whitford (Fln), Bd17
Whithorn (Wig), Ke4,11,21,f5–6,g2
Wickham, West (Ken), Fb51
Widnes (Lan), Hg89
Wilderspool (Lan), Bc163
Wilton (Wil), Dd4
Wiltshire, Bb88,180; Eg19,27,k48; Fg26;
   Ga9,i50
Winchelsea (Ssx), Gb29
Winchester (Ham), Bb126,205; Eb9;
   Hk7; bishopric, Ef63
Wisbech Barton (Cam), Ek46
Witham (Som), priory, Ee20
Wolford, Great (War), Ef62
Wollaton (Ntt), Fe57
Wolverhampton (Sts), Hc22
Woodlands, Calne (Wil), Hc20
Woolaston (Gls), Ca21
Wootton (Nth), Bd39
Worcester, Bb11,102; Dc22,f17; Fg31;
   cathedral priory, Ee4; diocese, Ee76;
   infirmary, Gj15
Wrexham (Den), Fe76
Wroxeter (Shr), Cb47; Dc3
Württemburg (Germany), He20

305

Index of Places

Yarborough (Lin), Ab10
Yarmouth, Great (Nfk), Ef62;
  Fb16,18,g22
Yeavering (Nbl), Df65
Yetminster (Dor), Fg4
York, Ac72; Bb84,153,c156–157; Dd21,
  f2,47–48; Ee15,17,f29,59,g7,15,17,
  k26,66; Fe80,h16; Gi3,10; Ke21;
  archbishopric of, Ek8; archdiocese,
  Aa128; diocese, Ee44; minster, Bb26;
  Ej16
Yorkshire, Bb32,c103,110; Cb13; Ee3,
  f29,44,j26; Fb8,e18,f23,h18; Hb35,
g46; Ih6,j2,28; east, Hc7; East Riding,
  Hi9; Ig7; North Riding, Fg27; west,
  He26; West Riding, Aa95; Ej54; Hh35,
  k43; wolds, Gf27
Youghal (Cor), Lf7
Ypres (Flanders), Ii20
Yugoslavia, Id52,63

Zaire, Nf29
Zambia, Nb21,f22
Zanzibar, Ne1
Zimbabwe, Na25,c39,d3,e12,f4
Zululand, Nb11,g7,20–23,25–27,39

INDEX OF SUBJECTS

Aborigines, Nf35,40,43,46
Accidents, Hf48; industrial, Hl21
Account books, Gi48
Accounting, Gf33; Hf9,75; If85
Accounts, Jf1
Act: Apothecaries (1815), Gj17; British
    North America (1867), Nc45;
    Canadian Constitution (Patriation),
    Nc45; Contagious Diseases, Hk58;
    External Relations (1936), Na5; Gin,
    Gc4; Great Reform, Hb49; Navigation,
    Gf59; Old Age Pensions, Mg3;
    Representation of the People (1918),
    Ib19; Salmon, Hc17; Stamp, Gd12;
    Work-house Test (1722), Ga14; and
    see Statutes
Actors, Hf31
Actresses, Fk38; Hf30–1
Addiction, Aa34
Administration, Gc12; Hh11; diocesan,
    Ge3,26; local, Hb60; naval, Gh10
Adoption, Jg8
Advowsons, Ee60
Aesthetics, Fe66; Gi46
Ageism, Fg53; discontent, Fh17; policy,
    Ne25; protest, Mb4
Agricultural Associations, Hf37
Agricultural Executive Committees, If28
Agriculture, Aa99,122–124,b34,c8,80;
    Ba2,40,90,96,b6,38,176,210,c204;
    Ca38,b32; Df24; Ef14,17,28,34,53;
    Ff17,25,g24,j33,k19,27,49,56,l14,23,
    25; Gc7,f9,17,27,32,57,66; Hf33–34,
    63,95,g12,45,68; If28,61,95,113,121,
    g41,h21,j78; Kf6; Lf10; Mf2,g12;
    Ne4,12; open field, Bb206;
    productivity, Ne10
Aid, humanitarian, Id30,78
AIDS, Ih1
Air forces, Ii54; operations, Bc20;
    pollution, Aa38
Alchemy, Fl7,19
Alien priories, Ee75
Aliens, Eb2
Allied Command, Ii15
Allies, Ia9
Almanacs, Fl15
Alphabet, Ogam, Li13
Anaesthesia, Ij52

Anatomy, Hk23,29
Ancien regime, Ga7,21,29
Anesthetics, Ei95
Anglicanism, Fe1; Ge18,50; He1,25,27,
    78; Ie9–10
Anglo-Irish, Mb17
Anglo-Jewry, Ib2,39,g17; and see Jews
Anglo-Saxon chronicle, Da7,b5,31
Annals, Ei11,68
Anthems, Fk63
Anthropology, Ac32,89,97; Ba98,c124;
    cultural, Nh8–9
Anti-Christ, Ge41
Anti-popery, Ge41
Antiquarianism, Aa169,c21; Fc19; Mi18
Antiquities, Ca48
Apartheid, Nb28
Appeasement, Ib72
Apprentices, Ff20,g46
Apprenticeship, Fg43
Arbitration, Ec2
Archaeology, Aa19–20,42,126,b18,55,
    63,c64; Ba143,148,b32,67,104,123,
    153,165,173,179,211,c41,53,64,96–
    97,106,113,129,137,146,156–158,
    160–172,183,186,191–193,211;
    Da25,d12,f44; Ea17,e84,f31,35,61,
    k18,21,23,49; Hl61,116; Ij71; Jh1;
    Kh1,i7; La4,f11; environmental, Bb84;
    garden, Ba62,135,184,b18,20;
    industrial, Bb9,21,24–25,51,82,107–
    108,112,120–121,127,129,133,151,
    155,160,163,216; Hk39; waterfront,
    Ef3,21,25,32,37,40,46,55,k11,26,36, 39
Archbishops, Ee15
Architecture, Aa104,c72; Ba7,203,b12,
    71,73,80,198,c104; Cb5; De23,f28,30,
    50; Ei77,j2,12,25,57; Fg31,k10,14,40,
    76; Gc14,i12–13,16,26,33,44,68;
    Hc18,k53,l33,42,83,121; Ij34,65,74,
    89–90; Ki11; La7,i5; Mi18; Nh24;
    church, He37,l8; ecclesiastical,
    Ee21,j16,36,42,62,64; He65; Hiberno-
    Romanesque, Li10; Romanesque, Ej4,
    10,22,29,59; vernacular, Bb210; Ej63,
    k38
Archives, Aa2,100; Ga3,d5,27; Id106;
    Ke3,5; diocesan, La9,e20; Vatican,
    Me10; and see Cartularies; Charters;
    Parish; Records; Rolls; and section Aa

Aristocracy, Df73; Ea7,11–14,24,30–31, 42,b5,c1–2,4,8,d12,e38,61,f1,g23,32, 37,44,50,h24–25,i77,83,k16; Hb78, g56; Ig41; La6; *and see* Class; Gentry; Nobility
Armada, Spanish, Fb30,i21
Armaments, If44
Armed forces, Aa142; Bc10; Ii37; Ng29
Armies, Eh9,21; Fi3
Arminianism, Fe20,66,92
Armour, Bc5; Df21; Eh1,6,22
Arms, Bc5; Df21
Army: Aa29,71,142,b37; Ba109,156, b91,192; Ca27,b3,9,17,21,47; Db37; Eh17; Gd20,h6,17,21,23–25,27,31, 33–36,43–44,47; Hj3,17,l84; Ii14,38, j72; Ld2,h6,11; Mh1–4; Ng20,28,32, 39,h5; Canadian, Ng9; Indian, Ng3; New Model, Fi10; reform, Hj9
Arson, Hc7
Art: Ba202,b16,c46,104,194; De27; Ej53; Fk36; Gi79; Hl18–19,118; Ij32, 81; collecting, Gi51; decorative, Aa156; Ba60,b80; design, Hl72; ecclesiastical, Ej15; institutions, Bb16; insular, Li9; military, Hl115; Romanesque, Ej44
Articles, of Church of England, Fe29
Artillery, Gh13
Artisans, travelling, Hb79
Artists, Ej51; Hl77; Nh17
Arts and crafts movement, Ij19
Astrology, Fl26
Astronomy, Fl15; Gj16; Hk32
Asylums, Ba153; Hh6,35,k1; Ij26
Atheism, He50
Atlases, Aa109,c13–14
Attainder, Fc26
Audio history, Aa135
Authors, female, Ba63
Automobiles, If101
Ave Maria, Fe82
Aviation, If86,i42,46,49

Bakers, Ga9
Balfour Declaration, Id65,92
Ballads, Ei72; Fg53; Mg18,20
Bands, military, Gi50
Bank: of England, Hf23,67; If17,49,60, 84
Banking, Ba83,160; If33,87
Baptism, Eg39
Baptists, Ge34; He16

Barbarians, Cb14; De12; Ld1
Barges, Ef38
Barony, court book of, Kc4
Barracks, Bb196
Basilicas, Bc167
Baths, Ca5
Battles, Dc23; of Assandun, Df35; of Brunanburh, Db12,38; of Culloden, Gh3; of Faughart, Lh3,13; of Gainsborough, Fi28; of Jutland, Aa101; Ii35; of lake Erie, Gh1,8,12– 13,16,22; of Long Island, Gh7; of Loos, Ii66; of Maldon, Db11,f74; of Mansurah (1250), Eh14–15; of Otterburn, Bc4; Ea22,h10,12,26–27, i72; of Ringmere, Db12; of the Holme, Db12; of Wakefield, Eh11,13; of Waterloo, Hl9; of Ypres, Ii20
Bayeux tapestry, De18; Ej20,39,45,k57
Bee-keeping, Fc22,j33
Belts, Ca27
Benefices, Bb102; Ee3,51
Benthamism, If80
Beowulf, Db32,34,e3,11,15,47
Berlin Control Commission, Ii2
Berlin wall, Id29
Bermuda conference, If74
Bestiaries, Ej20
Bible, Fe4,j45; He34; Ji9; New Testament, De42; Old Testament, De25
Bill of Rights, Fb36
Bi-metallism, Hf103
Birth control, Mi9,k3; rate, Hg32,78,93
Bishoprics, Ef9
Bishops, Ed5,e29,43,73,h10; Fi12; Ke27; Le15–16
Bishops' palaces, Je18
Bismarck, the, Ii36
Blasphemy, Ib54
Blitz, Ii6
Blockade, Gh42
Boat building, Ef31
Boats, Gh28; Jf3
Bolshevism, Ib39
Bomber Command, Ii10; Ng4
Bombing, Ih6,i60
Book of Llandaff, De54
Bookbinding, Ab47
Bookland, Db35
Books, Ee24,i54,73,79,85,89–91,94,101; of hours, Ji7; office, Dc14; service, Ee59
Borders, Kc14

Boroughs, Gb29,c8; parliamentary, Hb20; *and see* Towns; Urban
Boundaries, Ca22; Df9; Hd3; borough, Hb60
Bounty, H.M.S., Gh11
Bowls, Ca15,47
Bretons, Ea30
Bretwalda, Db27
Brewing, Ef50
Bricks, Ca16
Bridges, Ba79,b201; Cb37; Db3; Ek39–40
Bridgewater treatises, Hk62
Brigandage, Ed2
Britain, Roman, Bc137,162–163,167, 184,215; Da17
Britannica, Fa7
British and Foreign School Society, Hi2
British Deaf Association, Ba74
British medical journal, Hk3; Ih2,15,j17
British Rail, Ab27
British Transport Historical Records Office, Aa12
British Union of Fascists, Ib2,22
Broadcasting, Ib37,j59
Brokerage, Fg41
Bronzes, Cb2
Brooches, Ca41; Df32; Jg2
Buildings, Ac35,40; Bb7,56,c172; Ca4, 39,b5; Ek38; Gf68; Ia30; farm, Ek46; Gf27; stone, Ek55; timber, Bc166; Ca24; Ek36; vernacular, Hl70
Bullion, Bc102
Burghs, Gb21; Kc18,20,e26,i5; *and see* Towns; Urban
Burial customs, Df6,18–19,27,34,45,62, 76; sites, Lg6
Burials, Bb90,126,c185; Cb7,18,29; Df21,52,70; Ek4,15,27; Ki1
Business, Aa76,83,108,110,114,130,b57; Ba41,55,66,b33,c33,37; Gf53; Ne17; history, Ab21; *and see* Industry

Cabinet Office, Ia11
Calendars, Ei58; Li12
Calvinism, He28; Kf7
Campaigns: Falklands, Aa142; Normandy, Ii16; Ng9; north African, Id49
Camps, Ca34,b45
Canals, Bb30; Gf65; Hf50
Capital: Gf60; flow, If62; merchant, Ne2
Capitalism, Ei17; Gf4,10,47; Hf14,100; If51; Kf7; Nc4

Cardinals, Ed5
Carpentry, Ej62
Cartels, If14; international, Id23
Cartography, Aa31,84,97,b6,19,48,51, c13; Ba165,c47; Fg32,k68,l5; Ga24, f11,26; Hc11,14
Cartularies, Ef13
Castles, Bb7,41,46,162; Ea11,35,e49,f48, h8,j6,12,k5,12–13,16–17,19,22,28–31,34,43–44,47,50,52,55,57–59,61, 63,65,67,68; Fi11,23; Ja6,9,e18,h1–12,i11; Kg1,h8,i7,9; La2,8,h1–2,5,7–9
Casualties' union, Ii11
Cathedrals, Aa64; Bb57,73; Ee14,26, i100–101,j1–2,8,12,18,23,25,37,41, 48,53,56–57,59–61; Fk73; Ie11,j89; Je6,13; La7
Catholic Emancipation, Gb7; Hb37
Catholicism, Aa148; Ba16,b171,c13; Fe7,41,92,k22; Gd20,24,e42; He1; Ie4–5,g35; Mb2,18,e1,3,6,8–9,i21; Anglo-, He21; anti-, Hb12; Irish, Ba64; Roman, Bb114; He14,18–19,41,53,62, 67; Ke3,5,19,22; Me10,13,i7,17; *and see* Church
Catholics, Fe94
Cavalry, Hj9
Caves, Ca46,b7
Celts, Ab28
Cemeteries, Bc160; Cb36; Df6,10,13,22, 26,46–47,49,58,60; Ek4,27; Je10; Ki1; Lg6; Nh11
Censorship, Ib54,c3
Census, Aa74,88; Gg1,22; Hg18,k43,l22; Ig18; bishops', Fg12
Ceramics, Ek37
Ceremonial, Ga15; civic, Gi77
Chancery, Ec16; Fb81
Channel tunnel, Hk63
Chantries, Ee17; perpetual, Ee12
Chapel Royal, Fk73
Chapels, Ge51; Catholic, Ge7
Charities, Ba9,17,34,142,b164; Ee13; Ga11,e1,j1,6,15,27; Hh1,k42,61; Ia4, i3; Mj5; medical, Bc26
Charity, Ee30,g8
Charters, Db6,8,15,17,f31; Ea2,23,35, 37,b10,c30,e39,f2,13,52,g11,i43,99, k8; Ka1; Lc4; Anglo-Saxon, Da16
Chartism, Hb2,56,64,84,g16,76,h21,i33
Chemistry, Gj10; Hk21,64,68; Ij4–5,35–36,38,47
Child guidance, Ih34

Childhood, Hk34,l54,104
Children, Bc57; Ea28,g16; Fg50,56; Gg14; Hh28,34,37,k20; Ih22,34–35; Mg7,23
Chivalry, Bc46,98; Ei91
Cholera, Hb25,k25,59
Christian Fellowship, Ia28
Christianity, Bc189; Cb14; He69; Nd1,15; African, Nd10
Christmas, Ba25
Chronicles, Ea8,d13,e4,h7,14,24,i11,25,28,57, 59,64,68
Chrysler company, If39
Church, Aa15,19,73,75,151,160; Bb13, c12,96,148,194,202; Ie8,15; Anglican, Nd8; art, Ie12; Baptist, Ac99; He60; building, He35; Celtic, Je2; colonial, Ge25,32; Congregational, He5,30; decor, He61; discipline, Ie12; furnishings, Ge7; in Wales, He40; Irish, Aa158; medieval, Aa16,c30; Bc51,70; of England, Aa91,103; Ge3,17,22,25– 26,30,32,39–40,44,46,50,g16; He10, 20,24,42–43,49,68,73–74,79–80,g86, l57; Ie9,10; Nd5,7; of Ireland, Ge25; Mi1; of Rome, He74; of Scotland, Ge25,43; He35; organization of, Dc3,40; Presbyterian, He39; reform, Ke10; Roman Catholic, Hb31,e2,4,20, 52,59,61,64; and see Catholicism; Christianity; Methodism; Protestantism; Puritanism
Churches, Ac72; Bb102,198,222; Df31, 59; Ea39,j15,40,49,64; Ge38; Hb23, e7,56,65,g65; Ie6; Je1,3–5,11–12,14– 17; Ke11; Le7,17; minster, Dc4,15; parish, Fg14
Churchyards, Df59; Je1
CIGS, Ii29
Cinema, Hj2; Ie6
Circumcision, Bd18; Nh29
City Philosophical Society, Gj13
Civil: rights movement, Mb21; servants, Ib10; service, He9; Ic6,f32; Nc40
Clans, Nf3
Clas, Je5
Class, Bc55; Hg16,51,72; Ib16–17; Nf3
Classes, Ga27; Hb71,g48,53; artisan, Hb14,g59; business, Hf6,85; gentry, Ba44; labouring, Gi19; middle, Gb14,33,g17–18; Hg5,39,83,85,l34, 81; social, Ac11,36,83; Ba128; Hf29,

45,g35,69,i7; Nf8–9,22; upper, Hg36, 43,66; working, Ac77; Bb69; Ga28, f29,57,g19; Hg8–9,13,15,42,45,47,52, 55,70,78, 93,h19,25,i21,32,j10,l21,73, 98,104; Ib89,g11,39; and see Aristocracy; Gentry; Nobility
Classis Britannica, Ca12
Clay pipes, Kj1
Clergy, Bb68,114,c202; Ee3,10,18,23,32, 34–35,46,f62,g30; Fe30,57; Ge26,44, i7,80; Le14,24; Catholic, Mb8; country, Ge22; parish, Ge17; parochial, Ge3; unbeneficed, Ee76
Climate, Ba101
Cliometrics, Ac112
Clockmaking, Gf12
Clocks, Ab49
Clothing, Bc53
Cluster analysis, Ef53,i32
Clydesdale bank, Ab45
Cock-fighting, Bb113
Cohabitation, Kc12
Coin hoards, Dd1,7–8
Coinage, Aa26; Ba35–36,57,181; Dd3, 14,17,19,26–27,30,32–33; Ef7,24,42, 59; Ff3; Gi26; If27; Kf2; and see Numismatics
Coins, Ab58; Bc139,215; Ca29,b6,16, 40–41; Df32; Ef60
Collectors, Hl61
Colonial Film Unit, Nh37
Colonial Office, Hk17
Colonialism, Ba164,c2; Nc4,29,f38
Colonies, Gd20,h10; Kd1; American, Fd20,26; Ga19,23,26,d12,31–32,g5
Colonization, Fd28–29; Mg4
Comedy, Fk58
Commerce, Ba33; Dd16; Gf53; Hf68; Ij46; Kf3
Commercialization, Ge27
Commissariat, Ca6
Commissions, peace, Ec29
Communalism, Nb23
Communism, Ib66
Communities, Ac2,3; Bc94; Db4; He7
Companies, Ab21; Bb78; If115
Computing, Aa51,94,137,c69,71,90; Ei32; Gb1,g11; Hg18; La3; and see Databases
Concentration camps, Ng24
Concerts, Gi81
Conciliar theory, Fj12
Confession, Ee34,44

Index of Subjects

Conscientious objectors, Ii58
Conservation, Ac72; Bb153; Ej31–32,50
Conservatism, Aa164; Gb6,d28; *and see*
  Parties; Politics; Thatcherism; Toryism
Consumerism, Ba152; Ga27,f38
Consumption, Gg24
Contraception, Ba126
Contracts, Fg10
Control Commission, Id51,54,72,93
Convention, Spanish, Gb16
Conventiones, Ec20
Conversion, Dc40
Convicts, Bc187; Hg63; Mg11; Nc15,21
Cooking, Bb101; Ei5; Fg16
Coronations, Ec23; Gi77
Corporatism, If55,116–117
Correspondence, Eg43; Fe65,k41; Ge44,
  i80; Ng37
Corrodies, Ee30
Cosmology, Lc3
Costume, Ef26,j14
Council: ecclesiastical, Ee36,53; of
  Clovesho (747), Dc11; Privy, Fb17,50
Countryside, Bc107; Fh13
Court: Chancery, Ec25; consistory, Ee11;
  Gg8; ecclesiastical, Ec4; law, Gc1; leet,
  Ec9; manorial, Ee6; of Chancery,
  Ab20; of High Commission, Fe40; of
  Session, Gc9; royal, Bc74; Fa3,b73,
  k47
Covenanters, Ke7,10,20,23
Craftsmen, Bb185; Ej51
Credit, Hf23,52
Cremation, Lg6
Creoles, Fd5
Cricket, He51
Crime, Aa85; Ba15; Ea38,c1,g28; Fc1,
  g45; Ga12,27,c1,11,13–14,f38; Hb39,
  c2,5–7,10,16,20,e9,h36,39,k33; Ih19;
  Lc2,e12; Mg11,17,19–20; juvenile,
  Hh26
Crofting, If61
Crown lands, Bc122
Crusades, Eh14–15
Cults, Dc33,41; Ei29
Culture, Ac2–3,49–50,106; Ba10,45,58,
  95,c94,121,222; Cb29; Ga8,f34,53,
  i46,80,j10; Li7; material, Kf6; political,
  Gc2; Ig10; popular, Ba100,176; Gi19,
  j3; Hg22,h18,36,j16,l31,119; Ig9,j16,
  58; Nh14,23,30; public, Nh20;
  Scottish, Ga6,26,32,i29
Custom, popular, Gi19

Dance, Gi32; Ij22
Danegeld, Dd22
Danelaw, Bc99; Db16,21; Ec14–15
Darwinism, Hk8
Databases, Aa137,c94; *and see*
  Computing
Dearth, Ba144
Death, Ba157,c185,216; Fg5; Li3
Debt, Ba190
De-colonization, Id88; Na8,10–11,18,
  b20,c22–25,43,52,54
Deer, Ef6
Defence: Ii9; national, Hj14; policy,
  Bc179; Id35,80,i29; Ng31
Defenderism, Mb26
Democracy, Ha3; social, Hb17
Demography, Ba82,166,188,200,b103,
  195,199; Cb35–36; Eg17,37,38; Fh6,
  8; Gg1–2,9,21; Hg47,78,93; Ig18,32,
  35,40; Mf13,g3
Demonism, Nd15
De-nazification, Aa100
Dendrochronology, Ba198
Depression, economic, If45; Great, Hg32
Design, industrial, Bb202
Devaluation, If22
Dialects, Ei2,4,47,84,96
Dialogue of the Exchequer, Ec17
Diaries, Bc178; Fb26,77,g33; Ga9,b27;
  Hg41,87,l39; Ii32; Na24;
  parliamentary, Fb82
Dictionaries, De60
Dictionary of national biography, Ab2
Diet, Bb101
Diocesan administration, Le3,10,17
Dioceses, Bb102; Dc2; Ke27; Le2
Diphtheria, Hk70
Diplomacy, Aa36; Fd24,27; Gd27; Md2;
  and see Foreign Office; Foreign policy
Diplomatic, Ea37,i43
Directories, Aa108
Disarmament, Nb17
Discipline, Gh33–35
Disease, Ba12,27,121; Cb45; Df39; Ei1,
  34,81–82; Gh47,j8,18,22,24; Hk31,
  40,69; Ih4; *and see* Health
Dissent, Ab36; Fe47; Ge9,18,48; Hb71;
  religious, Bd30; Fe60; *and see*
  Nonconformity
Dissenters, Ga19
Disturbances, popular, Hb49
Divorce, Gg15
Docks, Bb30,63; Hc1; If114

Dockyards, Ig27
Doctors, Gj17; Hk60
Domesday Book, Ab22–23,25; Bc109–111,120; Db37; Ea4,31,c3,11–13,15, 33,d12,f12,44,54,g49,i20,k42,53
Douglas, Ge43
Drama, Ec5,i71; Gi23; Hl75; *and see* Theatre
Drawing, De43
Dress, Ba111,b39
Dreyfus affair, Hl90
Drink, Hc4
Dublin review, Mi21
Dyestuffs, If112
Dykes, Cb20; Offa's, Df24; Jh1

Earthworks, Bb67
East India Company, Hl76; Nc17,f16, h13
Easter rising, Ib96
Ecclesiastical administration, Ke15; Le24
Ecclesiology, Ge29,46
Economic: decline, Ia25; growth, Gf13; If51; history, Ac31; institutions, Bc142; policy, Gf64,66; Hf28,41; Ia39,d14,61,f8; theory, Gf1,8,34,37, 42,46,49,51–52,58,60,62,64,i40,66; thought, Gf56; warfare, Id47
Economics, Aa113; Bc29–30,52,144; Hf3,11,l91; Ma1
Economists, Ab4
Economy, Ba150,190; Jf2; Mf7,g12; international, If67; Ne7; political, Nb8
Edinburgh review, Hg62
Education: Aa59,115,117,b50,c69,84; Ba73,141,b44,65,98,105,145–146, 150,164,175,178; De37; Ee68; Gi31, 58; Hb83,h9,l106; Ib20,54,f2,21,57, g11,h4,7,18,27,30–31,35,j3,50,61, 78,82; Li7; Mg9,i2; Nh3, 22,40; adult, Ij28; Catholic, Nd11; higher, Hi29; legal, Gc3; medical, Gj25; Hk60; military, Ii38; scientific, Hi30,k44,62; technical, Hi4,28–29; university, Ge19; women's, Bb141; *and see* Schools; Education; Effigies, Ee32,j35; Li6
Egyptology, Ij71
Ekumeku movement, Nb22
Elections, Ba80; Fb18,40; Gb1,17,21,23, 28–29; Hb62,65; Ib7,14,28,34,57,75; Mb32; Nb5; Canadian, Nc6; county, Gb32; parliamentary, Hb11; *and see* Poll books

Electoral systems, Hb28
Electorate, Hb4
Electrical, Electronic, Telecommunications & Plumbing union, If72
Electrical Trades union, If72
Electricity, Hk57
Electronics, Ij6
Electrotherapeutics, Hk50
Elites, Ba4; Hg36,43; Ig28
Emergency, Malayan, Nb6
Emigration: Fe96,h3; Ga26; Hf61,g17, h38; Kd1; Mf6,g4,9,14,16,18,21–23, i16; Nf2,5–6,17–20,26; female, Nf44; Irish, Nf12,28,31–32,50; Scottish, Nf23–24,33–34,36; Welsh, Nf27
Emigrés (Russian), Hl53
Empire, Ba81; Ga18–19,23,b9–10,d17, 19,24; British, Aa97,c1; Ba120; Fa1,d9,14; Id103; Mg4; informal, Na19; *and see* Section N
Employers' organizations, If73
Employment, Hf31; Ib61; children, Hh30
Enclosure, Bb6,38; Gf11,14,67; Hf21
Encyclopedias, Ac115
Engadine, the, Ii35
Engineering, Bd24; Hf50; civil, Hf53,90; electrical, Hk66; water, Ef19; Gf61
Enlightenment, Ac25; Ga6,e15,20,f34, i29,59,j10,24; Hl96; Scottish, Gf62,i87
Entertainers, Hf31
Entertainments, Fk45; Hf31; royal, Fk13
Entomology, Hk19
Entrepreneurship, Ne23
Environment, Ca46; Gj26; Hh4; Ia23,j36
Epidemics, Hk52; Nc10
Epidemiology, Ba177,c25; Ih1,4
Epigraphy, Cb47
Episcopacy, Ec1; He23; Ke10,27
Episcopal: authority, Le22; registers, Ee18,72,76
Episcopalians, Kc16
Episcopate, He24; American, Ge32
Epitaphs, Ab64
Erotica, Fg40
Eschatology, Fj26
Espionage, Aa171; Bc18; Fb38,c11; *and see* Intelligence
Estate management, Da6; Ef1,9,49; Gf22, 26; Mf11
Estates, Aa31; Ek8; Fh11; Gg6; Kg7; crown, Fc14,f26–28; multiple, Df43
Ethics, De40; Gf51

Index of Subjects

Ethnicity, Ja5
Eucharist, Ee66–67
Eugenics, Mi9
Europe, Df11
European: Community, Bc201; Ib59,62,
    c2,7,d31–32,94; integration, Id12,36;
    monetary union, Id37
Evacuation, Ig8,h9
Evangelicalism, Ge4,13,27,31,34–35,37,
    39,45,47,49,51; He13,17; Nh5
Evangelism, Hh28
Evolution, Hb55,e31,38,k8,23,46,73;
    Nd12
Ewe question, Nc26
Examinations, medical, Hk17
Excavations, Ca19
Exchequer, Aa7; Ec17; Fc7
Exhibition, London, Hf97
Exiles, Ld2
Exploration, Ab19; Ed11; Gd13,21; Na7

Fables, Ba192; Ej20
Factories, Gf41
Fairytales, Ei8
Falconry, Ef47
Families, Aa144,145; Eg32,42; Gg10,14;
    Hg47; Ig3,21; Mg2
Family, Ba78; Ig26; reconstitution,
    Ba166; Gg11,25
Famine, Ba12; Fg49; Mb31; Nf50
Farmers, Mj3
Farmhouses, Hl70
Farms, Hg79
Fascism, Ib2,35
Fashion, Gf38
Feasts, Ei13
Federalism, Mc1
Feminism, Bc56,63,197; Fk32; Gg8,i74;
    Hb1,18,g7,21,24,27–29,31,33,61,77,
    81,h36,l51; Ia15,h22,j56; Nh32; and
    see Women
Ferranti company, If115
Ferries, Ff5
Fertility, Hg47; Ig32,35
Festivals, Ba25,38; Hd7,l58,88; Ki3
Festivities, Ba17
Fetter Lane Society, Ge37
Feudalism, Ac78; Ei17; Gi66; bastard,
    Eg44
Field surveys, Ca22; systems, Df24; Ek7,
    32
Films, Hj2; Ij43,69,93; Nh37
Finance: Aa11; Bc95; Gf35,46,57; If62;

Mb28; external, Id4; government,
    Gb26,f47,64; international, Ne13;
    public, Ne21; war, Ea24
Financial crisis, (1931), Ib94
Financiers, Hf29,41,45
Fire Brigades union, Ia2–3,19–20,36,
    b73,80,f3,11–12,34,36,50,65,68,92,
    94,g13,36
Fire-fighting, Bb8,c7; Mf3
Fish, Bb116; Ef20; farming, Ba49
Fisheries, If96; Mf15; Ne2
Fishing, Ef39; Hc17; communities, Bb195
Fishweirs, Ef58
Fleta, Ec18
Folklore, Aa131
Folktales, Ei8
Fonts, Ej44
Food, Ba144; Cb27,31; Ei5,13; supply of,
    Bb182
Foreign Office, Aa6,10; Id39; and see
    Diplomacy
Foreign policy, Ac102; Gb2,26,d18–19,
    h4–5; Id89; and see Relations
Forestry, Bb191
Forests, Ef6,k6; Fg29; royal, Ff14
Forfeda, Li13
Forgeries, Ee37; Fb46; Ij23
Fortifications, Bb115,159; Ca10,b20;
    Df66; Eh5,23; Jh1; Kh1,6; Lh1,4,10;
    Roman, Bc186
Forts, Ca5,17,31,33,b33; Jg9,h9
Fosterage, Jg8
Franchise, Hb4; Ib19
Frankalmoign, Ef51–52
Franks, Da26
Fraud, Ba159
Freehold Land Society, Hh24
Freemasonry, Bb86
Freemen, Bb47
Freethinkers, He69
Frontiers, Ea22,h26,k16; Lh10
Funerals, Ba71; Fg5
Furniture, Gi84

Gaelic revival, Lg5
Galloglass, Lh6
Gambling, Ba39
Gardening, Gi55; Hl50
Gardens, Ab18; Ba62,104,135,148,184,
    201,203,b15,18,20,28,58,152,165,
    c64; Ca52; Ek37; Fk19,48,l14,23;
    Gi44,65; Hl108; Ij92
Gardens, landscape, Gi6

Index of Subjects

Gender, Ac36; Ba161,c223; Fk9; Gg23; He22,81,g14,16,51,59,62,72,80,85, l18,27,40,68,71,87,99–100; Ib16–17, 22,g9,20,34,36,h12,22,32–33,j15,55; Nc5,f21,45,h12,21

Genealogy, Aa143–147,172–174,b16; Db32; Ea6,g34; Fh1; Jb3; Lg5,i4

General Strike (1926), Ib12,e3

Gentry, Ea7,b3,5,e61,g1,5–6,23,28,33, 42,50,h4; Fh4; Gg6; Hl80; *and see* Class

Geography, Aa109; Ba113; Hl55; Ma4

Geology, Aa169; Gj4; Kh8

Girl Guides, Ia42

Glass: painting, Gi10; stained, Bb26,36; Fk88; Hl25

Gleaning, Gc7

Glider Pilot Regiment, Ii55

Glossaries, Aa87

Gods, Ca3

Gold Coast Cocoa Marketing Board, Ne9

Gold standard, If45

Goldsmiths, Ab24; Fk28

Gospels, De43

Gothic revival, Hl32

Gout, Gj7

Government: Bc69; Fe26,j4,19,29,46,k1, 31,70; Gd20; Burmese, Nc19; Indian, Nc16; Labour, Bc131; Ia17,b52,55,71, 92–93,d35,88,f8,78; local, Db10,24, 33; Ec15,f5; Fb5; Gc6,8; Hc3,g65; Ib56,c1,7–8; Mj7; municipal, He56, f66,83; national, Ib78,94; parish, Bb10; *and see* Parties; Politics

Graffiti, Li9

Grammar, Gi3

Grand tour, Gd2,5,9,26,i12,51

Granges, monastic, Ef35

Gravestones, Ej54

Graveyards, Ki1

Greater London County Council, Bb3

Guilds, Eg11,18,31; craft, Ef57

Gulliver's travels, Gi37

Haberdashers' Company, Ba5

Hadrian's Wall, Ca14,42,49,51,b22

Hagiography, Bc229; Dc10,22,30,34,39, e35–36; Ee2,83; Ke25; *and see* Saints

Halls, Ek16; Ja4; aisled, Ek21

Handwriting, Ei32

Hanseatic League, Ed7

Harbours, Ff1; Gh37; Kh1

Hawking, Ef47

Head-dresses, Li9

Health: Ab53,c10; Ba27,c211; Ei34; Gj26; Hg51; Mg1; care, Hk20,51; children's, Hk70; mental, Hc21,g71, h6,16–17,35,37,k4–5,14,22; policy, Ih15; public, Hb25,g50,h19,k2,7,26, 31,52–53,55–56,58,71; Ig21,h5; services, Ih23,24; *and see* Disease

Helmet Anglian, Bc156

Heraldry, Ea11,h1,j43; Ja7

Heresy, Fe35

Heritage, Bc149; Ij24–25,62

Hermits, Le9

Highland clearances, Hh38

Highlanders, Scottish, Hg91

Historians, female, Ac9

Historical writing, Gi83

Historicism, Aa17

Historie of the arrivall of King Edward IV, Ed13

Historiography, Aa177,c16–17,27,29, 34,42,52,58,67,73,85,103,109–111; Ba95,c8–9,45,48,152; Da9,11,13,15, 17,20; Ee4; Fa6,e58,j11,20,k50; Ga6, 22,32,b25,e44,f5,30,50,58,i39,47,70, 87; Hl32,68; Kb3,7,e10,f3; La5,10; Mb13,16,i4–5,8,15,17,20; Nh1,5,18, 25,33

Historiometry, Ac98

History, local, Ab29,61,c8,80,93; Fe93; La5

Hoards, Ca25,b6

Holinshed's *Chronicles*, Fk21

Holocaust, Ac46

Holy Grail, Ji10

Home: front, Ig19; Rule, Hb51–2; Mb28, c1

Homiletics, Ge1

Homilies, Dc37,43; Ee81

Homosexuality, Ba140; Gc13

Honors, Ea31

Honours, colonial, Nb15

Hood, the, Ii36

Horses, Bb144

Hospital services, Ih29

Hospitals, Aa56; Ba142,b167,189,194, 196; Eg7,36,41,i75; Fl16; Ga11,i8,j1, 6,15,24,27; Hk30,42; Ih8; children's, Hk70; cottage, Bb40

House of Lords, Gb18,35,e2; Hb38,67, 78

Households: Cb23; Ef1; Mg6; noble, Ec8,f18,66; Gg24; royal, Ea9

Houses, Ba7,168,178; Ek48; Fa16;
    Hh32; Ja4; Nh16; country, Bb125;
    Fh12; Gg24,i22,44; manor, Db40;
    Eg5,k49; medieval, Je18
Housing: Bb199,212–213,c38; Hh3,25,
    l62; Ig31,h3,10,37; Mg1; council,
    Hh7; policy, Hh20; public, Hh29;
    slum, Fh5
Hudson's Bay Company, Nc14
Huguenots, Bd35; Fe6,16,f4
Humanism, Fj40
Humanitarianism, Gj1
Hunger, Ba144
Hussars, German, Gh23
Hussites, Ei42
Hygiene, Hk2
Hymns, Fe19

Ibstock Johnsen company, Bb33
Iconoclasm, Fe25
Iconography, Ba185; De2,28,41,61,63;
    Ee15,i29,j18; Fe25,k33
Idealism, Hl96
Illegitimacy, Gg25
Illuminations, Ej65
Illustration, biblical, Fe4
Immigration, Fe16; Ig24; Nf13
Immunology, Ba175
Impeachment, Fc26
Imperial preference, Ne27
Imperialism, Ba187,c10–11,130; Hi23,
    l125; Ij72; Ld1; Na13,g29; and see
    Section N
Impressment, Gh36
Income, Bc34; Hf102
Indians, Fd19; Gh16; American, Fd20
Individualism, Hb82
Industrial: conflict, Ga28; estates, If48;
    growth, Gf31; policy, If39,63,78;
    relations, Hf2,49,56; Ia2–3,19–20,36,
    b10,73,80,88,f3,11–12,21,34,36,50,
    65,68,72,92,94,114,118,g13,17,36,
    h11,36; revolution, Bb74,c151; Gf5–6,
    13,24,29–31,50,53–54
Industrialization, Ba92,197,b14; Gf6;
    Hf10,20,27,46,51,59–60,74,89,g48,
    90,l2; Kf3; Mf4; proto-, Mg5
Industry: Ba54,136,b131,142,c131; Dd2,
    11; Ef3,40; Gf25,57,j21; Hd8,f1,58,
    82,g65; If9; aeronautical, If102;
    aircraft, If43; aviation, If74; banking,
    Ab45; Hf14,24,26,28,36,41,51,67,74,
    100,105; If26,62; basket-making,

Hg34; biscuit-making, If1; brewing,
    Bb161; Ne26; brick-making, Bb33;
    Hf84; building, Ef48; car, If39,52,107,
    119; chemical, Ba30,c36; Hf33,80,g89;
    If38,40,47,75–76,79,88,112,121,j35,
    42; coach-building, Bb124; coal-
    mining, Bb188,197; Gf7; Hf40,42;
    101,g49,Id84,f30,89,123,g6,h14;
    Nc44; computer, If59,115; coprolite
    mining, Hf69; cotton, Hf49,54,71,g61;
    If42,111; cutlery, Hf96; dairy farming,
    Hf92; distilling, Aa120; dye, Hk36;
    electricity, Bb200; electronic, Ij6;
    engineering, Id45,f102; film-making,
    Ij1; finance, Hf14,100; fishing, Ba169;
    Ef4,61; Ff10; Gf44; Hc8,f25,93; If105;
    funerary, Ej47; glass-making, Lf3;
    glove-making, Hf7; gun-making, Bb50;
    hop, Ab34; Bb161; hosiery, Ba191; in
    India, Ne23; insurance, Gf48; Hf52,
    71; iron, Bb170,d34; Gf19–20; Hg94,
    h15; Id84; iron-making, Ca21; iron-
    working, Bb81; jute, Bb208; lace,
    Bb221; linen, Mg6; malting, Bb161;
    manufacturing, Gf5; match, Hf19;
    metal-mining, Aa83; Hf16–17; metal-
    working, Ab32; Dd24–25,f4; Ef26,43;
    Ij4; mining, Ab32; Ff23; Hf15,39;
    munitions, If91; pharmaceutical, Ab31;
    Hf72; If88; pin-making, Ba31; plastics,
    If75; pottery, Bb75,92; Dd21; Ef10;
    Hg6; printing, Aa52; Fe89; publishing,
    He34,76,l15,44; quarrying, Bb204;
    railway, Hf101; If16; retail, Ab33;
    If17; salt, Ca7; service, Ab38; If101;
    ship-building, Ab54; Ef32; Fi9; Gj2;
    Hf56–57,81; If64,73; shipping, Ba137,
    b111,134,217; Df29; Ef16,31,k40;
    Ff8; Hf4,8,61,65,88; If4,6–7,85; Lf5;
    shoe-making, Ba87; slate-quarrying,
    Bb53; Hf70; smelting, Ff23; steel, Hf9;
    If14; textile, Bb52,149,166; Ef25;
    Fc18,f22; Gf24; Ij47; tin-plating,
    Bb133,143; tin-working, Ff11;
    tobacco, If13; transport, Aa12,b27,c7,
    87; If41; watch-making, Ab49;
    wholesale, Ab33; wine-importing,
    Ff24; and see Business
Infection, Hk40
Inflation, Bc144; Ef7; If5
Information technology, Ac70; Ij42,60
Inheritance, Eg37; Fh1; Gc5
Innovation, Hf91

Inoculation, smallpox, Gg2
Inquisitiones post mortem, Ef14
Insanity, Hk37; Nf16
Inscriptions, Ca11,30,37,42,b1,47
Inspectorate, schools, Ba73
Insurance, If4; textile, Gf48
Insurrection, Id97
Intellectuals, Nh34
Intelligence, Aa171; Fc11; Hj1; Ib66,
    d19,27,i12,17,27,30; Nb24; *and see*
    Espionage
International Monetary Fund, If22
Inventions, Ba127; Gf24; Hk57
Inventories, Fh14,22; Gi53
Investment, Hf20; If29
Investors, Hf8
Irish: Hg3; Free State, Mb30,c2,f4;
    Housewives' Association, Mb29; in
    Britain, Hb6,g58; in Scotland, He2;
    Republican Army, Mb21
Irreligion, He48,50
Islam, Nd1
Italians, Ba43,b106

Jacobitism, Fb66; Ga15–16,b15,e42,h3, 27
Japan Society, Hl29
Jazz, Ij20
Jewellery, Ca41,50; Ef26
Jews, Aa106; Ba1,d18; Eb9,c31,e1,30,
    45,70,g2,9,i59,103; Fe24,39,48,88;
    Hb34,e36,g58; Ih25; Nf41; *and see*
    Anglo-Jewry
Journal of Economic History, Ac112
Journalism, Ba154,c28; Hb2,j18; Id44,
    52,e5; Mi1; Na24,h7,36; medical,
    Hk41; *and see* Newspapers
Journals, Hk18,l3,81
Judaism, Bc189; He17,25,43,69,78,l56;
    Ie1,j51
Judges, Ec25
Judicial combat, Ec27
Judiciary, Ic10
Juries, Ec9
Jurisdiction, ecclesiastical, Ke21
Jurisprudence, Gc3
Justice, Db28; Fc21
Justices of the Peace, Ec32

Kikuyu, Nf39
Kilns, Ca13; Ek20
Kingdoms, Fc27
Kings, Df73; *and see* Index of personal
    names

Kingship, Db26,34; Ec7,22,i63; Fb2; Jb2;
    Ke14; Lg5,i4
Kinship, Hg17; Jg8
Knight service, Ef52
Knighthood, Bc98
Knights, Ea9,11,16,h24
Kulturkampf, Hl117

Labour: Ac19; Ba174,b61,c117,176,219;
    Gf8; Hb14; Ib43,61; Ne2,f10,21,45,
    h21; African, Ne12; agricultural,
    Hg26,h13; Ig7; black, Nf4; child,
    Hk34; Ne11; conditions, Hf7,35,37;
    female, Ef29–30,33; Hf62,g61; force,
    Ig4,h12; history, Aa32,35,119; Mb22;
    markets, Ba88; Ne6; militant, Ac92;
    Hf2; Ib9,51,68,g11; movement, Ab8;
    Bb69,c175; Hg3,13; Ig5–6,22,27;
    organization, Hf82,h13; penal, Nc15;
    rural, Bc116; Ne22; urban, Bc116
Labourers, agricultural, Gi4
Ladies' Land League, Mb27
Land: Aa27,31; Nc46; native, Ne4;
    ownership, Hf34; policy, Nf4;
    question, Ba163; reform, Ih21;
    settlement, Nf37; tenure, Db13,19,35;
    Ef47; Hh23; use, Gg4
Landed estates, Aa27
Landholding, Bb128,c122; Ea13,31,40,
    e16,75,77,f14,34,51–52,54,63,g19,37,
    49; Fh11; La6,g7; Mf12
Landowners, Fh12; Jg1
Landownership, Gc5; Mf10,11
Landscape: Aa84,c8,80; Ba72,118,b152,
    c169–170,181; Ca9,b4; Df69; Ek8,
    24–25; Ff7; Gf9; Hf27; Ia27,f100,j92;
    industrial, Bb156
Language: Ba8,91,c58; Cb19; Ei52,74;
    Gi41; Hl114; foreign, Hi3; Hebrew,
    Ei103; Latin, Cb1; Middle English,
    Bc228; De4,8,26,30–31,33,57,59; Old
    English, Aa87; Bc227; De12,17,24,
    29,34,44,46; Ei6–7,16,21,30,49,62;
    political, Gb6,33,36; Welsh, Bb5;
    De54; Hg50
Latitudinarianism, Fe3,59; Ge50
Law: Aa1,3,175,b20,41,c48; Ba22,102,
    170,b72,c128,195; Da3,b1,28–29,39,
    f8; Ea19,c21–22,g28,33,i43; Fb61,c18,
    25; Ga12–13,b29,c2,4,11,e8,i9; He80,
    g8; Ic5; Kc1–2,5–8,10–12,15,17,19;
    Lc2,e4; Nc33,37; canon, Bc125; Ec31,
    e68; Jc1; codes, Db29; common,

Index of Subjects

Bc199; Ec6,26; company, Gf10; Hc9; criminal, Ba24; Hb39; Kc13; Nc5; ecclesiastical, Ba20,84; Le18; land, Ba3; marcher, Kc14; maritime, Kc3; natural, Fj17; New Poor, Hh11; penal, Nc50; Poor, Ba179; Ga10; Hh2,8; Nc13; vagrancy, Gc12; Welsh, Jc1; *and see* Court; Judges; Judiciary; Juries; Justices of the Peace
Laws: Corn, Hb52
Lawsuits, Ec33
Lawyers, Ba22,b103; Ee11
Legends, Ba176,c213; Da24; Eh3,14–15,i31,37,45,77; Fj34; Ji2
Legions, Cb47
Leisure, Ia38,g9,j94
Letters, Bc178; Ee80; Na24
Levellers, Fe21
Leviathan, Fj15
Leyrwite, Eg24
Liberalism, Ac19; Bb147,c16,219; Ha3, b10,15–17,29,66,69,76,l64; Ib24,g22; *and see* Parties; Politics
Liberation, Hb44
Libraries: Aa12,30,69,105,121,134,153, 155,168,b7,17,29–30,35,39–40,42, 59,c4, 108; Ba67,b64,105,c61; Ei54, 73,100–101; Fk16,64; Hl,46; Ke3,5, 19; college, Gi57; public, Hl10
Life of St Katherine, Ee82
Lighthouses, Hf22
Lighting, gas, Hf43
Limekilns, Gf43
Linguistics, De39,52; Mi19
Linnean Society, Fl2
Literacy, Ei19; Gg3; Hg63,j10
Literature: Aa92,161,c17; Bc39–40,56, 58,62,180,204,220; Db34,e5,14,16, 22,38,45,58,62; Ei13,41,45,67,70; Fa11,b29,g56,k25,50; Gi9,17,24,41, 52,62,73–74,78,80,87; He48,g1,j8, l16,27,34,36,46,71,93,97,99–100, 114; Ij29,37,56; Lf9; Mf14,i5,14; Nh6,14; American, Nf18; Arthurian, Ei65,77; Ji2,10; autobiography, Fk18, 91; Kh7; biography, Ab2,26,c107; Fk74; Gi35; Hb18; Ib32; criticism, Hl40; French, Ei83; Hiberno-Latin, Li3; Irish, Li2; juvenile, Ij15,50,72; medieval, Aa90; military, Hj14; novels, Fg8,52; pastoral, Ee34; scientific, Hk62; travel, Id104; *and see* Drama; Poetry

Liturgy, Dc9,16–18,25–26,32; Ee59,74, i71,j56; Je9; Li8
Livery companies, Ba5
Livestock, Ef63; Lf10
Lloyds of London, If4
Lobbying, Fb21; Ga18–19
Logistics, Cb3
Lollards, Fe33
London Committee for the Abolition of the Slave Trade, Gb24
London Revolution Society, Gb8
Lordship, Db40; Ec14,h24
Los Alamos project, Ij86
Luftwaffe, Ii6
Lunatic asylums, Hk37

Machiavellianism, Fj14
Macmillans (publishers), Ba50
Magistrates, Hc22
Magna Carta, Ea27
Maidstone (ship), Gh29
Management, Hf89; If58,66
Manors, Ec14,f12,k8; Fc13,17,g6,26,h11
Mansiones, Ca5
Manuscripts: Aa5,24,47,105,121; Bc220, 222; De6,20–21,48; Ei100; Fk55; Ji9; illuminated, Ji7; liturgical, De13; Middle English, Ei2,4,9,47,60,76,84, 96; surgical, Ei48
Maoris, Nc37,49,h2
Maps, Aa31,84,97,b6,9,48,51,c13; Bc47; Ei40,k14,32; Fg4,32,k68; Ga24,f11, 14,22,26,39; Hc11,14; Kh6; estate, Gf3
Marketing, If115
Markets, Bc167; Ef41; Gf1,57,63; capital, Gf47; corn, Gf66; domestic, Hf73; stock, Gf47
Marriage: Aa128; Ba38,182; Ec10,f30, g4,15,48; Fg39,48,51,k45; Gg8,15; Hc5,g23; Ig15,41; Kc12,i5; Li4; clandestine, Gg16; clerical, Fe13
Marxism, Aa177,c27,29,34,52,67,109–110,111; Bc9; Hb9,g4,14,l119; Nb1
Masons, Gf68
Masons' marks, Je13
Masques, Fk24,93
Mathematics, Aa23,50; Bc87; Fl6,9,27; Gi21,75,j5,11,14
Mau Mau, Nb14,f39
Mausoleums, Gi13
Measurement, Df5,25,41,56; Ef65
Mechanics' Institute, Hi27

317

Index of Subjects

Media, Ac91; Hf79; Ii61
Medical relief, Hj11
Medicine, Aa28,56,152,b53,c10,26; Ba9,
  28,145,151,154,175,194,b40,c23,26,
  28,57,90; Df39,51,68; Eg16,41,i3,34–
  36,75,95,103; Fl1,13,26; Gg2,i8,j3,6–
  8,12,15,19,22,24,26–27; Hk13,43;
  Ih2,8,i39,j9–10,18,26,52,57; Nf29,47;
  preventive, Hk26; psychological, Ij80;
  tropical, Bb45; Hk18; veterinary, Hf72
Medievalism, Aa17; Bc225; Hl12,35,37,
  119
Memorials, Ii8,j64; church, Ej24; war,
  Hl83; Ia21
Mendicants, Le23
Mental illness, Ba153
Mentalités, Fg58
Mercantile marine, Ff2
Mercantilism, Ff13
Mercenaries, Kh3–5,9
Merchants, Ba37,139,c31; Fd7; Ga25,f15
Metallurgy, Ba199; Dd18
Metalwork, Ca15,25,35,50; Dd6,f2–3,
  61,72; Hg94; Jg2
Meteorology, Gj23
Methodism, Fe58; Ge10,12–13,27,31,35,
  37–39,45,47–49,g14; He8,26,g88;
  Calvinistic, Ge51; early, Ge5;
  Primitive, He29,57
Métis, Nc14
Mfecane, Nb7,11,32
Migration: Eg15,25,40; Fg51,h8; Gg5,
  22; Hb79; Kh3; Nf42; female, Nf30;
  Irish, Bc182; Fd16; Nf7,25; Scots,
  Fd16; Nf7
Militia, Bb95–96; Fb5; Gi50
Millionaires, Ig41
Mills, Ef27,36; water, Bb193,c158; Ff5;
  Lf8
Millstones, Ef27
Mines, Bb22–23,54,160; coal, Bb168;
  gold, Bb42
Miniatures, Ej5
Mining, Ba199,b54
Ministry of Supply, If44
Mint, Royal, Bc226
Mints, Aa26; Ba35–36,57,181; Dd4,15;
  Ef42; Ff3; If27
Miracles, Ee2,g16; Fg17
Misericords, Ej40,61
Missals, Je9
Missionaries, Aa103; He39; Baptist,
  Ba180

Missions, He10,g42; Nd2–4,6,8–9,13–
  14,17; Scottish, Nd16
Moated sites, Ek35,49
Monarchy, Ba65,b35; Ea36; Fj8; Ga8;
  Ha2,b40
Monasteries, Aa81; Dc13,23,28,f28,44;
  Ee4,16,20–21,29,31,33,35,40,42,49,
  52,57,63–64,71,77,84,f9,49,51–52,
  54,g32,h10,i101,k54,56,60; He44;
  Je7,15,19; Ke2; Le8; dissolution of,
  Fe9,f19; double, Dc1; *and see*
  Nunneries; Orders
Monasticism, Dc19,42; Ge42; He45;
  Ke1,16; Li7
Monetary: policy, Hf103; system, Id37
Money, Cb41
Moneyers, Dd15,29; Ef59
Monopolies, If77
Monumental: brasses, Ej47; inscriptions,
  Bd9,16,20,27,31
Monuments, Ac40; church, Ej35
Morality, Hl43,64; Ij77
Morant Bay rising, Nc8
Moravians, Ge37
Mortality, Ba11,71,b126,c216; Cb18,36;
  Df52; Eg14,17; Fh10; Ig18,40; infant,
  Ba110,116; Hk71; maternal, Ba115–
  116
Morte d'Arthur, Ec27
Mosaics, Ca40
Motherhood, Hg80
Mottes, Ek13,47; Lh8
Mountaineering, Hl60
Mughals, Fd13
Murage, Lh15
Murals, Ej3
Murder, Fg15; Gc13
Museums, Aa102; Bb54,112; Hi1,l59;
  Ij76
Music: Ab39; Ba29,c77,78; Ei10,26–27,
  33,50,61,86–88,97–98; Fk2–4,12,30,
  62–63,66,85; Gi49,80–81; Hl14,17,
  20,31,56,89,109–111; Ie7,j3,63,88;
  Ki4,6,8; Church, Gi42; Hl57; Ie13;
  dramatic, Fk5; early, De56; halls,
  Hj16,l74; military, Ba69; popular, Ij16
Musical instruments, Hl38,112
Musicians, Ba14
Mutiny, Ii20,53
Myntling register, Eg24–25
Mysticism, Ee7,i54
Mythology, De16
Myths, Ba32,c213; Fa7

Index of Subjects

Names, Ef59; personal, Ab12; Ba68,189; Eg39–40,i20,56; Jg6
National Buildings Record, Ac35
National Gallery, Hl118
National Health Service, Ig43,h13
National: identity, Ac82; Bc190; Ha6; origins, Ba149
National Register of Archives, Aa4
National Trust, Bb121
National Unemployed Workers' movement, Ib18
Nationalism, Hb69,i8; Mb18; African, Nb21,c26; English, Hb51; Scottish, Ij45
Nationality, Ba42,b119; Eh9
Nationalization, If83
Natural history, Hk72
Naturalists, Hk48
Nature, Fg44; Gi65
Naval and Military Emigration League, Nf17
Naval: bases, Ng15; policy, Hj7,15
Navigation Ab19; Ba6; Na7
Navy, Aa101; Ba26,52,56,c173; Ff2,i1, 16; Gh2,4–5,10–11,14–15,28–30,37–38,40–42,47,j2,19; Hd12,j1,12,k59; Id25,i28,68; Australian, Ng16; Canadian, Ng12,33; merchant, Hf32
Neoplatonism, Fe54
Newspapers, Aa116,b46; Ga4; Hb35,53, f79,l22,45; Ib50; government, Gb3; and see Journalism; Press
Nineteenth century, Hl15
Nobility, Ac23; Bc49; Ea11,21,42,b5, g42; Fb10,34,43,62,84,f29; Gg24; and see Aristocracy
Nomadry, Lf10
Non-conformity, Aa95,125,b36; Bb154, c198; Fe31,94; Ge8,23–29; He6,12; and see Dissent
North Atlantic Treaty Organisation, Id6
North West Company, Nc14
Nuclear: energy, Ij5; weapons, Id76
Numismatics, Aa26,b58; Ba35–36,57, 181,c139; Cb16,40; Dd1,23,31; Ef42–43,60; Ff3; If27; and see Coinage
Nunneries, Ee79; Fe9
Nuns, Fg23
Nurses, Gj25
Nursing, Hk61

Office for Strategic Services, Ii34
Ontology, Fl3

Opera, Hl17,109–110
Optics, Hk12
Oral history, Aa135–136
Orangeism, Hg3; Ib89; Nc6
Orders: Benedictine, Ee24,54; Ge42; Capuchin, Fe90; Carthusian, Ee20,69, i54; Cistercian, Ab7; Ee16,42,f35,i3; Le22; Dominican, Li5; Me1; Franciscan, Bb37; Ee62; Fe11; Me3; Gilbertine, Ab7; Ek60; Jesuit, Aa60; Bb100; Ke23; Mc2; military, Ee48,i64, k23; preaching, Fe33; Premonstratensian, Ab7; religious, Ee22,33; Ke2; Servite, Aa70; and see Monasteries; Nunneries
Ordinances, Fc20
Ordination lists, Ee18
Ordnance office, Fi24
Ordnance Survey, Ab9,48; Hc14
Organists, Fk73
Organs, Hl56,112
Orientalism, Ba117,129; Nh4–5,13,26
'Original contract', Gi18
Ornithology, Ij2
Orthography, Ei20
Osteoarchaeology, Ee50,i81–82
Outlaws, Ea26,h3
Oxfam, Ia4

Pacificism, Id33,i58
Packmen, Bb130
Paganism, Cb14; Df42,55,75
Pageantry, Fk35,44; Hl84
Pageants, Ki3
Painters, Hl77
Painting, De43; Ej3; Fk14,89; Gi1,4–5, 48,60,82; Hl18,37,52,78,85–86,115; portrait, Ba183
Palaces, Ek19; archiepiscopal, Ek41; episcopal, Ek41
Palaeography, Ei68,99
Pale, Lh10
Pamphlets, Bc203; Fj1,22,48
Pantomine, Aa157
Papacy, Ed1,5–6,e3; He64; Ke2,8
Papal provisions, Ee71
Parades, Nh10
Paramilitaries, Mb5
Parasitology, Ij18
Paris Peace conference, Id100
Parish: records, Aa95; registers, Aa172–174; Bd1–8,10–15,17; Fh2; Gg25; revenues, Ef62

319

Parishes, Bc100,119; Dc2–3,5–6,35; Ee9,k64; Gg13,20; Ke18,26; Mj6

Parks, Ba104,148,b152; Hh4; deer, Ef6; national, Bb25,51,129

Parliament, Bb147,c80; Eb5,c18,i63; Fa2,8,b1,7,17,20–21,27,39,42,67,77–78,82,87,c4,26; Gb2,12,16,27,e2; Hb21,47,62,67,77–78,c18; Ib6,13,45, 58,76; Kc16,20; Lc1; Mb17, 32; Australian, Nb33; Cavalier, Fb41; records of, Fb26; representation, Hb70; Rump, Fb4

Parliamentary: debates, Gb33; reform, Hb24; Ib38

Parties: British Communist, Ib86; Communist, Ib11,73,g6; Conservative, Aa164; Ba171; Hb26,38,54,67,75,83; Ib20,30,72,74,83,94,d22,h32; Labour, Ab8; Ia41,b1,8,15,17,26–28,34,48,55, 63,65,77–78,82,84,88,90–91,93,f39, 63,65,g22,h33; Liberal, Hb8,10,33,85; Ib81; Liberal Unionist, Hb42; Nazi, Ib50,d103; political, Gb22; Hb21,63; Ib4,14,16,36,45,75,g27; Scottish Nationalist, Ib29; Social Democrat, Ib81; Tory, Hb27; Whig, Fb63; Gb19–20,22,c9,h27; Hb13,23–4,e80; Kb3; and see Conservatism; Liberalism; Politics; Socialism

Partition, of India, Nc38

Party organisation, Ib19

Paston letters, Eg43

Pastoral care, Dc6,11,35,42; Je15; Le4–5, 21,23

Patents, Ba127

Paternity, Nf10

Patriotism, Ed2; Ga8,b6; Ie7

'Patriots', American, Gb9,10

Patronage: Ee40,i80,83; Fb32,58,g41, k82; artistic, Gi51; ecclesiastical, Ee51; Gb30,e52; literary, Gi34; political, Fb1

Patrons, art, Hl2

Peace movement, Hd14; Ib48

Peace of Paris, Gd19

Peace Societies, He16

Peasantry, Eb1,c9,d2,e67,g19,24–25,29, 34,39–40,42,45,51; Hg68; Indian, Nc20,f14

Peasants' Revolt (1381), Ec29

Peerage, Fg30

Peninsular and Orient Company, If85

Penitentials, Lc2

People's budget (1909–1910), Hb43

'Percentage agreement', Id3

Periodicals, Gi85; He11,l15,26,44,90, 113,123; and see Journalism; Press

Pesticides, Ij35

Petition of Right (1628), Fb14

Petitions, Gb24

Pewter, Ej6

Philanthropy, Bb167; Gi8; Hg6,42,h33; Ih14

Philately, Ia33,i18

Philosophy, Bc14,60; Ei51; Fe49; Gf70, i7,24,28,71,75; Hb1,80,k47,49,l1,48, 96; Mi11,21; moral, Gi40

Photography, Aa133; Hk65,l73,101; If79; aerial, Ab63; Ba72,201,c181; Cb4; Df69; Ek7,24,25; Ff7; Gf9; Hf27; If100,h6

Physicians, Ei35; Gj20,25

Physics, Hk16; Ij30

Physiocrats, Hf18

Physiology, Hk28

Piano, Hl38

Picts, Da10; Ki10

Pietism, Fe69; Ge49

Piety, Ee27,38,65,i80

Pilgrimage, Ee8; Je20

Pilgrimage of Grace, Fb13

Pilotage, Ia24

Pins, Df32

Pipe rolls, Ea24

Piracy, Fd8,33,i17

Place-names, Aa19,66,72,b10–11,13–15, 43,52,62,c47; Bb117,c96; Ca14,b19, 26; Df9,14,16,63; Ea17,f2,i20,k42; Hl125; Je16,g6; Ka2,j3; Le7,11,17; Scandinavian, Aa66

Plague, Eg12; Nc10

Plainsong, Fk62

Planetology, Hk32

Planning, Ij13

Planters, Fd5

Plants, De50

Plate, ecclesiastical, Fe76

Plays, Hg1; mystery, Ek62; passion, Ei60

Plot, Gunpowder, Ba46

Poetry: Aa137; De9,53; Fk34,42,55; Gi4, 63,67,86; Hl107,122; Ij44,67; Je20, i1,3,5,8; Ki2; Li8; Mi14; Anglo-Norman, Eh15; bardic, Li1; Latin, De1; Middle English, Ee25; provencal, Li1

Poisoning, Hc2

Polemic, religious, Fe15,44

Police, Ba15,61,b89,169,180,c11; Gc12; Hc1,19,k56; Ih19; Mc3,e4; Nc1–2,27, 35,42,52,53
Policing, Bb10; Ib44
Political: economy, Ga10,f62,70; Hf98; Mi2; ideas, Ib30; language, Gi83; philosophy, Ei55; science, Ac58; thought, Ac18; Gb11,19,25,31,36,f56, i18,59,69; Kb4
Politics: Aa164; Ba13,c14,16; Fe26,j4, 19,29,46,k1,31,70; Gh2; Ma1; Association, Bc200; Labour, Ba51; Liberal, Hb73; parliamentary, Ge8; popular, Ga30,b21,23,e8; Ib26; Mb4, 26; and see Parties
Poll books, Gb1
Polychronicon, Ei96
Polyphony, Ei10,27,33
Poor: Gc12,j6; relief, Ga14,g13,20; Hh10,14; Mf8
Popular: culture, Ga27; movements, If10; protest, Ga12,27,f2; Hh18
Population, Aa74; Ba200; Eg37–38,i2, k1; Fh6; Gf57; Hg92; Mg3
Portraiture, Aa25; Bc93; Ej5,65; Fk20; Gi1,27,53,60,82; Hl86; Ij84
Ports, Bc207; Cb46; Dd10; Ff24; Jg5; Lf7
Post office, Hc15
Postal: history, Ba167,b59; systems, Ii23
Postan thesis, Ba40
Postcards, Nh27
Potters, Bb92; Ja3
Pottery, Aa20,c64; Bc191,215; Ca6,18, 28,32,b13; Dd12,f48; Ef39,j9,13,17, 21,26,33–34,38,46,52,58,k20,37; Fg1, 54; Jf4,i11; Kg1; Lg4; Iberian, Lf6
Poverty, Ac114; Ga10; Hg20,64,h5; Ig9,42; Mb25,f8
Prayer books, De43; Fe12
Prayers, Je8
Preaching, Dc42; Ee81; Ge10,33; He8
Predestination, Fe70
Pre-Raphaelite movement, Aa33,89; Hl19,79,82,101,118
Presbyterianism, Fe10,72; He28; Ke10
Press, Ab46,c91; Gd8; Hb2,53,57,k33, l3,26; Ib50,d52; medical, Hk15; scientific, Hk10; and see Journalism; Newspapers; Periodicals
Pressure groups, Ib5
Prices, Ef27
Priests, He19
Printing, Aa150; Bc88,d37; Fl18; Hl36

Prisoners of war, Ii32,52,64
Prisons, Ba67; Ec30,k44; Gc10,14,e12
Probate, Fg7; accounts, Ff6
Profanity, Ba91
Professional Association of Teachers, If21
Professionalism, Hl81
Professions, Fc12,f9; medical, Gj6,17; Hh2,k3,6,10,15,24,41; playwrights, Hl105; teaching, Ih9
Proofs of age, Eg22
Propaganda, Eb6; Fb45; Gb34; Ib53,93, d96,i61,j1; evangelical, Nd17
Prophecy, Fe71; Ge20,i52
Prosopography, Ee46
Prostitution, Hg5,h36
Protection, Id14
Protestant League (1725), Ge6
Protestantism, Fe78; Ga8,e24,49; Hb12; Ib89,f93; Mb9–10,14,e5
Psalmody, Gi42
Psalters, Ej65
Psychiatry, Bc145; Hc21,g71,k4–5,22, 27,35,54,67; Ig43,h34,i39,j8–9,12,41, 48,68,70,80,87,91; Mk1,2; Nf16
Psychology, Ac98; Ij10
Public opinion, Ba185; Gb24; Hd3; Ib59
Public works, Hc3
Publishing, Ba19,28,50; Ge44,f53,h6,i12, 32,38,42; Hl89,113; Ij58
Punishment, Hb39
Puritanism, Fe28,37,42,51,58,85,87,92, 96–97,k82

Quakers, Fd6,e17,38,46,50,55,63,67,77, 80–81,86,93; Ga17,e28,36,i3,58; Me14
Quarter Sessions, Bb76
Quays, Cb46
Queens, Ba65; Ea44; Lc3
Queenship, Ec24
Queensland-British Food Corporation, Ne18

Race, Hg14,30; Ib16,d28; Nc25,31,f8
Racism, Gi43; Ij7
Radar, Ii28
Radicalism, Bc153; Fe31,g11; Ga8,b11, 34,36,e8,11,i39,52; Hb30,50,68
Radicals, Scots, Nb4
Radio, Ac91; Id96
Radio-navigation, Ii51
Railways, Ab27; Bb204; Hf48,g57,k63, l16; Great Western, Hf12

Ransoms, Eh25,28
Ranters, Fj28
Rebellions: Ef23,g19,51; Lb1; Mc3;
  Cade's, Eg29; of 1641, Mb24; of 1798,
  Mi5
Reconstruction, If87
Record linkage, Ac71; Ba166; Gb1,g11;
  Hg18
Record Offices, Ab29
Records: Chancery, Ec16,25; criminal,
  Aa85; ecclesiastical, Aa81; household,
  Ef18,66; manorial, Aa165; public, Aa8
Red Cross, Ii23
Reflections on the revolution in France,
  Gb25,i15
Reform: Hb13; army, Ng1; ecclesiastical,
  Le2; legal, Hb39; liturgical, Ge18;
  parliamentary, Hb63,77; social, Hb81
Reformation: Ac79; Ee35,61; Fa6,b85,
  e13,18,40,62; Ke6,10,12–13,24;
  Counter-, Le6; Irish, Le18; of manners,
  Ga11
Refugees, Id62,g24
Regalia, Df20
Regality, court book of, Kc4
Regiments, Mh2
Regions, Bb14,135; Ib43
Registers, Bd18
Regularis concordia, Dc12
Relations, Anglo-: American, Gh45;
  Hl44; Id5,9,11,23,27–28,41,55,61,67,
  73–76,90,108–109,f74; Argentine,
  Bc136; Hd16; Id44,66,69; Australian,
  Id102; Austrian, Gd6; Hd1,11;
  Brazilian, Ba47; Bulgarian, Id17;
  Canadian, Ia32; Chinese, Gd30;
  Id101,108; Danish, Gd1; Dutch, Ff13;
  Finnish, Id39; French, Ed1,9; Gd3,17;
  Hk9,63; Id49,64,85,i47; Md2; Na10;
  German, Bc140; Hd8,12,l117; Ib50,
  d34,60,71,83; Iberian, Ed4; Iranian,
  Ba93; Iraqi, Id68; Irish, Mb28;
  Japanese, Hd4,9,15,j13,l5,11,13,24,
  30,42,67,93; Id2,8,10,13,16,20,26,38,
  42,45,57,77,98,105,j44; Mexican,
  Hd6; Ottoman, Bc147; Papal, Df56;
  Fd1; Gd16; Russian, Ib39,d55–56;
  Scandinavian, Id86; Scottish, Db25;
  Soviet, Id3,29,39,99; Spanish, Hd17;
  Id81; Swedish, Gd25; Id84; Swiss,
  Id47; Wittelsbach, Gd4; Yugoslav,
  Id62; *and see* Diplomacy; Foreign
  policy
Relations: British-Saudi Arabian, Id59;
  Cambro-Norman, Ja5; foreign, Aa138;
  international, Id21; Italo-British, Id18
Relics, Ee63,80
Religion, Aa18,75,81,93,125,140,148,
  c79,99; Ba13,32,76–77,b27,57,214,
  c13,206; Ca36–37,b29; Fe36; Ge9;
  Hg42,l49,114; Ib17,d39,e2,7,g17;
  Ma1; evangelical, Ba157; Jewish, Hb34
Renaissance, Bc221; Fa12,14,k32
Republicanism, Fj24; Gb19,i39; Hb57;
  Mb33
Research: aeronautic, Ij39; methodology,
  Ab5; Ba174
Resistance, black, Nb22,c12; Indian,
  Nc20
Resorts, Hg54
Restoration, Bc79; Fa9
Reuters, Hf79
Revenge, Fi6
Revolt, Peasants, Eg51
Revolution: Hk25; American, Gb31,d32,
  e8,f59; French, Bc132; Ga22,b4,8,11,
  31,i23,52,63,72,83; Glorious, Bc48;
  Fb19,45,52,60,69,c2,d2,9,11,14,j51;
  Ga22,b8; Kb3–4,e10; Mb3; Russian,
  Ib11,39
Rhetoric, Eh7,i24; Nh38
Rheumatology, Ij85
Richard III, Ek62
Riddles, De1
Rights of man, Gb25
Ringforts, Lh8,14
Rings, Ca50; Ji4
Riots, Ga12,30,b21,e14,53,f2; Hb12,
  e58,g10; Kb7
Ritual, Ec24; Gb23
Ritualism, Hl57
Rivers, Ef58
RMT company, If69
Roads, Ca2,45,b25–26,37; Gf40; Hf13,
  38; If96; turnpike, Bb132
Rolls, court, Aa129; Exchequer, Aa7;
  plea, Eh17
Romance, Ei77
Romanticism, Hj8
Roofs, timber, Ej63
Royal Africa Company, Ga25
Royal Agricultural Collge, Ij78
Royal Air Force, Aa176; If43,i3,10,12,
  22,30,42,49,62; Ng4
Royal Commission on Historical
  Manuscripts, Aa5,24

Index of Subjects

Royal Commission on the Ancient & Historical Monuments of Scotland, Ac40
Royal Corps of Signals, Ii25
Royal family, Bd41
Royal Irish Constabulary, Me4
Royal Literary Fund, Hl92
Royal Mail Steam Packet Company, If7
Royal prerogative, Gb18
Royal Society, Fl20; Ge24; Hk45,49
Rubber, Ne5
Rural society, Aa102; Ba12,c54; Cb10; Df73; Ff25,h17,19; Gg19; Hb84,f21, 34,62,g12,26,41,45,67,79,h31,39, i10; Ig2

Saints, Bc229; Da24,c8,24,38,41; Ec23,e2,8,25,54,58,78,82–83,g16; Je20; Ke4,25; lives, De51; *and see* Hagiography
Salvation Army, Hc13,e33,58,81
Sandemanians, Ge9,j13
Sanitation, Fg34; Hk7,55
Satire, Fe91
Scholarship, De37
Schoolboys, Gi73
Schools: Gi21,31,58; Hg65,h9,26,i11,15,19,23,k34,l75; attendance, Hi9,35; boarding, Hi24; cathedral, Bb44; Catholic, Bb150; charity, Ga11,14; elementary, Hi26; nautical, Hi5; navigational, Hi6; *and see* Education
Science, Aa23,41,50,67,98,b1,c44,51,95, 98,104; Ba54,b207,c60,87; Fe49,l7; Ge24,i24; Hb55,e55,l114; Ij40; practitioners of, Fg47
Scots, Id70
Scribes, Ea35,e57,i76,99–100
Scrofula, Gj8
Sculpture, Ca3; De10; Ej30,32,42,50; Fk23,53; Gi51; Hl61; Li11; Anglo-Saxon, Ej28; Romanesque, Bc105; Ej7,28,31,55; wood, Ej27
Scurvy, Gj19
Seals, Ee15,37,j9,19
Seamen, Gd11,13
Secret societies, Mb26
Sectaries, Fj21
Sects, religious, He72
Secularisation, Ie2
Secularism, Fe83; Hb59
Security, Ib44,d19

Security services, Id24
Serfdom, Eg19; Jewish, Ec31
Serjeanty, Ef47
Sermons, Aa151; Ee23,81; Fe23–24,73; Ge1; Ke9; Le23
Servants, Fg21
Settlement, Ab9; Bb62,c168,210;Ca1,20, b4,7,17,19,21,24; Da8,f15,23,33,37, 43,54,57,69; Eg10,27,k1–2,8–9,24,35, 42; Ja2,g7,9; Kg3,6–7; Lf11,g1,9–10; Mg15,24; history, Gg4
Settlements: Bb1; Ca29,33,44; Df66; Ek45; Anglo-Norman, Lf1,2; British, Df63; deserted, Bb66; disabled officers, Nf15; laws, Gg13,20; nucleated, Jg4; rural, Jf2; Scandinavian, Df38
Seven Oaks massacre, Nc14
Sex, Eg24; Hg35,h36
Sexual behaviour, Ig20
Sexuality, Aa18,b56; Gg23,i73–74,79,j1; Hg87,l95,97; Ig15
Share-cropping, Ef34
Sheep, Hc9
Shell-shock, Ij8
Sheriffs, Bb31; Ea24,f41; Ff19
Shields, Df21
Ship money, Fi25
Ship owners, Hf8
Shipping, Hl107; Ia24
Ships, Bc207; Hk11; Lh12; Ng8,17; model, Gh20
Ship-service, Db37
Shires, Db18; Kc20
Shopkeepers, Hg44; Mj3
Shops, Hf73,g55
Sieges, Eh8; Fi2; Kh6; Lh1
Signal towers, Ca51
Silver, Ca25; Dd18
Silversmiths, Ab24; Bb122
Slave trade, Aa54; Fd21; Ga25,b24,e28, f55; Hf44; Lf4
Slavery, Aa54; Ba63,132; Fd3,f15; Na9, f48–49
Sleeping sickness, Nf29
Slum clearance, Ih37
Smallpox, Gj24
Smuggling, Gf21; Hf99
Sobriquets, Ab12
Social: history, Ac60,100; Bc138; Nd1; inequality, Ac114; life, Hg38; policy, Ac10; Ga14,c4,11; Hb36,74; Ih20; Mb25; reform, Hh33; relations, Ga13; structure, Ac11,23,39,83; Bc43; Fj9;

Index of Subjects

Gi62; Mg5,12; theory, Ac22; unrest, Hc13
Socialism, Bc176; Hb7,32,50,74,76,80, g4,l66,103; Ib15,47,65,90,96; Mb22; *and see* Parties; Politics
Societies: Co-operative, Hf78; Friendly, Ih25; popular, Hi34; religious, Mf1
Society for the Promotion of Christian Knowledge, Ga14
Society for the study of addiction, Aa34
Society of Arts, Gh20
Society of Genealogists, Aa43
Sociology, Ac32,67; Hl103
Sokes, Ec14
Solar energy, If76
Soldiers, Gh9; Hj8,16; Kh2–5,9; Ld2; Scottish, Bc21; Ng2,19
Songs, Hl111
South Sea Company, Gf47
Southcottians, Ge40
Sovereignty, Aa73; Bc12; Fj43
Special Air Service, Ii31
Spies, Eh2; Fc11
Spiritualism, Ie14
Sport, Aa55; Ba21,186,b113; Ei18; Gi25; Nh28; cricket, Ba23; horse racing, Hl80; rowing, Ba196; tennis, Ba138
Standard of living, Ef62; Ga27,f29; If5
State, Gf64; papers, Aa111
Statistics, Fl10
Status, Db30,36; Fh20
Statutes: Eb2; of Carlisle (1307), Ee16; *and see* Act
Steam engines, Gf36; Hf43
Stereotypes, Nh30
Sterling area, Ne15
Stewards, estate, Fg24
Stone carvings, Li6
Stones, inscribed, Ji6
Strategy, Gh41; Ii67; military, Gh19
Strikes, Bc175; Hc1,f54; Ib80,f30,114,118
Submarine warfare, Ii33
Subscription lists, Gi38
Suez: canal, Id80; crisis, Ib72,d22; Nb12
Suffrage: Hb46; female, Hb19,32,45, h22; Ke22; question, Hb3
Suffragettes, Ib11,95
Suffragism, Aa159; Bc50
Sugar, Ne24
Suicide, Hg19
Surgeons, Ei35
Surgery, Ac63; Ba105,c24; Ei44,48,95; Fl17; Gj8,12,20,27; Hk29

Surveying, Gf26,39
Swedenborgianism, Ge20
Swords, Eh18
Syphilis, Ei82

Tactics, naval, Ii59
Tale of a tub, Gi24
Tanks, Ii63
Tapestries, De18; Ej20,39,43,45,k57
Tartan, Bb39
Taxes: Cb38; Dd22; Ec3,e75; Fb42; Gf57; Hf18; Ic1; Kf1; assessment of, Ec11–13; cider, Gb35; clerical, Eg30; hearth, Fg2,27; land, Ba70; Gf16; lay subsidies, Eg39,40
Tea, Ba60; Ne24
Teachers, women, Hb45
Teaching, Hi3
Technology, Ba54–55,92,103,188,c22, 32,86,134,208; Gf5,19,25,36,j21; Hf1,11,58–60,82,91,i1,k11,36; If43, j66,75; telephone, Ia12
Telephones, Ia12
Television, Ac91
Temperance, Hh22; Me7,11
Temples, Ca26,36
Terrorism, Id82
Testaments, Kc11
Textiles, Bc53,174; Ki9
Textus Roffensis, Ee77
Thatcherism, Ib46
The battle of Maldon, Df74
The Builder, Hk53
The Canterbury tales, Eg2,i74
The faerie queene, Fc21
The Guardian, Aa116
The Intelligencer, Gi85
The merchant of Venice, Fc25
The parlament of women, Fk83
The Times, Ib50
The way to the temple, Gi38
Theatre, Ba158,b136,c63; Fk38,43,67, 78,87,92; Ge43,i54,61,67; Hg2,7,21, 24,27–29,31,33,77,81,l51,105; *and see* Drama
Theatres, Hg22
Theology, Bc60; Dc27; Ee1,i38; Fe20, j26; Ge41,50; He20,32,63; Mi21
Theosophy, Hb81
Tides, Cb46
Tiles, Ca12,16; Ek20
Time: Ei58; Ie1; reckoning of, De55; series, Gf37

324

Tithe surveys, Hc11
Tithes, Aa99; Gf2; Le5; herring, Gf44
Tithing-penny, Eg38
Tokens, Ab34; trade, Gi26
Toleration, Fe95,j35; Gi59; Kc16
Toll houses, Bb132
Tombs, Ej8,48
Topography, Df1,53; Lg2,4,9–10
Toryism, Gi70; *and see* Parties; Politics
Tourism, Fd32; Ga17,d9,23,i56; Ia38,
   j24–25,62
Tournaments, Ki5
Tower houses, Ki11; Lh1,4,7
Town: councils, Kc18; planning, Gi33;
   Ih16,j53
Towns, Aa16,61,77,127,b6,44,c30,84,
   86; Ba11,121,b2,11,131,140,148,205,
   c55,70,107,126,212,214,216; Cb10,
   12,28,30,48; Dd5,13,f7,17; Ec2,9,29,
   e12–13,17,26,65,f3,5,21–23,25,27,32,
   37,40,45–46,48,55,57,64,g11,18,21,
   31,35,47,h26,k2–4,11,14–16,22,26–
   27,53; Ff7,h13; Gf63,g12; He56,f66,
   76,g11,37,57,65,l52; Ib87,g27,j13;
   Jf4,g1,3; Kg4; Mf5,g1,j1,6; walled,
   Lh15; *and see* Boroughs; Burghs;
   Urban
Tractarianism, Hl57
Tractarians, He53
Trade: Ac1; Ba37,130,139,147,b75,
   111,c31,209,215; Dd2,9–10,13,16,20,
   23; Ed7,f45; Ff2,i1; Gf21,66; Hf80;
   Kf4–5,8,h1; Ne16; Anglo-Australian,
   Ne15,27; Anglo-Canadian, Ne19;
   Anglo-Portugese, Ef15; associations,
   Hg44; If55; Atlantic, Ff21; Gf45,59;
   Baltic, Gf18; book, Aa170; Fl18; Gc1,
   f28,i38,42; He76; cloth, Ef8; clothes,
   Gf38; cotton, Ba111,b139; East India,
   Ff13; food, Ia32; foreign, Cb15; Ga23,
   d19,f15; free, Hf47,68; Id14; gin, Gc4;
   maritime, Ef21,46; Lf7; overseas,
   Ed11; Gf23; policy, Ne3; retail, Ac96;
   Bb218,c35; Hf73,78,86–87; second-
   hand, Gf38; unionism, Ig25; Trade
   unions, Ba107,b183,c7; Hf48,56,94,
   g76; Ia36,b10,73,77,83,88,f72–73,92,
   94,g36,h36; Mf3; West African, Hf57;
   wool, Ef8
Trades Union Congress, If65
Tramps, Ba48
Tramways, Hf75,77
Translations, Fk22

Transport: Ba130,c117; Gf40,54,57,69;
   Hf5,38; Id23,f69,82; Mf9; maritime,
   If46; railways, Hf4; river, Ef38,55;
   urban, Hf77
Transportation, Hc6; Mb11,g9,11,17,
   20,i14; Nc3,18,28,32
Travel, Fd7,32,k65; Ga17,20,d2,5,10,
   26,30; Hf12,g38,i21,l9,39,47; Ia22;
   Nh31,39; writing, Mi22
Travellers, Ld4
Treason, Ea38; law of, Gc2
Treasure, Ca25; Dd7,8
Treasury, Aa11; Ic3,f32; Mb28
Treaties: ANZUS (1952), Nb18; of Union
   (1707), Kb2,6; of Waitangi, Nb29,c37
Trial by ordeal, Ec19
Tribute, Eh16
Tropers, De56
Turnpikes, Gf40; Hf13
Two treatises on government, Bc155;
   Fe26,j4,19,29,46,k1,31,70

Unemployment, Ia7,14,f10,g12
Union, Anglo-Scottish, Kb2; Malayan,
   Nc30; Scottish, Ga26
Unionism, Hb42; Mb5,10,16
Unitarianism, Bb154; Ge14,16,53;
   He70–71,77,g42,h5
United Distillers, Aa120
United Irishmen, Mb23
United Nations War Crimes Commis,
   Id58
Universities, Aa59,118,121,134,175,c30;
   Ba4,b207; Ee24,i15,22–23,78; Fg16,
   37,k26,l18; Ge19,f33,i57,j25; He47,
   i12,16,20,25,31,k21,28; If57,j33;
   Kc10; Mi2,12,16; *and see* Education
Urban: development, Bb11; Lc4,f1–2,12,
   g2; economy, If70; history, Aa77,c38,
   84,88; Ba143,c42,126; Gf63,i77;
   Ke17; planning, Nh15; society, Bc55;
   unrest, Ih11
Urbanism, Bc214
Urbanization, Cb30; He54,g5,11,53,90–
   91,h19; Ig2; Lf7,g8; *and see* Boroughs;
   Burghs; Towns
Urns, Cb11
Usury, Ec4,g2
Utopianism, Fd18,30

Values, Victorian, Bc196
Vanity fair, Nh7
Vestments, He27

Veterans, Ii11
Vichy government, Ie5
Victoria county history, Aa62
View of the present state of Ireland, Fc21
Views, topographical, Ek66
Vikings, Bc108,159; Da12,15,b2,23,f23; Lf11,g2,8,h12
Villages: Bb29; Ek45; deserted, Ek1; medieval, Ab9
Villas, Ca21,40,43,b24,43–44
Vindication of the rights of woman, Gi74
Vision of Leofric, Dc17
Visionaries, Dc8
Visions, Dc8
Visitations, episcopal, Ee29
Voluntarism, He56
Voting, Hb62
Voyages, Nd2

Wage rates, Ba114
Wages, Ef48; family, Ig4
Walls, Cb20
War: air, Hj14; Ii5,22,26,40,42,46,49, 56–57,62,65; crimes, Id34,58; debts, Id102; economy, Bc188
War Office, Aa97; Ii29
Warfare, Aa13,c20,74; Bc141,150; Cb8; Eh25; Fi3,7,19; Kh7; Ld1; Mh3; colonial, Ng10,11
Warhorses, Dd6
Wars: Bb94,c69; Fb80; Hj11; Afghan, Ng38; Anglo-Boer, Hj5–6; Anglo-Mysore, Gh31; Anglo-Scottish, Bc4; Ea22–23,f46,h3–4,10,12,16–17,26–27,i72; Anglo-Spanish, Gh47; Boer, Hj2,3; Ng24,34,h17,27; cod, If105; cold, Id19,29,71; Nb19; colonial, Hj18; Ng18; Crimean, Hk59; Egyptian, Ng1; English civil, Bc79,107,203; Fb7–8,56,74–75,c20, d15,e2,21,38,78,f25,29,g36,h4,7,9,11, 13,15–17,21,i4–5,11–12,14,20,22,29, j1,20,22–23, 27–28,38,45,48; Gi47; Falklands, Ii68; first world, Aa101, c56; Hb33,j3,7; Ia18,21,42–43,b53, 79,95,d100,e7,14,f4,28,70,113,i8,13, 17,19–20,35,39,43,47–48,54,63,66–67,j8,26,64; Mh1,2; Ng35; French revolutionary, Gh21,42; Gulf, Ii61; hundred years', Ea1,21,d2,6,h3–4,9, 19,21,28; Korean, Id84,109; Napoleonic, Gd1,25,h19,21,23; north-west frontier, Ng32; of 1812, Bc85;
Gh1,8,12–14,16,22,45; of American independence, Gb9–10,d31,e12,h2,7, 26; of the Austrian succession, Gh10; of the Roses, Ea34,b6,d3,13,h2,11,13; Peninsular, Gh9,21,46; Scottish, Ea8, 38; second world, Aa29,100; Bc20; Ia9,29,b7–8,28,49,d3,6,9,27–28,30, 33,47–49,52,55–56,58,78,96,f91, 116–117,g1,8,19,36–38,h6,9,16,i4–7,10,12,15–16,22–23,26–28,30–34, 36–37,40–42,46,49–51,53,56–58,60, 64,j1,81,86; Md1; Nb3,g4,6,9,13,30; seven years', Gb26,d14–15,19,h24,30, 44; Spanish civil, Ib10,d70,79; thirty years', Fb87; Vietnam, Nb6; world, Ba81; Zulu, Ng7,20–23,25–27,39
Water power, Bb186
Waterfronts, Aa42; Bc41,166
Wealth, Hf6,85,g74–75,84; Je14
Wealth of nations, Gf42
Weapons, nuclear, Ij86
Welfare: Hg60,k20; Ih22–24,26,29; Mb25; child, Hh30,34; Ih28; policy, Ig42; state, Aa107,c6; Bc44; Hb36, h12,27; Ih10,13,33
Western European Union, Id12
Western front, Ii63
Wetlands, Ca1
Whiggism, Scottish, Gi67
Whisky, Aa120
Widows, Ec6; Fg6
Wife beating, Gg8
Wills, Bb184; Db20,37; Fb46,g7,13; Gg10; Kc11
Witchcraft, Fg18–20,h18
Witches, Lg3
Women, Aa18,c9,16,36,55,65,73,85,99, 104; Ba82,88,129,132,158,b19,27,c8, 39–40,44,62,66,68,91–92,135,177; Cb23; Dc21,e49,51; Ea42–44,c6,21, e7,27,83,f1,29–30,33,60,g3–4,6,8–9, 15,20,24,46,48–49,i31,80; Fe2,50,69, 71,86,g3,9,28,42,h21,j21,27,41,k25, 46,52,57,72,94; Ga8,13,31,e10,g8,i9, 39,69,74,78,j3; Ha5,b3,18,46,74,c4–5,12,e39,41,55,61,f30,104,g1–2,7,19, 21,24–25,27–29,31,33,35,38,49,58, 72–73,77,81,87–88,h15,18,22,33, k45,l51,87,123; Ib13,34,g4,15,26,30, 37–39,j14,43; Lb1; Mb27,29,g6,i4,6; Nd9,f10–11,21,30,39,45,g24,h1,18, 21,32; gentry, Ba128; working-class, Ib79,g29; *and see* Feminism

Women's movement, Ig33
Woodland, Ef44
Work, Eg20; Gf8; Hf30; female, Ba88; Ig26
Work-houses, Bb196; Ga11; Hh8; Mg7, 23
World Economic Conference, If33
Worship, Ge7

Writing, Ei19

Yachts, Hl72
Young Ireland, Mb11

Zamindari, Nc46
Zionism, Id53,65,91
Zodiac, Ej44
Zulus, Nb11